D0138354

ELECTRICITY & ELECTRONICS

Thomas E. Newman

GLENCOE
McGraw-Hill
New York, New York   Columbus, Ohio   Mission Hills, California   Peoria, Illinois

Cover Photograph: The Stock Market/Clayton J. Price 1994

Photo credits appear on pages 559–560, which are hereby made part of this copyright page.

Library of Congress Cataloging-in-Publication Data

Newman, Thomas E.
Electricity & electronics / Thomas E. Newman.
p. cm.
Includes index.
ISBN 0-02-801253-4
1. Electric engineering. 2. Electronics. I. Title. II. Title: Electricity and electronics.
TK146.N64 1995
621.3—dc20 94-37494
CIP

Electricity & Electronics

Copyright © 1995 by The Neville Press, Inc. All rights reserved. Except as permitted under the United States Copyright Act, no part of this publication may be reproduced or distributed in any form or by any means, or stored in a database or retrieval system, without the prior written permission of the publisher.

Send all inquiries to:
Glencoe/McGraw-Hill
936 Eastwind Drive
Westerville, OH 43081

ISBN 0-02-801253-4

Printed in the United States of America.

2 3 4 5 6 7 8 9 0 RRD-W 00 99 98 97 96 95

CONTENTS

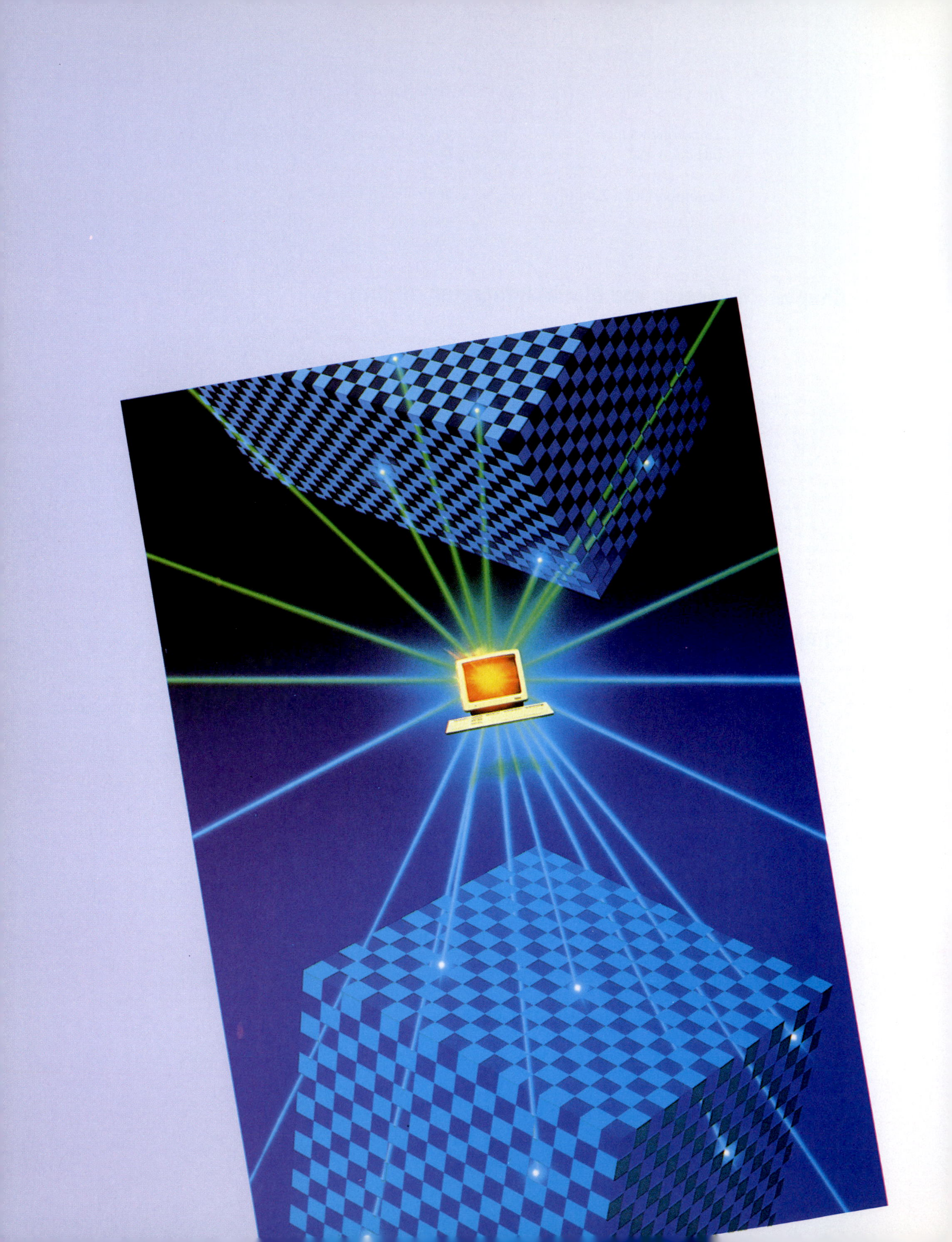

PREFACE

You are about to take the first step as you prepare for a career in the exciting field of electronics in the twenty-first century. All around us we see—and use—electronic products that weren't in existence for much of this century. We can only imagine what products will be developed—perhaps with your help—for our use in the next century. *Electricity & Electronics* has been written and designed to help you develop an understanding of the basic concepts and principles you'll need to know to become a well-prepared electronics technician.

Organization

The first chapter of the text provides some background information on the works of the early pioneers and their studies of electric charges and electric fields. Chapter 2 provides a three-dimensional view of atoms and matter and Chapter 3 covers the necessary topics of units, definitions, and notations. The three basic quantities of voltage, current, and resistance are introduced in Chapter 5. Basic electric circuits, how they function, and how to troubleshoot them are covered in the next five chapters. The subjects of magnetism and magnetic devices are discussed in Chapters 12 and 13. Alternating current (AC) is introduced in Chapter 14. The next eight chapters continue the study of alternating current through the development of various types of circuits. Chapters 23 to 26 introduce diodes, transistors, and integrated circuits. These are subjects that you will study in greater detail in later courses. The final chapter provides information on career opportunities for technicians.

Features

Design The text's vivid photography, dynamic graphics, and computer-generated art are especially designed to help you understand DC/AC theory as well as to enhance the text's appeal. Some very creative art has been included which we hope you find interesting.

Chapter Opener Each chapter starts with an introduction of what is to be covered and a list of performance objectives.

Chapter Ending At the end of each chapter is a summary of the main points covered and a list of the most important vocabulary words. Following those features are *Questions, Problems,* and *Critical Thinking* questions. The *Questions* and *Problems* help reinforce the material covered in the chapter, while the *Critical Thinking* questions are intended to build your analytical skills.

Review Quizzes To assure that you have understood the material covered, each chapter contains *Review Quizzes*. The answers are provided for you at the end of each chapter.

Other Features You will find a feature called *Student to Student* throughout the text. It is intended to provide you with helpful hints from someone who has already taken the course.

Two other features, *Notes* and *Electronic Facts*, provide additional insights into the history of electronics and some important electronics breakthroughs.

Supplementary Material

Tutorial Software A Windows-compatible software program has been developed to supplement the text. The software includes questions and circuit problems for every chapter except the chapter on careers.

Experiments Manual A lab manual directly correlated with the text has been developed. It provides you with hands-on learning activities for the theories covered in the text.

Problems and Exercises Manual This manual has been specifically developed to correlate with the text. It provides an abundance of extra practice material for you.

Mathematics Manual Intended as a math refresher, this manual has been developed to help you with the required electronics math. It includes both traditional solutions and the use of calculators to solve problems.

Instructor's Management System Also available is an instructor's package which includes a computerized-test generator, transparency masters, study guides, and answers to the text and supplements.

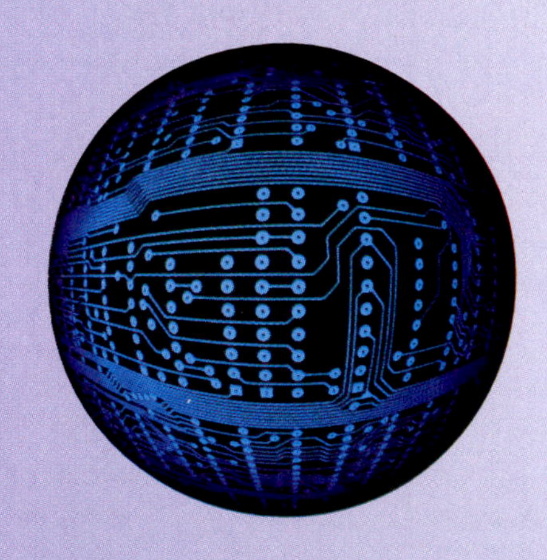

DEDICATION

In
memory
of
Richard S. Neville,
founder of The Neville Press.

HIS SPIRIT LIVES ON . . .

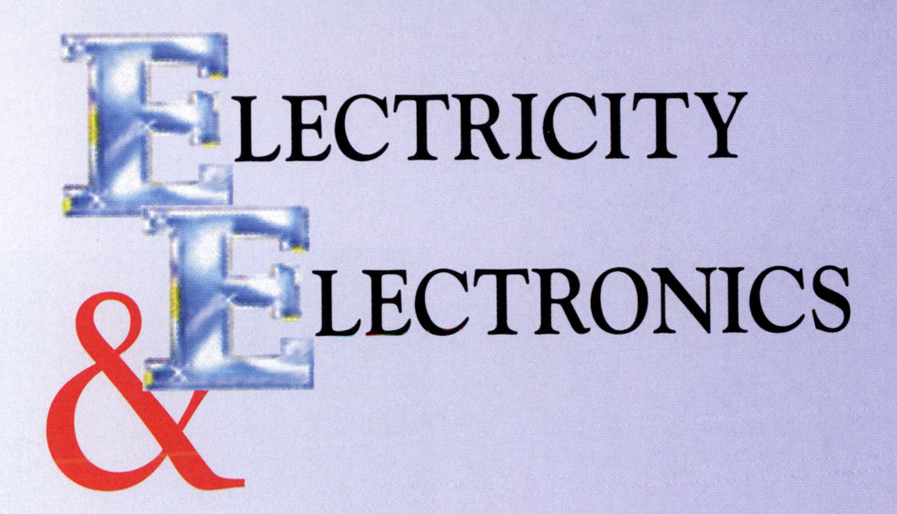
ELECTRICITY
&
ELECTRONICS

Chapter 1
Electrostatics

ABOUT THIS CHAPTER

As you read this chapter, you will retrace the steps of electrical pioneers who brought electricity out of the realm of the supernatural and into their laboratories. From the development of the first electric machine in 1672 to the development of the battery in 1800, the characteristics of electric charge and electric fields were identified and measured.

After completing this chapter you will be able to:

- Describe how various experiments established the basis for our modern electrical theory.
- Cite how electric charge is measured and how its strength varies with distance.

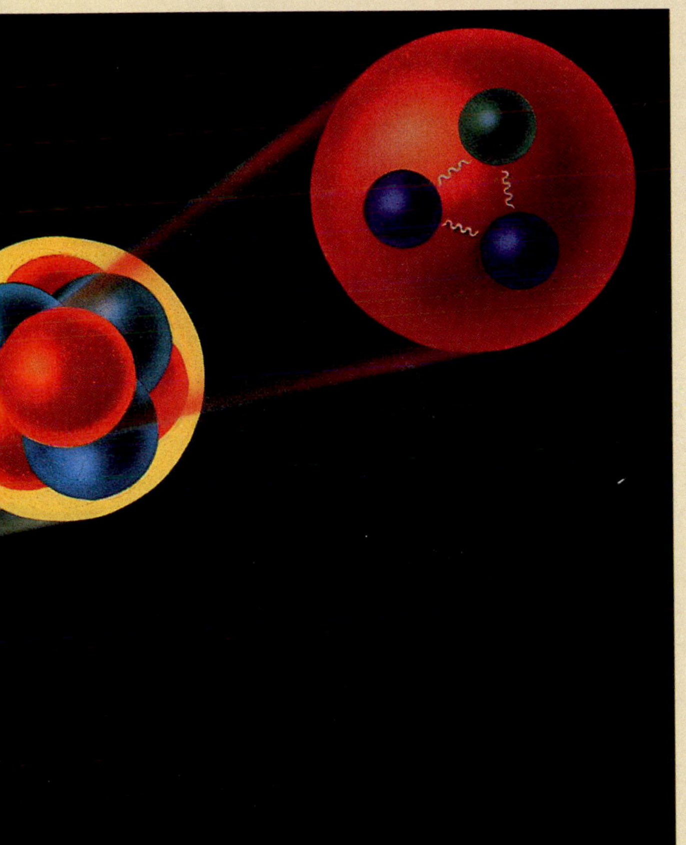

The knowledge gained between these important milestones became the foundation for the technological explosion that makes today's sophisticated electronics possible. This chapter examines the properties of electric charge and electric fields.

- Explain the general nature of electric fields and the necessity of creating models.
- Define *electric flux* and how it relates to charge.

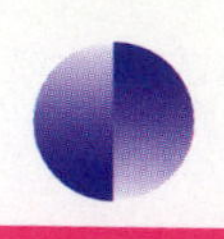

STUDENT
to
STUDENT

When I took this course, I learned that almost every electrical unit is someone's name.

1-1 Electrical Observations

Over two thousand years ago the Greeks discovered that rubbing a fossilized resin, called amber, with wool caused the amber to attract small pieces of straw and lint. The ancient Greeks thought that the invisible force of attraction was a supernatural phenomenon. Some believed that because amber had once been the sap of a tree, the tree's imprisoned soul was reaching out for and attracting objects in the world of the living.

Amber's ability to attract other materials after being rubbed with wool has nothing to do with the supernatural. What the ancient Greeks discovered was **static electricity.** The word *static* means to stand without motion. Static electricity is produced whenever dissimilar nonmetallic materials are rubbed together.

The Word *Electricity*

English scientist William Gilbert (1540–1603) used the word *electrica* in his renowned book on magnetism, *De Magnete*, in 1600. The word described all substances that, like amber, have the property of attracting matter. In 1646, Sir Thomas Browne (1605–1682) used the word **electricity** for the first time in his book *Pseudodoxia Epidemica: Inquiries into Many Commonly Received Tenets and Commonly Presumed Truths.*

Static Electricity Machines

Static electricity remained a generally ignored curiosity until German physicist Otto von Guericke[1] (1602–1686) developed the first static electricity machine in 1672. The heart of von Guericke's machine was a sulphur ball about the size of a large grapefruit with a shaft placed through its center. The shaft was equipped with a crank that allowed the ball to be turned when mounted on a stand. While turning the crank with one hand, von Guericke held his other hand against the rotating ball. This process caused sparks to emerge from his fingertips and caused the ball to become electrically **charged.** After the ball was charged, paper and small metal shavings were attracted to it. A recreation of von Guericke's machine is shown in Fig. 1-1.

Von Guericke continued his experiment by removing the charged sulphur ball from its stand shaft and placing it near a feather. At first the feather was attracted to the ball, but then it suddenly flew away after touching it. Von Guericke chased the feather around the room with the ball until the feather came into contact with another object. After that contact, the feather was again attracted to the ball and the action repeated itself, although not as vigorously as before. Von Guericke had discovered the phenomenon of **electrical repulsion.**.

Von Guericke was also the first to document **electrostatic induction.** He found that light bodies suspended close to his charged ball acquired the ball's electrical properties. After suspending a linen thread over the ball, making certain that the two did not touch, von Guericke noted that the thread moved away when he reached for it. The thread had not been in contact with the ball and had not previously been charged, but it seemed to be charged when in the ball's presence. Von Guericke also observed that if the thread was allowed to touch the charged ball, the thread became electrified to a distance of several feet.

Figure 1-1 The heart of von Guericke's static electricity machine was a sulphur ball that became charged when von Guericke held his hand against it while turning the crank.

English experimenter Stephen Gray (1696–1736) determined that some materials are electrical **conductors** and others are **nonconductors.** He performed an experiment similar to but more elaborate than von Guericke's. Gray fastened an ivory ball to a long piece of twine and tied the other end of the twine to a glass tube. He then hung the ball over the edge of a balcony 26 feet high and found that when he rubbed the glass tube the ivory ball would attract small particles.

Gray became very curious about how far the charge could be carried by the twine and added more. To gain additional height, he connected long canes into each other and to the glass tube, similar to the way a fishing pole is assembled in sections. The charge continued to be carried by the twine. Unable to gain more height, Gray attempted to increase the distance by running part of the twine horizontally to a nail driven into a wooden beam. The remaining portion of the twine hung down with the ivory ball at its end. To his amazement, the electric charge stopped at the nail. Gray concluded that the charge went into the beam and in the process he discovered what might have been the first **short circuit.**

Two Electricities

Charles Du Fay (1698–1739), a member of the French Academy of Sciences, observed two kinds of electricity. He called one **vitreous** and the other **resinous.** *Vitreous* refers to glass, and *resinous* refers to any plasticlike material made from or containing resin. Du Fay discovered that unlike kinds of electricity attract and that like kinds repel. This interaction is known today as the **law of charges.**

Just over 10 years after Du Fay's terming of vitreous and resinous electricity, American writer, scientist, philosopher, and statesman, Benjamin Franklin (1706–1790), along with his friend Ebenezer Kinnersley, also recognized two types of electricity and called them **plus** and **minus.** Their theory was similar to Du Fay's, but there is no evidence that Franklin and Kinnersley had any knowledge of Du Fay's work at the time.

Franklin believed in the existence of a single ethereal[2] electrical fluid. He imagined that it was distributed throughout the universe and present in all matter. In a now classic experiment, Franklin rubbed sealing wax, a resinous material used to seal letters, with wool and found that both became **electrified.** Afterward, he observed the electric effects of attraction and repulsion. Franklin felt that these ef-

[1]Von Guericke was mayor of the German city of Magdeburg. He is best known as the inventor of the air pump (vacuum pump) and for his experiment of the "Magdeburg hemispheres," which he performed for the King of Prussia in 1654. In that experiment two metal hemispheres were placed together to form a sphere. Then, using his air pump, von Guericke removed the air from the sphere through a valve, which he then closed. Two teams of eight horses could not pull the hemispheres apart. After opening the valve von Guericke was able to separate the hemispheres by hand. The king was very impressed and awarded von Guericke a lifelong pension.

Von Guericke was the first to demonstrate that light could pass through empty space. He did this by showing that objects placed in glass containers could still be seen after the air was evacuated from the containers. Using a similar technique, he proved that sound required a medium for its propagation.

[2]Something that lacks material substance, often considered unworldly or spiritual in nature.

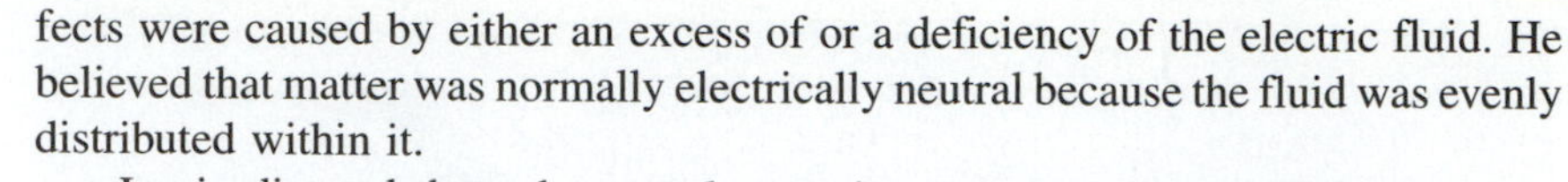

fects were caused by either an excess of or a deficiency of the electric fluid. He believed that matter was normally electrically neutral because the fluid was evenly distributed within it.

Logic dictated that when wool came into contact with a liquid, the liquid would be absorbed by the wool. This logic prompted Franklin to label the wool **positive** because he believed it had gained additional electric fluid. Conversely, he labeled the sealing wax **negative** because he believed that it had lost an equal amount of fluid. The identification of **electric polarities** allowed Franklin to observe that similar types of electricity repel and opposite types attract, just as Du Fay had done before him.

Electricity in a Jar

The rapid and uneven discharges common to static electricity made accurate measurements difficult, if not impossible. A partial solution to that problem was found around 1746 by the Dutch doctor and scientist Pieter van Musschenbrock (1692–1761).

While at the University of Leyden in Holland, van Musschenbrock experimented with ways of storing electric charge. He developed what would later become known as the **Leyden jar**. The jar was made from a glass bottle that was partially filled with water. The top of the bottle was sealed with a cork. A wire was pierced through the cork's center and immersed in the water. The free end of the wire was connected to an electric machine. The person operating the machine placed a free hand around the outside of the bottle, forming an outer conductor. This unfortunate choice of an outer conductor caused van Musschenbrock to receive what may have been the first near-lethal electric shock ever recorded.

The operator's hand was soon replaced by a metal coating, as suggested by Dr. John Bevis (c. 1695–1771). Additional improvements abandoned the use of water and replaced it with metal foil.

In the improved Leyden jar, shown in Fig. 1-2(*a*), the opening of a glass jar was covered with a wooden lid equipped with a metal rod running through its center. A small chain hung down from the rod and touched the foil inside. The foils inside and outside of the Leyden jar were connected across an electric machine, which was then activated. This system allowed a great quantity of electricity to be stored and delivered on demand. The discharge from these jars was so great that it could melt an iron wire placed across them. Bevis replaced the glass jar with a flat glass plate covered by a sheet of foil on each side, as shown in Fig. 1-2(*b*). That arrangement performed the same function as the Leyden jar and later became known as a **condenser** because of its ability to concentrate electricity.

(*a*)

(*b*)

Figure 1-2 The Leyden jar (*a*) and Bevis's condenser (*b*) could store large amounts of electric charge.

In America, Benjamin Franklin became interested in the Leyden jar and employed it in numerous experiments. The mechanism for the storage of charge mystified Franklin, and he was determined to solve it. He disassembled one of the jars and carefully examined each part, concluding that the electric force must be stored in the glass.

Franklin believed that the necessity of having foils both inside and outside the jar was to give electricity to one side and take it from the other. He later proved that the positive charge on one side of the glass was exactly matched by a negative charge on the other side. He also discovered that the charges could be increased by enlarging the surfaces of the foils facing one another inside and outside or by using a jar with thinner walls.

The invention, and subsequent improvement, of the Leyden jar was a very significant discovery. It allowed experimentation with true electric circuits and provided the mechanism for the rudimentary evaluation of electrical quantities, such as the conductivity of various materials and the relative speed of electricity.

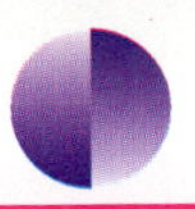

STUDENT *to* STUDENT

The condenser became one of the most common electrical components. It is now known as a capacitor and is only called a condenser in automotive electrical systems.

REVIEW QUIZ 1-1

1. Electricity produced by rubbing dissimilar nonmetallic materials together is called ______________.
2. Objects close to an electric charge appear to be charged through a process called ______________.
3. Benjamin Franklin observed two types of electricity, which he called ______________ and ______________.
4. After taking apart a Leyden jar, Benjamin Franklin decided that electric force was stored in the ______________.

1-2 Measuring Electric Charge

Experimenters observed electrical interactions and developed theories to explain them. Those theories often pale in light of today's technology, but a few of them were brilliant, if not entirely accurate. Franklin's theory of an electric fluid evenly distributed throughout all matter is an excellent example because it could explain the two types of electric charge and why matter is normally electrically neutral.

The Electroscope

The electrical nature of matter is clearly demonstrated with an **electroscope**. An electroscope is a simple instrument that detects the presence of electric charge. It can also be used to compare the polarity of one charge to another. The detection mechanism consists of two very thin metal strips, called leaves, that are held vertically facing one another. The leaves are usually made from gold leaf because it is so thin, but aluminum works well and is more durable. The upper ends of the leaves are held together, but the lower ends are free to move. The leaves are fastened to a vertical metal rod with a metal ball on the upper end. This assembly is then mounted into a supporting structure, often a glass container [Fig. 1-3(*a*) on the next page] or a metal ring placed on end. If a metal ring is chosen, the assembly must be insulated [see Fig. 1-3(*b*)].

(*a*)

(*b*)

Figure 1-3 Electroscopes detect the presence of electric charge and can be used to compare the polarity of one charge to another.

Figure 1-4 The leaves of a charged electroscope repel each other because they are connected at the top and therefore have the same charge.

The leaves of an electroscope separate when they become charged (Fig. 1-4). The action is identical for both positive and negative charges. The metal ball at the top of the rod serves as a charge collector. For the moment, assume that Franklin's **fluid theory** is correct. When a positively charged object, such as a glass rod that has been rubbed with silk, is brought near the metal ball, electric fluid would be pushed into the leaves. The leaves become positive and repel each other in accordance with the law of charges. The closer the rod is to the ball, the farther apart the leaves move. When the glass rod is withdrawn, the leaves return to their normal resting position.

When rubbed with wool a hard rubber rod becomes negatively charged. Bringing the rod near the metal ball pulls electric fluid into the ball and away from the leaves. The leaves again move apart because they are both negative. If the rod is allowed to touch the ball, the leaves will not collapse when the rod is withdrawn because electric fluid has been removed from the electroscope's main assembly. The loss of fluid leaves the entire assembly negatively charged. If the positively charged glass rod is again brought near the metal ball, the leaves will move closer together. Electric fluid is again pushed into the leaves, but this time the addition of the fluid brings the leaves into electrical balance, accounting for the collapse.

Franklin's single fluid theory left several questions unanswered. It was reasonable to assume that the fluid could repel itself and be attracted to an area of less concentration. But the theory did not account for the observed repulsion between two objects deficient in fluid. Some, even before Franklin, believed there were two electric fluids. The French priest Jean Antoine Nollet (1700-1770), who had worked with Du Fay, proposed such a two-fluid theory. Nollet called one of the fluids **affluent** (to flow toward) and the other **effluent** (to flow out). However, neither of the fluid theories could explain the action at a distance, so clearly demonstrated by electricity.

The inability of either fluid theory to provide a complete explanation of the actions observed through experimentation caused opinions on the theories' validity to change frequently. The one point of agreement between the two theories was that electricity exists within all materials in some form of balance—the even distribution of a single electric fluid or equal quantities of opposite electric fluids.

The early experimenters made clever devices to measure electricity.

Coulomb's Balance

How electric charge varied with distance was accurately described in 1785 by the French natural philosopher Charles Augustin Coulomb (1736-1806). He measured the force of attraction and repulsion between two electrically charged spheres with a device he developed called a **torsion balance**, shown in Fig. 1-5. It was similar to one used later by a leading eighteenth-century English scientist, Henry Cavendish (1731-1810), to measure the force of gravity. It is possible that Coulomb and Cavendish collaborated on the development of both devices.

Coulomb's torsion balance was once a familiar instrument to every student of physics. It consisted of a glass cylinder 12 inches in diameter and 12 inches tall. It had a glass cover 15 inches in diameter with two holes, one in the center and one toward the outside edge. A glass tube 24 inches long was cemented into the center hole. The upper end of the tube supported a torsion micrometer that was fitted with a graduated disk to read the angular motion of the head. A thin silver wire fastened to the torsion head supported a light horizontal rod made from straw or silk coated with shellac. The rod had a pith ball (pith is a light spongy material found in the center of plant stems) on one end about the size of a pea and a varnished paper disk on the other end. The paper disk balanced the pith ball and dampened any oscillations that might develop in the suspended rod. A small cylindrical weight was also attached to the lower end of the wire to keep it taut.

The second hole allowed a charged pith ball to be inserted into the balance. In that way the force between the two charged pith balls could be measured. A paper scale divided into 360 degrees was pasted onto the outside of the glass cylinder at the height of the pith balls. The torsion balance could accurately measure the relative force of repulsion or attraction at different distances between the stationary and moveable pith balls. Coulomb used this device to establish the **law of inverse squares** and to define the fundamental unit of charge quantity known as the **coulomb.**

Several electrical investigators sought to determine how the strength of electric charge varied with distance. In 1773 Cavendish performed an experiment to determine the rate of falloff of electric charge. He used an inner metal ball surrounded by a larger metal ball made from two hemispheres. Once the balls were charged, Cavendish surmised that the electric attraction or repulsion between the balls varied inversely as the square of the distance between them. Several other experimenters came to the same conclusion at approximately the same time, but credit for devising a system for the accurate measurement of charge relative to distance belongs to Coulomb.

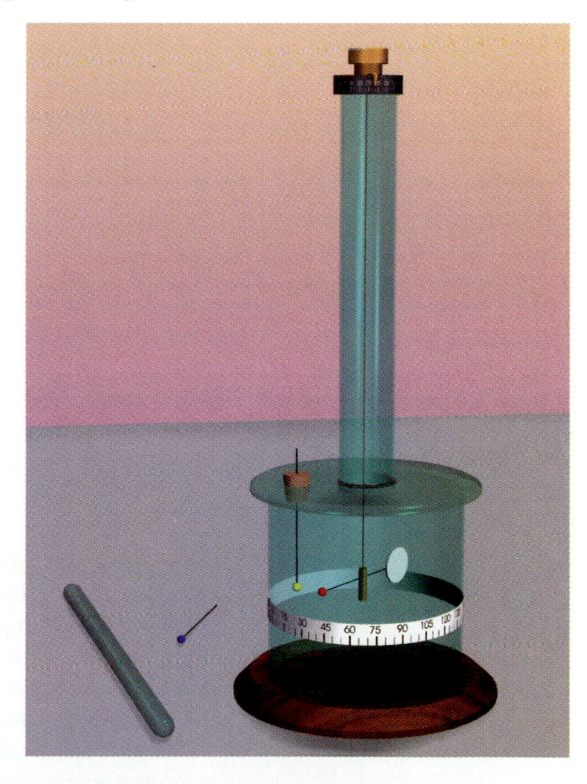

Figure 1-5 Coulomb's torsion balance accurately measures the relative electrical force of attraction or repulsion between two electric charges.

Coulomb's Law

With his balance, Coulomb correctly identified the relationship between the force of attraction and repulsion between two charges and the distance between them. The result of his observations is stated as **Coulomb's law**: *The electrical force of repulsion or attraction between two charged spheres is directly proportional to the strength of their charges and inversely proportional to the square of the distance between their centers.* Doubling the distance between the balls causes the charge to drop to one-fourth of its previous value. Cutting the distance between the spheres in half causes the force to increase four times.

Although Coulomb adopted the two-fluid theory for his experiments, he did not seek to understand the nature of electricity or to engage in the one-fluid/two-fluid controversy. He only sought to measure the interactive force of electric charge. Coulomb conducted many experiments on the distribution of charge on conducting bodies. He showed that the electrical distribution on a charged object is dependent solely on its shape and dimensions, not its mass or the material from which it is made.

REVIEW QUIZ 1-2

1. The electroscope is an instrument that measures the presence of electric ______________.
2. Franklin's one-fluid theory did not adequately account for the repulsion between ______________ charges.
3. Coulomb used his torsion balance to develop and verify the law of ______________ .
4. Coulomb proved that the charge on an object is not dependent on its ______________ or the material from which it is made.

1-3 The Electric Field

Neither of the fluid theories could adequately explain electricity's action at a distance. In that respect, the invisible electric force seemed similar to magnetism and gravity. While analyzing the previously unexplained movements (perturbations) of the orbital paths of planets, English philosopher and mathematician Sir Isaac Newton (1642–1727) coined the term **field.** He said: *When an action occurs at one point in space that is caused by an action at another point in space, with no physical connection between those two actions, they are connected by a field.* The cumulative electric field around an electrified object is the embodiment of electric charge.

There are three types of fields with known properties. They are **electric, magnetic,** and **gravitational.** Fields do not interact directly with matter, only with other fields and then only with other fields of their own type—electric with electric, magnetic with magnetic, and gravitational with gravitational. Fields have no physical structure and cannot be seen or touched. The only energy available from a field is that which was put into it. The field itself cannot be consumed.

Modeling the Electric Field

Electric fields surround charged objects like an invisible atmosphere. They are modeled as **lines of force** that radiate away from charged objects. The number of lines in a given area represents the strength of the field. If the charge is spherical, the lines of force radiate equally in all directions, becoming less dense as the distance from the charge increases.

The inverse-square relationship of the electric charge on a spherical object, as measured by Coulomb, is demonstrated using two parallel squares. One of the squares is twice the distance from the center of the charge as the other. Four lines of force are followed from the center of the charge across the four corners of the first square and past the second square. The size of the second square is adjusted so that its corners touch the same four lines. The area of the second square is four times greater than the area of the first square. This indicates that the strength of the charge has dropped to one-fourth because the same number of lines of force are spread over four times the area, as shown in Fig. 1-6.

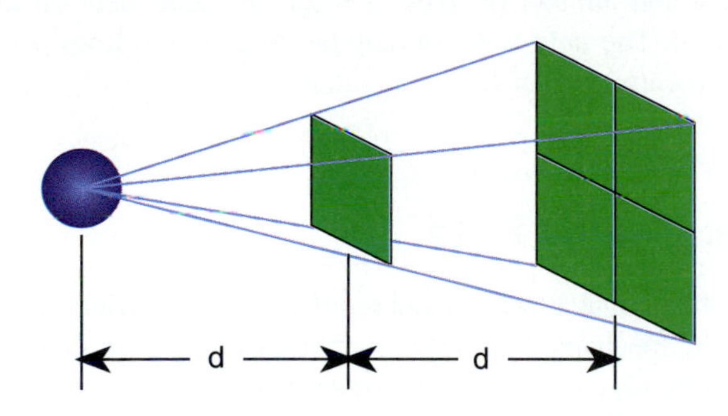

Figure 1-6 The strength of the electric field around a spherical charge drops to one-fourth when the distance from its center is doubled.

To graphically distinguish a positive charge from a negative charge, the lines are assigned arrowheads that point away from the charge if it is negative [Fig. 1-7(*a*)] and toward the charge if it is positive [Fig. 1-7(*b*)].

If a mobile negative charge is placed between a fixed positive charge and a fixed negative charge, the mobile charge is attracted toward the positive and

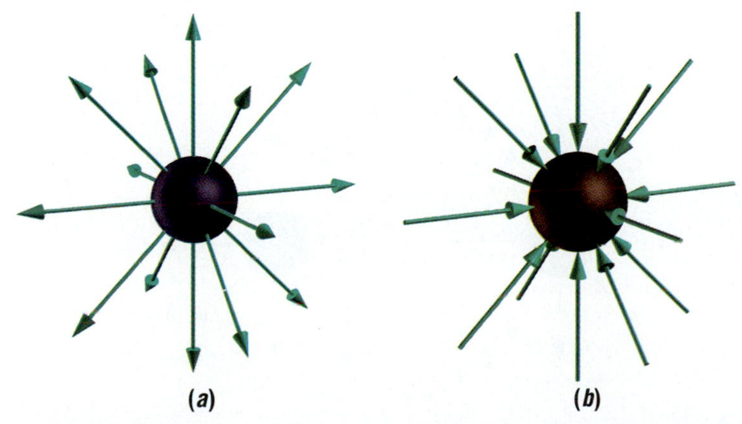

Figure 1-7 Arrowheads are drawn on lines of force pointing (*a*) away from a negative charge and (*b*) toward a positive charge.

repelled from the negative in compliance with the law of charges. When the focus is placed on the mobile charge, the arrows assigned to the lines of force should correctly identify the direction of its expected movement. This is an important point because in almost every electric circuit it is the negative charge that moves. Scientists and engineers, for example, concentrate on the forces that create movement. Electronic technicians, however, are more concerned with the effect of those forces on physical objects.

The lines of force model accurately represents the observed characteristics of electric fields. In the long run, the direction of the arrowheads assigned to the lines of force is unimportant, because each model is valid in its own frame of reference. The field is not actually composed of lines and there is no evidence that it is in motion, as might be suspected from the presence of the arrowheads. The arrowheads simply show that there is a difference between the two types of electric fields and that those fields produce opposite effects.

The major advantage of the lines of force model is that it allows a field to be described mathematically. The total number of lines of force represents the charge quantity; the lines of force in a given area represent the strength of the charge. The area and number of lines of force are dependent on the system of measurement used. The field strength may be measured in lines per square inch, lines per square centimeter, or lines per square meter.

The Elastic Electric Field

The reason for the mutual attraction and repulsion between electric fields remains a mystery, but the mechanics of those interactions have been observed for many decades. Electric fields are elastic and change their shape easily when in the presence of another field. When two oppositely charged objects are brought close together, their lines of force bend toward each other. This action concentrates the field between the objects. The field becomes increasingly concentrated as the objects are brought closer together. Another charged object near the pair will only be slightly affected because there are so few lines of force available for interaction (Fig. 1-8). When the particles have similar charges, the lines of force have a different action. Instead of a mutual attraction, they push away from each other. As the particles are forced closer together their lines of force form a right angle to the imposed movement.

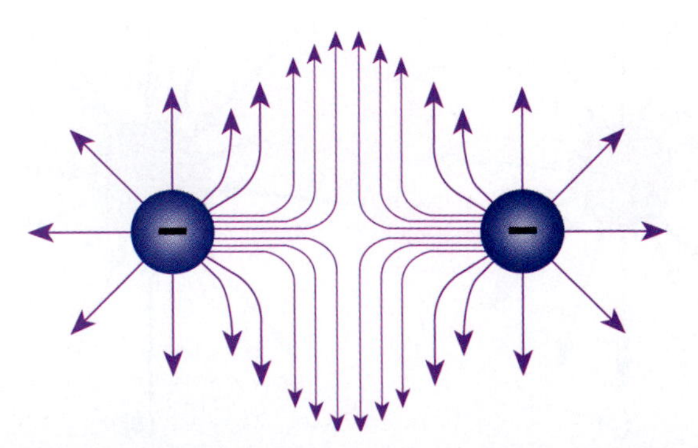

Figure 1-8 Every electrical charge has a finite number of lines of force. The density of those lines of force changes when influenced by other electric charges.

The interactions of electric fields can be seen by suspending many short fine threads in a light oil and applying a strong electric charge to two metal rods immersed in the oil. The threads align in the electric field, forming the illusion of lines of force (Fig. 1-9).

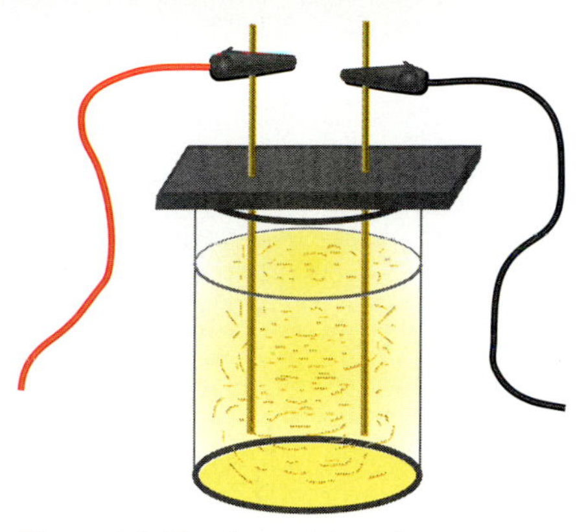

Figure 1-9 The lines of force between charged metal rods immersed in a jar of oil are visualized by suspending fine pieces of thread in the oil.

REVIEW QUIZ 1-3

1. In many respects electric fields are similar to ____________ and ____________ fields.
2. From an electronic technician's point of view, the lines of force surrounding a negatively charged object point ____________ from the object.
3. The lines of force surrounding oppositely charged objects bend ____________ each other.
4. A mobile negatively charged object will move ____________ a fixed positively charged object.

1-4 Electric Flux and Gauss's Law

While Coulomb's law provides a basis for calculating the force between two electric charges, it does not give any insight into the relationship between field and charge. The first step toward discovering that relationship is to define a quantity called **flux.** Flux is the number of lines of force that pass through any closed surface, real or imaginary; and flux is directly proportional to the amount of charge enclosed by that surface. That general relationship is called **Gauss's law,** named after German mathematician Johann Karl Friedrich Gauss (1777–1855). Strictly interpreted, the word *flux* implies motion. In the case of fields, that motion often remains implied and is not always evident. Flux is the rate at which something passes through a surface. The amount of air passing through a cross section of a heat duct in a specified unit of time could be described as its flux.

STUDENT *to* STUDENT

Try to imagine what the electric field might look like and how it behaves. These electric fields will be covered again in other chapters.

Electric flux can be described in the same way as the air in the previous example. Consider a uniform electric field passing perpendicularly through a surface. The field is represented by lines of force. Its uniformity is indicated by the even distribution of those lines [see Fig. 1-10(*a*) on the next page]. The flux can be increased by using a larger area [Fig. 1-10(*b*)] or a stronger field. The stronger field is indicated by the closer spacing between the lines of force [Fig. 1-10(*c*)]. The flux is also dependent on the incident angle between the surface and the field. When the area is not perpendicular to the field, the flux is reduced [Fig. 1-10(*d*)]. The flux is zero when the surface and the field are parallel [Fig. 1-10(*e*)]. Any surface, real or imaginary, used to implement Gauss's law is called a **Gaussian surface.**

Gauss's law verifies that a charge placed on a conductor resides entirely on its surface. (The effect was discovered by Benjamin Franklin long before Gauss's law came into existence.) When any conductor is charged, the charge distributes itself over the surface until it reaches equilibrium. The result is that the internal electric field becomes zero. Even a thin conductive coating can be used to surround an object and shield it from external electric charges.

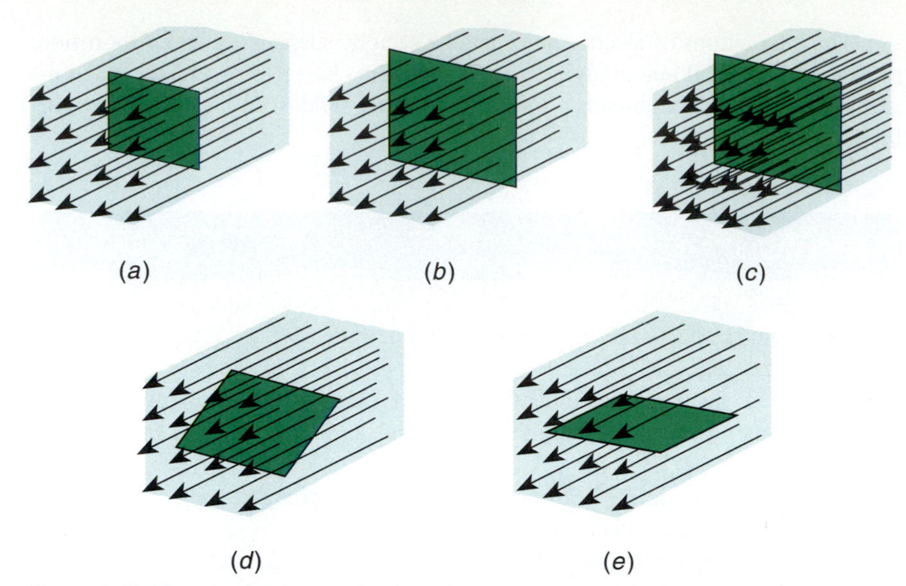

Figure 1-10 The perceived strength of an electric field is determined by the number of lines of force passing through a surface and the angle of those lines to that surface.

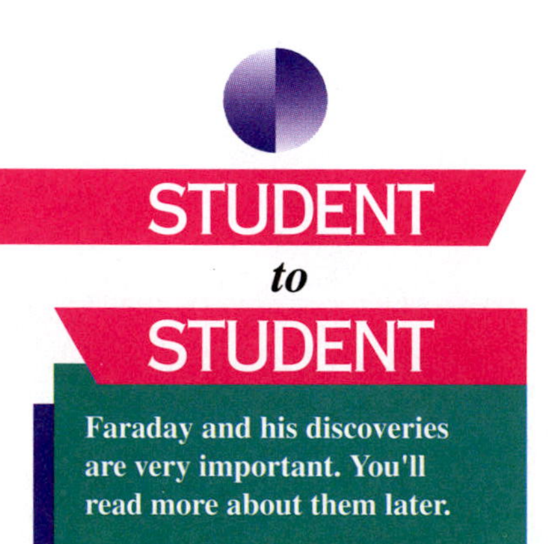

English chemist and physicist Michael Faraday (1791–1867) performed extensive experiments on shielding. As a result of those experiments, electrostatic shields, commonly made from a wire coil or mesh, are frequently called **Faraday's cages** or **Faraday's shields.** Electrostatic shields are so effective that a person sitting inside a wire mesh sphere is safe from the effects of an electrical discharge as strong as a lightning bolt (Fig. 1-11).

An application of shielding that is more practical than protecting someone from lightning bolts is the use of plastic bags or other plastic containers with conductive coatings to store and protect static-sensitive electronic components. The coatings often look metallic and are usually pink, green, blue, or clear.

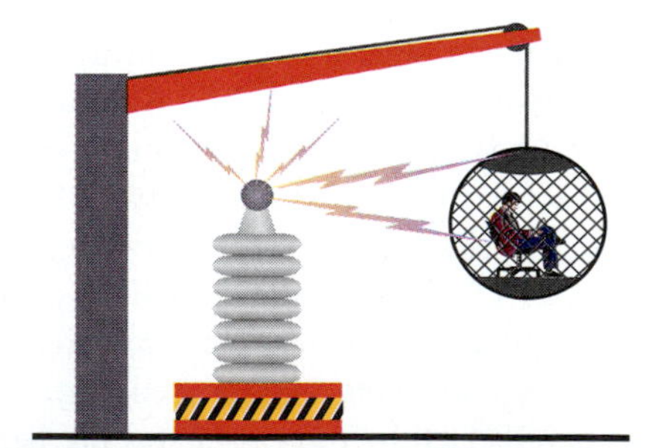

Figure 1-11 The internal electric field of a conductor is zero. Thus objects enclosed inside conductors are protected from the effects of external electric charges.

REVIEW QUIZ 1-4

1. Electric flux is the number of lines of force passing through a(n) ______________ surface.
2. The ______________ electric flux occurs when the lines of force are perpendicular to the Gaussian surface through which they pass.
3. Electric charge resides entirely on the ______________ surface of a hollow metal sphere.

SUMMARY

For just over one and a quarter centuries, between 1672 and 1800, experiments with electricity were centered around electric machines. Several major discoveries resulted from those experiments:

1. There are two types of electricity called positive and negative.
2. Similar types of electricity repel and opposite types attract.
3. Objects can be electrified simply by being near another electrified object through a phenomenon called electrostatic induction.
4. Charge is determined by the surface area of an object and not its mass or the material from which it is made.
5. Charge can be stored for later use.
6. Charge can be transferred through some materials and not others. Materials through which charge can readily move are labeled conductors. Materials that cannot support the movement of charge are labeled nonconductors.

Electric fields, like magnetic and gravitational fields, are without substance and must be modeled in order to describe them. The model represents electric fields as lines of force. The total number of lines of force is called the electric flux. The electric flux density represents the strength of the field. The electric charge surrounding any object is in reality an electric field emanating from that object.

NEW VOCABULARY

affluent
charge or charged
condenser
conductors
coulomb
Coulomb's law
effluent
electric field
electric polarities
electrical repulsion
electricity
electrified
electroscope
electrostatic induction
Faraday cage
Faraday shield
field
fluid theory
flux
Gauss's law
Gaussian surface
gravitational field
law of charges
law of inverse squares
Leyden jar
lines of force
magnetic field
minus electricity
negative
nonconductors
plus electricity
positive
resinous
short circuit
static electricity
torsion balance
vitreous

Questions

1. What types of materials are rubbed together to produce static electricity?
2. Why is static electricity produced more easily in winter than in summer?
3. Who developed the terms *plus* and *minus*?
4. What is implied by the term *static electricity*?
5. What modern component was inspired by the Leyden jar?
6. What name is given to a wire mesh used to shield something from external electric charges?
7. What is stated in the law of charges?
8. How does the electric field around a charged sphere vary with distance?
9. Does the electroscope behave differently for positive charges than for negative charges?
10. How are electric fields modeled?
11. What action was not explained well by either the one-fluid or two-fluid theory?
12. Does the volume of an object affect its electrical charge?
13. What was the major purpose of a Leyden jar?
14. What name was given to materials that allowed charge to move freely through them?

Problems

1. What would be the increase in the strength of an electric field around a spherical electric charge if the distance between the measuring point and the center of the charge were cut in half?
2. What would happen to the force of attraction between two spherical charges of opposite polarity if one of the charges were doubled?

Critical Thinking

1. In von Guericke's experiment with a charged sulphur ball and a feather, why was the feather repelled by the ball after it came into contact with it?
2. Are objects that appear charged because of electrostatic induction truly charged, or do they appear charged because their internal electrical charges have been redistributed? Explain your reasoning.
3. If a charged object were to change in size without gaining or losing charge, would the amount of flux change? Explain your reasoning.

Answers to Review Quizzes

1-1
1. static
2. electrostatic induction
3. plus and minus (or positive and negative)
4. glass

1-2
1. charge
2. negative
3. inverse squares
4. mass

1-3
1. magnetic and gravitational
2. away
3. toward
4. toward

1-4
1. closed
2. maximum
3. outside

Chapter 2

The Nature of Electricity

ABOUT THIS CHAPTER

Alessandro Volta's development of the battery in 1800 was the key to discovering the composition of atoms. Batteries gave both scientists and the curious the reliable source of electrical power needed for their experiments. This chapter follows the most significant developments of early atomic research. It details the structure of atoms, and

After completing this chapter you will be able to:

- Describe the development of batteries as a practical source of electricity.
- Explain the general structure of matter and its components.

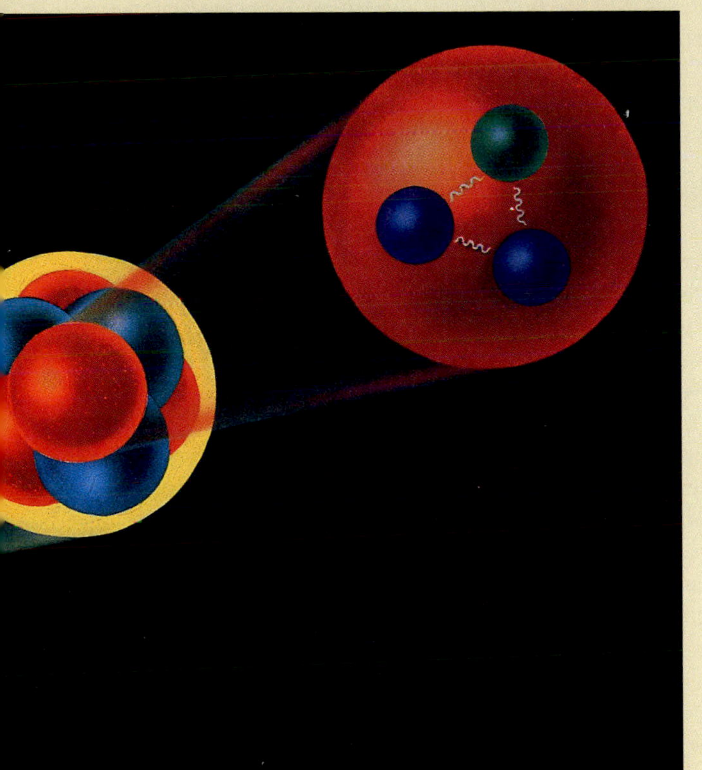

it illustrates the three-dimensional orbitals of electrons. Thus the chapter provides a solid foundation for understanding today's sophisticated gaseous and solid state components. The sources and uses of electricity are also presented and it is shown how the two are related.

- ▽ List the ways that electricity converts energy from one form to another.
- ▽ Describe the sources and uses of electricity and their relationship to each other.

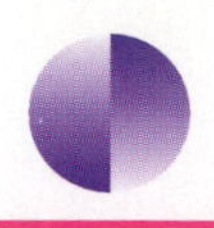

STUDENT *to* STUDENT

It's interesting that pulling the disks apart made the charge on them appear stronger.

2-1 Practical Electricity

The nature of electricity was the subject of much debate throughout the 1700s. Experimentation had raised many questions that could not be answered by the theories of the time. In 1796, however, an Italian physicist named Alessandro Volta (1745–1827) performed an experiment that would ultimately change the course of electrical science. For this experiment Volta used two metal disks about 6 inches in diameter. One disk was made from copper and the other was made from zinc. A glass handle was attached to each of the disks. Volta found that when the faces of the disks were brought into contact with each other and then suddenly separated a charge could be detected by a sensitive electroscope. From this Volta concluded that the initial charge on the disks resulted from chemical action and that their sudden separation significantly multiplied the effectiveness of the charge.

The Battery

Four years later Volta developed the first chemical **battery.** It became known as a **voltaic pile** and it consisted of disks made from silver and zinc that were stacked alternately with pieces of moistened cardboard between them. Each pair of disks formed a **cell.** On March 20, 1800, Volta wrote a letter to Sir Joseph Banks, then president of the Royal Society of London, in which he described his battery and claimed that the device repeatedly gave a shock when touched (Fig. 2-1).

Batteries provided the first practical source of electricity. Although the potentials of batteries were much less than those of electric machines, the quantity of charge that could be moved was generally much greater. Of more importance, however, was the fact that batteries maintained a constant output over long periods of time. Having a steady, predictable source of electricity allowed experimenters to advance their knowledge of electricity more rapidly than ever before.

The success of Volta's battery inspired him and other scientists to experiment with different combinations of metals and methods of construction. Soon batteries were commonplace and were routinely used in a variety of scientific investigations. One of the most significant was changing water into hydrogen and oxygen through a process called **electrolysis.** Electrolysis produces a chemical change when electricity is passed through a liquid. When the gases are collected in separate glass tubes, the volume of hydrogen collected is twice that of the oxygen, but the oxygen has eight times more mass. These ratios are consistent with observed chemical activity and support the early Greek theory that matter is composed of fundamental building blocks called **atoms.**[1]

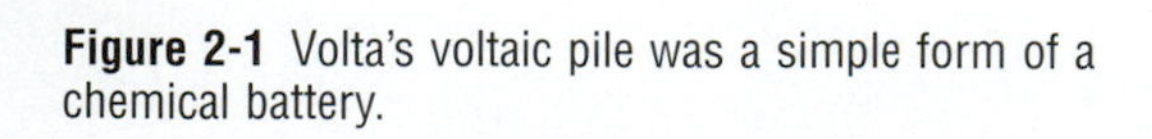

Figure 2-1 Volta's voltaic pile was a simple form of a chemical battery.

REVIEW QUIZ 2-1

1. Alessandro Volta developed the first chemical ______________.
2. A process that produces a chemical change when electricity is passed through a liquid is called ______________.

2-2 The Structure of Matter

The seemingly limitless number of materials in the universe are made from 92 different atoms called **elements**. Each element has its own unique atomic structure. Copper, gold, hydrogen, oxygen, and mercury are examples of naturally occurring elements. Nineteen additional elements have been created artificially by bombarding atoms with high energy atomic particles. All 111 elements are arranged according to their characteristics on a chart called a **periodic table.** A typical periodic table is shown in Fig. 2-2.

Materials made from combinations of different atoms are called **compounds.** Some common compounds are sucrose, table salt, ammonia, and water. The smallest amount of a material that can exist and still maintain the properties of that material is called a **molecule.** Molecules are usually made from combinations of different atoms, but not always. The molecules of most gases are formed from pairs of like atoms. For example, the chemical representations of hydrogen and oxygen molecules are H_2 and O_2 respectively. A water molecule, which is formed from two parts hydrogen and one part oxygen, is represented by the symbol H_2O.

[1]As early as the fourth century B.C., Greek philosophers believed that matter was composed of indestructible building blocks that were so small they could not be seen. They called these building blocks atomos, meaning indivisible. The word *atom* is derived from the Greek word *atomos.*

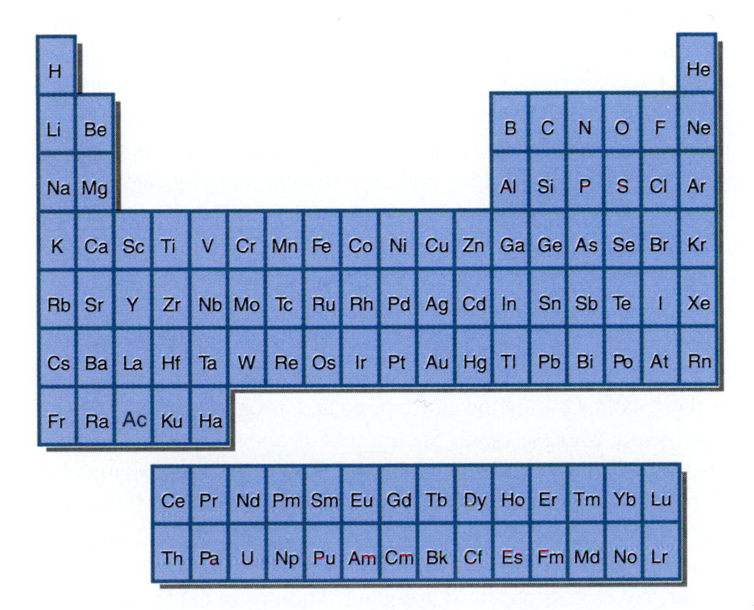

Figure 2-2 The periodic table organizes the elements according to their characteristics. These charts usually provide additional information about each element, such as atomic mass, thermal conductivity, and electron configuration.

Figure 2-3 The amount of charge is determined by the number of lines of force passing through a closed surface.

Atomic Structure

Atoms are composed of three basic particles: **electrons, protons,** and **neutrons.** Electrons and protons represent the basic units of negative and positive electric charge. Atoms are electrically balanced when they have the same number of protons and electrons. The number of protons in the nucleus (core) of an atom is its **atomic number.**

Mass-to-Charge Ratio of Electrons and Protons

Although the charges on electrons and protons are exactly equal and opposite, their masses are vastly different. A proton is 1836 times more massive than an electron. However, the difference in mass does not equate to as large a difference in size as one might expect. That's because electrons and protons are spherical. A sphere has the greatest ratio of volume to surface area of any geometric shape. Thus, the diameter of a proton is only 12.25 times greater than the diameter of an electron.

The similar charges and different masses of electrons and protons can be demonstrated by placing an electron inside a sphere the size of a proton. The lines of force that pass through the electron's surface also pass through the sphere's surface. The charge is unaffected because the number of lines of force does not change (see Fig. 2-3). Remember that flux is the number of lines of force that pass through any closed surface. The area of that surface is unimportant.

Electrons

Around 1889, British scientist and inventor Sir William Crookes (1832–1919), developed a method for producing a vacuum that was far superior to the one created by von Guericke's pump (see Chapter 1). Crookes, who was interested in the behavior of electrical discharges in a vacuum, created what would later become known as the **Crookes tube.** Crookes placed a positive electrode (called an **anode**) and a negative electrode (called a **cathode**) in a glass tube and then evacuated the air from it. When a strong electric potential was applied across the tube, a green glow became visible at the positive end. Fluorescent materials were also placed inside the tube to enhance the effect. Crookes believed that these materials glowed when the tube was electrified because of "rays" emitted from the negative electrode. Thus, the rays emitted from the cathode were called **cathode rays.**

Some scientists proposed that cathode rays were similar to light; others thought they were particles. Cathode rays were observed to travel in straight lines and to cast shadows on the interior of the tube when small metal objects were placed in their path. One of the most popular Crookes tubes contained a metal piece shaped like a Maltese cross. Such a tube is shown in Fig. 2-4.

The particle theory was given a substantial boost by an experiment performed by French physicist Jean Baptiste Perrin (1870–1942) in 1895. Perrin showed that cathode rays were attracted to a positive charge and repelled by a negative charge, thus proving that the rays were themselves negative. His experiment was refined two years later by British physicist Joseph J. Thompson (1856–1940). Thompson built a variation of the Crookes tube that contained internal deflection plates (Fig. 2-5). He used this tube to establish that cathode rays were particles. He also determined the mass-to-charge ratio of cathode rays.

Thompson found that the particles composing the cathode rays could be deflected by connecting a battery to the metal deflection plates. Through a series of

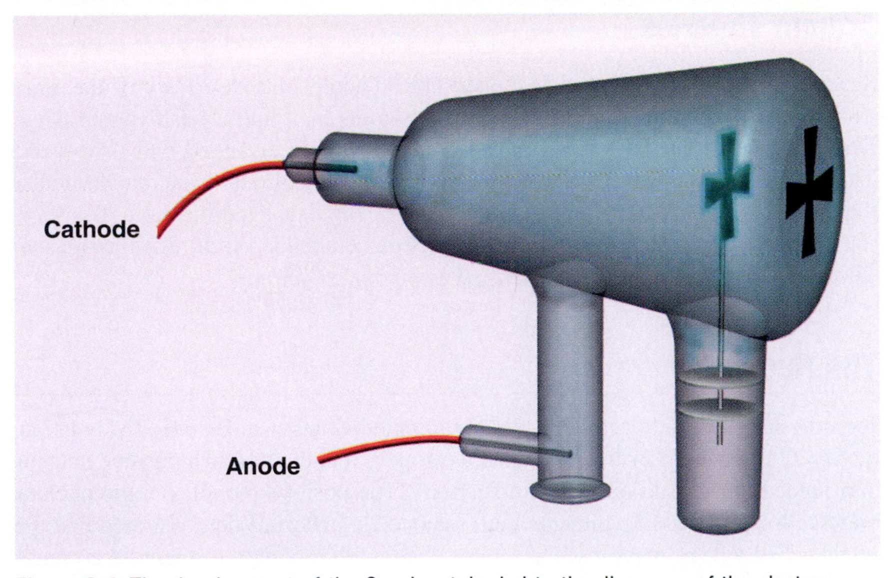

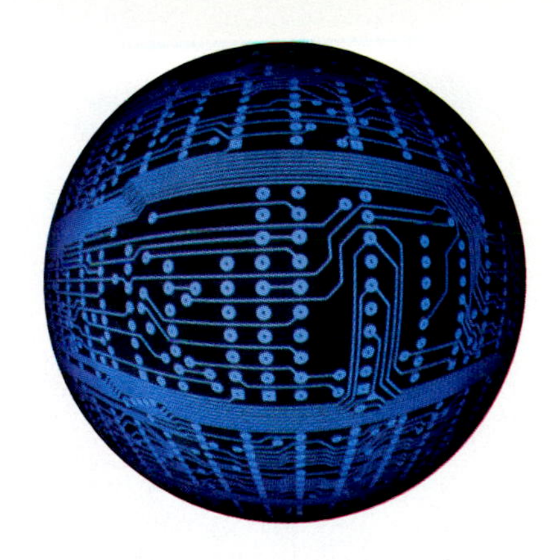

Figure 2-4 The development of the Crookes tube led to the discovery of the electron.

calculations based on measurements of the deflection of the particle beam, Thompson determined that the mass of each particle was less than one-thousandth the mass of a hydrogen atom, which was known to be the smallest of all atoms. From his experiments, Thompson concluded that:

1. Atoms are not indivisible because negative particles could be removed from them.
2. The particles composing the cathode rays all have the same mass and the same electrical charge.
3. The charge on the particles was equal to the smallest known charge in electrochemistry.

Thompson called the particles *corpuscles,* which later became known as electrons. Because of his careful investigations, Thompson has become known as the discoverer of the electron. He was awarded the Nobel prize in physics in 1906 in recognition of that accomplishment.

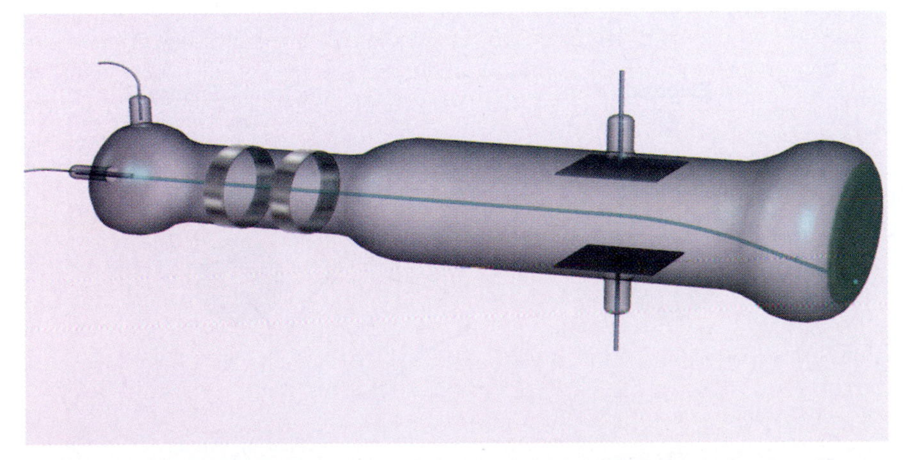

Figure 2-5 Thompson found that cathode rays could be deflected by an electric charge applied to metal plates inside a modified Crookes tube.

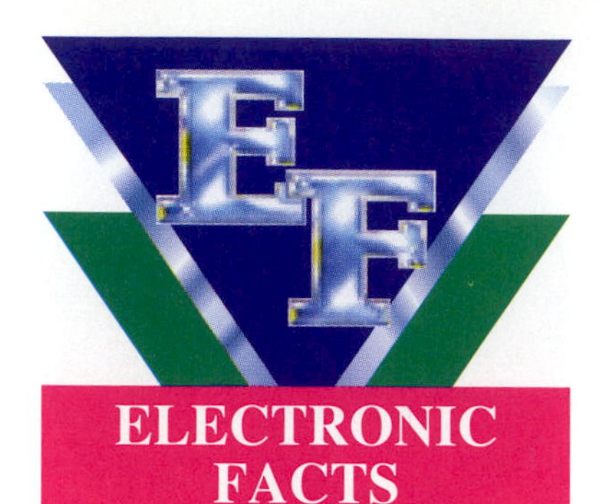

Spectroscopy is a process that divides light into its component colors by passing it through a prism. The resulting bands of color are then used to identify the specific energies being radiated from the light source.

Atomic Theories

Several atomic theories were developed after Thompson's discovery of the electron. Experimentation clearly identified that atoms had both a positive and negative component, but how those opposite charges were arranged was a mystery. Thompson had proven that the negatively charged electrons were much smaller than the positively charged portion of the atom. He visualized the atom as a positively charged spherical mass with the electrons embedded in it. For that reason, Thompson's model was often described as a *plum pudding.*

The Nucleus

Experiments with radioactive materials prompted Ernest Rutherford (1871–1937), a New Zealand–born British physicist, to propose that atoms had a positive nucleus surrounded by a cloud of negative electrons. The positive particles in the nucleus were called protons. Experiments with radioactive materials demonstrated that the nucleus had a large mass-to-charge ratio. It was felt among the scientific community that the positive charge of some protons in the nucleus was canceled by the presence of electrons. Rutherford did not agree and theorized that the nucleus contained neutral particles of about the same mass as a proton. It was difficult for Rutherford to prove his theory because the equipment available at that time could only detect charged particles.

Whereas theories such as Thompson's and Rutherford's were very creative, they could not answer all of the questions raised through experimentation and observation. In 1913, however, a revolutionary new atomic model was developed by Danish physicist Niels Bohr (1885–1962). Bohr had worked with both Thompson and Rutherford for about a year and incorporated some of their ideas into his own atomic theory. Bohr proposed that hydrogen's single electron revolves around its nucleus in a stable circular orbit, as shown in Fig. 2-6(*a*). The diameter of the orbit is dependent on the amount of energy possessed by the electron. Bohr believed that the electron must be in motion to prevent it from being "captured" by the positively charged nucleus.

Experimental evidence suggested that the electron's energy only existed in certain increments. Bohr was not sure why that should be, but the evidence through **spectroscopy** was undeniable. Limiting the electron's energy to specific levels au-

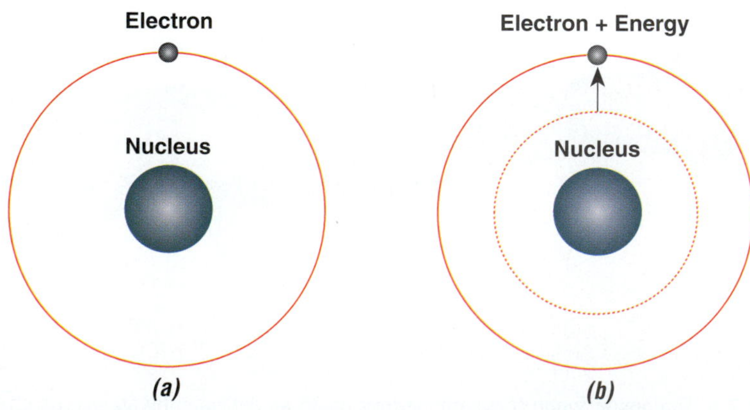

Figure 2-6 An electron has more energy when it is farther from the nucleus of an atom.

tomatically decreed that only certain orbits were allowed. Electrons in orbits farther from the nucleus contain more energy than those closer to it, as indicated in Fig. 2-6(*b*). The amount of energy is greater in the more distant electron orbits because energy must be added to pull an electron away from the nucleus.

Spectroscopy and Atoms

It had been known for many years that the color of a flame could be changed by inserting various materials in it. For example, copper turns a flame green. Around 1750 a Scottish experimenter named Thomas Melville used a prism to spread the light from a flame into a spectrum of colors. He found that only certain bands of color were produced. Those bands were dependent on the material held in the flame. The areas between the bands were dark. This differed greatly from the apparently continuous rainbow spectrum produced by passing sunlight through a prism, as shown in Fig. 2-7(*a*).

Assorted gases confined in glass tubes at low pressure glowed when a strong electric potential was applied to the electrodes. Each gas produced a different color. When the light from the glowing gas was passed through a prism, only certain bands of color were produced. Atomic theories before Bohr's could not explain these color bands; it is on this phenomenon that Bohr based his atomic theory.

When passed through a prism, the light emitted from glowing hydrogen gas produces four visible bands of color. The areas between the bands are black, as shown in Fig. 2-7(*b*). The production of only certain colors of light implies specific electron energy levels. Bohr concluded that the emission of light from an atom occurs when an electron drops to a lower energy level. When an atom absorbs a specific quantity of energy, an electron jumps to a higher energy level. The amount of energy radiated or absorbed is exactly equal to the difference in energy between the two levels. That difference in energy is called an **energy gap.**

Bohr calculated the energies required for electrons to maintain stable orbits and found that they matched the energies in hydrogen's emission spectrum. These findings prompted Bohr to believe that he had discovered a universal model for the atom. Bohr's work, based partly on speculation, was so revolutionary that in 1922, at the age of 37, he was awarded the Nobel prize for physics.

In the mid-1920s Bohr's atomic model was applied to other atoms and shown to be incorrect. Some attempts to explain the observed phenomena in atoms other than hydrogen were made by using elliptical orbits, but they also failed. At that point a new model became necessary. Despite the widely known fact that Bohr's model works only for hydrogen, the orbital paths of electrons for all atoms are

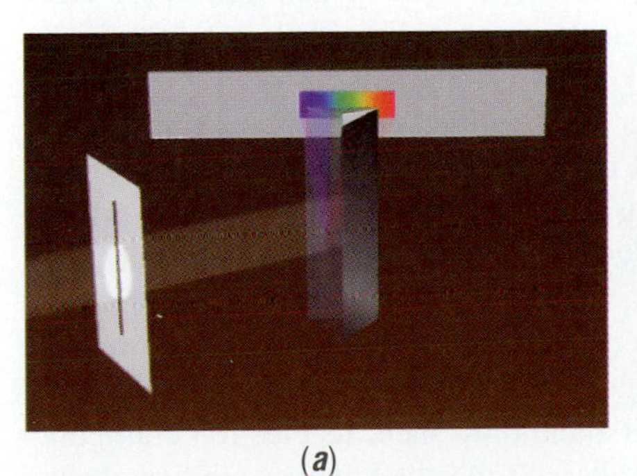

(*a*)

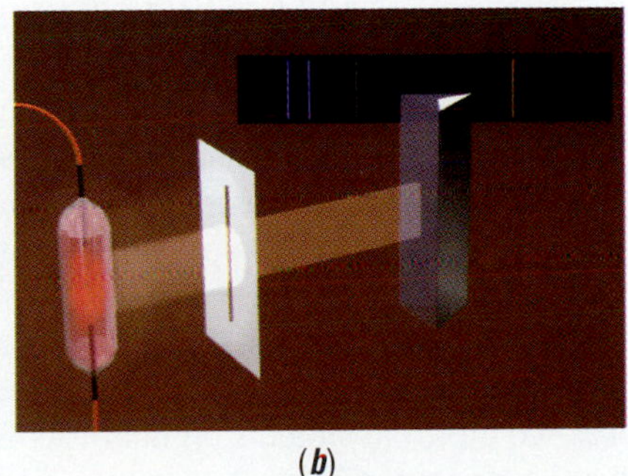

(*b*)

Figure 2-7 The various color components of visible light can be identified by passing light through a transparent glass prism.

often— although incorrectly—identified as circular. As we will see, however, electrons do not move in circular orbits around the nuclei of atoms.

A French physicist named Louis de Broglie (1892–1987) was the first to propose that electrons move about the nucleus as standing waves. Erwin Schrödinger (1887–1961), an Austrian physicist, pursued this idea and focused on the electron's wave properties. He and de Broglie began working on a wave-mechanical model for the atom.

Schrödinger developed a complex equation to define an electron's motion in three-dimensional space. A mathematical analysis of the equation is extremely complicated and extends well beyond the scope of this text, but it is sufficient to know that the equation can produce many solutions. These solutions are dependent on the energy present in the electron, which must exist in specific increments to maintain stable orbits. Graphic plots of the equation produce the appearance of three-dimensional shapes. These shapes may change dramatically as the various numbers representing the allowed energy levels for the electrons are plugged into the equation. It is easier to visualize these three-dimensional shapes if electrons, like a skywriting airplane, are thought to leave a trail to mark their path. The trail forms the surface of an imaginary object called an **orbital.** The wave functions, which form a key portion of Schrödinger's equation, cannot calculate the exact path of an electron. Rather, they are used to calculate the orbitals because orbitals provide probability maps of where electrons might be. They provide a great deal more information about the shape and structure of atoms than the positions of their electrons at any instant of time.

Electron Configurations

The electron configurations of the known elements are well-defined and are usually listed on the periodic table of elements. Bohr's proposal that electrons orbit the nucleus at specific distances was well founded—but as orbitals not circles. The orbitals occupy zones around the nucleus called **shells.** The term *shell* applies because the high velocity of the orbiting electrons causes their paths to seem almost solid. The known elements may contain as many as seven shells. The shells are numbered innermost to outermost *1, 2, 3, 4, 5, 6,* and *7.* Sometimes the shells are lettered *k, l, m, n, o, p,* and *q.*

Orbitals bear the designations *s, p, d,* and *f* and divide the shells into specific zones. These zones are often called **subshells.** The *s* orbital in any of the seven shells is nearest the nucleus and therefore represents the lowest energy level for that shell. The *p, d,* and *f* orbitals represent successively increasing energy levels within a shell. The introduction of energy into an atomic system may cause electrons to jump to a higher level orbital. Electrons occupying their natural orbitals are said to be in their **ground state.**

The electron configurations of atoms are identified by groups of three terms. The first is the shell number, the second is the letter designation for the orbital type, and the third is an exponent representing the number of electrons in the orbital. This designation system does not identify cases where two or more similarly lettered orbitals exist in the same shell.

Hydrogen, which is atomic number one, has only a single *s* orbital in shell one. Its electron configuration is designated as $1s^1$. Helium, atomic number two, has two electrons in the *s* orbital of its first and only shell. Its electron configuration is designated $1s^2$. All *s* orbitals, regardless of the shell they occupy, may contain up to two electrons. Since a shell can contain only one *s* orbital, the *s* orbital is limited to two electrons and is said to be filled when both are present. Hydrogen has a partially filled *s* orbital in shell one because it has only one electron. Helium, however, has a filled *s* orbital in shell one because it has the maximum two electrons.

Lithium, atomic number three, has a filled *s* orbital in shell one and a partially filled *s* orbital in shell two. The electron configuration for lithium is $1s^22s^1$. The fourth element on the periodic table, beryllium, has filled s orbitals in both shells one and two. The electron configuration for beryllium is $1s^22s^2$. The progressive movement of electrons to a new shell stops until the second shell contains eight electrons. Obviously, because any *s* orbital can contain only two electrons, the additional electrons must occupy a different type of orbital.

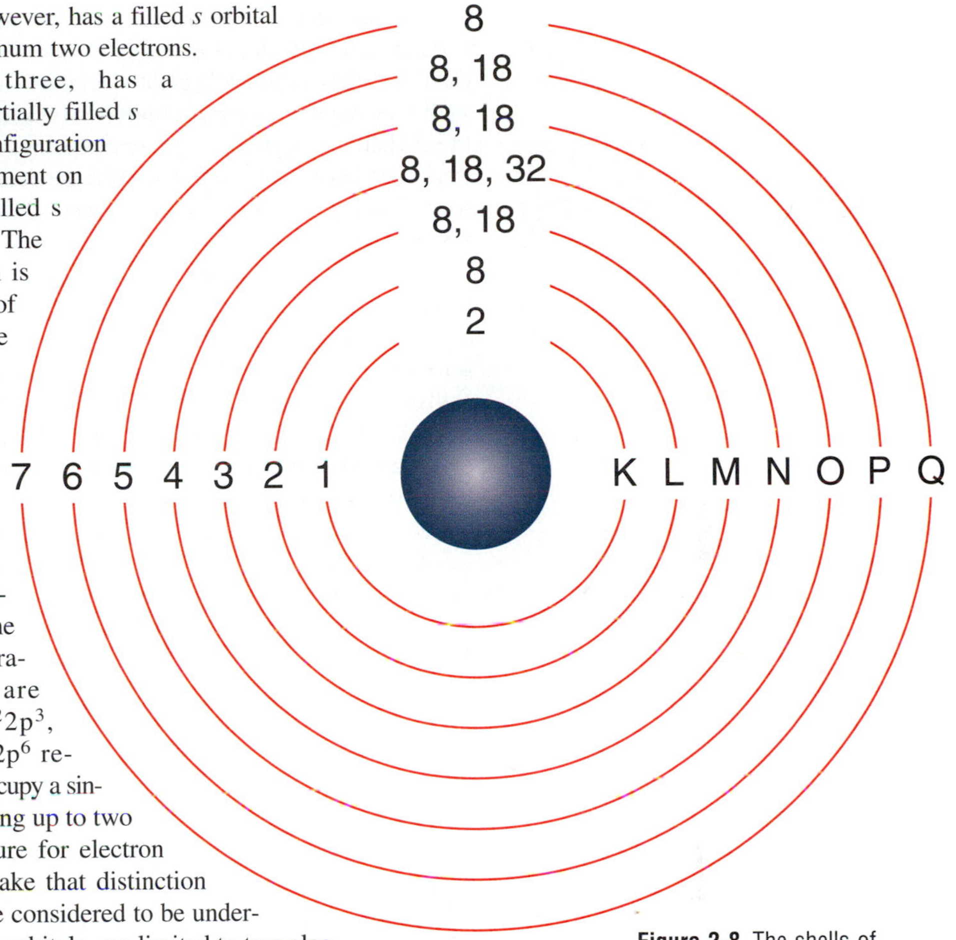

The next six elements on the periodic table—boron, carbon, nitrogen, oxygen, fluorine, and neon—have electrons occupying *s* and *p* orbitals in the second shell. The electron configurations for these six elements are $1s^22s^22p^1$, $1s^22s^22p^2$, $1s^22s^22p^3$, $1s^22s^22p^4$, $1s^22s^22p^5$, and $1s^22s^22p^6$ respectively. Up to three *p* orbitals may occupy a single shell, with each *p* orbital containing up to two electrons. (The standard nomenclature for electron configuration, however, does not make that distinction and the true nature of the orbitals are considered to be understood.) Since the *p* orbitals, like the *s* orbitals, are limited to two electrons, the *p* orbitals of any shell may contain no more than a total of six electrons. However, since shell two may contain one *s* orbital and three *p* orbitals, it can carry a maximum of eight electrons.

Figure 2-8 The shells of an atom have limits on the number of electrons they contain.

The first and second shells of all atoms after neon are identical. To simplify identification of the electron configuration, the elements from sodium through argon use the chemical symbol for neon, Ne, encased in brackets to represent the first two shells. For instance, the electron configuration for sodium ($1s^22s^22p^63s^1$) becomes $[Ne]3s^1$. The first three shells are identical for all atoms after argon. For those elements from potassium through krypton the chemical symbol for argon, Ar, replaces that of neon within the brackets and represents the first three shells. The first four shells are identical for all elements after krypton, so Kr replaces Ar for the elements rubidium through xenon. The first five shells are identical for all elements after xenon, so Xe replaces Kr for the elements cesium through radon. The first six shells are identical for all elements after radon, so Rn replaces Xe for the remaining known elements, francium through unnilhexium.

The boundary elements—neon, argon, krypton, xenon, and radon—are all chemically inert gases containing eight electrons in their outermost shell. No atom has more than eight electrons in its outermost shell. Technically, helium also marks a transition point because the first shells of all atoms after helium are identical. However, because there is only one atom before helium little would be accom-

Scientists often have to develop the equipment necessary to conduct their experiments.

plished by abbreviating the electron configuration at that point. For consistency's sake, helium is also a chemically inert gas.

As has already been hinted, there is a limit to the maximum number of electrons in each orbital and therefore in each shell. Shell one may contain up to two electrons; shell two, up to eight electrons; shell three, eight or eighteen electrons; shell four, eight, eighteen, or thirty-two electrons; shell five, eight or eighteen electrons; shell six, eight or eighteen electrons; and shell seven, in theory, eight electrons (no element known today has more than two electrons in its seventh shell). A simplified representation of the number of electrons in each shell is shown in Fig. 2-8 on page 27.

Each successive orbital may contain four electrons more than the one before it. The *s* orbitals contain up to two electrons, the *p* orbitals up to six electrons, the *d* orbitals up to ten electrons, and the *f* orbitals up to fourteen electrons. The reason for the variation in the maximum number of electrons in shells three, four, five, and six is that *d* orbitals appear in shell three for all elements after calcium, shell four for all elements after strontium, and shell five for all elements after barium. Shell four also contains *f* orbitals for all elements after lanthanum, and shell five contains *f* orbitals for all elements after thorium. Figure 2-9 shows how the shapes of the *s, p, d,* and *f* orbitals change with the addition of electrons.

The outermost shell of all atoms bears the special name **valence,** and the electrons in that shell are called **valence electrons.** All electrical activity occurs in the valence shell. The geometry of its electron orbitals determines how a material behaves chemically, or structurally, and indicates whether or not it is an electrical conductor. That is, all of the physical properties of matter are dictated by the electrical forces within its atoms.

The Neutron

In 1931 Sir James Chadwick (1891–1974), an English physicist, created an experimental apparatus where helium nuclei, called **alpha particles,** were allowed to bombard a beryllium target. The result was the emission of a previously unknown radiation, which was unaffected by electric charge or magnetic fields. Chadwick then allowed the unknown radiation to strike a paraffin (wax) sheet, which then emitted numerous protons (hydrogen nuclei). Paraffin was chosen because it contains a high percentage of hydrogen. Through a series of energy calculations Chadwick proved that the unknown radiation was a neutral particle with a mass similar to that of a proton as Rutherford had suspected. This particle was given the name **neutron,** and in 1932, Chadwick was awarded a Nobel prize in physics for its discovery.

REVIEW QUIZ 2-2

1. All materials in nature are formed from 92 natural atoms called ______________.
2. Materials made from combinations of different atoms are called ______________.
3. The smallest amount of a material that can exist and maintain the properties of that material is called a(n) ______________.
4. Electrons in the outermost shell of an atom are called ______________ electrons.

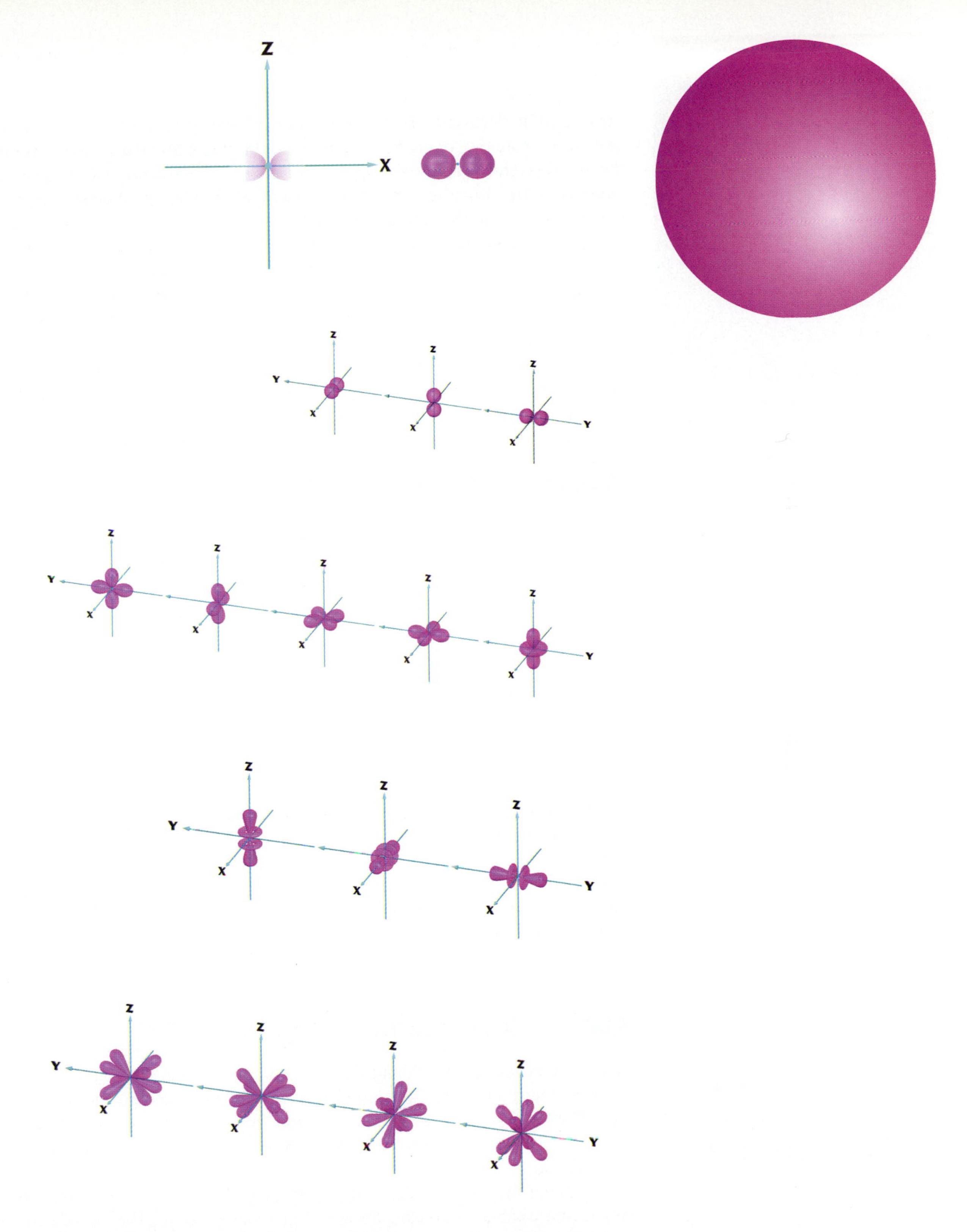

Figure 2-9 A computer-generated plot of Schrödinger's equation provides a much different picture of an atom than the traditional, but inaccurate, "solar system" model proposed by Bohr.

Figure 2-10 Generators convert a rotational force into electricity and are used for most commercial power production.

2-3 Energy Conversions

Very often in electronics, energy becomes the focus of attention as it moves from one place to another or changes from one form to another. An example would be the sound produced by a compact disc player. The sound represents energy that is being converted from light to a digital code, then from a digital code to an analog electric signal, and then back into sound.

Electricity is a medium through which energy can be converted from one form to another. The energy can be moved over great distances without any apparent mechanical system. Electricity is present in all matter and contains no energy other than that which was placed in it by an outside source. Four forms of energy produce electricity. They are: mechanical energy, chemical energy, heat energy, and light energy. Because it was produced from one of these four forms, electricity can be converted back into its original form of energy or it can be converted into any of the remaining three forms.

Mechanical Energy to Electrical Energy

The mechanical energy of friction was the first source of electricity. Although it was not a very practical source, it was all that was available for over two thousand years. Today, **generators** convert mechanical energy into electricity (Fig. 2-10).

Chemical Energy to Electrical Energy

Volta's battery is a prime example of how chemical energy can be converted into electrical energy. Batteries continue to be an important source of electricity and are available in a wide variety of shapes and sizes (Fig. 2-11). They are used in automobiles, trucks, buses, trains, telephone systems, countless portable entertainment products, cameras, flashlights, and so on.

Living organisms also convert chemical energy into electricity. In fact, "animal electricity" was once thought to be a separate form of electricity apart from that of metallic contact. The concept had been skirted around by many experimenters but was most strongly promoted by Luigi Galvani (1737–1798), an Italian scientist who discovered that the leg of a dead frog would move when two pieces of dissimilar metals were brought into contact with the leg and with each other. Belief in this "alternative" electricity was called **Galvanism.**

Heat Energy to Electrical Energy

Figure 2-11 Batteries come in a variety of shapes and sizes and are often designed for specific applications.

In 1821 Thomas Seebeck found that two dissimilar metals in contact with each other produced an electric potential when heated. He called this the **thermoelectric effect.** A device called a **thermocouple** converts heat energy into electrical energy. Seebeck's thermoelectric effect later became known as the **Seebeck effect** in his honor.

Twisting two wires of dissimilar metals together produces a simple thermocouple, but welding or bonding the metals is more effective. The use of two junctions in series, heating one and cooling the other, as shown in Fig. 2-12, improves the thermocouple's output. Although thermocouples cannot effectively compete with other sources of electric power, they are useful for measuring moderately high temperatures when they are connected across an electric meter.

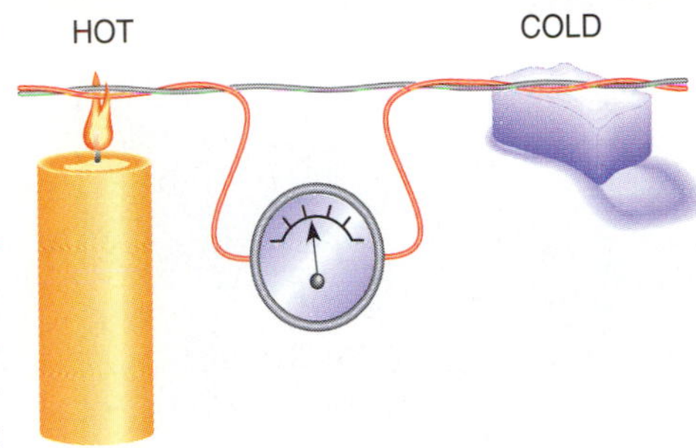

Figure 2-12 Thermocouples produce electricity from heat or differences in temperature.

Light Energy to Electrical Energy

Light energy is converted directly into electrical energy with a device called a **solar cell,** which operates on a principle called the **photovoltaic effect.** Solar cells are a source of electric power for portable calculators and for battery-charging systems in communications networks. Improvements in efficiency and reduced cost have opened new markets for solar cells in homes, automobiles, and commercial power production. Solar cells are a primary power source in satellites, where their direct exposure to the sun's light produces the maximum power obtainable (Fig. 2-13). Although they will work with any light source, only the sun provides enough light to produce practical amounts of power. Solar cells often contribute large amounts of power to local utilities in areas that have a constant supply of sunlight.

Figure 2-13 Solar cells are a primary source of power in satellites and are also used in solar farms where a constant supply of sunlight is available.

Hydroelectric Power

Electric utilities produce electricity from generators. The generators are driven by hydroelectric power, oil-powered engines, or steam turbines. The steam is produced in oil or coal-fired boilers, or from heat produced in nuclear reactors.

Hydroelectric power begins by building dams across rivers. The water backs up behind the dams, creating artificial lakes. Releasing water from these lakes provides power to turn the shafts of the generators. Hydroelectric power is a very clean and economical means for producing electricity on a commercial scale.

Hydroelectric power generation is really a solar-powered system as shown in Fig. 2-14. Heat from the sun converts water from the earth's surface into vapor. As the vapor rises into the upper atmosphere, it cools and condenses into small droplets, forming clouds. Eventually, these droplets gather and produce larger droplets that fall back to earth as precipitation. The energy used to produce hydroelectric power comes from falling water. Clouds may hold that water in suspension for a long time, in essence storing the sun's energy by opposing the pull of gravity.

REVIEW QUIZ 2-3

1. Generators convert ________________ energy into electrical energy.
2. Batteries convert ________________ energy into electrical energy.
3. Thermocouples convert ________________ energy into electrical energy.
4. Solar cells convert ____________ energy into electrical energy.

Figure 2-14 Hydroelectric power is a solar-powered system and an excellent example of how energy is converted from one form to another.

2-4 Uses of Electricity

Electricity is produced by only four forms of energy: mechanical energy, chemical energy, light energy, and heat energy. The uses of electricity for the transfer of energy, however, are numerous. Electricity can transfer power, sound, light, and information over distances limited only by the length of the wires that carry it.

Electrical Energy to Mechanical Energy

Electricity can be easily converted into mechanical energy with a **motor** or **solenoid.** An electric motor usually converts electrical energy into a rotational force, but some motors produce a linear (straight line) force. A solenoid produces a pulling or pushing force and could be considered a simple form of linear motor.

Motors and generators perform exactly opposite functions. The electricity produced by a generator can be used to power an electric motor over considerable distances. A generator's shaft rotating miles from the motor will rotate the motor's shaft as though a mechanical connection existed between them. The "mechanism" for that apparent connection is the invisible electrical energy flowing through the wires that connect the generator to the motor.

Electrical Energy to Chemical Energy

Electrical energy can also be converted into chemical energy. That conversion is most evident in **secondary batteries** like the lead-acid batteries used in automobiles. Volta's voltaic pile was a **primary battery.** The conversion of chemical energy to electrical energy in a primary battery consumes the materials from which it was made. Therefore, a primary battery cannot be recharged. Common batteries, such as carbon-zinc, alkaline, and mercury are primary batteries.

Secondary batteries do not consume the materials from which they were made. As they are used, energy is released by a chemical process. The energy that is stored chemically was placed there by the electrical energy used to charge the battery. Secondary batteries rely on reversible chemical reactions that do not consume the materials of the battery. Common rechargeable batteries are lead-acid and nickel-cadmium.

The electrolysis of water is another means of converting electrical energy into chemical energy. The energy used to separate water into hydrogen and oxygen is returned when the combination of gases is ignited. The energy is released in much less time than it takes to store it.

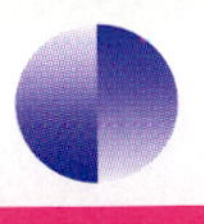

STUDENT *to* STUDENT

I've read that new types of rechargeable batteries are being developed for electric cars.

Electrical Energy to Light Energy

One of the earliest applications of electricity was the production of light. Thomas Edison spent hundreds of hours perfecting his **incandescent** lamp. *Incandescence* is a process that produces light from heat. Light is also produced by raising the energy levels of the valence electrons in gases in order to free them from their atoms. Removing electrons creates a positive **ion.** An ion is an atom that has become electrically unbalanced because it has gained or lost an electron. The energy used to remove the electron is often radiated as light when the electron returns to its natural orbit.

Electrical Energy to Heat Energy

Electricity flowing through conductors generates friction. The friction in return creates heat. Conductors that oppose the flow of electricity heat more rapidly than those that do not. This principle is used in electric heaters, ovens, toasters, irons, and numerous other small appliances.

REVIEW QUIZ 2-4

1. Motors convert electrical energy into ______________ energy.
2. Secondary batteries convert electrical energy into ______________ energy.
3. Incandescent lamps convert electrical energy into ______________ energy.
4. Electric ovens convert electrical energy into ______________ energy.

SUMMARY

The characteristics of atoms determine the electrical behavior of all materials. Atoms were once thought to have electrons orbiting the nucleus in circular orbits, but that model has been proven to be incorrect. Electrons do not travel in circular orbits. The orbital path that most closely resembles a circle is the spherical *s* orbital. It is relatively unimportant where an electron is at any instant. The important thing is the three-dimensional shape of the orbitals created by mapping where an electron might be. The orbitals determine how atoms bond together and define their electrical characteristics. Ultimately, only the valence orbits are important because they predict how an atom behaves electrically and chemically.

Atoms provide the means for energy to pass through various materials. Materials classified as conductors become a medium through which energy can be converted from one form to another or transferred from one place to another. Only four forms of energy produce electricity, and once produced, the only use for electricity is to produce energy in one of those four forms. The sources of electricity and the uses for electricity are the same.

NEW VOCABULARY

alpha particles
anode
atoms
atomic number
battery
cathode
cathode rays
cell
compounds
Crookes tube
electrolysis
electrons
elements
energy gap
Galvanism
generators
ground state
incandescent
ion
molecule
motor
neutrons
orbital
periodic table
photovoltaic effect
primary battery
protons
secondary battery
Seebeck effect
shells
solar cell
solenoid
spectroscopy
subshells
thermocouple
thermoelectric effect
valence
valence electrons
voltaic pile

Questions

1. What was the original name of Volta's battery?
2. What is the name of the chart used to organize the elements according to their characteristics?
3. What two gases combine to make water?
4. What are the three basic particles of an atom?
5. Are the charges on electrons and protons exactly equal?
6. Which is heavier, the electron or proton?
7. Were cathode rays emitted from the positive or negative electrode of a Crookes tube?
8. Cathode rays are composed of which atomic particles?
9. What was used to separate light into its component colors?
10. What name is given to the electrons in the outermost shell of an atom?

Problems

1. What three terms are used to identify the electron configurations of atoms?
2. Copper is atomic number 29 on the periodic table. That means that it has 29 protons and 29 electrons. Those 29 electrons occupy four shells. How many electrons are there in each shell?

Critical Thinking

1. Other than the high velocity of orbiting electrons, why would the orbitals be described as shells?
2. Would light be emitted by a hydrogen atom as it is ionized or deionized?
3. Why did Bohr feel that electrons had to be in motion around the nucleus of an atom?
4. How is work stored in electricity?
5. Was Galvani's "animal electricity" really any different from that produced by Volta's battery?
6. How are primary and secondary batteries different?
7. Would the absorption of energy by an atom cause an electron to jump to a higher orbital or leave the atom completely?
8. Knowing that light is emitted from atoms only at specific electron energies, does it seem likely that light from the sun, which is composed only of hydrogen and helium, is really a continuous spectrum?

Answers to Review Quizzes

2-1
1. battery
2. electrolysis

2-2
1. elements
2. compounds
3. molecule
4. valence

2-3
1. mechanical
2. chemical
3. heat
4. light

2-4
1. mechanical
2. chemical
3. light
4. heat

Chapter 3

Numbers, Conversions, *and* Units *of* Measure

ABOUT THIS CHAPTER

This chapter begins with a discussion of numbers and how they are used in the decimal system. It introduces and explains many mathematical terms as a prerequisite to the introduction of scientific notation and the conversion of fundamental units from one to the other.

After completing this chapter you will be able to:

- ▽ Describe the decimal system of numbers.
- ▽ Explain the functions of fractions, logarithms, and scientific notation.
- ▽ Use scientific notation to multiply and divide numbers and explain when it is used.

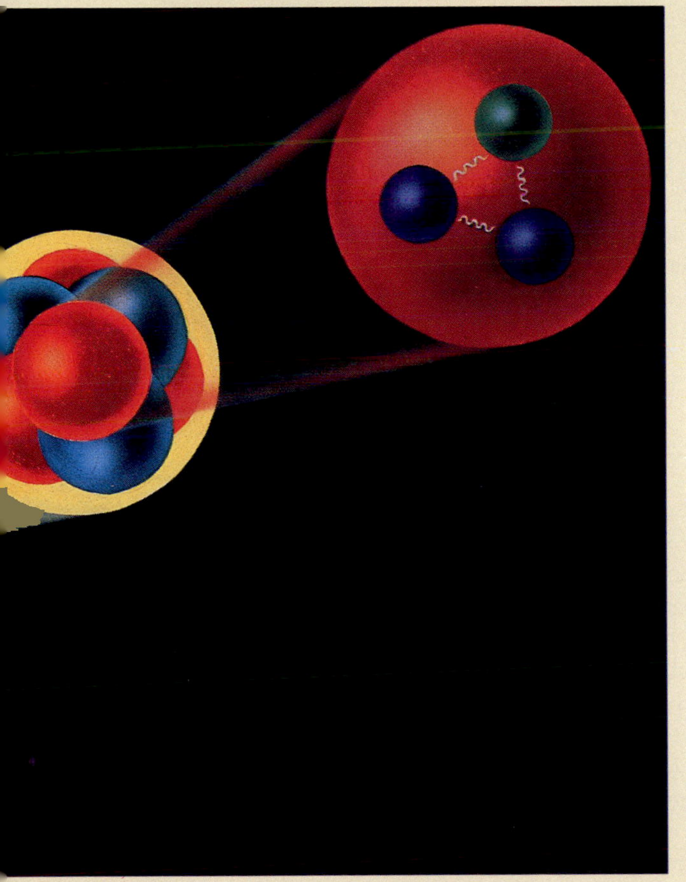

Numbers encountered in electronics range from the very small to the very large. These numbers are divided into named groups every thousand units. These fundamental units often become the center of attention because of the inconvenience of referencing large and small numbers to one. How these units relate to each other and how numbers are converted from one unit to another is presented.

There are three major systems of measure in common use: metric, American, and the SI system. Charts are provided to show the interrelationships of the three systems using units of length, mass, force, temperature, energy, and time.

▽ State the basic units of physical measurement and the common systems of measure.

▽ Convert numbers from unit to unit or system to system.

3-1 Number Systems and Usage

The most commonly used number system is **decimal.** The word *decimal* means based on 10. The decimal system is composed of the ten familiar symbols: 0, 1, 2, 3, 4, 5, 6, 7, 8, and 9. Each of these symbols is called a **digit.** The word *digit* comes from the Latin word *digitus,* meaning finger. The popularity of a system based on ten symbols is not surprising considering the convenience of counting on one's fingers.

Number systems may be based on any number of symbols, but only a few are commonly used. The number of symbols used in a number system defines its **base.** Decimal is a **base ten** system because it uses ten symbols. Sometimes the base of a number is shown as a subscript after the number. For example, 1001_{10} is identified as a base ten number by the subscripted 10 shown after the last digit.

STUDENT *to* STUDENT

Pay attention to the definitions in this chapter. You'll need to know these terms later.

Counting

Counting is a matter of using the symbols of a number system as place markers. In the decimal system, counting sequentially steps through the symbols 1, 2, 3, 4, 5, 6, 7, 8, and 9. Notice that the symbol 0 has not appeared yet. The decimal system begins counting with 1, assuming the count to be zero until there is something to count. When all the symbols have been used, but the count has not been completed, the symbols are repeated. The number of repetitions must also be counted. After the first set of symbols, the number of times the set is repeated is placed before each symbol in the set. The 0 appears in all subsequent repeats of the symbol set: 1, 2, 3, 4, 5, 6, 7, 8 9, 10..., 20..., 30..., 100..., 200..., 300..., and so on.

Leading Zeros

Certain types of numbers, such as calendar dates, have practical limits on how many decimal places are shown. Dates contain three general divisions: months, days, and years. The months can range from 1 to 12, the days within a month from 1 to 31, and the years, using only the last two digits, from 00 to 99. Each date element requires no more than two decimal places.

Computerized equipment often requires that all decimal places be occupied by a symbol. A date, such as January 5, 1996, becomes 01/05/96. This example displays the **leading zeros** associated with numbers displayed using a fixed number of decimal places. Television remote control systems often require a leading zero before channels 2 through 9 can be selected. An alternative system waits to see if a second digit is entered within a predetermined time period. If it is not then the single digit channel is selected, unless it was 0 or 1, which is ignored.

Leading zeros in the decimal system are normally suppressed. This practice is so common that it is easy to forget their existence. Other number systems display leading zeros far more frequently.

Integers

Numbers that represent complete units are called **integers.** Integers are also called **whole numbers.** They include all of the **natural numbers,** the negatives of those numbers, and zero. Natural numbers include the number 1 and all numbers created by adding 1 to another natural number.

Fractions

A **fraction** is a form of **ratio** written as a division problem that establishes the equality or inequality of two numbers. The number used as a reference is written below the first number with the two numbers separated by a horizontal line or a slash. Regardless of the value of the bottom number, it represents one unit. The number on top can be greater than, equal to, or less than the number on the bottom, but its value will always be in relation to the bottom number, as shown in Fig. 3-1.

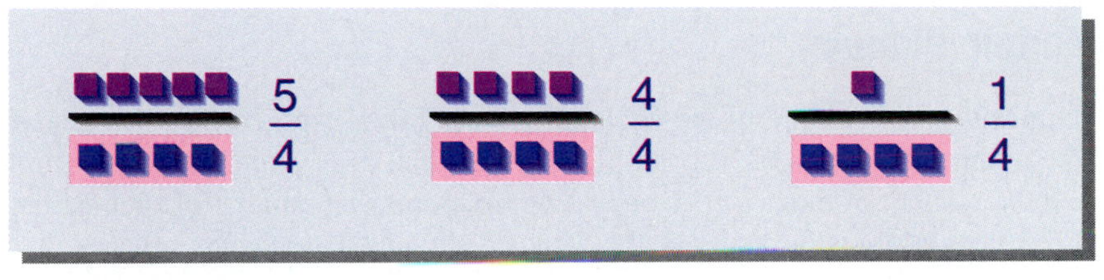

Figure 3-1 Fractions are written as the ratio of incomplete units to whole units.

The top part of a fraction is called the **numerator** and the bottom part of a fraction is called the **denominator.** The answer obtained when the numerator is divided by the denominator is called a **quotient.**

A fraction whose numerator is smaller than its denominator is called a **proper fraction.** A fraction whose numerator is larger than or equal to the denominator is called an **improper fraction.** The quotient of a proper fraction is called a **decimal fraction.** A decimal fraction has a denominator that is understood to be ten or a multiple of ten. The denominator is expressed with a **decimal point.** For instance: $0.035 = {}^{35}\!/_{1000}$. (See the discussion under "Number Places" on page 40.)

The nonzero quotient of an improper fraction is called a **mixed number.** A mixed number is composed of an integer and a fraction such as $4\,{}^{5}\!/_{8}$. When a mixed number is written as a decimal, the integer is placed to the left of the decimal point. In decimal form $4\,{}^{5}\!/_{8}$ becomes 4.625.

Placing a zero before the decimal point in a proper decimal fraction clearly identifies the number as a fractional value, but the zero is not mandatory. The leading zero can be helpful because there is always the possibility that what appears to be a decimal point is just a blemish or accidental mark on the paper. That possibility is eliminated when the leading zero is used.

The Real Number Line

To help visualize how numbers relate to each other, imagine they exist along a horizontal line called a **number line,** as shown in Fig. 3-2 on the next page. The numbers that reside on that line are called **real numbers.** The line is called the **real number line** for that reason. The real number line is separated into two halves by the digit 0. Zero and all numbers to its right are positive. All numbers to the left of 0 are negative. Although 0 is positive, it separates the positive and neg-

-9 -8 -7 -6 -5 -4 -3 -2 -1 0 1 2 3 4 5 6 7 8 9

Negative Positive

Real Number Line

Figure 3-2 The 0 and all numbers to its right are positive. All numbers to the left of 0 are negative.

ative numbers because it is not a **counting number.** The counting numbers are 1, 2, 3, 4, 5, 6, 7, 8, 9, . . . , and the negatives of those numbers.

Number Places

Each position to the right of the decimal point is called a **place.** The first place is the number of tenths, the second place is the number of hundredths, the third place the number of thousandths, and so on for an infinite number of times. Figure 3-3 demonstrates the order of the values assigned to the decimal places.

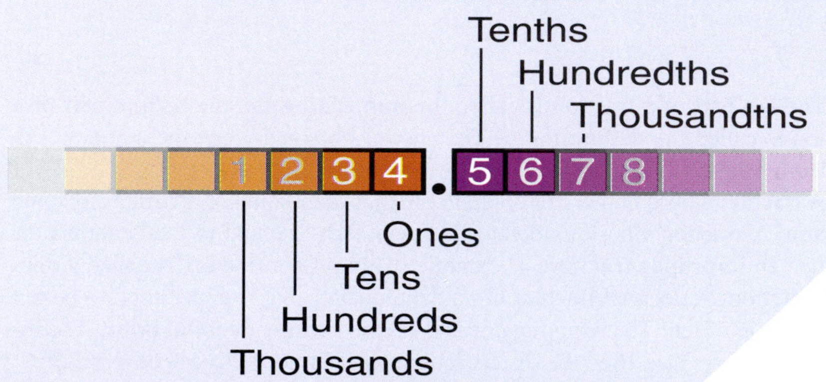

Figure 3-3 Each decimal place is one-tenth the value of the place to its left when a number is written in decimal form.

Place Accuracy

The fractional portion of 1.5 indicates that the number is larger than the integer 1 by five-tenths. Placing a zero in the hundredth's place (1.50) indicates that the fractional portion is larger than the integer by 50 hundredths, and placing a second zero in the thousandth's place (1.500) indicates that the fractional portion is larger than the integer by 500 thousandths. This process could be repeated any number of times. It may seem redundant, but there are many situations where additional digits must be shown for the sake of accuracy. Increasing the number of digits after the decimal point, even though they may be zeros, expresses the overall number with greater precision.

The precision to which a number is expressed is identified by how many places in the fractional portion are occupied by a symbol. An infinite number of invisible places are considered to exist anytime a decimal point is placed after an integer, but it is not necessary or practical to place digits in all of them.

Rounding Off

How many fractional digits a number needs is determined by the required accuracy. The accuracy of a fractional number is established through a process called **rounding off**. The number 1.5 is expressed to one place accuracy because only the first place to the right of the decimal point is occupied by a digit. The same number at two place accuracy is 1.50. At three place accuracy it becomes 1.500. The extra zeros may seem unnecessary, but they are required when the number must be accurately stated. For example, when the number is taken to four place accuracy it becomes 1.4996. The reason for the apparent change in value is that the 1.5, 1.50, and 1.500 resulted from rounding off the number's original value.

A number is rounded off by first determining the desired place accuracy. If the digit in the next highest place is five or greater, the digit in the desired place is rounded up. For instance, to express 2.683 to one place accuracy, look at the digit to the right of the 6 and determine whether it is 5 or greater. Since 8 is greater than 5, the 6 is rounded up to 7 and all digits beyond one place are thrown away. Thus the number becomes 2.7. If the digit to the right of the desired place is less than five, it and any subsequent digits are discarded and the number in the desired place remains unchanged. For example, the rounded off value of 2.64 to one place accuracy is 2.6.

STUDENT *to* STUDENT

Be careful! Round-off errors can cause you to think you have made a mistake when you haven't.

Floating Point

The default mode of most calculators is **floating point.** Calculator displays are limited to a specific number of digits, usually eight, ten, or twelve. The decimal point moves (floats) along the visible display to accommodate the varying size of the integer portion of the displayed numbers. The last digit in the display is rounded off. Most calculators carry many more places internally than shown in the display. This improves the accuracy of the answers while dealing with the necessity of rounding off numbers because of physical limitations in the calculator's circuitry.

Fixed Point

Some calculators allow the display to be fixed at any number of places from zero to the limits of the display. This mode of operation is called **fixed point.** A beneficial feature of using fixed point is that the calculated results are easier to read. It can be very confusing to look at eight or more digits. Most calculations performed in determining the parameters of electronic circuits result in answers that are adequately displayed using one- to four-place accuracy. Even though the calculator display is rounded off to a fixed number of decimal places, the number is maintained internally to as many places as the circuitry will allow.

Round-Off Error

Calculations involving rounded off numbers can produce unexpected results. For example, using one of the most sophisticated hand-held scientific calculators available to find the square root of 2 produces 1.41421356237. The calculator also has a function to square a number. Taking the square root of 2 and then squaring the result should return the 2, but it does not. The answer in the display is instead 1.99999999999.

The calculator did not give the expected result because of **round-off error.** The square root of 2 produces what is called an **infinite series.** That means that it cannot be completely expressed regardless of the number of places allowed. Even the extensive number of places carried internally by this expensive calculator is inadequate to produce accurate results.

If the mode of the calculator is changed from floating point to fixed point at ten place accuracy, the correct and expected result is displayed. When the square root of 2 is calculated, the answer to ten places becomes 1.4142135624. The last digit at eleven places is 7. The tenth digit is then increased from 3 to 4. The numbers that the calculator carries internally do not change, but limiting the number of places displayed by the calculator produces the expected result. Round-off error is an integral part of mathematics that must be accepted and understood. It can be reduced when using a calculator by avoiding writing down a calculated result and manually reentering it later. Whenever possible, maintain all numbers within the calculator.

REVIEW QUIZ 3-1

1. The decimal is a base ______________ system.
2. Numbers that represent complete units are called ______________ .
3. The top part of a fraction is called the ______________, and the bottom part is called the ______________ .
4. Numbers with many digits after the decimal point must be ______________ to make them easier to work with.

3-2 Scientific Notation

STUDENT *to* STUDENT

Scientific notation is very helpful once you understand it.

Very large and very small numbers are common in scientific studies such as electronics. Using such numbers increases the possibility of introducing errors because they contain many zeros either before or after the decimal point. These zeros can make large and small numbers difficult to read. Another problem is that their length often exceeds the display capability of most calculators. Converting such numbers to **scientific notation** overcomes these problems. Even many inexpensive calculators convert to scientific notation when a calculated result becomes too large or too small for the display. Scientific notation is a form of numeric shorthand that eliminates the troublesome zeros by expressing numbers as a **power of 10.**

Logarithms and Scientific Notation

The actual mechanism behind scientific notation is based on the use of logarithms. The **logarithm** of a number is the **power** to which 10 is raised to equal the number. *Power* is another word for the **exponent** of 10. A decimal point sometimes divides logarithms into two parts. The part to the left of the decimal point is called the **characteristic,** and the part to the right of the decimal point is called the **mantissa.**

With 10 or an exact multiple of 10, its logarithm contains only the characteristic and does not have a mantissa. For example, the log of 100 (which is 10^2) is 2, and the log of 1000 (which is 10^3) is 3. It follows, then, that a number larger than 100 but smaller than 1000—for example, 625—has a log that is greater than

2 and less than 3—in this case the log is 2.79588. The mantissa represents that portion of the number that lies between exact powers of ten.

In scientific notation, 625 would be written as 6.25×10^2. The 6.25 is the mantissa in **antilog** form, the 10 is the base, and the exponential 2 is the characteristic. Antilogs are the opposite of logs because they are the number that raising 10 to some power represents. Like logs, antilogs are looked up in tables or found using a scientific calculator. The log of 625 is 2.79588, with the mantissa being 0.79588. The antilog of 0.79588 is 6.25. When 6.25 is multiplied by 100, also written as 10^2, the 625 reappears.

Scientific notation is just as useful for small numbers as it is for large ones. The logarithms of numbers less than one, but larger than zero, have negative logarithms. For example: $\log 0.005 = -2.301$.

The Mechanics of Scientific Notation

Nonwhole numbers use a decimal point to separate the integer portion from the fractional portion. Integers do not require a decimal point, but it would not change their value if one were present, it would only change their implied accuracy.

The first step in converting an integer to scientific notation is to place a decimal point after the integer. This decimal point is then moved leftward as many times as necessary until only a single counting number remains to its left. The convention of maintaining one *counting* number to the left of the decimal point is called **normalizing.** Remember that zero is not a counting number. Each place the decimal point was moved leftward divides the number by 10. It is this repetitive division that makes the number small and manageable, and allows the zeros to be eliminated.

Figure 3-4 The decimal point at the end of a number does not change its value.

The zeros cannot be simply thrown away because the resulting number no longer represents the original value. The divisions by 10 are compensated for by multiplying the number by 10 raised to some power. The power is written as an exponent that indicates how many times the number was divided by 10.

The following example demonstrates the process of converting an integer to scientific notation.

Example 3-1 Convert 4384 to scientific notation.

1. Place a decimal point at the end of the number as shown in Fig. 3-4.
2. Move the decimal point in the mantissa to the left until only one counting number remains, as shown in Fig. 3-5. This will make the number smaller and more manageable.
3. Combine the mantissa with $\times 10^3$ to obtain 4.384×10^3, which reads as four-point-three-eight-four times ten to the third power.

4.384

Figure 3-5 Each place the decimal point is moved to the left divides the mantissa by 10. Three divisions of 10 lead to a total division of 1000. The mantissa is restored to its original value when it is multiplied by 10 three times. Scientific notation indicates this as $\times 10^3$.

Notational Shorthand

Expressing whole multiples of ten in scientific notation does not require the use of the $1\times$ before the power of 10. It is understood that the 1 is there.

10	=	1×10^1	=	10^1	10,000	=	1×10^4	=	10^4
100	=	1×10^2	=	10^2	100,000	=	1×10^5	=	10^5
1000	=	1×10^3	=	10^3	1,000,000	=	1×10^6	=	10^6

Fractional Number Conversion

A fraction must be converted to a decimal before it can be expressed in scientific notation. Converting a decimal number to scientific notation is similar to converting whole numbers. However, the movement of the decimal point is to the right instead of the left. The goal of having a single counting number to the left of the decimal point remains the same. The value of the exponent decreases by a factor of 10 each time the decimal point is moved to the right. Leading zeros in the mantissa resulting from the movement of the decimal point are discarded.

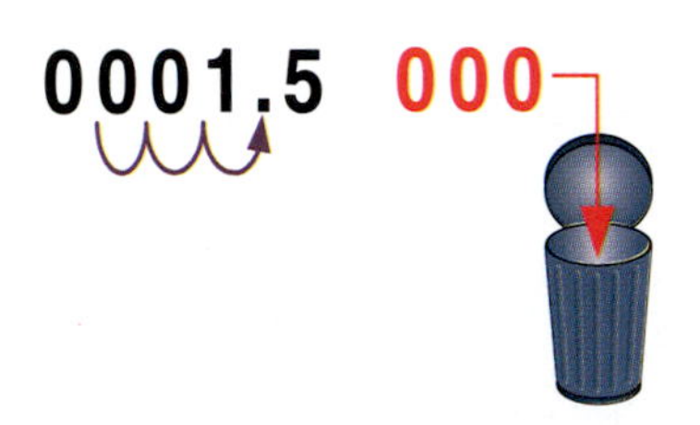

Figure 3-6 When converting decimal values to scientific notation the decimal point is moved to the right of the first counting digit. Any leading zeros created in the process are discarded.

Example 3-2 Convert 15 ten-thousandths to scientific notation.

1. Write 15 ten-thousandths as a fraction and convert it to decimal form by dividing it by 10,000:

$$\frac{15}{10{,}000} = 0.0015$$

2. Move the decimal point to the right past the first counting number and stop. In this example, the decimal point is moved three places. The leading zeros created in this process are discarded as shown in Fig. 3-6.
3. Moving the decimal point to the right multiplied the mantissa by 10 three times, making it 1000 times larger. Dividing the number by 10 three times restores its original value. That division process is indicated by $\times\ 10^{-3}$. The result is written as 1.5×10^{-3}.

Conversion to Standard Notation

The sign of the exponent is the key to converting scientific notation to standard notation. A positive exponent directs you to repeatedly *multiply* the mantissa by 10 the number of times indicated by the exponent. For example, with 4.384×10^3, moving the decimal point to the right three places accomplishes the necessary multiplication. Discard the $\times$ 10. If the number is an integer discard the decimal point as shown in Fig. 3-7.

A negative exponent indicates that the mantissa is *divided* by 10 the number of times indicated by the exponent. The division is accomplished by moving the decimal point to the left. The empty places between the start and end positions of the decimal point are filled with zeros as shown in Fig. 3-8. A leading zero is added for clarity.

Basic Math Using Scientific Notation

Scientific notation simplifies the multiplication and division of large and small numbers. Although it may also be applied to addition and subtraction, the numbers must be adjusted so that their exponents are equal. That requirement usually cancels any benefit scientific notation has to offer if it is being done without a calculator.

Multiplication

Only the numbers before the $\times$ 10 are multiplied when using scientific notation. The exponents are added **algebraically** as shown in the following examples. Do

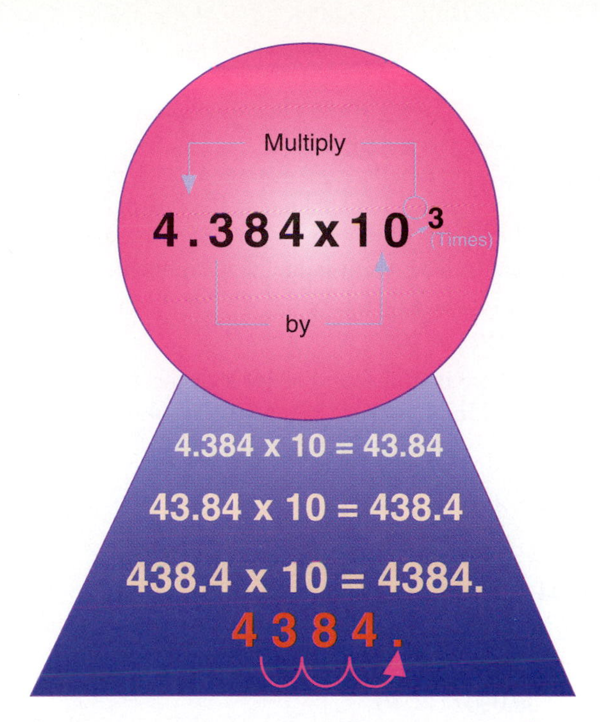

Figure 3-7 A positive exponent indicates that the mantissa is to be multiplied by 10 as many times as indicated by the exponent.

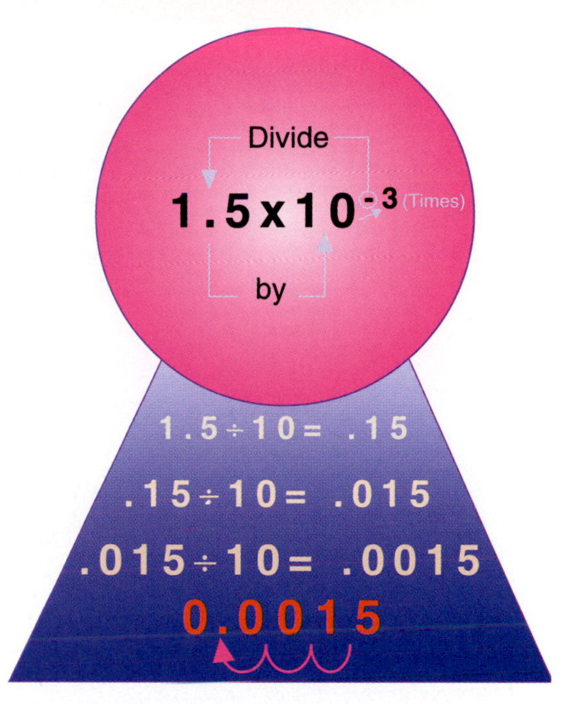

Figure 3-8 A negative exponent indicates that the mantissa is to be divided by 10 as many times as indicated by the exponent. The places between the start and end positions of the decimal point are filled with zeros.

not forget to *add* the exponents. It is very tempting to multiply them, but it doesn't work that way. To add numbers algebraically means to consider their positive and negative signs. Any number without a sign is understood to be positive.

Example 3-3 Multiply (2.5×10^3) by (2×10^3).

1. Multiply the mantissas.

$$2.5 \times 2 = 5$$

2. Insert a $\times$ 10 after the 5.

$$5 \times 10$$

3. Add the exponents algebraically. The result becomes the new exponent.

$$(+3) + (+3) = 3 + 3 = 6$$

4. The scientific notation for $(2.5 \times 10^3) \times (2 \times 10^3)$ is 5×10^6.

Example 3-4 Multiply (3.15×10^6) by (1.6×10^{-8}).

1. Multiply the mantissas.

$$3.15 \times 1.6 = 5.04$$

2. Insert a $\times$ 10 after the 5.04.

$$5.04 \times 10$$

3. Add the exponents algebraically. The result becomes the new exponent.

$$(+6) + (-8) = 6 - 8 = -2$$

4. The scientific notation for $(3.15 \times 10^6) \times (1.6 \times 10^{-8})$ is 5.04×10^{-2}.

Example 3-5 Multiply (1.256×10^4) by (3.2×10^{-4}).

1. Multiply the mantissas.

$$1.256 \times 3.2 = 4.0192$$

2. Insert the $\times$ 10 after the 4.0192.

$$4.0192 \times 10$$

3. Add the exponents algebraically. The result is the new exponent.

$$(+4) + (-4) = 4 - 4 = 0$$
$$4.0192 \times 10^0$$

4. When the exponent is zero there is no need to show it or the $\times$ 10.
5. The scientific notation for $(1.256 \times 10^4) \times (3.2 \times 10^{-4})$ is 4.0192.

Example 3-6 Multiply (6.354×10^{-9}) by (5.68×10^2).

1. Multiply the mantissas.

$$6.354 \times 5.68 = 36.09072$$

2. Add the $\times$ 10 after the 36.09072. Do not normalize the number at this time.

$$36.09072 \times 10$$

3. Add the exponents algebraically. The result is the new exponent.

$$(-9) + (2) = -9 + 2 = -7$$
$$36.09072 \times 10^{-7}$$

4. Normalize the mantissa by moving the decimal point one place to the left. This divides the mantissa by 10.

$$36.09072 \text{ becomes } 3.609072$$

5. Round off the mantissa to three places.

$$3.609$$

6. Change the exponent to compensate for dividing the mantissa by 10. This is accomplished by increasing the exponent by 1 for each time the mantissa was divided by 10.

$$-7 + 1 = -6$$

7. The scientific notation for $(6.354 \times 10^{-9}) \times (5.68 \times 10^2)$ is 3.609×10^{-6}.

Division

Because only the mantissas are divided, division using very large and very small numbers is greatly simplified in scientific notation. The exponents are algebraically subtracted. Algebraic subtraction is similar in most respects to algebraic addition. The signed exponents are placed within parentheses. The parenthesized values are then separated by a minus sign to indicate subtraction. For the sake of protocol the numerator should be placed ahead of the denominator as shown in Example 3-7.

Example 3-7 Determine the mass ratio between a proton and an electron.

Given: The mass of a proton is $1.6726231 \times 10^{-27}$ kg.
The mass of an electron is $9.1093897 \times 10^{-31}$ kg.

Do not round off the mantissas at this point.

1. Divide the mantissas.

$$1.6726231 \div 9.1093897 = 0.183615275566$$

And round off the mantissa to four places.

$$0.1836$$

2. Algebraically subtract the denominator from the numerator. The result is the new exponent.

$$(-27) - (-31) = -27 + 31 = 4$$

$$0.1836 \times 10^4$$

3. Normalize the mantissa by moving the decimal point one place to the right. This multiplies the mantissa by 10.

0.1836 becomes 1.836

4. Add the × 10.

$$1.836 \times 10$$

5. Change the exponent to compensate for multiplying the mantissa by 10. This is accomplished by decreasing the exponent by 1 for each time the mantissa was multiplied by 10.

$$4 - 1 = 3$$

6. The mass ratio between a proton and an electron is 1.836×10^3

Example 3-8 Divide (2×10^3) by (2×10^{-3}).

1. Divide the mantissas.

$$2 \div 2 = 1$$

2. Add the × 10.

$$1 \times 10$$

3. Algebraically subtract the denominator's exponent from that of the numerator. The result is the new exponent.

$$(+3) - (-3) = 3 + 3 = 6$$

4. The scientific notation for $(2 \times 10^3) \div (2 \times 10^{-3})$ is 1×10^6.

Example 3-9 Divide (9.838×10^{-6}) by (1×10^{12}).

1. Divide the mantissas.

$$9.838 \div 1 = 9.838$$

2. Add the × 10

$$9.838 \times 10$$

3. Algebraically subtract the denominator's exponent from that of the numerator. The result is the new exponent.

$$(-6) - (12) = -6 - 12 = -18$$

4. The scientific notation for $(9.838 \times 10^{-6}) \div (1 \times 10^{12})$ is 9.838×10^{-18}.

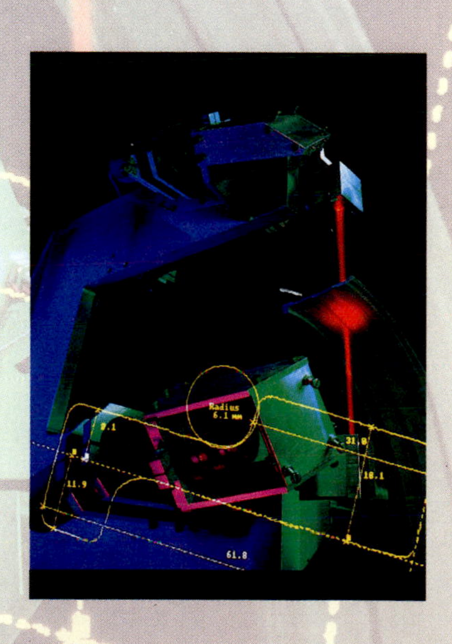

Example 3-10 Divide 1 by (6.2415×10^{18}).

1. Divide the mantissas.

$$1 \div 6.2415 = 0.1602$$

2. Algebraically subtract the denominator's exponent from that of the numerator. The result is the new exponent. In this case the numerator does not have an exponent. The exponent is then understood to be zero.

$$0 - 18 = -18$$

3. Normalize the mantissa by moving the decimal point one place to the right. This multiplies the mantissa by 10.

0.1602 becomes 1.602

4. Add the $\times$ 10

$$1.602 \times 10$$

3. Change the exponent to compensate for multiplying the mantissa by 10. This is accomplished by decreasing the value of the exponent by 1 for each time the mantissa was multiplied by 10.

$$-18 - 1 = -19$$

6. The scientific notation for $1 \div (6.2415 \times 10^{18})$ is 1.602×10^{-19}.

REVIEW QUIZ 3-2

1. When a number is expressed in scientific notation the portion ahead of the $\times$ 10 is called the ____________.
2. The portion of a number written in scientific notation at the upper right of the $\times$ 10 is called the ____________.
3. Scientific notation expresses numbers as a power of ____________.
4. When numbers expressed in scientific notation are multiplied their exponents are ____________.

3-3 Unit Conversion

Standard Fundamental Units

In day-to-day life, quantities generally revolve around the number 1—How many children do you have? How many dollars per hour do you make? How many cars do you have? As such, 1 serves as the reference point for evaluating other quantities.

Numeric values in electronics are usually cataloged into units that increase or decrease by a factor of 1000. Each unit has its own name and symbol. Table 3-1 lists the units and symbols from smallest to largest.

Conversion of Units

Conversion of units is often necessary in electronics and should eventually become second nature. The conversion from one named unit to another is not always obvious. It requires an understanding of the relationship between the two quantities. Converting common units, such as feet to inches, may not seem difficult but is often done incorrectly. For example, if 3 feet were converted to inches, the usual

Table 3-1 Powers of 10

Prefix	Symbol	Pronunciation	Power of 10	Number
atto	(a)	one-quintillionth	10^{-18}	0.000000000000000001
femto	(f)	one-quadrillionth	10^{-15}	0.000000000000001
pico	(p)	one-trillionth	10^{-12}	0.000000000001
nano	(n)	one-billionth	10^{-9}	0.000000001
micro	(m)	one-millionth	10^{-6}	0.000001
milli	(m)	one-thousandth	10^{-3}	0.001
one	(1)			1
kilo	(k)	one thousand	10^{3}	1000
mega	(M)	one million	10^{6}	1,000,000
giga	(G)	one billion	10^{9}	1,000,000,000
tera	(T)	one trillion	10^{12}	1,000,000,000,000

method is to multiply 3 feet by 12 inches for a total of 36 inches. The unwanted unit of feet simply gets discarded. In fact, the procedure used gives an answer of 36 feet-inches, which is confusing, although correct. This method may not be a problem with units encountered daily, but when unfamiliar units appear it can be unnecessarily confusing.

The proper procedure employs a **conversion factor** that relates what you want to know to what you already know. An equality must exist between the two. The following problems are examples.

Example 3-11 Determine how many inches in 3 feet.

Given: 12 inches = 1 foot

1. Set up the units of feet and inches as a division problem. The unit you want in the answer must be in the numerator. This establishes a ratio between the units called a conversion factor. Although the units are different, the quantities they represent must be equal.

$$\frac{12 \text{ inches}}{1 \text{ foot}}$$

2. Multiply the conversion factor by the variable quantity whose units are in the denominator. Here the conversion factor is multiplied by 3 feet. The units of feet and foot cancel.

$$3 \cancel{\text{feet}} \left(\frac{12 \text{ inches}}{1 \cancel{\text{foot}}} \right) = 3 \times 12 \text{ inches} = 36 \text{ inches}$$

Example 3-12 How many kilometers are there in 5 miles?

Given: 1.609 kilometers = 1 mile

1. Set up the conversion factor.

$$\frac{1.609 \text{ kilometers}}{1 \text{ mile}}$$

2. Multiply the conversion factor by the variable quantity whose units are in the denominator. Here the conversion factor is multiplied by 5 miles. The units of miles and mile cancel.

$$5 \cancel{\text{miles}} \left(\frac{1.609 \text{ km}}{1 \cancel{\text{mile}}} \right) = 5 \times 1.609 \text{ km} = 8.045 \text{ km}$$

Example 3-13 How many megaseconds are there in 24 hours? (This problem cannot be done in one step because the unit asked for in the question is not included in the information.)

Given: 60 minutes = 1 hour.
60 seconds = 1 minute.

1. Set up a conversion factor that relates minutes to hours.

$$\frac{60 \text{ minutes}}{1 \text{ hour}}$$

2. Convert the number of hours given to minutes. The units of hours and hour cancel.

$$24 \cancel{\text{hours}} \left(\frac{60 \text{ minutes}}{1 \cancel{\text{hour}}} \right) = 24 \times 60 \text{ minutes} = 1440 \text{ minutes}$$

We now know the number of minutes in 24 hours.

3. Set up a conversion factor that relates seconds to minutes.

$$\frac{60 \text{ seconds}}{1 \text{ minute}}$$

4. Convert the number of minutes calculated in step 2 to seconds. The units of minutes and minute cancel.

$$1440 \cancel{\text{minutes}} \left(\frac{60 \text{ seconds}}{1 \cancel{\text{minute}}} \right) = 1440 \times 60 \text{ seconds} = 86{,}400 \text{ seconds}$$

The number of seconds in 24 hours has now been found, but the question asked for the number of megaseconds in 24 hours. One more conversion factor is necessary.

5. Set up a conversion factor that relates megaseconds to seconds.

$$\frac{1 \text{ megasecond}}{1 \times 10^6 \text{ seconds}}$$

6. Convert the number of seconds calculated in step 4 to megaseconds. (It is advisable to convert the seconds from standard notation to scientific notation because the numbers being dealt with are so large.)

$$8.64 \times 10^4 \cancel{\text{seconds}} \left(\frac{1 \text{ megasecond}}{1 = 10^6 \cancel{\text{seconds}}} \right) = (8.64 \times 10^4)(1 \times 10^{-6} \text{ megaseconds})$$
$$= 8.64 \times 10^{-2} \text{ megaseconds}$$

REVIEW QUIZ 3-3

1. A(n) ______________ relates two quantities that are equivalent but expressed in different units.
2. Conversion factors are written as ______________ problems.
3. Conversion factors allow the ______________ of unwanted units.

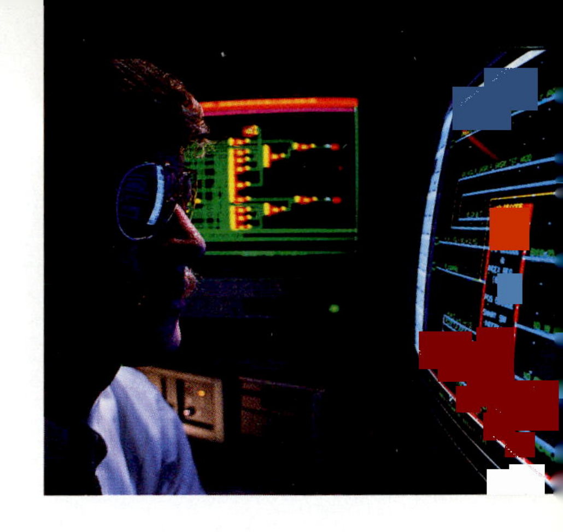

3-4 Physical Measurement

The units of measure used in electronics continue to be a mixture of metric, American, and **SI** (*Système International*) values, despite efforts to consolidate them under the SI system. To make matters even more confusing, the metric system has two major divisions, **cgs** (centimeter-gram-second), and **mks** (meter-kilogram-second). The SI system uses mks units. The use of the different systems is often perpetuated by manufacturers of electronic components and equipment who are reluctant to change to the SI system.

The International Bureau of Weights and Measures, located at Sèvres, France, has hosted the General Conference of Weights and Measures, a meeting that is attended by representatives from virtually all the world's nations. In 1960, the General Conference adopted a system called *Le Système International d'Unités* (International System of Units), which has the international abbreviation SI. The SI system was adopted by the Institute of Electrical and Electronic Engineers, Inc. (IEEE) in 1965 and by the United States of America Standards Institute as a standard for all scientific and engineering literature in 1967.

Nonelectrical units of measure fall into one of six general categories: length, mass, force, temperature, energy, and time.

Length Measurements

Units of length are **scaler quantities** that primarily revolve around the SI/metric unit of **meters** and the American standard unit of inches. The values of the other units in the two systems are based on the meter and the inch. If there is any doubt, consider the following questions: What is a yard? A yard is 3 feet or 36 inches. What is a foot? A foot is 12 inches. The metric system is simpler to learn because all of its multiple and submultiple units are based on the meter and show their origins in their names. For example, a centimeter is one-hundredth of a meter, and a kilometer is one thousand meters. Length units encountered in electronics are multiples and submultiples of the meter in the SI/mks system and miles, yards, feet, inches, and mils in the American standard system. The concept of a fundamental unit, around which other units are based, is very important to a complete understanding of electric circuits.

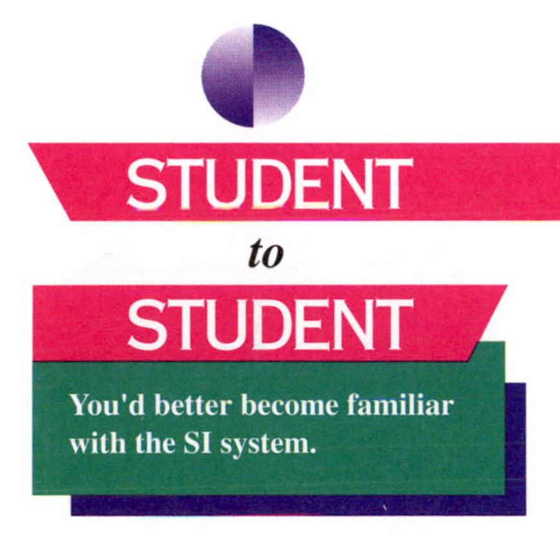

Mass

Mass is the measure of a body's resistance to acceleration. It is different from weight but is proportional to it. Weight is the force that gravity exerts on mass. If the gravity were changed, the weight would change, but the mass would remain the same, even in space. The major units of mass are the **kilogram** (which is an SI/mks unit) and the **gram** (which is a metric cgs unit). The American standard unit of mass is the **slug** and is of little or no importance in scientific studies. It is the amount of mass that is accelerated at a rate of one foot per second per second when acted upon by a force of one pound weight. The chart in Fig. 3-9 on the next page provides the necessary conversion factors to convert units of mass from one system to another.

Force

Force can create motion. It is referred to as a vector quantity because it has both magnitude and direction. One **newton** of force is an SI/mks unit that can accelerate a one kilogram mass one meter per second per second. That means that in the

Mass Conversion Chart

System	Units	SI MKS	CGS	Am. Stan.
SI MKS	Kilogram (kg)		1000 g	0.0685 slug
CGS	Gram (g)	0.001 kg		0.0001 slug
Am. Stan.	Slug (slug)	14.5939 kg	14593.9 g	

Figure 3-9 This conversion chart provides the conversion factors necessary to convert mass units between systems of measure.

first second it would have moved 1 meter, in the second second 2 meters, in the third second 3 meters, and so forth. The cgs unit of force is the **dyne**, which can accelerate one gram one centimeter per second. It would take 100,000 dynes to produce the force of one newton. In the American standard system force is measured in pounds. The pound unit is the force exerted by a one pound mass in the earth's gravitational field. The chart in Fig. 3-10 provides the necessary conversion factors to convert units of force from one system to another.

Force Conversion Chart

System	Units	SI MKS	CGS	Am. Stan.
SI MKS	Newton (N)	1 N	1×10^{5} dynes	0.2248 lb
CGS	Dyne	1×10^{-5} N	1 dyne	2.248×10^{-6} lb
Am. Stan.	Pound (lb)	4.448 N	4.448×10^{5} dynes	1 lb

Figure 3-10 This chart provides the conversion factors necessary to convert units of force between systems of measure.

Temperature

Heat is nothing more than the motion of atoms and molecules. When all molecular motion stops there is no heat whatsoever. The temperature at which that occurs is called **absolute zero**. Two systems of temperature measurement, **Fahrenheit** and **Celsius** (which is called **centigrade** in the cgs system) have an arbitrary range of values based on the freezing and boiling point of water. Fahrenheit is an American standard system where water freezes at 32°F and boils at 212°F. The Celsius/centigrade system sets the freezing point of water at 0°C and the boiling point at 100°C. The SI unit for temperature is **Kelvin**. The Kelvin scale starts at 0 K, called absolute zero, and increases in the same size units as the Celsius/centigrade system. Occasionally it is necessary to convert from one system to another. The chart in Fig. 3-11 provides the conversion factors necessary to do so.

Energy

Energy is the ability to do work, which is force multiplied by distance. In the SI/mks system the unit for energy is the **joule**; in the cgs system it is the **erg**; and

Temperature Conversion Chart

System	Units	SI	MKS	CGS	Am. Stan.
SI	Kelvin (K)		°C + 273.15	°C + 273.15	$5\left(\frac{°F + 459.67}{9}\right)$
MKS	Celsius or Centigrade (°C)	K - 273.15		Units are equal	$5\left(\frac{°F - 32}{9}\right)$
CGS	Centigrade (°C)	K - 273.15	Units are equal		$5\left(\frac{°F - 32}{9}\right)$
Am. Stan.	Fahrenheit (°F)	$\left(\frac{9(K)}{5}\right)$-459.67	$\left(\frac{9(°C)}{5}\right)$+32	$\left(\frac{9(°C)}{5}\right)$+32	

Figure 3-11 This chart provides the conversion factors necessary to convert units of temperature from one system to another.

in the American standard system it is the **foot-pound**. The joule is the amount of work done in moving one kilogram one meter; an erg is the amount of work done in moving one gram one centimeter; and the American standard foot-pound is self-explanatory. Notice that the amount of time used to move the mass is not indicated. The work done in moving something is not dependent on the time used to move it. The chart in Fig. 3-12 identifies the relationships between the various units.

Energy Conversion Table

System	Units	SI MKS	CGS	Am. Stan.
SI MKS	Joule (J)	1 J	1×10^{7} ergs	0.738 ft-lb
CGS	Erg [Dyne-centimeter]	1×10^{-7} J	1 erg	7.376×10^{-8} ft-lb
Am. Stan.	Foot-pound (ft-lb)	1.356 J	13.558×10^{7} ergs	1 ft-lb

Figure 3-12 This table identifies the relationships between units of energy in various systems of measure.

Time

The units of time are the same in all systems: hours, minutes, and seconds. The second is often divided into the much smaller units of milli, micro, pico, and nano.

REVIEW QUIZ 3-4

1. One meter is equal to ______________ yards.
2. One newton is equal to ______________ foot-pounds.
3. One inch is equal to ______________ centimeters.
4. The standard measurement system in science and engineering is the ______________ system.
5. The ______________ measurement system is still widely used in the United States.

SUMMARY

Electronics is a field of study that makes extensive use of numbers. Numbers range from the very small to the very large with both extremes commonly encountered simultaneously. A valuable tool for dealing with such numbers is scientific notation. Certain aspects of scientific notation will become mechanical to you as you use them. Scientific notation offers a powerful solution to the multiplication and division of large and small numbers, but it is not as convenient when applied to addition and subtraction. A scientific calculator is a useful tool for those studying electronics.

The wide range of numbers encountered in electronics is divided into named units. Each unit has a symbol that acts as an abbreviation for its name. Electrical and physical quantities must often be converted from one unit to another. Before that can happen, an equality must be established between the two units involved in the conversion. The equality is established by using a conversion factor.

Electronics also involves a great deal of measurement. Not all measurements are electrical. Length, mass, temperature, energy, and time are measured in American standard, metric, and SI units. The SI system is the official standard in science and engineering.

NEW VOCABULARY

absolute zero
algebraically
antilog
base
base ten
Celsius
centigrade
cgs
characteristic
conversion factor
counting number
decimal
decimal fraction
decimal point
denominator
digit
dyne
erg
exponent
Fahrenheit
femto
fixed point
floating point
foot-pound
fraction
gram
improper fraction
infinite series
integers
joule
Kelvin
kilogram
leading zeros
logarithm
mantissa
meters
mixed number
mks
natural numbers
newton
normalizing
number line
numerator
place
power
power of 10
proper fraction
quotient
ratio
real number line
real numbers
round-off error
rounding off
scaler quantities
scientific notation
SI
slug
whole numbers

Questions

1. What number system is based on 10 symbols?
2. What numbers are integers?
3. What is another name for integer?
4. What is a proper fraction?
5. Is a number written in decimal form a type of fraction?
6. Is zero a counting number?
7. Can rounding off numbers introduce errors into an equation?
8. Is the zero before a proper decimal fraction mandatory?

Problems

1. Multiply (3×10^5) by (2.04×10^{-2}).
2. Multiply (4.15×10^0) by (4×10^6).
3. Multiply (3.18×10^{-4}) by (1.5×10^{-5}).
4. Divide (6.28×10^{-4}) by (2×10^6).
5. Divide (1×10^3) by (4×10^3).
6. How many feet are in 1 kilometer?
7. How many pico units are in 1 micro unit?
8. How many meters are in 1 mile?

Critical Thinking

1. (a) Is it feasible that the counting system in number systems other than decimal is similar to decimal except for the number of symbols used? (b) Could the count start at 0 instead of 1? (c) Show how a count of 10 would be written in a base 3 number system using standard symbols and starting with zero.
2. How could you indicate that 1001 is a binary (base 2) number instead of a base 10 number?
3. How are weight and mass different from each other?
4. Why can't anything be colder than 0 K?

Answers to Review Quizzes

3-1
1. ten
2. integers (whole numbers)
3. numerator, denominator
4. rounded off

3-2
1. mantissa
2. exponent
3. ten
4. added

3-3
1. conversion factor
2. division
3. cancelation

3-4
1. 1.0936
2. 0.2248
3. 2.54
4. SI
5. American standard

Chapter 4

Voltage, Current, and Resistance

ABOUT THIS CHAPTER

Voltage, current, and resistance are often called "the big three" in electricity. This chapter defines and examines each quantity. It also discusses how charge is measured and how other quantities are related to it. Current and charge are contrasted using graphs that show how closely related quantities can produce vastly different results when

After completing this chapter you will be able to:

- Define basic electrical quantities, such as charge, current, voltage, and resistance.
- Understand the color-coding system used on resistors.
- Describe the numerical equivalent of the standard EIA colors.

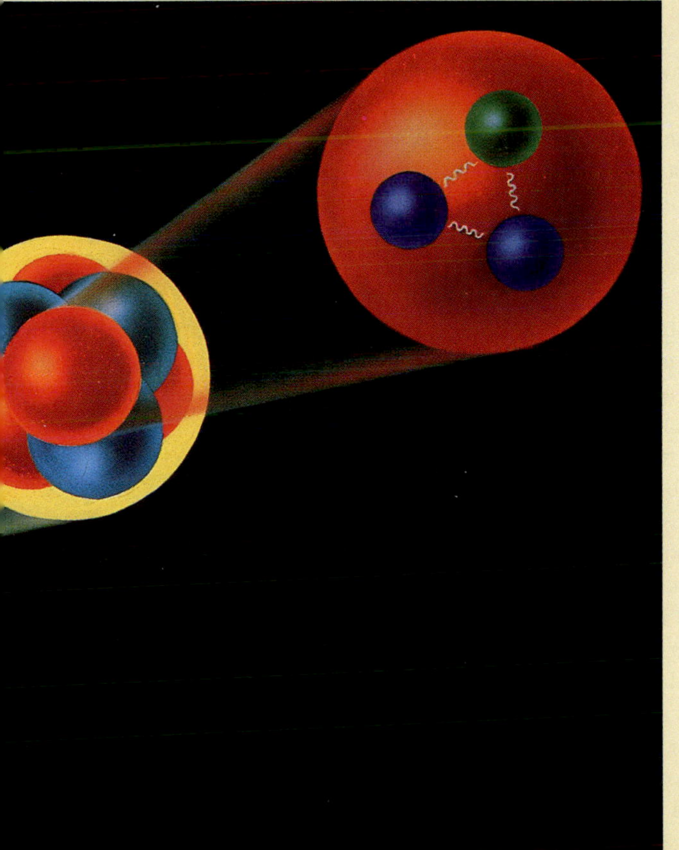

viewed over time. In addition, the chapter presents the categories of color-coded resistors and shows how resistor manufacturers use color codes to encrypt resistance values. One of the most fundamental relationships in electricity is Ohm's law. The chapter discusses how voltage, current, and resistance are mutually defined using Ohm's law rather than the physical quantities upon which they are based. Finally, the first of five circuit types, the simple circuit, is presented and examined.

- ▽ Explain the interrelationship of voltage, current, and resistance using Ohm's law.
- ▽ Describe how a simple circuit operates.
- ▽ Explain how circuit values vary from the OFF to ON condition.

4-1 Electrical Quantities

The work required to move an individual electron is very small and seldom of interest. A practical amount of work is achieved when millions of electrons are moved between points of dissimilar charge. An understanding of the mechanism behind the conversion of work into electricity and the reciprocal conversion of electricity into work is made possible by defining the following quantities:

1. A quantity of charge at rest.
2. The quantity of charge moved in a given period of time.
3. The electrical pressure that causes the charge to move.
4. Any opposition to the flow of charge that may exist.

These quantities become named electrical units.

The Coulomb

The **coulomb** is the basic unit of electric charge. It is equal to the combined charge on 6.24×10^{18} electrons or protons. The charge on an electron or proton is often called a **unit charge**; therefore, the coulomb is also defined as 6.24×10^{18} unit charges. The coulomb can be thought of in the same sense as a gallon or liter of liquid, in that all are quantities that may be at rest or in motion. A capital C is normally used to abbreviate the word *coulomb*. An italicized capital C represents capacitance. An italicized capital Q or an italicized lowercase q may be used in equations to represent a quantity of charge in coulombs.

The nature of matter is such that objects become electrically charged because electrons are added to or taken from them. The quantity of electrons moved in the process is normally measured in coulombs. When Benjamin Franklin disassembled a Leyden jar to discover where the charge was stored, he concluded that it was in the glass. He also determined that both inside and outside foils were necessary to allow charge to be taken from one side of the glass and given to the other.

Removing 1 coulomb of charge from the foil on the inside of the Leyden jar caused it to become positively charged. Depositing that same coulomb on the outside foil caused it to become negatively charged. The absolute difference in charge between the inside and outside foils is then 2 coulombs because the inside foil is $+1Q$ and the outside foil is $-1Q$. The total difference in charge of $2Q$ is exactly twice the quantity of charge that could be moved by discharging the Leyden jar.

An electron-moving force is developed between any dissimilarly charged objects. The objects do not have to be charged equally and oppositely. The difference in charge can be between objects that are both positive, but one is more positive than the other; between objects that are both negative, but one is more negative than the other; between a positive object and an object that has zero charge; between a negative object and an object that has zero charge; between positive and negative objects of unequal charge; and between equal and opposite charges. Examples of differently charged objects are shown in Fig. 4-1.

French physicist André Marie Ampère (1775–1836).

Current

Electric current is charge in motion—any charge, electrons, protons, or ions. The unit for current is the **ampere,** which is named in honor of French physicist André Marie Ampère (1775–1836). In electrical formulas, current is represented by an italicized capital letter I, meaning intensity. The type of charge I is to represent

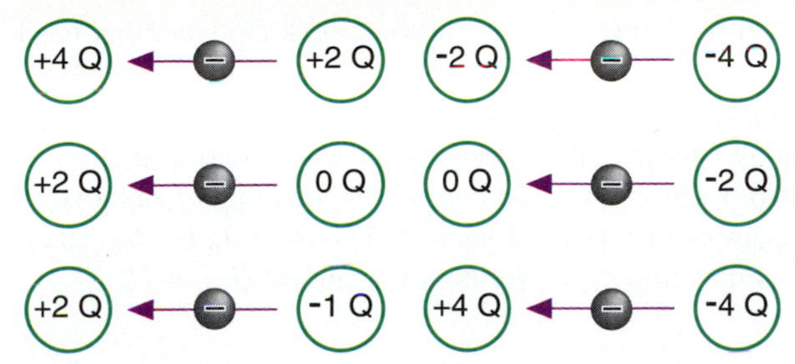

Figure 4-1 An electron moving force is developed between any difference in charge, not just between positive and negative charges.

must be established before the rules of circuit operation can be developed and adopted. Unless specifically noted otherwise, I will represent electron flow.

Ampère suspected that two parallel, current-carrying wires exert a force on each other because of interactions between their magnetic fields. Through several innovative experiments he discovered that two parallel wires attract each other when currents flow through them in the same direction and that they repel each other when the currents are made to flow in opposite directions. During 1822 and 1823, Ampère developed the mathematical formulas necessary to calculate the forces between the two current-carrying wires.

The ampere unit was not defined by Ampère, but by German physicist Wilhelm Eduard Weber (1804–1891), in cooperation with German mathematician Johann Karl Fredrich Gauss (1777–1855). Gauss defined the **unit magnetic pole** and worked with Weber to establish a consistent set of interrelated electrical units. In 1849 Weber defined current in terms of a current-carrying wire of unit length exerting a unit force on one of Gauss's magnetic poles at a unit distance. The name *ampere* was given as the unit of current at the Paris International Congress in 1881.

The modern definition of an ampere is more closely related to Ampère's original experiments. The Ninth General Conference on Weights and Measures adopted the ampere unit for electric current in 1948, with the following definition:

> The ampere is that constant current which, if maintained in two straight parallel conductors of infinite length, of negligible circular cross section, and placed 1 meter apart in a vacuum, would produce between these conductors a force equal to 2×10^{-7} newton per meter of length.

Although technically the ampere is defined in terms of the physical force developed between two current-carrying wires, it is more common to declare the ampere as the rate of flow of electric charge. Traditionally, it is equivalent to 1 coulomb of charge moving past a given point in 1 second:

Equation 4-1

$$I = \frac{\Delta Q}{\Delta T} \Rightarrow \text{Current} = \frac{\text{change in charge}}{\text{change in time}} \Rightarrow 1 \text{ ampere} = \frac{1 \text{ coulomb}}{1 \text{ second}}$$

where ΔQ = the change in the number of coulombs
I = the average current in amperes
ΔT = the time interval current flowed in seconds

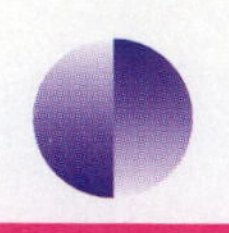

STUDENT *to* STUDENT

The slope of a line is positive if the line is uphill, zero if the line is flat, and negative if the line is downhill.

The symbol Δ (**delta**) is a letter of the Greek alphabet that has been adopted by mathematicians to mean change in or difference in. The symbol ⇒ represents the word *then*.

Interestingly enough, the coulomb, which was used to define current, is often defined in terms of current. One coulomb is the quantity of charge transferred when a current of 1 ampere flows for 1 second. The formula for the quantity of charge moved is a manipulation of the formula for current (Eq. 4-2).

Equation 4-2

$$Q = I \times T \Rightarrow \text{Charge} = \text{current} \times \text{time} \Rightarrow 1 \text{ coulomb} = 1 \text{ ampere per second}$$

where Q = the charge in coulombs
I = the current in amperes
T = the time in seconds

It is not uncommon for electrical units, such as the coulomb and ampere, to define each other. This seemingly ambiguous system of circular references is generally more convenient than referring to the absolute physical constants on which the units are based. The graph in Fig. 4-2 provides an example of how charge and current are related.

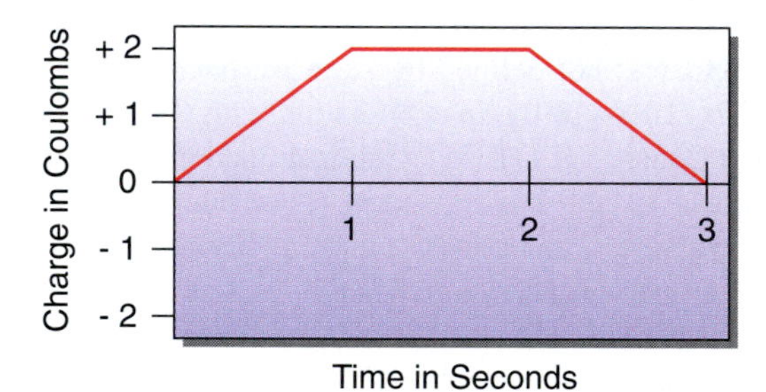

Figure 4-2 The relationship of charge and current.

The graph shows the charge-versus-time **waveform** for a particular circuit. A waveform is any graphic display of some quantity in relation to time. During the first time interval, the charge builds from 0 to 2 coulombs. During the second time interval, the charge remains at 2 coulombs and does not change. During the third time interval, the charge drops from 2 coulombs back to 0 and stops.

Prior to doing the calculations necessary to plot a graph of the resultant current, the slope of the line used to trace the quantity of charge over the given time period reveals what the current is doing. For the first time interval, the slope is positive because the value of charge is increasing. Slope indicates that the current is positive, whatever direction positive is agreed to be. During the second time interval, the slope is 0 because the charge did not change. There can be no current because there is no change in charge. Current is not dependent on the presence or absence of charge; it exists only when charge moves. During the third time interval, the slope of the line is negative because the charge is declining. The current, therefore, will be negative, indicating that it has changed directions.

The graph in Fig. 4-3 shows the current-versus-time waveform resulting from the charge-versus-time waveform of Fig. 4-2. The values on this graph are constant from one time interval to the next. The current for each of the three time intervals is found using Eq. 4-1. The change in charge during the first time interval was +2 coulombs. The change in time was 1 second. The change in charge during

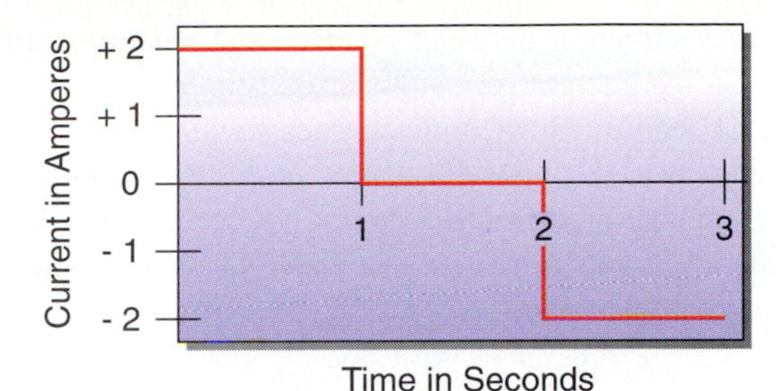

Figure 4-3 A 2-amp current flowed in one direction for 1 second, stopped for 1 second, then a 2-amp current flowed in the reverse direction for 1 second.

STUDENT *to* STUDENT

You'll be seeing lots of schematics, so it's important to memorize the symbols used for electrical quantities.

the second time interval was 0 because the charge began and ended at 2 coulombs. The change in time was again 1 second. Zero divided by any number is 0, so no current flowed during the second time interval. The change in charge during the third time interval was −2 coulombs because the charge went from 2 coulombs to 0. The change in time was once again 1 second. Minus 2 coulombs divided by 1 second equals −2 amperes. Two coulombs divided by 1 second equals 2 amperes.

Waveform analysis is an important tool for technicians. The graphs in Figs. 4-2 and 4-3 show that two closely related electrical quantities can have markedly different waveforms.

Voltage

Voltage is the force that causes charge to move. A voltage is created with any difference of electric charge, but it is normally considered to be the electron moving force between a positive and negative charge. Voltage is often described as a **difference of potential** or **electromotive force,** abbreviated **emf.** The unit of electromotive force is the **volt** and is named for Alessandro Volta. If the positive and negative polarities of a voltage source do not reverse at regular intervals, the source is said to be **DC,** which is an abbreviation for **direct current. AC**, which is an abbreviation for **alternating current,** will be discussed later.

Voltage is represented in mathematical formulas by the capital letter *V*, or sometimes the letter *E*. The letter V is an abbreviation for *volt* and the E is an abbreviation for *electromotive force.* The volt unit, like the ampere, was adopted at the Paris International Congress in 1881 and was also the result of efforts by Weber and Gauss to standardize electrical quantities.

For simplicity's sake, DC voltage sources in this text are represented using the **schematic** symbol for a battery, as shown in Fig. 4-4. A schematic is a wiring diagram that uses a standard set of symbols to represent components and the wiring that connects them.

The basic battery symbol consists of a pair of parallel lines, one long and one short. The long line is the positive terminal, and the short line is the negative terminal. Any number of line pairs may be used to represent a battery or DC voltage source, but there must always be an even number of lines beginning with a long and ending with a short. The voltage cannot be determined by the number of line pairs used in the symbol, but generally speaking more line pairs indicate higher voltages.

The basic units of charge—electrons and protons—coexist within matter in equal numbers, forming an electrical balance. To create an electrical imbalance

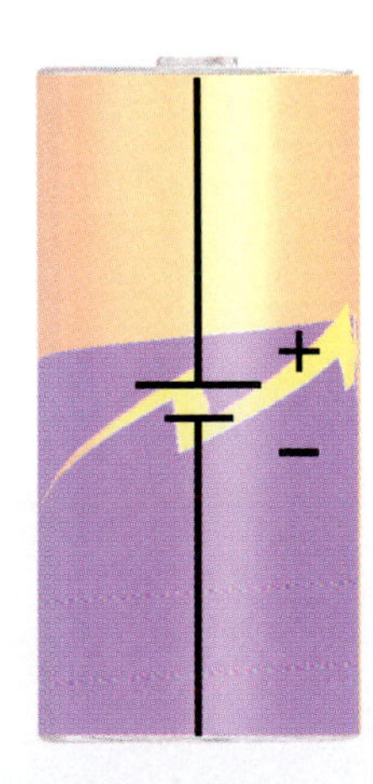

Figure 4-4 A long and short line in the battery symbol represents a single cell. The actual voltage of a battery is not indicated by the number of cells portrayed in its symbol. It is printed alongside the symbol.

(voltage), the mutual attraction between positive and negative particles must be overcome by introducing energy from an external source. The energy used in taking the charges apart is returned when the charges are allowed to go back together.

The mutual attraction of opposite charges represents a force that can create motion. The sheer application of force does not represent work, only the potential for work. In electrical circuits, the force is created by voltage and the movement is electric charge, usually electrons.

Resistance

Resistance is an opposition to current and is caused by a lack of available charge carriers or a difficulty in moving charge carriers through a material. The unit of resistance is the **ohm,** which is named in honor of Georg Simon Ohm (1787–1854), a Bavarian-born German physicist. The Greek capital letter omega (Ω) is often used as an abbreviation for the word *ohm.* In mathematical equations, the ohm is represented by the capital letter *R.*

REVIEW QUIZ 4-1

1. The force that causes charge to move is called ______________ .
2. Any charge in motion is called ______________ .
3. Opposition to current is called ______________ .
4. The unit for quantity of charge is the ______________ .

4-2 Color-Coded Resistors

There is often a need to limit the current in an electronic circuit. That need is met using prepackaged resistance elements called **fixed resistors.** Resistors are manufactured in a wide range of values designed to meet the general needs of circuit designers. The **carbon film** resistor is one of the most common electronic components. The desired resistance is created by coating the outside of a small ceramic tube with a thin coating (film) of carbon. Then metal caps with wire leads welded to them are placed over each end to make a mechanical and electrical connection to the carbon element. A protective coating of epoxy resin is applied, then bands of color are painted around the resistor to identify its resistance value and tolerance. A cross section of a carbon film resistor is shown in Fig. 4-5(*a*), and a full view is shown in Fig. 4-5(*b*). Resistors are manufactured in various sizes depending on the power they are required to handle.

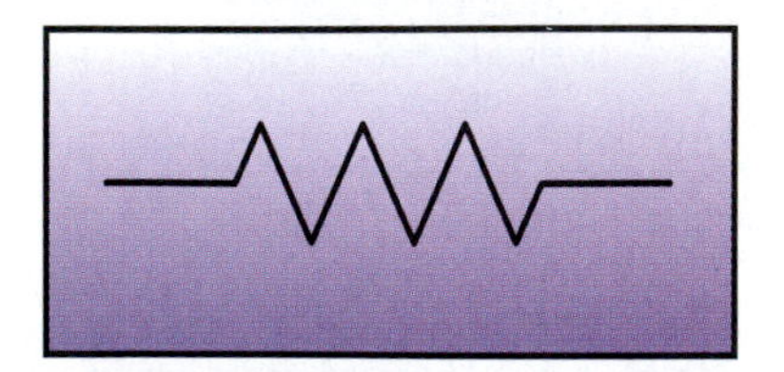

Figure 4-6 The schematic symbol for a resistor is a zigzag line that usually has three points on each side. The length or size of the symbol does not alter its meaning.

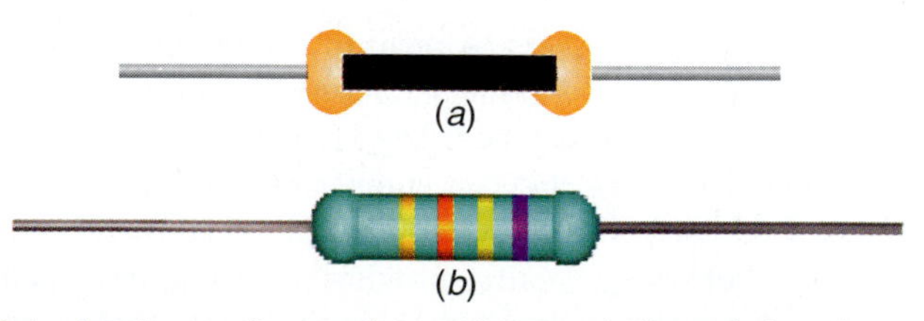

Figure 4-5 Color bands are often used on resistors to indicate their resistance values. Resistors of different sizes may have identical color bands because the size is an indication of how much heat a resistor can dissipate without being damaged.

The schematic symbol for a resistor is a zigzag line that usually has three points on each side (Fig. 4-6). However, sometimes the symbol may be drawn with more than three points. The actual number of points or the size at which the symbol is drawn does not affect its meaning. A resistor's value is written next to its symbol. If the value is followed by k, it is in kΩ (1 kΩ = 1000 Ω). If the value is followed by MΩ, it is in megaohms (1 MΩ = 1,000,000 Ω).

Figure 4-7 Standard EIA colors represent the numerals 0 through 9.

EIA Color Code

Color-coded resistors use three, four, or five color bands to identify their value and accuracy. The specific colors and their meanings have been standardized by the **Electronic Industries Association** (EIA) and are universal.

The EIA has adopted 10 colors to represent the numerals 0 through 9. They are as follows: black = 0, brown = 1, red = 2, orange = 3, yellow = 4, green = 5, blue = 6, violet = 7, gray = 8, and white = 9. The colors are shown in ascending order from top to bottom in Fig. 4-7.

The following whimsical saying may help you memorize the colors and the numbers they represent:

> *Black Birds Ravaged Our Yellow Grain Because Violet Got Witty.*

The bands on all color-coded resistors are painted closer to one end than the other. The end to which the bands are closest should be viewed so that it is on your left. Although there are necessarily differences in the way three-band, four-band, and five-band resistors are read, generally they are very similar.

Three-Band Resistors

Resistors with only three bands are very obsolete and would be found only in older equipment where the accuracy of its components was not critical. Three-band resistors have a tolerance of ±20% and were available in only six **decade values**,[1] shown in Table 4-1. Aside from having broad tolerances, these resistors were often unstable under extremes of thermal and physical stress. That instability seemed most pronounced for resistors of 1 MΩ or more. The EIA has classified these resistors as type E-6.

Table 4-1

10	15	22	33	47	68

The colors of the first and second bands represent literal numbers and are called the *first significant figure* and *second significant figure*, respectively. The third band is called the *multiplier*. It represents the number of zeros to be placed after the number obtained from the first two bands. For example, assume that the colors on a resistor are red, violet, and orange, as shown in Fig. 4-8. The value of the resistor would be: red = 2, violet = 7, orange = 3, but in the multiplier band it is three zeros (000). Therefore, the value of the resistor is 27,000 Ω, or 27 kΩ. The tolerance of the resistor is ±20% because there is no fourth band.

[1]Decade values are those resistors from 10 Ω to the last value under 100 Ω that are available in a specific group. In addition to the decade values multiples and sub-multiples of those values complete the available selection.

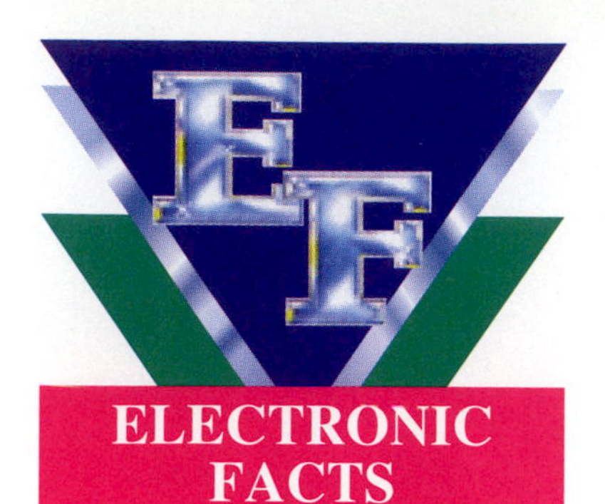

ELECTRONIC FACTS

Often on schematic diagrams, the omega sign is not used if the resistor value is in kilohms. For example, a 10-kilohm resistor is shown as 10K on some diagrams, rather than 10 kΩ. Both methods are considered acceptable.

Figure 4-8 The first two bands of a three-band resistor are read literally. The third band is the number of zeros. If the third band is gold or silver, the value represented by the first two bands is divided by 10 or divided by 100. With no fourth band, the tolerance is ±20%.

Four-Band Resistors

Resistors with four color bands are available in tolerances of ±10%, ±5%, and ±2%. The 10% tolerance resistors were considered standard for many years. Now, like the three-band resistors, they have become a thing of the past. The four-band, 10% tolerance resistors were available in 12 decade values and were classified by the EIA as type E-12 (Table 4-2).

Table 4-2					
10	12	**15**	18	**22**	27
33	39	**47**	56	**68**	82

Today the standard resistor has a tolerance of ±5%. These resistors are manufactured in 24 decade values and are classified by the EIA as type E-24 (Table 4-3). The same resistor values are also manufactured to the tighter tolerance of ±2%, but at a slightly higher cost. It is very helpful to commit these values to memory because it makes recognizing the most common color-coded resistor values much easier.

The only difference between the way three-band and four-band resistors are read is that the fourth band specifies the tolerance of the resistor as ±10% if it is silver, ±5% if it is gold, and ±2% if it is red.

Table 4-3							
10	11	**12**	13	**15**	16	**18**	20
22	24	**27**	30	**33**	36	**39**	43
47	51	**56**	62	**68**	75	**82**	91

The use of silver and gold is not limited to the fourth band. These colors sometimes appear in the third (multiplier) band when values more than 1 Ω and less than 10 Ω must be indicated. If the third band is gold, the value obtained from the first two bands is divided by 10. If the third band is silver, the value obtained from the first two bands is divided by 100. For example, the bands on a resistor are brown, black, gold, and gold, as shown in Fig. 4-9. The value of the resistor would be: brown = 1, black = 0, gold = 10, gold = ±5%. The resistor value is 10/10 or 1 Ω ±5%.

Five-Band Resistors

Five-band resistors are available in ±1%, ±0.5%, ±0.25%, and ±0.1% tolerances. The large number of available values allows the closest possible match to design specifications. These resistors are found only where the need for very tight tolerances

Figure 4-9 The first two bands of a four-band resistor are read literally. The third band is the number of zeros. If the third band is gold or silver, the value represented by the first two bands is divided by 10 or divided by 100. If the fourth band is silver, the tolerance is ±10%; if it is gold, the tolerance is ±5%; and if it is red, the tolerance is ±2%.

can justify their higher cost. The five-band resistors with 1% tolerance are available in 96 decade values and are classified by the EIA as type E-96 (Table 4-4). The five-band 0.5%, 0.25%, and 0.1% tolerance resistors are available in 192 decade values and are classified by the EIA as type E-192 (Table 4-5).

Table 4-4

10.0	10.2	10.5	10.7	11.0	11.3	11.5	11.8	12.1	12.4	12.7	13.0
13.3	13.7	14.0	14.3	14.7	15.0	15.4	15.8	16.2	16.5	16.9	17.4
17.8	18.2	18.7	19.1	19.6	20.0	20.5	21.0	21.5	22.1	22.6	23.2
23.7	24.3	24.9	25.5	26.1	26.7	27.4	28.0	28.7	29.4	30.1	30.9
31.6	32.4	33.2	34.0	34.8	35.7	36.5	37.4	38.3	39.2	40.2	41.2
42.2	43.2	44.2	45.3	46.4	47.5	48.7	49.9	51.1	52.3	53.6	54.9
56.2	57.6	59.0	60.4	61.9	63.4	64.9	66.5	68.1	69.8	71.5	73.2
75.0	76.8	78.7	80.6	82.5	84.5	86.6	88.7	90.9	93.1	95.3	97.6

The first three bands of a five-band resistor are read literally. The bands are called the *first significant figure*, the *second significant figure*, and the *third significant figure*. The fourth band is now the *multiplier*, which is read in the same way as the multiplier band on four-band resistors. The fifth band is the tolerance. If it is brown, the tolerance is ±1%; if it is green, the tolerance is ±0.5%; if it is

Table 4-5

10.0	10.1	10.2	10.4	10.5	10.6	10.7	10.9	11.0	11.1	11.3	11.4
11.5	11.7	1.8	12.0	12.1	12.3	12.4	12.6	12.7	12.9	13.0	13.2
13.3	13.5	13.7	13.8	14.0	14.2	14.3	14.5	14.7	14.9	15.0	15.2
15.4	15.6	15.8	16.0	16.2	16.4	16.5	16.7	16.9	17.2	17.4	17.6
17.8	18.0	18.2	18.4	18.7	18.9	19.1	19.3	19.6	19.8	20.0	20.3
20.5	20.8	21.0	21.3	21.5	21.8	22.1	22.3	22.6	22.9	23.2	23.4
23.7	24.0	24.3	24.6	24.9	25.2	25.5	25.8	26.1	26.4	26.7	27.1
27.4	27.7	28.0	28.4	28.7	29.1	29.4	29.8	30.1	30.5	30.9	31.2
31.6	32.0	32.4	32.8	33.2	33.6	34.0	34.4	34.8	35.2	35.7	36.1
36.5	37.0	37.4	37.9	38.3	38.8	39.2	39.7	40.2	40.7	41.2	41.7
42.2	42.7	43.2	43.7	44.2	44.8	45.3	45.9	46.4	47.0	47.5	48.1
48.7	49.3	49.9	50.5	51.1	51.7	52.3	53.0	53.6	54.2	54.9	55.6
56.2	56.9	57.6	58.3	59.0	59.7	60.4	61.2	61.9	62.6	63.4	64.2
64.9	65.7	66.5	67.3	68.1	69.0	69.8	70.6	71.5	72.3	73.2	74.1
75.0	75.9	76.8	77.1	78.7	79.6	80.6	81.6	82.5	83.5	84.5	85.6
86.6	87.6	88.7	89.8	90.9	92.0	93.1	94.2	95.3	96.5	97.6	98.8

Figure 4-10 The first three bands of a five-band resistor are read literally. The fourth band indicates the number of zeros to be placed behind the number obtained from the first three bands; the fifth band is the tolerance.

blue, the tolerance is ±0.25%, and if it is violet, the tolerance is ±0.1%. For example, assume that the bands of a resistor are red, orange, violet, yellow, and brown, as shown in Fig. 4-10. The value of the resistor would be: red=2, orange=3, violet=7, yellow=4, but in the multiplier band it is four zeros (0000). The value of the resistor is 2,370,000 Ω. The fifth band is brown, making the tolerance ±1%.

Cautions and Special Cases

Care must be taken not to read the colors in reverse order when reading five-band resistors or four-band resistors with a 2% tolerance. The colors used in the tolerance band make it less identifiable than gold or silver. Occasionally the color bands are placed in the center of the resistor by mistake. Knowing the standard decade values, or having a list nearby, will eliminate mistakes because resistors read in reverse order will usually equate to nonstandard values.

REVIEW QUIZ 4-2

1. Three-band resistors have a tolerance rating of ± ______________%.
2. The third color band on a four-band resistor is called the ______________.
3. If the third band on a four-band resistor is gold, the number represented by the first two bands is divided by ______________.
4. The most accurate five-band resistor has a tolerance of ______________%.
5. The color bands on a 220-Ω, ±5% resistor are red, red, ______________, and gold.
6. The color bands on a 634-kΩ, ±1% resistor are blue, orange, yellow, ______________, and brown.

Physicist Georg Simon Ohm (1787–1854).

4-3 Ohm's Law

Voltage, current, and resistance are mathematically related by three simple formulas collectively known as Ohm's law. Each formula solves for one of the three quantities. If any two of the quantities related through Ohm's law are known, the third may be found using the proper variation of the Ohm's law formula.

Voltage is current × resistance:

Equation 4-3

$$V = I \times R \Rightarrow 1\ \text{V} = 1\ \text{A} \times 1\ \Omega$$

Current is voltage ÷ resistance:

Equation 4-4

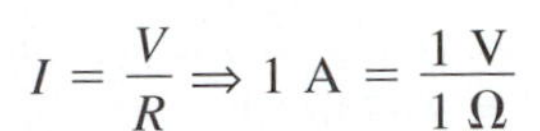

$$I = \frac{V}{R} \Rightarrow 1\ \text{A} = \frac{1\ \text{V}}{1\ \Omega}$$

Resistance is voltage ÷ current:

Equation 4-5

$$R = \frac{V}{I} \Rightarrow 1\ \Omega = \frac{1\ \text{V}}{1\ \text{A}}$$

STUDENT *to* STUDENT

A reminder! Ohm's law is extremely important. Make sure you understand it.

A knowledge of Ohm's law is extremely useful when working with electrical circuits because the quantities that most affect circuit operation become easily identified. Examples 4-1, 4-2, and 4-3 show how voltage, current, and resistance may be found in a simple circuit. All three examples refer to the schematic diagram in Fig. 4-11.

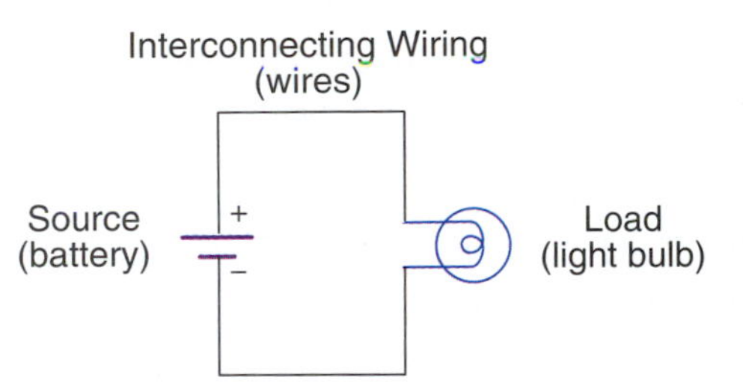

Figure 4-11 This circuit contains a voltage source, a load, and interconnecting wiring. Ohm's law may be used to find the source voltage, the current in the circuit, or the load resistance.

Example 4-1 The circuit in Fig. 4-11 contains a battery connected across a light bulb with a filament resistance of 12 Ω. The current in the circuit is 0.25 ampere. Using Ohm's law, determine the battery voltage.

$$\begin{aligned} V &= I \times R \\ &= 0.25\ \text{A} \times 12\ \Omega \\ &= 3\ \text{V} \end{aligned}$$

Example 4-2 The circuit in Fig. 4-11 contains a 3-volt battery connected across a light bulb with a filament resistance of 12 Ω. Using Ohm's law, determine the current in the circuit.

$$I = \frac{V}{R} = \frac{3\ \text{V}}{12\ \Omega} = 0.25\ \text{A}$$

Example 4-3 The circuit in Fig. 4-11 contains a 3-volt battery connected across a light bulb with an unknown resistance. The current in the circuit is 0.25 ampere. Using Ohm's law, determine the resistance of the bulb.

$$R = \frac{V}{I} = \frac{3\ \text{V}}{0.25\ \text{A}} = 12\ \Omega$$

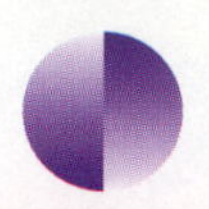

STUDENT *to* **STUDENT**

The circuit of a flashlight is not as simple as it looks. A lot of new students have trouble identifying all the voltages in the ON and OFF conditions.

REVIEW QUIZ 4-3

1. According to Ohm's law, voltage is current times ______________ .
2. The current through a 1-ohm resistor with 1 volt across it is ______________ ampere(s).
3. Increasing the resistance in a circuit ______________ the current.
4. Resistance is found by dividing ______________ by ______________ .

4-4 The Simple Circuit

An electric circuit must contain a minimum of three components: a **source,** a **load,** and **interconnecting wiring.** The source develops power for the load, the load consumes power from the source, and the interconnecting wiring provides a path for the power. For a circuit to be practical, a fourth component called a **controlling device** must be included. The simplest form of controlling device is a **switch.**

A small flashlight employs a simple four-component circuit. A battery supplies the power, an electric lamp represents the load, and wires, in conjunction with a portion of the case, connect the battery to the lamp through a switch. A schematic diagram of a flashlight is shown in Fig. 4-12. The circuit of Fig. 4-12(*a*) shows the flashlight in the OFF position, and the circuit of Fig. 4-12(*b*) shows the flashlight in the ON position. The position of the switch changes the voltage across two of the four components.

The voltage across the battery remains constant for as long as the battery is in good condition. It should not change when the circuit is switched on. The wiring in the circuit should never have voltage across it. Its purpose is to get power from the battery to the light bulb, not to consume any of the power. The light bulb will have voltage across it only when the switch is turned ON. The switch will have voltage across it only when it is turned OFF.

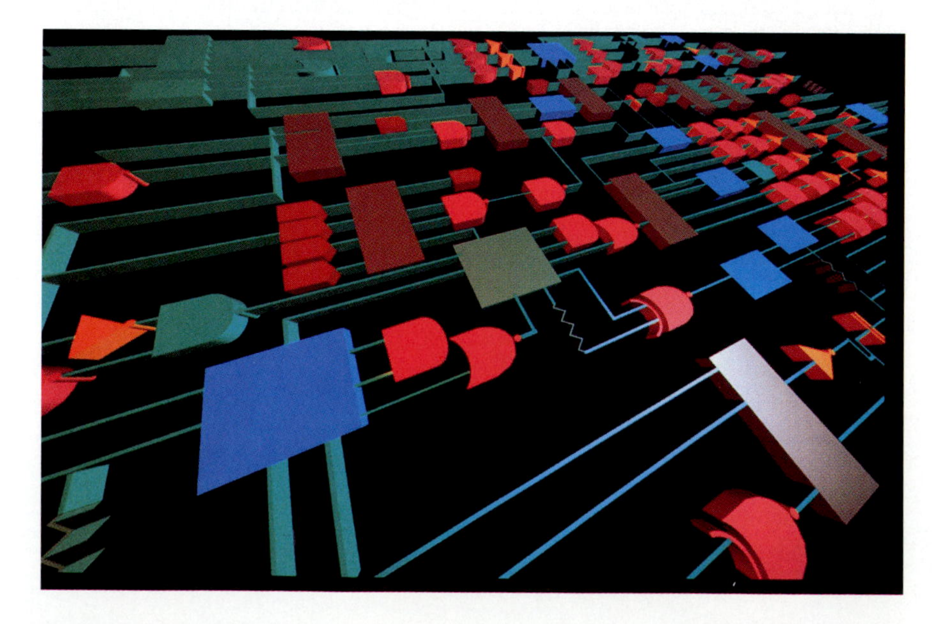

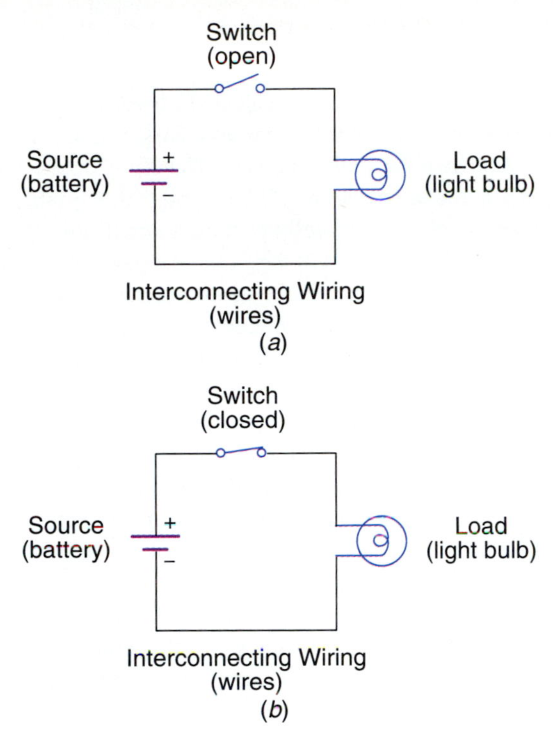

Figure 4-12 The schematic diagram of a flashlight is shown with the switch in both the off (*a*) and on (*b*) conditions. The voltage across some components will change in response to the switch setting.

REVIEW QUIZ 4-4

1. A closed switch in an operating circuit should have the ______________ voltage across it.
2. In the simple circuit, the light bulb represents the ______________ .
3. The load consumes ______________ from the source.
4. The voltage across the light bulb in the simple circuit can be calculated by multiplying the circuit current by the ______________ of the light bulb.

SUMMARY

Electricity is present in all matter as positive and negative charges, called electrons and protons. The electron is the moving particle in electricity because it is far less massive than the proton, but also because the proton is held captive in the nuclei of atoms. Therefore, objects become electrically charged by adding or subtracting electrons. The coulomb was established as a practical measurement because the electron represents such a small unit of charge. The coulomb is a quantity of charge equivalent to 6.24×10^{18} unit charges.

The movement of charge while objects are being charged or discharged is current. In fact, current is charge in motion. The force created by the separation of charge, causing charge to move, is called voltage. The opposition to current, because of a lack of charge carriers or a restriction on the flow of charge within a material, is called resistance. Voltage, current, and resistance are known as "the big three" in electricity.

One of the most basic electronic components is the resistor. The common type of resistor uses a color code to identify its value. The colors and their numerical equivalents are established by the Electronic Industries Association (EIA). All electronics professionals are expected to commit the color code to memory. EIA colors are black, brown, red, orange, yellow, green, blue, violet, gray, and white.

The simplest practical circuit must contain a source, a load, connecting wiring, and a controlling device. The voltage across the wiring and the source should remain the same under all conditions of circuit operation, but the voltage across the controlling device and the load transpose as the circuit is turned on and off.

NEW VOCABULARY

- alternating current (AC)
- ampere
- carbon film
- controlling device
- coulomb
- decade values
- delta
- difference of potential
- direct current (DC)
- electromotive force (emf)
- Electronic Industries Association (EIA)
- fixed resistor
- interconnecting wiring
- load
- ohm
- resistance
- schematic
- source
- switch
- unit charge
- unit magnetic pole
- volt
- voltage
- waveform

Questions

1. What is the basic unit of electric charge?
2. What does the quantity 6.24×10^{18} electrons represent?
3. What happens to an object when electrons are removed from it?
4. What do the initials *emf* stand for?
5. What is defined as charge in motion?
6. In what unit is an electric difference of potential measured?
7. What is the basic unit of resistance?
8. How many color bands are there on a $\pm 1\%$ tolerance resistor?
9. What was the controlling device in the flashlight circuit in Fig. 4-12?
10. What part of a simple circuit consumes power?

Problems

1. What is the current in amperes if 1 coulomb of charge moves past a given point in 2 seconds?
2. How long would 3 amperes have to flow to move 6 coulombs of charge?
3. How much current would flow through a 12-Ω resistor if a 6-volt battery were connected across it?
4. If a car headlight draws 10 amperes from a 12-volt battery, what is the resistance of the lamp?
5. How much voltage is required to push 1 ampere through a coil of wire with a 30-Ω resistance?

Critical Thinking

1. What color bands would be on a 1-MΩ, $\pm 5\%$ resistor?
2. If 2 mA flowed through a resistor having brown, black, brown, and gold color bands, what are the maximum and minimum voltages that could be measured across the resistor?
3. Why would there be no voltage across a closed switch in an operating circuit such as a flashlight?
4. If the slope of a charge waveform were zero (horizontal), why would the current be zero?
5. Could a $\pm 5\%$ resistor be used to replace a $\pm 10\%$ resistor if the first three colors were the same?

Answers to Review Quizzes

4-1
1. voltage
2. current
3. resistance
4. coulomb

4-2
1. $\pm 20\%$
2. multiplier
3. 10
4. 0.1%
5. brown
6. orange

4-3
1. resistance
2. one
3. decreases
4. voltage, current

4-4
1. source
2. load
3. power
4. resistance

Chapter 5

Conductive *and* Resistive Properties *of* Materials

ABOUT THIS CHAPTER

An understanding of electrical conductors and their characteristics is essential to understanding electric components, circuits, and distribution systems. This chapter begins by examining electrical conductivity. It shows how materials conduct electricity rather than oppose it. Current is charge in motion, but the charge is not always carried by electrons. This chapter identifies other types of

After completing this chapter you will be able to:

- Describe electrical conductivity and how it is measured.
- Identify the different charge carriers that represent electric current.
- Explain the different electrical classifications of materials.

charge carriers and the media that supports them. Moreover, it describes how materials are classified according to their electrical conductivity and compared to each other, and it examines the systems for measuring specific resistance. These systems allow the electrical conductivity of one material to be compared with another because materials do not conduct equally under all environmental conditions. Finally, the chapter discusses how components make use of their nonlinear characteristics.

- Show how specific resistance allows electrical conductivity of materials to be compared.
- Define linear and nonlinear resistance and identify which components use these properties.
- Explain how the resistance of materials varies with temperature.

[1]An older unit of conductance is the *mho*. *Mho* is *ohm* spelled backwards. *Mho* is represented by the Greek capital letter omega turned upside-down .

5-1 Electrical Conductivity

Conductance is a measure of how easily current can flow through a material, component, circuit, or device. The unit[1] of conductance is the **siemens.** The name *siemens* honors Ernst Werner von Siemens (1816–1882), a German technologist who did research on electrical conductivity. Conductance is represented by a capital G in math equations. The word *siemens* is abbreviated to a capital S. Electrical conductance is calculated by dividing current by voltage:

Equation 5-1

$$G = \frac{I}{V} \Rightarrow \text{Conductance} = \frac{\text{current}}{\text{voltage}} \Rightarrow 1 \text{ siemens} = \frac{1 \text{ ampere}}{1 \text{ volt}}$$

Conductance and resistance are **reciprocals** of each other. Reciprocals are pairs of fractions where one is created by inverting the other. For instance, the reciprocal of $^4/_5$ is $^5/_4$ and the reciprocal of $^5/_4$ is $^4/_5$. The conductance of a resistor is equal to the reciprocal of its resistance value, $G = {}^1/_R$. The opposite is true for converting conductance to resistance. Resistance is found by taking the reciprocal of conductance, $R = {}^1/_G$.

Scientific calculators have a special key labeled $^1/_X$ that calculates the reciprocal of X. The $^1/_X$ key is called the *reciprocals key*. On virtually all calculators, X is the number seen in the display. When the $^1/_X$ key is pressed, the number in the display is divided into one.

REVIEW QUIZ 5-1

1. Conductance is measured in ______________ .
2. Conductance is current divided by ______________.
3. The $^1/_X$ key on a scientific calculator calculates the ______________ of a number.

5-2 Types of Charge Carriers

Electric current is usually conducted through solids, liquids, or gases. It may also consist of charged particles projected through a vacuum. Current is charge in motion; the type of charge carrier depends on the current-carrying medium. Electric current may be a flow of electrons, positive ions, negative ions, or protons. These charge carriers may exist in any combination depending on the current-carrying medium.

Conduction in Solids

Current through a solid is caused by the movement of electrons. Because most electrical conductors are solid, electrons are the most common charge carrier. Electric current cannot exist without mobile charge carriers. Solid electrical conductors are usually made from metal. The high electrical conductivity observed in metals results from the ease with which its valence electrons can move within the confines of the metal's atomic structure. A metal's valence electrons can move so freely that they are called **free electrons.** The interactions of these free electrons with any atom, or with other electrons, is not particularly important if the temperature is not elevated excessively. A metal's free electrons can be thought of as a cloud distributed evenly throughout the space that the metal occupies.

[2]Italian scientist (1737–1798) noted for his experiments in bioelectricity. Galvani believed that there was an inherent animal electricity, the effect of which became known as *Galvanism*. A sensitive ammeter is sometimes called a galvanometer.

Conduction in Liquids

An important milestone in understanding electrical phenomena was the discovery that electricity could be conducted by seawater. Around 1804, Giovanni Aldini (1762–1834), a nephew of Luigi Galvani,[2] performed an experiment to demonstrate the conductivity of seawater using a recently killed animal and a voltaic pile. The animal carcass was connected to the voltaic pile by completing an electric circuit using brass wires inserted into a portion of the English Channel. When the circuit was completed, the muscles of the dead animal were forced to contract because of the applied voltage. At that time, physicists did not believe that electricity would flow through the sea because it was so vast.

The conductivity of seawater can also be proved in an experiment using a battery, connecting wires, a container of pure water, table salt, an electric lamp designed to operate from the same voltage as the battery, and two metal rods. The first step is to connect a wire from a terminal of the lamp to one of the battery terminals. Another wire connects the other terminal of the battery to one of the metal rods. The remaining metal rod is then connected to the unused lamp terminal with a third wire. This circuit forms a simple continuity tester [Fig. 5-1(*a*)]. The lamp should light if the rods touch each other.

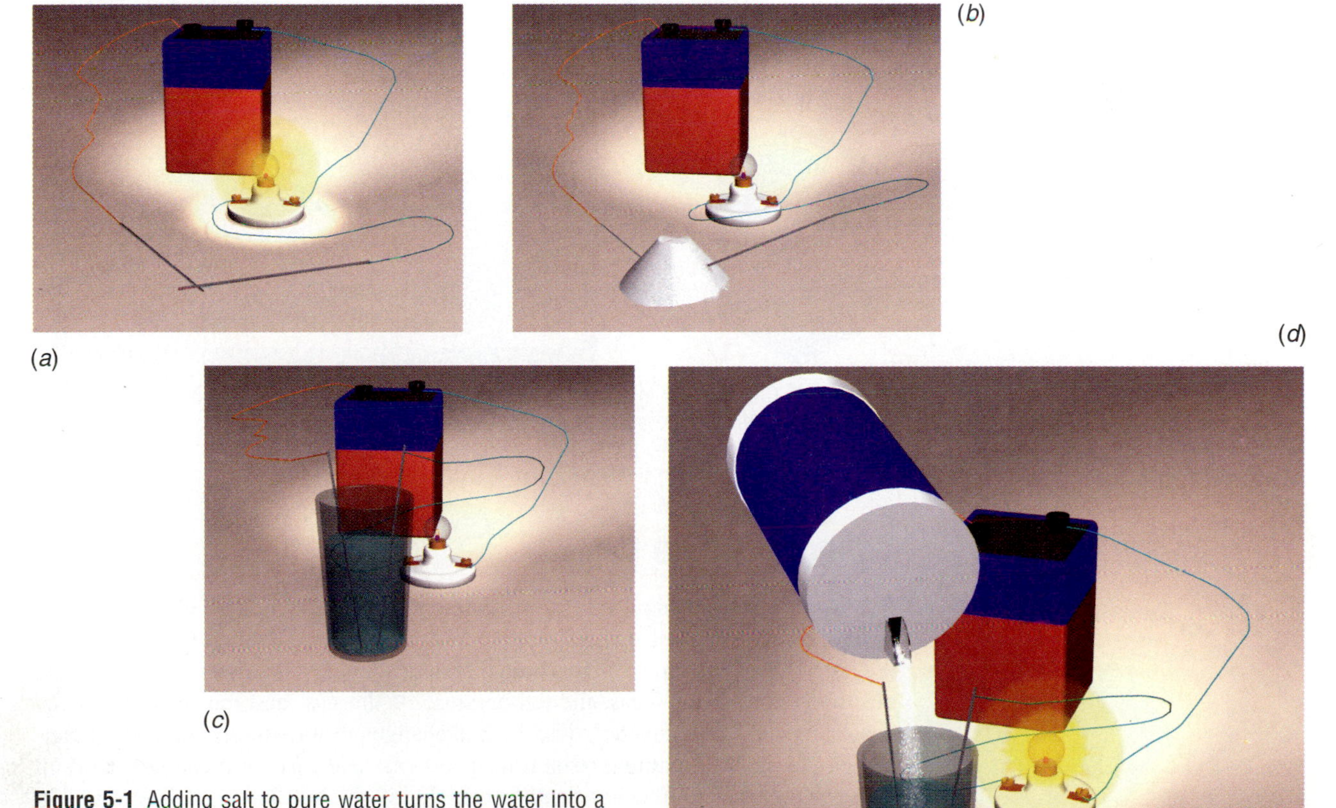

Figure 5-1 Adding salt to pure water turns the water into a good conductor by supplying positive and negative ions.

To determine what is a good conductor, begin by placing the dry metal rods (not touching) into the salt. The bulb should not light because dry salt is not a conductor [Fig. 5-1(*b*)]. After removing the rods from the salt, and cleaning off any residual salt, place the rods into the water. Again, the bulb should not light [Fig. 5-1(*c*)]. Thus pure water is not a good conductor. (For the sake of safety, water should not be thought of as an insulator.) Then immerse the rods in the water. Adding salt to the water will cause the lamp to light brightly once the salt is dissolved [Fig. 5-1(*d*)]. The addition of the salt caused the water to become a good conductor.

The chemical formula for water is H_2O. Water is a combination of hydrogen (H) and oxygen (O). The chemical formula for salt is NaCl. Salt is also composed of two elements, sodium (Na) and chlorine (Cl). Hydrogen, oxygen, and chlorine are gases. Sodium is a metal.

Sodium has only one valence electron and will give it up easily. Chlorine's valence shell is almost complete with seven electrons and will readily accept one more. The valence characteristics of sodium and chlorine allow them to bond tightly together. That bond is broken when the salt is dissolved, but the chlorine retains sodium's valence electron, which forms a positive ion with a singular positive charge (Na^+). Having gained an electron, the chlorine becomes a negative ion with a singular negative charge (Cl^-). These ions are free to move within the liquid solution, where they form the bulk of the mobile charge carriers.

Conduction in liquids is primarily caused by the presence of equal numbers of positive and negative ions. The negative ions flow from negative to positive and the positive ions flow from positive to negative. The flow of these two charge carriers is shown in Fig. 5-2. A liquid conductor is called an **electrolyte.**

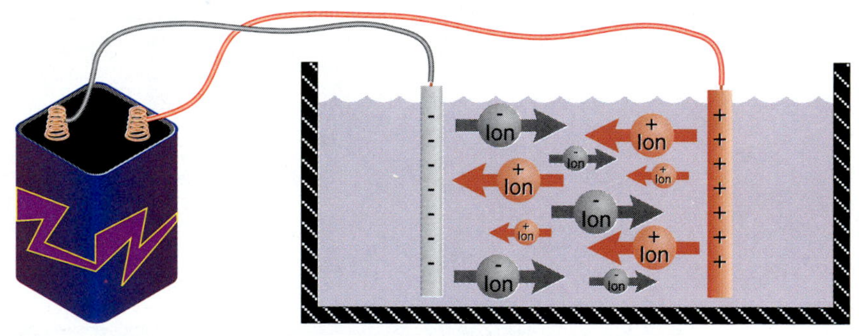

Figure 5-2 Current flow in a liquid is from positive to negative and negative to positive because there are two types of charge carriers.

Conduction in a Gas

Gases are normally thought of as insulators and will not conduct electric current until placed between electrodes that have enough voltage applied to produce ionization. The actual value of the voltage is dependent on the distance between the electrodes, the type of gas, the temperature of the gas, and the pressure of the gas. The electrical characteristics of a gas change abruptly under the correct conditions. In nature, lightning strikes are good examples. Electric charges build up between clouds and the Earth, different clouds, or hot and cold layers of air.

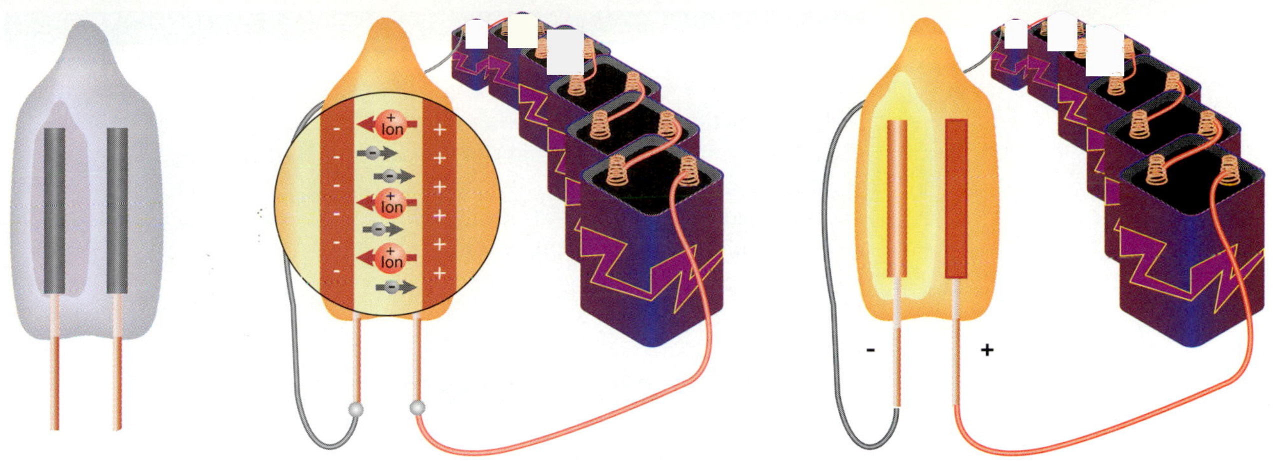

Figure 5-3 When a gas is electrically ionized, only the negative electrode of the voltage source produces light.

These charges dissipate when the ionization potential of the air has been reached, producing dramatic electrical discharges.

The effects of ionizing a gas can be examined using a **neon glow lamp.** The construction of the lamp is simple. The lamp consists of two parallel metal electrodes inserted into a glass bulb filled with neon gas [Fig. 5-3(*a*)]. Neon is tenth on the periodic table, where it takes its place as one of the elite elements known as the *noble gases*. The valence shell of neon is full at eight electrons, and it will not accept additional electrons, nor will it easily give any up.

When voltage is applied to the lamp's electrodes, the positive nuclei of the neon atoms are attracted to the negative electrode, and the electrons are attracted to the positive electrode. The potential on the electrodes places the neon atoms under stress and can suddenly ionize them. Mobile charge carriers are created in the process. In an ionized gas, current flow consists of positive ions and electrons [Fig. 5-3(*b*)]. As in a liquid, current flow is positive to negative and negative to positive. Once ionized, the applied voltage will have to be lowered well below the ionization potential before the gas returns to a stable state. Less energy is required to maintain the ionization of a gas than was initially required to ionize it.

The positive gas ions reacquire their lost electrons when they reach the negative electrodes. The energy initially used to ionize the gas is radiated by these electrons as they fall into stable orbits, usually as visible light [Fig. 5-3(*c*)]. The polarity of the applied voltage is determined by observing which electrode is glowing. Only the negative electrode produces light. The light from a neon lamp is orange. The color of ionized air can range from violet to a bright white, but is very often blue.

Understanding the nature of conduction in a gas is useful when working around high voltages. Many technicians, upon receiving an unexpected shock, have said, "I never even touched it." Their statement is most likely true. You do not have to touch a source of high voltage; it will touch you. The human body represents a large conductive mass with a normally neutral charge. The air between the body and a high voltage source is placed under an electrical stress because any difference of potential represents voltage. If that stress reaches the ionization potential of the air, there may as well be a wire connecting the person to the voltage source.

REVIEW QUIZ 5-2

1. Conduction in a solid is owing to ____________ .
2. Conduction in a liquid is owing to ____________ and ____________ .
3. Conduction in a gas is owing to ____________ and ____________ .
4. A liquid conductor is called a(n) ____________ .

5-3 Electrical Classifications of Materials

Materials are divided into three general categories according to their ability to support the flow of electric current. The three categories are **conductors, insulators,** and **semiconductors.** Conductors allow electric charge to move through them easily. Insulators are materials that provide significant opposition to electric current. The conductivity of semiconductors lies somewhere between that of conductors and insulators.

A fourth class of materials called **superconductors** promises to move from a laboratory curiosity to a practical reality. That reality could revolutionize electrical devices and the transmission of electric power. Superconductors allow an absolutely unobstructed flow of electric current—but at a price. Superconductors are normal materials that take on their new dimension only when they are cooled to temperatures near absolute zero (−273.15°C, −459.67°F, 0 K). This requirement for such cold temperatures limits their practicality.

Conductors

Electrical conductors are usually made from metal. The choice of which metal to use is dependent on cost, conductivity, malleability, and resistance to corrosion. The four metals used most often in electrical work are silver, copper, gold, and aluminum.

The best natural conductor is silver, but it has several disadvantages. It is expensive, prone to oxidation, requires special connectors, and does not bond well using conventional tin/lead solders. Despite its disadvantages, silver plays a significant role as an electrical conductor. Switch contacts that commute large amounts of current depend on silver to provide the contacts with maximum conductivity. Silver is also important in communications electronics, where wire coils in tuned circuits are often silver-plated to reduce their surface resistance.

The most practical conductor is copper. It is second only to silver, having 94.5% of silver's conductivity. Copper is abundant, affordable, malleable,[3] solderable, and easy to protect against corrosion. Malleability is an important property of a conductor because many electrical components require wire to be wound into a coil without breaking.

The third most conductive metal is gold. It has 77.75% of silver's conductivity. The price and availability of gold prohibit its widespread use in electrical circuits, but it has some advantages that override any cost disadvantage. Gold is extremely resistant to friction and corrosion. Both properties are important for the metal contacts of electrical connectors. Such connectors must often undergo

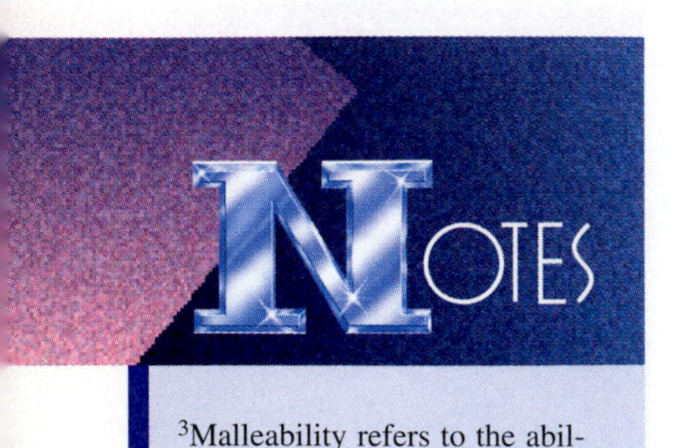

[3]Malleability refers to the ability of a material to be shaped by pressure.

many cycles of insertion and removal that would cause other metals to fail quickly. Fortunately, gold need only be plated onto the contacts in an extremely thin layer to do the job. The base metal is often made from brass or copper.

Another advantage of gold is its solderability. Gold and solder form an excellent bond, resulting in more reliable connections than offered by other metals. Gold is also one of the most malleable metals, but its high cost and relatively large electrical resistance prohibit its use in coils, where it would otherwise be a prime candidate.

A conductor that finds more applications in electrical circuits than in electronics is aluminum. It is very inexpensive when compared to other metals used in electrical applications, but it has some serious disadvantages. The electrical resistance of aluminum is high, with only 58% of the conductivity of silver and 61% that of annealed[4] copper. Aluminum is chemically active and corrodes easily, especially if it is in physical contact with a dissimilar metal, such as copper. Therefore, it requires special connectors when used in electrical wiring.

[4]Annealed means to be heated and then slowly cooled to toughen and reduce brittleness.

Aluminum cannot be soldered in the conventional sense. The lack of solderability requires that connections to aluminum be mechanical. While mechanical connections are customary in electrical distribution, most electronic components are soldered. There are special solders and fluxes available that claim to be compatible with aluminum, but their value and performance are questionable.

However, aluminum is not without its virtues. It is very light and it is an excellent conductor of heat. It is easily machined and is extremely malleable. This last feature is advantageous when making metal housings for various components. These housings, because they have intricate shapes created in a press, could not be made as easily with other metals.

Not all metals are chosen because they are good conductors of electricity. Some applications, such as the filaments of incandescent lamps, require high electrical resistance combined with a high melting point. Incandescence involves heating something until it produces light. Most metals melt before producing an acceptable amount of light, but not tungsten. It has a melting point of 6164°F (3406°C), which is more than twice that of iron at 2797°F (1536°C).

Another requirement often encountered in electrical circuits is the need for a material that has a stable electrical characteristic over a wide range of temperatures. In addition, these materials are often required to carry moderate to large amounts of current. That requirement is met by a metal alloy called **nichrome,** which is made of copper, nickel, and iron. Almost all heating elements are made from nichrome because it changes resistance very little with changes in temperature.

Insulators

Insulators do not allow charge to move through them easily. That is not to say that insulators can totally prohibit current flow when exposed to an electromotive force, but any current that does flow is negligible. Common insulators are glass, plastic, rubber, paper, mica, oil, dry air, most ceramics, and aluminum oxide.

Insulators are sometimes called **dielectrics.** The dielectric strength on an insulator is based on the maximum voltage a material of some specified thickness can withstand without breaking down. That thickness is usually specified in mils (0.001 inch), millimeters, or meters. Insulators vary considerably in their ability to withstand voltage. Dry air at sea level is able to withstand approximately 76 volts/mil before a spark will occur. On the other end of the scale, aluminum ox-

Table 5-1 The dielectric strengths of some common insulators

Insulating Material	Dielectric Strength in Volts/mil
Dry Air	76
Porcelain Ceramic	200
Polystyrene	300
Oil	375
Rubber	700
Wax Paper	1300
Teflon	1500
Glass	3000
Mica	5000
Aluminum Oxide	20,000

ide can withstand 20,000 volts/mil. Table 5-1 lists the dielectric strength of some common insulators.

As a general rule the dielectric strength of an insulating material is greatly reduced with an increase in temperature. One of the first accountings of this was written by English chemist Joseph Priestley (1733–1804) in 1767. In Section IV of his book, *History and Present State of Electricity*, Priestley states that red-hot glass is a conductor and not an insulator. This phenomenon was verified by several scientists of the period. It is important to keep in mind that electrical experiments prior to 1800 were performed using electric machines that produced fairly high voltages. The concept of a good conductor, in terms of these high-voltage, low-current circuits, is different from that of a good conductor by today's standards. The resistance of red-hot glass is in the tens-of-thousands of ohms, which would not necessarily represent a good conductor, but it does represent a resistance that is vastly lower than the almost immeasurably high resistance of glass at **room temperature.** Room temperature is 20°C (68°F), but it is occasionally defined as 25°C (77°F). Many unexplained insulator failures can probably be attributed to excessive heat. When high temperatures and high dielectric strength are required at the same time, the choice of insulators becomes fairly limited. Old stand-bys are mica and ceramics. In recent years, however, compounds employing **silicones** have become increasingly popular. Silicones are organic compounds, such as rubber, made synthetically using silicon to replace carbon in their atomic structures.

Semiconductors

Semiconductors have conductivities that lie between those of good conductors and good insulators. The most common semiconductors are the elements silicon (Si) and germanium (Ge). Other semiconductor materials are compounds such as gallium arsenide (GaAs), indium antimonide (InSb), cadmium sulfide (CdS), and zinc sulfide (ZnS).

All semiconductors have crystalline structures that confine the valence electrons to a specific geometric structure. The ideal semiconductor has very few mobile charge carriers and should behave like an insulator. However, this is not the case because the amount of energy required to remove an electron from the valence shell of a semiconductor is much less than in an insulator. Thus, modest amounts of electrical pressure, normal thermal activity, and exposure to visible light are adequate to provide mobile charge carriers by freeing valence electrons from their atoms and crystalline prison.

Each electron that is removed from the valence shell leaves a **hole** that another electron can occupy. The hole is often thought of as a positive charge equal to that of a proton. The positive charge actually belongs to the nucleus of the involved atom. The apparent movement of a hole is caused by an electron filling one hole as it creates another. The actual moving charges in semiconductors are electrons, just as in any other solid. The apparent movement of holes is shown in Fig. 5-4.

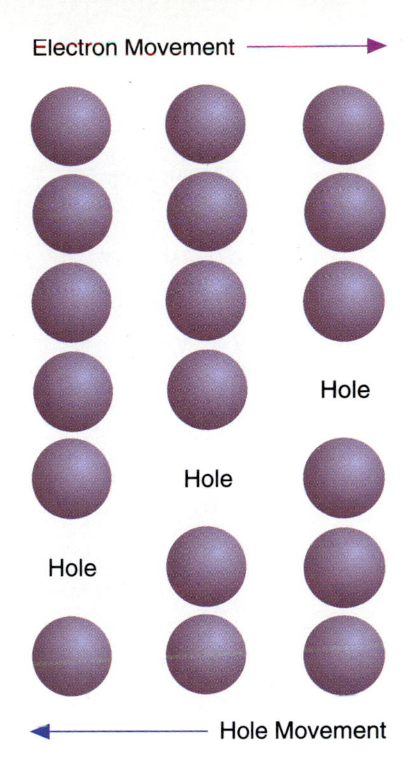

Figure 5-4 The holes in a semiconductor appear to move oppositely of electrons.

Pure semiconductor materials, often referred to as **intrinsic,**[5] generally have four valence electrons. The conductivity of these materials must often be increased by the addition of certain impurities, called **dopants.** The dopants fit into the crystalline structure of the semiconductor, but they have a different number of valence electrons—usually three or five. Once the impurities are introduced, the material is referred to as **extrinsic.**[6] Dopants with five valence electrons are called **donor** impurities and those with three valence electrons are called **acceptor** impurities. A semiconductor containing a donor impurity is **N-type** because the majority charge carriers are negative electrons. Those semiconductors containing acceptor impurities are **P-type** because the majority carriers are holes. In any event, the conductivity of the semiconductor is vastly improved by adding acceptor impurities, and, more important, controlled.

Interest in the conductive properties of semiconductor materials increased greatly with the invention of the transistor late in 1947. At that time, and for many years to follow, the hole was represented as a positively charged, massless particle and was given credit for being a principal charge carrier. This inaccurate representation proved very confusing to those attempting to make the technological transition from vacuum tubes to transistors without a formal scientific background. Sadly, many never made that transition and were forced to stand and watch as the tide of technology passed them by.

Superconductors

Superconductors are materials that lose all electrical resistance at some critical temperature, denoted by T_C. Many believe superconductors will revolutionize electric motors, power distribution, and computer logic circuits, if they can be made to operate at high enough temperatures. However, there are some dark clouds on the horizon that may destroy the hopes for some superconductor applications and open doors to new ones not previously considered.

In 1911, Dutch physicist Heike Kamerlingh Onnes (1853–1926) discovered that when mercury (Hg) was cooled to below 4 K (−269°C, −452.5°F), its resistance suddenly dropped to zero. It had become a superconductor. Onnes, who was awarded the 1913 Nobel Prize in physics for his discovery, was able to create such a cold temperature because he had succeeded in liquefying helium gas. Helium was the last and most difficult gas to be liquefied. In subsequent years other elements, such as aluminum (T_C = 1.2 K), lead (T_C = 7.2 K), niobium (T_C = 9.3 K), and compounds, such as niobium-tin (Nb_3Sn, T_C = 18 K) and niobium-germanium (Nb_3Ge, T_C = 23 K) were identified as being superconductors.

In 1933, German physicists Walther Meissner and Robert Ochsenfeld discovered that superconductors expel magnetic fields when their temperature is taken below the transition temperature. The complete expulsion of magnetic fields is known as the **Meissner effect.**

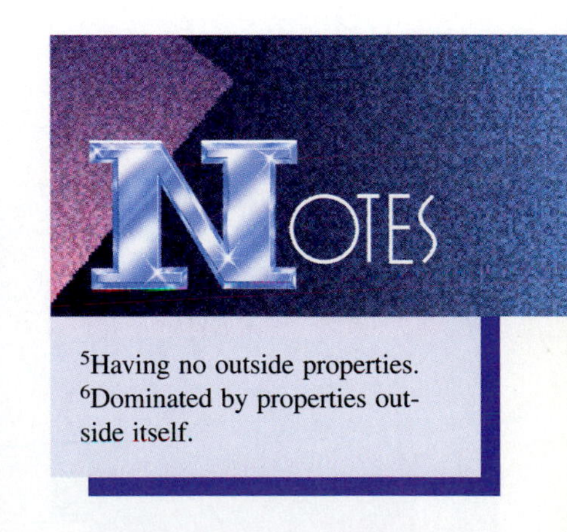

[5]Having no outside properties.
[6]Dominated by properties outside itself.

The atomic models of the 1930s could not account for superconductivity. This left researchers without a sound theoretical model. A slight glimmer of hope emerged in 1950 when two Russian scientists, Vitaly L. Ginzberg and Lev D. Landau, formulated a series of equations to describe superconductivity. However, they could not explain why superconductivity occurred.

In 1957, American physicists John Bardeen (one of the trio who developed the transistor), Leon N. Cooper, and J. Robert Schrieffer developed the **BCS theory** (**B**ardeen, **C**ooper, **S**chrieffer) to explain superconductivity. According to the BCS theory, the mobile electrons travel in pairs, called **Cooper pairs.** These electron pairs are formed because they interact with mechanical vibrations in atoms that form the crystalline structure of the material. The vibrations are caused by what little heat is present. The vibrations neutralize the normal repulsion between electrons and actually seem to create a small attracting force between them. At temperatures above the **transition temperature,** when a material loses its superconductive properties, the vibrations of the material's atoms destroy the Cooper pairs.

The major problem with all superconductors is that they must be kept at extremely low temperatures to remain below their transition points. With the materials previously listed, the required temperature could only be obtained with the use of liquid helium. Liquid helium is very expensive and difficult to work with, prompting the search for materials with higher transition points, preferably above 77 K, the boiling point of liquid nitrogen. Liquid nitrogen is less expensive than liquid helium and easier to handle.

In 1986, J. George Bednorz and K. Allen Müller, employees of IBM, discovered an oxide compound of barium, lanthanum, and copper that became superconducting at 35 K. This discovery earned them the 1987 Nobel Prize in physics and sparked a resurgence in superconducting research. In February 1988, a compound with a transition point of 125 K was discovered. In August 1988, a thallium compound achieved a transition point of 162 K. Thallium is a very toxic material, which may ultimately limit its use. The graph in Fig. 5-5 identifies the

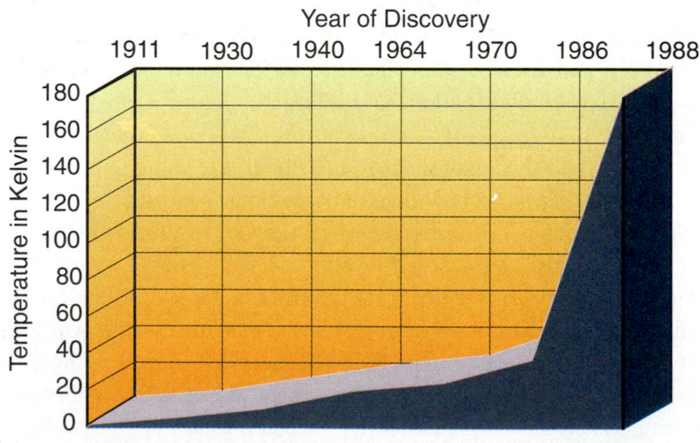

Figure 5-5 Superconductor research has developed materials with progressively higher transition temperatures in recent years.

tremendous progress that has been made in developing superconducting materials in recent years.

One of the greatest hopes for superconductivity was a perceived improvement in electric motor technology. Linear motors and related levitation devices

You'll be hearing a lot about superconductors. Their potential uses could significantly change the way we live.

were planned for high-speed futuristic trains, floated on and propelled by magnetic fields, but a problem has surfaced. Exposure to external magnetic fields causes superconductors to lose their superconducting properties. The external magnetic fields interact with electrons in the superconductor to form circulating eddies of current that form columnar vortices (Fig. 5-6). The applied current adds

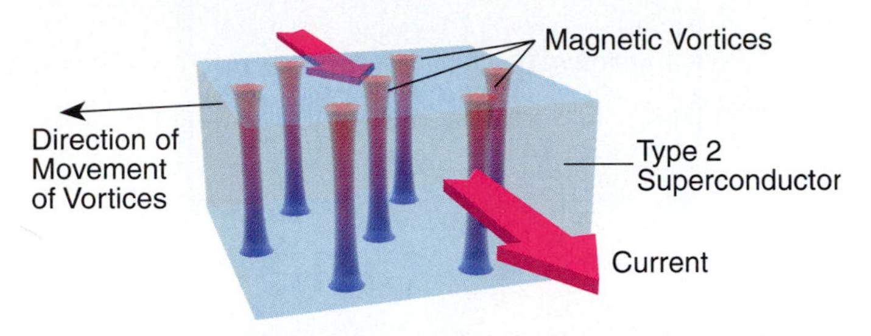

Figure 5-6 Current flow through a superconductor can be impaired by magnetic vortices created when the superconductor is exposed to an external magnetic field.

to one side of each vortex and subtracts from the other. As a result, the vortices are pushed at a right angle to the applied current. This removes energy from the applied current and effectively increases the resistance of the superconductor. Although this phenomenon dims hopes for superconductor applications that rely on magnetic interactions, it provides a new vision of high-speed switching devices that could revolutionize computer technology.

REVIEW QUIZ 5-3

1. ______________ is the best natural conductor.
2. Glass is a good ______________ .
3. N-type semiconductors are made by adding ______________ impurities.
4. The temperature of superconductors must be kept near ______________ .
5. The superconducting properties of a material can be destroyed by exposure to an external ______________ .

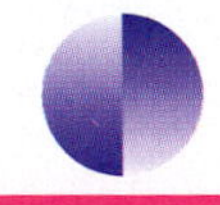

STUDENT *to* STUDENT

A circular-mil is a relative area found by squaring the diameter of a round wire, a technique used to find the area of a square. Although the circular-mil is not the true area of the circular cross-section of a wire, it allows the area of one wire to be accurately compared to another with a minimum of mathematics.

5-4 Specific Resistance

The **specific resistance** of a material, also called **resistivity,** is the ohmic value of a sample at room temperature that conforms to a set of predetermined dimensions. The symbol for the resistivity of a material is the Greek letter ρ (**rho**). The resistance of a conductor is found by dividing its length by its area and multiplying the result by ρ:

Equation 5-2

$$R = \rho \frac{l}{a} \Rightarrow \text{Resistance} = \text{rho} \frac{\text{length}}{\text{area}}$$

When calculating the resistance of wires, ρ is in ohms/circular-mil-foot and the cross-sectional area of the wire is given in circular mils with the length in feet. The value for ρ is usually taken from a table like the one shown in Table 5-2 on page 84.

Table 5-2 Specific resistance of metals at room temperature

Metal	Specific Resistance Ω/cir-mil-ft @ 20°C
Silver	9.8
Copper	10.37
Gold	13.66
Aluminum	17.02
Tungsten	33.2
Iron	61.0
Constantan	294.0
Nichrome	660.0

When comparing the resistance of one material to another, it is necessary to establish some standard size for that comparison. As usual, the standard depends on the measurement system being used. In the American Standard system the unit of measure for metals is the **circular-mil-foot.** That standard is based on a wire of circular cross section 1 foot in length and 1 mil (0.001 inch) in diameter. Wires of the correct dimensions are prepared from each type of metal used in the electrical industry, and their resistances are compared to each other. Because some metals produce more extreme changes in resistance than others in response to temperature changes, it is important that resistance comparisons be made at the same temperature. Normally, the resistance of each wire is measured at room temperature.

The resistances may also be evaluated against each other by declaring their conductivity compared to silver. The best natural conductor, silver, is assigned a value of 1. Because all other metals are poorer conductors than silver, they will have a value less than 1 as indicated by the graph in Fig. 5-7. When this value is multiplied by 100, it can be expressed as a percentage.

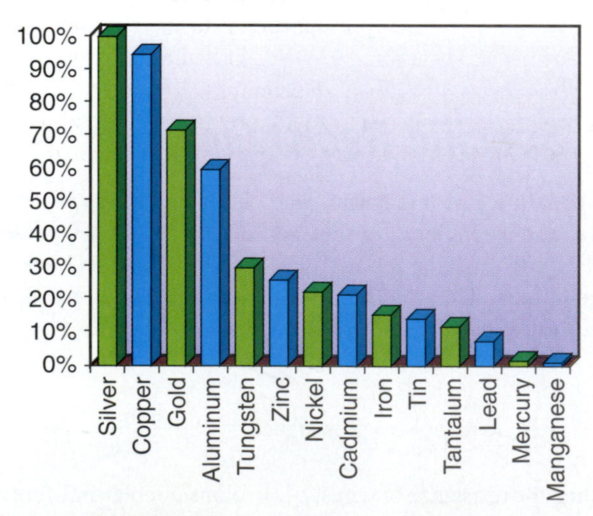

Figure 5-7 The electrical conductivity of all metals is compared to that of silver.

Specific Resistance in Nonmetals

Many materials other than metals are used to conduct electric currents, and not all of them can be made into wires. To accommodate such materials, and also allow a comparison to metals, resistance samples are formed into a unit cube. In the SI system this cube is 1 meter on each side. The goal of standardization sought by proponents of the SI system does not always lend itself to practical values. Using a cubic meter as a practical resistance sample is such a case. The unit of Ω/m^3 results in very low values of resistance because the samples have such a large cross section. The problem can be masked by depicting the resistance of the sample in scientific notation, or by using its conductance value, which becomes as large as the resistance was small. A more appropriately sized sample is the cubic centimeter (cm^3).

A common representation for ρ is the Ω-m and Ω-cm. This may seem incomplete because only a length is specified, but it is a derivative of the specific resistance formula presented earlier in this chapter and it is based on a material sample of 1 m^3 or 1 $cm.^3$ The formula is rearranged to solve for ρ instead of R, as indicated in Eq. 5-3 and Eq. 5-4. The three dots forming a triangle represent the word *therefore*.

Equation 5-3

$$R = \rho \frac{l}{a} \therefore \rho = \frac{RA}{l} = \frac{\Omega \cdot cm^2}{cm} = \Omega \cdot cm$$

Equation 5-4

$$R = \rho \frac{l}{a} \therefore \rho = \frac{RA}{l} = \frac{\Omega \cdot m^2}{m} = \Omega \cdot m$$

Wire Sizes

Wire sizes are standardized throughout the United States and numbered in **American Wire Gauge** (**AWG**) units from AWG 40 to AWG 0000 [pronounced "four aught" (4/0)]. The gauge of the wire is inversely proportional to the size of the wire. Number 40 is the smallest and four-aught (4/0) is the largest.

For every drop of three gauge numbers, the area of the wire doubles, and for every drop of 10 gauge numbers the area increases by a factor of 10. Because electrical conductivity is directly proportional to the conductor's area, decreasing the gauge of a wire by a factor of three cuts the resistance in half, assuming the length of the wire is unchanged. Reducing the gauge of wire by 10 causes a reduction in resistance to one-tenth of its previous value. Wires from gauges 6 to 4/0 are made from several strands of smaller wire for flexibility.

Table 5-3 provides information on the AWG sizes for solid copper wire. The areas of the wires are given in circular-mils and the resistance is measured in ohms/1000 ft @ 20°C.

REVIEW QUIZ 5-4

1. In the SI system, the standard size for a sample conductor is 1 cubic ______________ .
2. Increasing a wire size by three AWG gauge numbers, without changing its length, reduces its resistance by ______________ .

Table 5-3 The diameter and resistance of wires from gauge 40 to 4/0

Gauge Number	Area in Circular-Mils	Ohms/1000 ft. at 20°C
4/0	211,600	0.0490
3/0	167,800	0.0618
2/0	133,100	0.0779
0	105,600	0.0982
1	83,690	0.1239
2	66,500	0.1559
3	52,441	0.1977
4	41,620	0.2492
5	33,120	0.3131
6	26,240	0.3952
7	20,740	0.5000
8	16,380	0.6331
9	13,000	0.7977
10	10,400	0.9971
11	8,230	1.260
12	6,530	1.588
13	5,180	2.002
14	4,110	2.523
15	3,260	3.181
16	2,580	4.019
17	2,060	5.034
18	1,620	6.401
19	1,290	8.039
20	1,020	10.17
21	812	12.77
22	640	16.20
23	511	20.29
24	404	26.67
25	320	33.41
26	253	40.99
27	202	52.34
28	159	65.22
29	128	81.02
30	100	103.7
31	79	131.3
32	64	162.0
33	50	207.4
34	40	259.3
35	31	334.5
36	25	414.8

5-5 Linear and Nonlinear Resistance

The electrical conductivity of almost any material varies because of some external influence. Changes in temperature, applied voltage, light, current, and pressure are often key factors. Circuit elements that change resistance in response to environmental changes are **nonlinear.** Resistance elements with values that remain constant under environmental changes, such as the fixed resistors introduced in Chap. 4, are **linear.**

The Volt-Ampere Curve

The electrical characteristics of any conductor, component, or circuit can be plotted on a graph called a **volt-ampere curve.** The applied voltage is placed on the horizontal axis and the resulting current is placed on the vertical axis. A line from

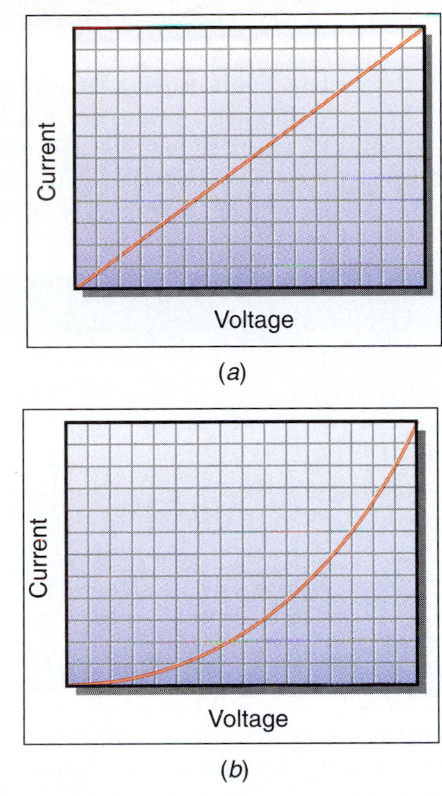

Figure 5-8 (*a*) If a volt-ampere curve is linear, it graphs to a straight line. (*b*) A nonlinear volt-ampere curve graphs to a curved line.

the graph's origin through all of the intersecting points of voltage and current defines the resistance. A graph that produces a straight line has a linear conduction characteristic [Fig. 5-8(*a*)]. If the graph produces a curved line, it is nonlinear [Fig. 5-8(*b*)].

Examples of Nonlinear Resistors

The schematic representation of a nonlinear resistance element uses the symbol for a fixed resistor enclosed by a circle, oval, or rectangle with rounded corners.

A resistive component that changes value with changes in applied voltage is called a **voltage-dependent resistor,** or **VDR.** The symbol for a VDR is that of a nonlinear resistance with the capital letter V inside the surrounding border. An alternate symbol for a VDR uses a line drawn through a shape that is similar to a tilted hourglass. The value of a VDR cannot be given in ohms because its resistance is dependent on the applied voltage. Instead, VDRs are rated by giving the current at some specified voltage. For instance, a VDR may be rated 20 mA at 100 volts. These components are generally unique to a specific piece of equipment and should only be replaced with an exact replacement or one recommended by the equipment manufacturer.

A resistive element that changes value when exposed to light is called a **light-dependent resistor,** or **LDR.** The LDR's dark resistance is very high, often in the millions of ohms. When exposed to light, its resistance drops dramatically. Most LDRs are made from cadmium-sulfide or silicon. They are commonly used as sensors for automatic lighting systems and full-featured cameras.

A resistor that changes value with changes in temperature is called a **thermistor.** The resistance of a thermistor is specified in ohms at some reference temperature, such as room temperature. For instance, a thermistor may carry the rating 120 Ω @ 20°C. It is easier to substitute thermistors than VDRs because they are often less critical, but only recommended substitutes should be used. The physical size of a thermistor is an indication of how quickly it responds to changes in temperature. Smaller thermistors respond more quickly than larger thermistors.

REVIEW QUIZ 5-5

1. Fixed resistors have a(n) ______________ volt-ampere curve.
2. Thermistors change resistance with changes in ______________ .
3. LDRs are sensitive to ______________ .
4. VDRs change resistance with changes in ______________ .

5-6 Temperature Coefficients

The resistance of all metals and most metal alloys increases with an increase in temperature. The change in a metal's resistance is constant over a very wide range of temperatures, but that resistance change diminishes and disappears near absolute zero. The graph in Fig. 5-9 plots the resistance of a typical conductor against temperature and emphasizes that no change in resistance occurs until the temperature reaches some nominal value, albeit very cold. Above that point the change in resistance is constant. Although technically the graph is nonlinear, the practical operating range of metal conductors lies on the linear portion of the curve.

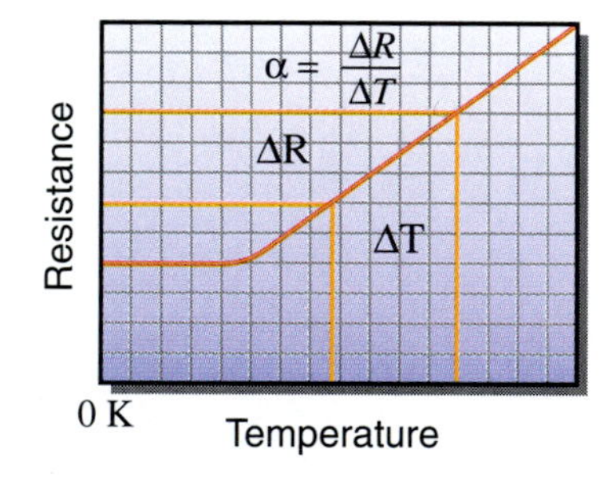

Figure 5-9 The temperature coefficient of most metals drops to zero as the metals' temperature approaches absolute zero.

A material has a **positive temperature coefficient** if its resistance increases with an increase in temperature. All metals and most metal alloys have a positive temperature coefficient. A material has a **negative temperature coefficient** if its resistance decreases with an increase in temperature. Electrolytes, semiconductors, and insulators have negative temperature coefficients. In mathematical formulas, temperature coefficient is represented by the lowercase Greek letter alpha (α). Table 5-4 compares the temperature coefficients of common metals used in the electrical and electronics industries.

Some thermistors have a positive temperature coefficient. This is indicated by the letters **PTC** inside the symbol's surrounding border. If a thermistor has a negative temperature coefficient, the letters **NTC** are found in the same location.

Conductivity in metals results primarily from the movement of free electrons. Heat energy, represented in matter as the random motion of atoms and molecules, impairs that movement, resulting in an increase in electrical resistance. For most metals the increase in resistance with temperature is linear over the expected operating temperature range for electrical equipment and transmission lines.

Knowledge of how temperature affects resistance is important in many electrical applications. An electric motor may produce adequate power when it is cold, but normal operation will cause its temperature to increase. When it does, the resistance of its windings will increase, lowering the amount of current through the motor. The lower current may result in the motor's being unable to

Table 5-4 The temperature coefficients of metals vary greatly

Metal	Temperature Coefficient °C/ohm @ 20°C
Silver	0.0038
Copper	0.00393
Gold	0.004
Aluminum	0.0039
Tungsten	0.0045
Iron	0.0055
Constantan	0.000008
Nichrome	0.00044

do its job properly. Thus, the maximum operating temperature must be considered in choosing the size of wire used in a motor's windings. This would also be true in the transmission of electric power over long distances, where the wires are exposed to extremes of environmental temperatures.

The tungsten filaments of incandescent light bulbs undergo drastic changes in resistances as they go from cold to hot, often increasing 20 to 50 times. When turning on a lamp, the low cold-resistance of the filament can lower the voltage in a home electrical circuit for an instant, as indicated by a blinking of other lights on the circuit. The current requirements of many components is far greater at the instant power is applied than after they have stabilized.

Conductivity in semiconductors depends on adding enough energy to the valence electrons to free them from the crystal lattice, generating hole-electron pairs in the process. Prior to the addition of that energy, there are not enough charge carriers to maintain any reasonable level of conduction. Because the bonding between atoms in semiconductor materials occurs in the valence shell, employing too many valence electrons as current carriers will destroy the structure of the material, which causes it to overheat and melt. As that process takes place, the temperature coefficient will reverse and eventually become positive.

Semiconductors that have had their conductivity increased by the addition of impurities are not so dependent on the breaking of structural bonds to support current flow, although the impurities form part of the crystalline structure. The process of adding impurity elements to semiconductors to improve their electrical characteristics is called **doping.** However, the resistance of doped semiconductors will also decrease with an increase in temperature. The properties of the intrinsic material will dominate over any thermal impairment of the charge carriers introduced by the impurity.

REVIEW QUIZ 5-6

1. Metals have ______________ temperature coefficients.
2. Semiconductors, electrolytes, and insulators have ______________ temperature coefficients.
3. Temperature coefficients are represented by the Greek letter ______________ .

SUMMARY

Electrical conductivity is the opposite of resistance; it deals with how well a material or device will conduct current. Conductance is measured in siemens or mhos and may be calculated taking the reciprocal of a resistance value.

Mobile electric charges may be electrons, protons, or ions. The type of charge that constitutes a flow of current is dependent on the current-carrying medium. Current is a flow of electrons in a solid; positive and negative ions in a liquid; and positive ions and electrons in a gas.

Materials are classified as conductors, insulators, semiconductors, and superconductors. Conductors allow the free flow of electric current and are generally metals. Insulators inhibit the flow of electric current and are represented by glass, plastic, rubber, cloth, and air. Semiconductors are crystalline materials that are not good insulators or good conductors. Naturally occurring semiconductors are silicon, carbon, and germanium.

The specific resistance of a material is the resistance of a sized sample at room temperature. Wires often use the circular-mil-foot as the standard, but the SI system defines specific resistance in terms of a sample one cubic meter in size.

The conduction characteristic of most materials does not remain constant under all circumstances. Materials whose current-to-voltage ratio is not constant are said to be nonlinear. Those with a constant current-to-voltage ratio, such as fixed resistors, are linear.

Most conductors change resistance with changes in temperature. The change in resistance divided by the change in temperature is called the temperature coefficient. The temperature coefficient of metals and most metal alloys is positive. The temperature coefficient of semiconductors, electrolytes, and insulators is negative.

NEW VOCABULARY

acceptor
American Wire Gauge (AWG)
annealed
BCS theory
circular-mil-foot
circular-mils
conductance
conductors
Cooper pairs
dielectrics
donor
doping
electrolyte
extrinsic
free electrons
hole
insulators
intrinsic
light-dependent resistor (LDR)
linear
Meissner effect
negative temperature coefficient (NTC)
neon glow lamp
nichrome
nonlinear
N-type
positive temperature coefficient (PTC)
P-type
reciprocal
resistivity
rho (ρ)
room temperature
semiconductors
siemens
specific resistance
superconductors
thermistor
volt-ampere curve
voltage-dependent resistor (VDR)

Questions

1. What is the standard unit of conductance?
2. How is conductance determined from resistance?
3. What are the three general types of charge carriers?
4. What are the charge carriers in a liquid?
5. What name is given to liquid conductors?
6. Under what circumstances does a gas become a conductor?
7. What is another name for dielectric?
8. What is the major advantage of nichrome over other wire conductors?
9. How is the resistance of a semiconductor affected by heat?
10. What name is applied to a semiconductor material that has been doped with an impurity?

Problems

1. What is the conductance of a 10-Ω resistor?
2. What is the conductance of a circuit element when 10 volts applied across it result in a current of 10 mA?
3. If copper has a resistivity of 10.37 ohms/circular-mil-foot at room temperature, what is the resistance at room temperature of 3 feet of #36 copper wire? Number 36 copper wire has a cross-sectional area of 25 circular-mils.
4. What is the resistance of a 20-siemens conductance?
5. If the wire size in Problem 3 is increased three sizes to number 33, what would be the resistance of the wire at room temperature if its length were increased to 6 feet?

Critical Thinking

1. How could you determine the effective conductance of a current-carrying vacuum device such as a Crook's tube?
2. Are free electrons one of the charge carriers in a gas? Explain your answer.
3. Could semiconductor materials be used to make thermistors? If so, what type of thermistor?
4. Why might the SI unit of Ω/m^3 be less convenient as compared to the Ω/circular-mil-foot when calculating the resistance of wires?
5. Is it possible that two nonlinear resistors could have the same rating, such as 20 mA @ 100 V, and have very different volt-ampere curves?

Answers to Review Quizzes

5-1
1. siemens; mhos
2. voltage
3. reciprocal

5-2
1. (free) electrons
2. positive ions; negative ions
3. positive ions; electrons
4. electrolyte

5-3
1. silver
2. insulator
3. donor
4. absolute zero
5. magnetic field

5-4
1. meter
2. one-half

5-5
1. linear
2. temperature
3. light
4. voltage

5-6
1. positive
2. negative
3. alpha (α)

Chapter 6

Energy *and* Power

ABOUT THIS CHAPTER

Energy is generally defined as the ability to do work. This chapter examines that ability from several points of view. The more important topics from our perspective are electrical energy and mechanical energy. The major use of electricity is as a source of power. Every electronic device requires power to operate. Power is the rate at which energy is expended or work is performed. The chapter also continues the topic of resistors and brings into

After completing this chapter you will be able to:

- ▽ Define energy and its units of measure.
- ▽ Describe how energy produces power.
- ▽ Explain the wattage rating system of resistors.

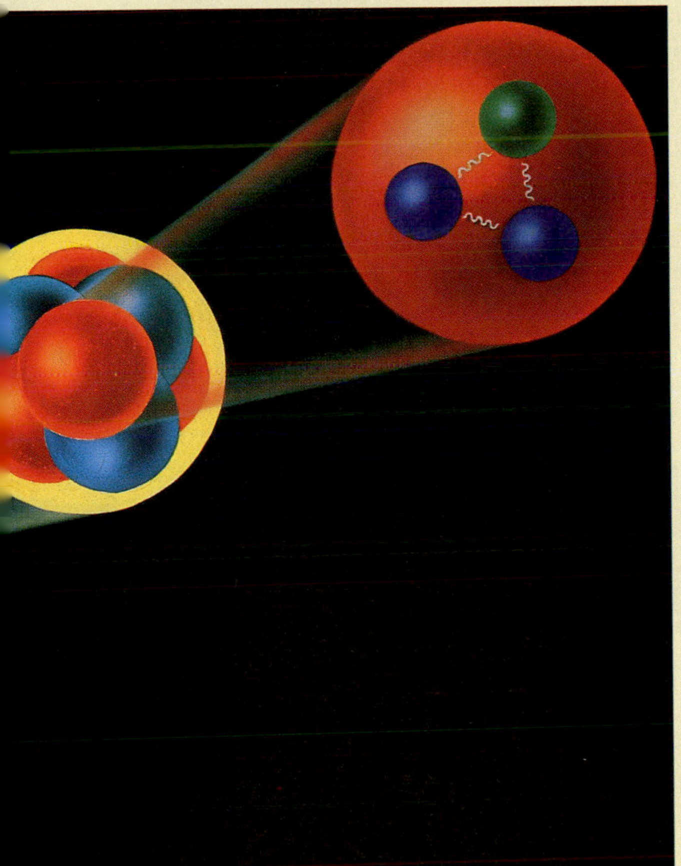

focus their power rating system. Color-coded resistors are made larger when they need to handle more power. The larger sizes are necessary to expel the heat that is developed within them. Moreover, the chapter presents a simple three-factor formula, similar in structure to Ohm's law that defines electrical power. This power formula is mathematically merged with Ohm's law to solve for voltage, current, resistance, or power with fewer steps.

- ▽ Discuss how power is distributed in a circuit and dissipated by components.
- ▽ Show how power, voltage, current, and resistance are combined in 12 interrelated formulas.

James Watt (1736–1819)

6-1 Energy

Chapter 2 revealed that electricity acts as a medium through which energy is moved from one place to another and converted from one form to another. The mechanism for creating electrical energy is the separation of charge. That separation creates a difference of potential measured in volts. Electricity releases its energy when the separated charges are allowed to reunite. The amount of power that is developed is dependent on how quickly that happens. Energy itself is timeless.

Energy is the ability to do work. There is little that is defined so simply and yet has a depth so great as to elude total comprehension. Energy is often described as **potential energy** or **kinetic energy.** Potential energy is stored energy. Electrostatic charges, unconnected batteries, stretched rubber bands, combustible fuels, water vapor masquerading as clouds, and even a book on a shelf represents potential energy. Kinetic energy is the energy of motion. A thrown ball continues to move after it has left the thrower's hand. The ball is physically the same as it was before it was thrown, but it now contains kinetic energy. When that energy is given to some other object, the ball again becomes lifeless.

An interesting experiment that demonstrates the transfer of kinetic energy involves placing several identical large steel balls (surplus ball bearings) adjacent to each other on a narrow track made from wood or metal, as shown in Fig. 6-1. A coved wooden molding is ideal for this. The balls should be able to roll freely as well as touch each other. The track must be as level as possible.

By carefully lifting one end of the track without disturbing the balls, a ramp is created. If one of the balls is lifted to the top of the ramp and released, several very important observations can be made. (1) When the rolling ball strikes those at rest on the level portion of the track, it immediately stops. (2) The instant the rolling ball stops, the end ball rolls away from the group with the same velocity as the ball that rolled down the ramp. (3) Only one ball will leave the group regardless of how fast the first ball is moving.

Repeating the experiment releasing two or more balls at the top of the ramp instead of one reveals that the number of balls entering the group is exactly matched by the number of balls leaving the group. The velocity of the rolling balls has nothing to do with the number of balls ejected. All the balls striking the group will instantly stop. This experiment demonstrates that the energy has been quantized, meaning it has been transferred in specific bundles. This quantization of energy is frequently observed when working with electricity. That should not be surprising because all electrons have identical mass and charge.

Energy cannot be created or destroyed; it may only be changed from one form to another. The meaning of that statement must be clearly understood. Energy will not just disappear. It must always have a place to go. Conversely, energy does not magically appear out of nowhere. It will always be given by something that previously possessed it—not a pleasant thought for those seeking to discover perpetual motion.

Figure 6-1 Energy often exists in exact quantities as demonstrated by the balls on the ramp.

[1]Be careful not to confuse ether with the liquid **ether** used in anesthesia.

Although energy seems to have a life of its own, from a mortal perspective it must always have a host to exist. That host can be matter in any of its four states or any of the three known fields. One of the most mystifying things about energy is that it can move effortlessly from matter to field and from field to matter. It is at once both matter and field and neither of them. It seems to require a medium to travel through, but when it cannot find one it makes its own. Such is the case with light waves and radio waves. Energy in these forms behaves in a wavelike manner, but waves must travel through a medium of some type, be it matter or field. That fact has caused scientists to believe that space, which is known to be about as void as anything can be, must be filled with some unperceived substance without substance.

The idea of space being filled with an invisible intangible substance of some kind is not new. The ancient Greeks had proposed that same idea over two thousand years ago, but they proposed it for a different purpose than to explain why light waves could pass through the apparent void of space. The Greeks called their contrived substance **aether,** which later became known as **ether**.[1] People of ancient times were not comfortable with the idea of stars and planets suspended in empty space. They were much happier believing that everything was suspended in the aether in the same way that fish are suspended in the sea.

The concept of space being filled with "ether" proved to be very convenient for those attempting to explain the transmission of light through space—that is, until Albert Michaelson (1852–1931) and Edward Morley conducted an experiment to determine whether the velocity of light was affected by the Earth's rotation. If the "ether" did exist, the velocity of light waves traveling with the Earth's rotation would be different from those traveling perpendicular to the Earth's rotation. Their experiment conclusively proved that the "ether" did not exist. Without a medium to travel through, light could not be a wave. Scientists then proceeded to define light as a massless particle called a photon, which was briefly mentioned in Chap. 2. Photons have properties of both matter and waves.

The Joule

Electrical energy involves voltage and a quantity of charge, measured in coulombs (Q). The SI unit of electrical energy is the **joule,** named after English physicist James Prescott Joule (1818–1889). Energy is symbolized by the capital letter J. One joule is equal to 1 volt-coulomb, as specified in Eq. 6-1. Voltage exerts a pressure on any electric charge, which will move if it is free to do so. When charge moves, work is performed. Without movement there is no work, regardless of the pressure voltage exerts.

Equation 6-1

$$J = V \times Q$$

Mechanical Energy

The SI unit of mechanical energy is also the joule. One joule is the amount of energy required to accelerate 1 kilogram 1 meter per second for every second that 1 newton of force is applied. The unit of time cancels in this equation, proving that energy is timeless. In equation form, 1 joule equals 1 newton multiplied by 1 meter, as shown in Eq. 6-2 on the next page. In the American Standard system, the unit of mechanical energy is the **foot-pound.** A foot-pound is the amount

of energy required to lift 1 pound 1 foot. One newton is a unit of force equivalent to 0.2248 foot-pounds.

Equation 6-2

$$1 \text{ joule} = 1 \text{ N} \times \text{m} = \frac{1 \text{ kg} \times \text{m}^2}{\text{s}^2} = 0.73756 \text{ foot-pound}$$

Work

Work is force multiplied by distance and is closely related to energy. The units for energy and work are the same. The major difference is that energy may exist without motion, whereas work cannot. Electrical force is voltage, and work is performed when that voltage causes charge to move. Anytime a voltage results in the flow of current, work is performed.

Work like energy is timeless. Assume that a small pile of bricks is to be moved from one place to another. The person moving these bricks could take one to the new location, set it down, and return for another. The process would be repeated until the job was completed. Another person, assigned to the same task, might take several bricks at a time, set them down, then return for more. A third person, being very strong, might lift the entire pile and take it to the new location.

Regardless of the method chosen to move the bricks, the work is the same. The first person exerted a small force to lift one of the bricks, but made many trips. The second person exerted a large force to move several bricks, but did not have to travel as far. The third person had to exert a very large force, but traveled only a short distance. The work produced by the three individuals was equal because the force times the distance moved is the same in all three cases. If a time limit had been imposed on our movers, things would have been different. Then the unit of power would come into play.

REVIEW QUIZ 6-1

1. Energy is the ability to do ______________ .
2. Electrical energy is measured in ______________ .
3. Electrical energy involves ______________ and ______________ .
4. The SI unit for mechanical energy is the ______________ .

6-2 Electric Power

When a time limit is imposed on work, the element of **power** is created. Power is the rate at which work is performed and always involves time. The unit of electrical power is the **watt.** In mathematical equations, power is represented by the capital letter P, and the word *watt* is abbreviated by a capital W. Initially, power is the energy in joules divided by the time in seconds. Although this equation is technically accurate, we do not have an energy or coulomb-meter. As a result, the energy-over-time relationship is inconvenient at best. The basic power equation states that the power in watts is the voltage in volts multiplied by the current in amperes. That relationship is distilled from the energy-over-time equation as shown in Eq. 6-3. The power equation ranks equally with the Ohm's law equations. Power is involved whenever work must be accomplished in a specified time period. Power is equal to energy divided by time. Energy is the volt-coulomb. Charge divided by time is current. The result of this symbolic slight of hand is that power equals voltage times current, as shown in this equation.

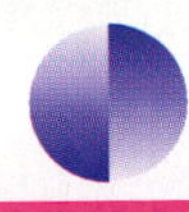

STUDENT *to* STUDENT

Memorize the power equation. You'll be tested on it.

Equation 6-3

$$P = \frac{J}{T} = \frac{V \times Q}{T} = V\frac{Q}{T} = V \times I$$

The Kilowatt-Hour

Electric utility companies do not charge for power; they charge for energy. Customers are charged for electricity by the **kilowatt-hour,** which, despite the name, is not a unit of power. A kilowatt-hour is 1000 watts of power used for 1 hour. The equation for determining kilowatt-hours is the power in watts divided by 1000, (which converts the power from watts to kilowatts); then the kilowatt-hour unit is multiplied by the time in hours.

Equation 6-4

$$kWh = \frac{P}{1000}T$$

Although the equation for the kilowatt-hour appears to involve time, multiplying power by time actually cancels time from the equation. Power is voltage multiplied by current, $V \times I$, and current is charge divided by time, Q/T. By substituting Q/T for I, power becomes $V(Q/T)$. Multiplying this equation by time causes the unit of time to cancel out, leaving $V \times Q$, which is the original definition of energy.

Electric utility companies charge for the ability to do work, not the work that is actually done. The efficiency of the customer's electrical equipment is not taken into account on the electric bill. It is for the work that could have been done that the customer is billed. In resistance, electrical power is completely converted into heat. Electric ranges, toasters, irons, and heaters make good use of electrical energy. Incandescent lamps, however, are not very efficient. The light they produce is a by-product of heat. A fluorescent lamp produces as much as four times the light per watt as an incandescent bulb. Electric motors are also relatively inefficient. They convert only 50% to 80% of the electrical power they consume into mechanical power.

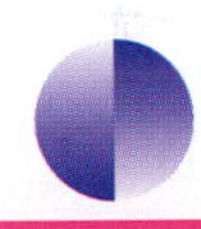

STUDENT *to* STUDENT

Look at an actual electric bill to see what your cost is for electricity. You can see this equation at work.

The cost of electricity varies, but in the United States, it is normally a few cents per kilowatt-hour. Assume that electricity costs 8¢ per kilowatt-hour and that you want to know the cost of keeping a 60-watt porch light on for a month. The power rating on the bulb is given in watts and must be converted into kilowatts by dividing the power in watts by 1000. Thus, 60 watts ÷ 1000 = 0.06 kilowatts. To determine the number of hours in a month, assuming a month is 30 days, multiply 24 hours by 30 days to get 720 hours. Multiplying 0.06 kilowatts by 720 hours gives 43.2 kilowatt-hours. At 8¢ per kilowatt-hour, the cost to operate the bulb for one month becomes \$0.08 × 43.2 kilowatt-hours, or \$3.46.

REVIEW QUIZ 6-2

1. Power is voltage times ______________ .
2. Power is energy divided by ______________ .
3. The kilowatt-hour is a unit of ______________ .

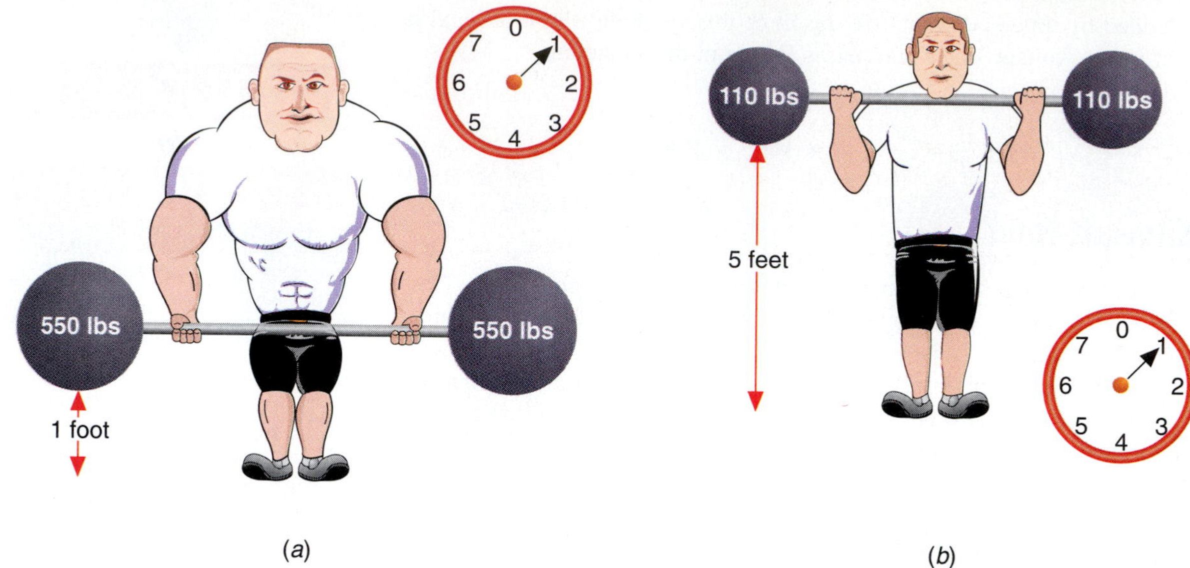

Figure 6-2 Developing the maximum power is a matter of matching the source of power to the job.

6-3 Mechanical Power

Mechanical power is often measured in **horsepower.** One horsepower is 550 foot-pound/seconds. That is, if you could lift 550 pounds 1 foot in 1 second you would have developed 1 horsepower. It might seem that a person would have difficulty developing 1 horsepower because only a few of the world's top weightlifters can actually lift 550 pounds [Fig. 6-2(*a*)]. The foot-pound unit is the product of feet times pounds. If 110 pounds could be lifted 5 feet in 1 second, 1 horsepower would also be developed [Fig. 6-2(*b*)]. This is a feat that could be accomplished by many individuals.

An ampere is 1 coulomb moving past a point in 1 second. In equation form, $I = Q/T$. If that coulomb is made to flow because of a potential difference of 1 volt, then 1 joule must equal 1 watt/second. That is,

$$1\text{ V} \times \frac{1\text{ C}}{1\text{ s}} \times 1\text{ s}.$$

Since 1 horsepower is equal to 550 foot-pound/seconds, any horsepower value can be converted into watts by dividing it by the conversion factor between joules and foot-pounds. This yields 745.7 watts per horsepower (550 foot-pounds/0.7376 foot-pounds = 745.7). Because electrical power is often converted into mechanical power, this is a useful conversion.

REVIEW QUIZ 5-1

1. One horsepower is equal to ______________ watts.
2. One horsepower is ______________ foot-pound/seconds.

6-4 Wattage Rating of Resistors

Resistance dissipates as heat. For that reason, resistors are available in several physical sizes (Fig. 6-3). The larger the resistor, the higher the power rating. Color coded resistors are available in sizes ranging from ⅛ watt to 2 watts in the following order: $^1/_8$ W, $^1/_4$ W, $^1/_2$ W, 1 W, and 2 W. Size being used as the power indicator is the reason why resistors having the same color code are available in different sizes. A higher wattage resistor can usually be substituted for a lower wattage one if space permits.

The wattage of a resistor is a free-air rating. If the resistor is located in an area where the air flow is restricted, the temperature buildup may be excessive and cause the resistor's value to change, although it was operated within its electrical power rating. Also, most resistors are mounted flush to a circuit board, preventing a uniform dissipation of heat. That is why resistors are often derated by operating them at no more than 50% of their rated value, or conversely, using a safety factor of two.

Resistors larger than two watts are normally **wire-wound** and operate at temperatures too hot to be touched comfortably. Because it is important that the resistor's value remains constant over its operating temperature range, the wire it is made from must have a very low temperature coefficient. This is accomplished using nichrome, which has a temperature coefficient of 0.00044. The temperature coefficient of copper (0.00393) is almost nine times that of nichrome. A wire-wound resistor made from copper would undergo a considerable change in value if it were heated or cooled to any degree. In addition, nichrome has about 58 times more resistance than copper at room temperature, so it does not take extraordinary lengths of it to get the desired resistance. The values of wire-wound

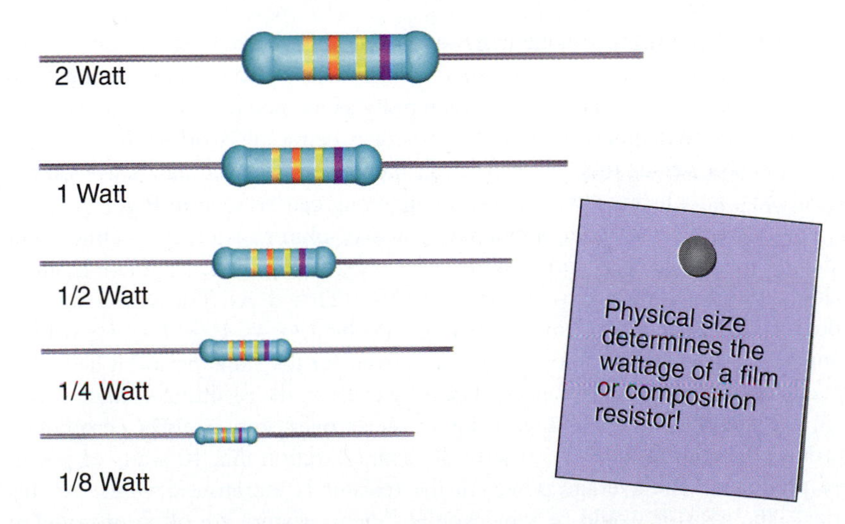

Figure 6-3 Only the size of a color-coded resistor determines its wattage.

5000 Ω± 10% 5 W

Figure 6-4 Wire-wound resistors have their value and wattage rating imprinted on their sides.

resistors and their wattage ratings are printed directly on the body of the resistors (Fig. 6-4). Heating elements in electric ovens, toasters, irons, and hair dryers are also made of nichrome.

Power in Pulsating DC Circuits

The electrical conductivity of any material is visually determined with a graph of the material's volt-ampere curve. Voltage is plotted on the horizontal axis and current on the vertical axis. For a fixed resistor, the current through the resistor is constant if the voltage across it is held constant. This forms an area under the graph that is rectangular. The area of this rectangle, as with any rectangle, is one side multiplied by its adjacent side, which here is voltage times current. The area under the volt-ampere curve is equal to the power dissipated by the resistor (Fig. 6-5).

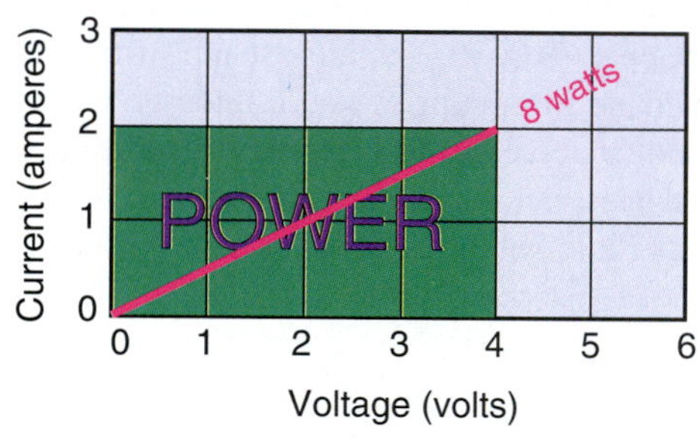

Figure 6-5 This graph tells the true story of how the power is being dissipated in the resistor.

James Prescott Joule (1818–1889)

In many circuits the voltage is not constant; rather, it pulsates or varies. The power dissipated by a resistor under such circumstances also pulsates, but the temperature of the resistor cannot change as rapidly as the power changes, and it may take some time to demonstrate that the resistor is being overworked. If a graph is made of power versus time, with time on the horizontal axis and power on the vertical axis, a realistic view of what is happening can be seen in Fig. 6-6.

Assume a voltage, when applied across a 10-ohm resistor, rises almost immediately to a value of 10 volts. While the voltage is applied, 1 ampere is made to flow through the resistor ($I = V/R = 10\ \text{V}/10\ \Omega = 1\ \text{A}$). Ten watts of power is produced in the resistor and converted into heat ($P = V \times I = 10$ volts x 1 ampere $= 10$ W). Then the voltage is removed for the same period it had been applied, and the process is repeated. The duty cycle of the resulting voltage waveform, a graph of what the voltage value has been over time, is 50%. Zero power is produced in the resistor for exactly the same duration that 10 watts of power were produced. The average power in the resistor is therefore 5 watts. At this wattage, the resistor would be wire-wound. Such resistors are often operated at or near their power rating and are much more forgiving than the color coded carbon resistors.

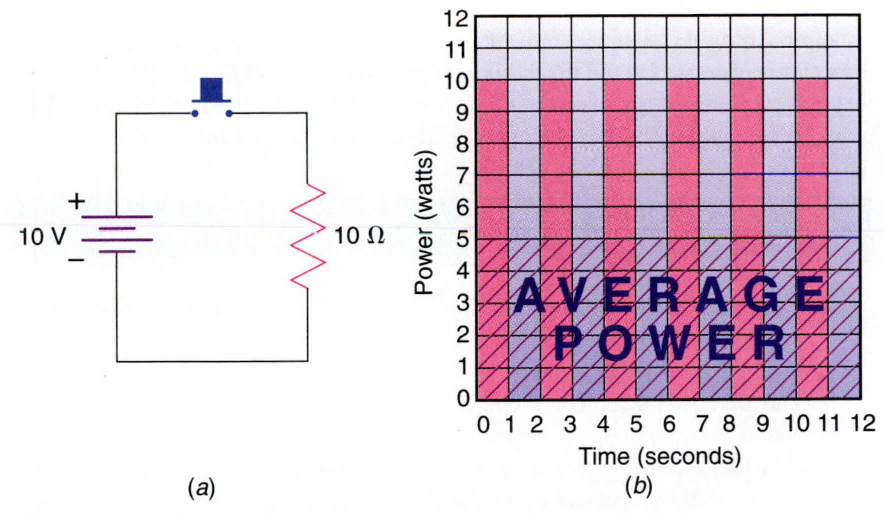

Figure 6-6 The average power value in a resistor may be deceiving and not indicate that the resistor is being abused.

The area under a square-wave power waveform can be found by multiplying the power by the time it is applied. The product of power multiplied by time is the watt-second, or joule. Therefore, the area under the active portion of the graph represents the amount of electrical energy used in transporting electrical charge from one place to another. This is true of any power-time graph regardless of how the power varies with time. The square-wave nature of the example provides an easy solution. The tools of integral calculus are usually required to calculate the areas of more complex waveforms.

If a 100-ohm resistor has a 100-volt square-wave pulse applied with a duty cycle of 1%, the current during the pulse would be 1 ampere ($I = V/R = 100\ \text{V}/100\ \Omega = 1\ \text{A}$). The peak power in the resistor is then 100 watts ($P = V \times I = 100$ volts $\times$ 1 ampere $= 100$ W). The average power, at 1 watt, is much less. The power is applied to the resistor for only one of a possible 100 time periods. During the remaining 99 time periods, no power is dissipated at all. This results in an average power dissipation in the resistor of 1 watt (100 watts/100 periods = 1 watt). Allowing a safety factor of two, a carbon resistor might mistakenly be used for this application. If only the average power were considered, it would most likely be a time bomb waiting to produce a predictable failure. The maximum energy a resistor could absorb and radiate as heat would be rated in joules or watt-seconds. Specific information on the upper limit of this value would be obtained from the resistor manufacturer. Some resistors are much better at handling temporary overloads than others. Wire-wound resistors are especially forgiving.

There are two factors to consider concerning the power rating of a resistor. The first is the amount of heat that can be removed from it in a specific period of time. The second deals with the types of structural changes that might occur when higher voltages are applied to the resistor. Excessive voltage applied to a resistor can also cause the voltage to jump across unintended paths in the resistor, thereby resulting in undesirable arcing.[2] Generally, the higher the wattage rating of a resistor, the higher the voltage rating. That is why resistors in high voltage circuits are often placed in series to get the desired value, although the value might be available in a single resistor.

[2]An arc occurs when electricity flows through space. It can be used constructively as in a carbon-arc lamp. However, it is often undesirable and destructive, as in the case described in the text.

Carbon resistors can be operated at temperatures up to 85°C (185°F), which is approaching the boiling point of water at 100°C (212°F). Wire-wound resistors can be operated at the much higher temperature of 300°C (572°F). Resistors may be cooled by the movement of air across them or the use of heat sinks. They may also be cooled by being immersed in a liquid such as oil.

REVIEW QUIZ 6-4

1. The wattage rating of a color coded resistor is indicated by its ______________.
2. Power is dissipated in resistance as ______________.
3. The highest power color coded resistor is usually ______________ watts.
4. Can the wattage rating of a resistor in a pulsating DC circuit be exceeded without exceeding its average power rating? ______________

6-5 Ohm's Law and Power Combined

Volts, ohms, amps, and power are generally the major units of concern in any circuit. The interrelationships among voltage, current, and resistance are expressed by Ohm's law, but often an Ohm's law unit must be found in terms of power. At that point the Ohm's law equations must be combined with the power equation. This combination produces 12 equations from the original four (Fig. 6-7).

Power		Current	
$E \cdot I$	$I^2 \cdot R$	$\frac{E}{R}$	$\frac{P}{E}$
$\frac{E^2}{R}$	P	I	$\sqrt{\frac{P}{R}}$
$I \cdot R$	E	R	$\frac{E}{I}$
$\frac{P}{I}$	$\sqrt{P \cdot R}$	$\frac{P}{I^2}$	$\frac{E^2}{P}$
Voltage		Resistance	

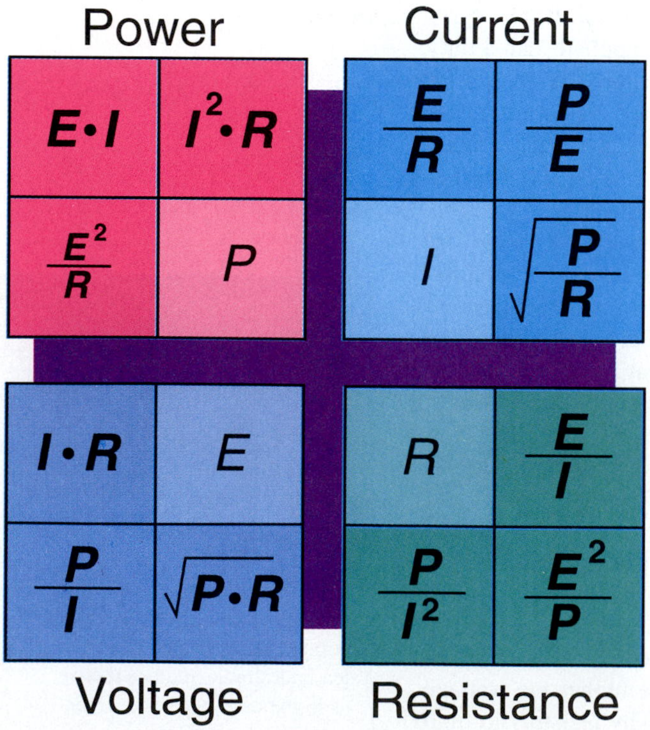

Figure 6-7 These twelve equations provide the key to solving many electrical problems.

Like Ohm's law, the power equation has three variations—one for each of the three factors in the equation ($P = V \times I$; $I = P/V$; and $V = P/I$). These equations can also be adapted to a mechanical aid (Fig. 6-8). The unknown quantity is placed at the top of the triangle, where it remains all alone. The other terms are placed in individual compartments in the lower half of the triangle. Cover the quantity that you want to know and proceed with the intuitive calculation. If the covered quantity is the lone one at the top of the triangle, the required operation is simply multiplication, as indicated by the adjacent algebraic variables. Remember, in algebra, unlike computer programming languages, only a single character may be used to represent a variable. If one of the bottom factors is covered, the unit at the top of the triangle is divided by the product of those left uncovered at the bottom.

The three Ohm's law equations—$V = I \times R$; $I = V/R$; and $R = V/I$—form some interesting combinations when mixed with the power equations. The basic power equation is $P = V \times I$ (power equals volts multiplied by amps). Suppose you do not know the voltage, but you know the resistance in the circuit. According to Ohm's law, $V = I \times R$ (voltage equals current multiplied by resistance). Since the resistance is known, the equation $I \times R$ can be substituted for voltage in the power equation ($P = I \times R \times I$). Collecting the common terms of I, the equation becomes $P = I \times I \times R$, or simply $P = I^2 \times R$. Assume you know the power and the resistance and want to find the current from our newly developed equation. Algebraic manipulation dictates that the equation now contains a square root. The solution would have yielded the equation $I^2 = P/R$, but we do not want to know the square of the current. The solution is to take the square root of both sides of the equation:

$$\sqrt{I^2} = \sqrt{P/R}$$

The equation then becomes current equals the square root of power divided by resistance:

$$I = \sqrt{P/R}$$

Ohm's law combined with the power equations forms squares and square roots in the resultant equations. This makes them more difficult to solve and even more difficult to remember. Generally, most electronic technicians remember only the equations they use daily and which can be done mentally. Most people do not use squares and square roots in mental calculations. Instead, they use charts that mechanically allow the selection of the appropriate equation. To use the chart you must be able to: Locate the quad that contains the quantity you are looking for—power (P), voltage (V), current (I), or resistance (R)—and then locate the equation in that quad containing the two factors you know. Only one equation will meet those conditions. That is the one to use.

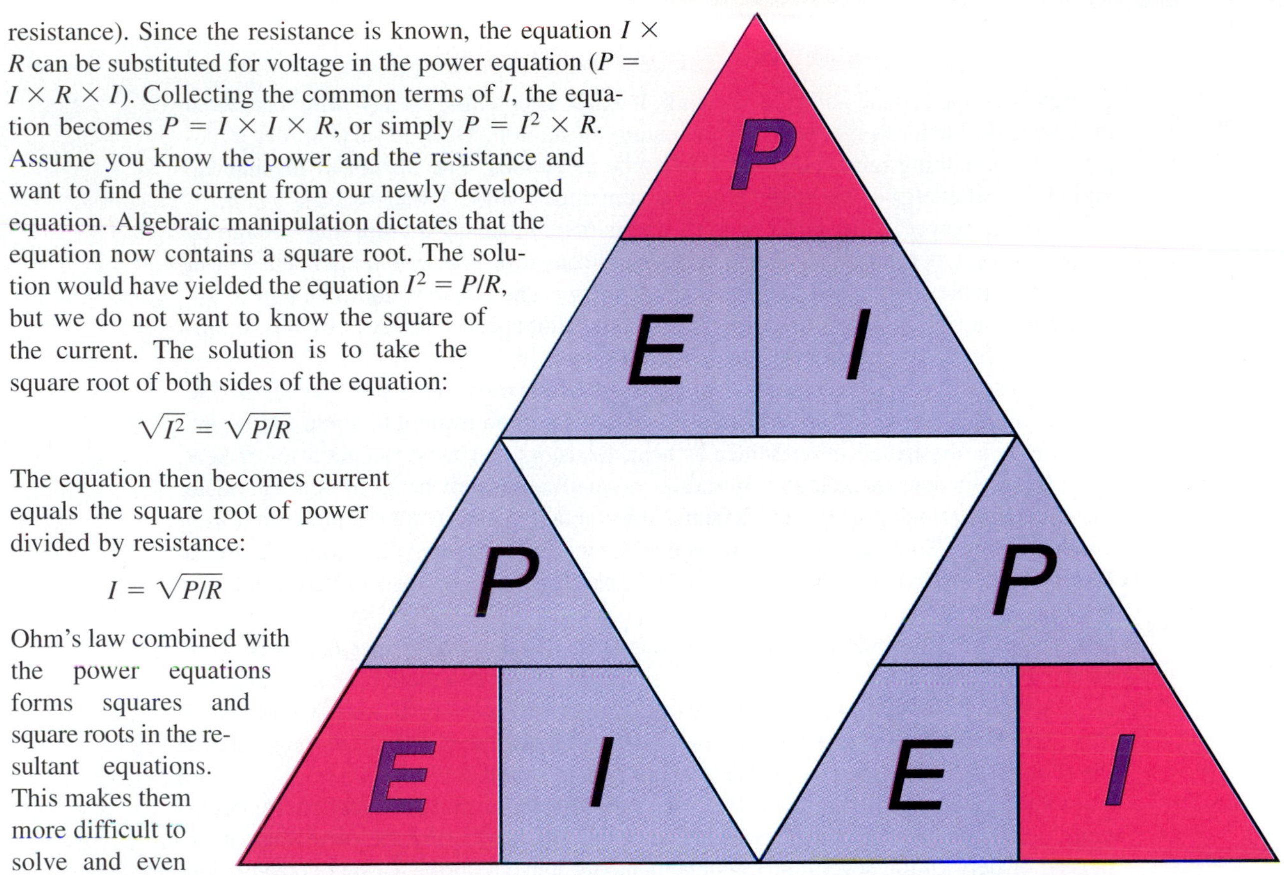

Figure 6-8 Mechanical solution guides can be very helpful when learning new equations.

REVIEW QUIZ 6-5

1. Two power equations other than $P = V \times I$ are ______________ and ______________.
2. ______________ equations are created by the combination of Ohm's law and the power equations.

SUMMARY

Energy is defined as the ability to do work. It exists as potential energy, which is stored energy, and kinetic energy, which is the energy of motion. Work is very similar to energy, but something must move for work to be performed. The SI unit of mechanical and electrical energy is the joule, which is sometimes called a watt-second.

Electric power is measured in watts and is determined by multiplying voltage by current. Power is the rate at which work is performed or energy is expended. Power utility companies do not sell power; they sell energy. The kilowatt-hour is a unit of energy because it multiplies power by time, which, although not always obvious, eliminates time from the equation.

Mechanical power is measured in terms of horsepower. One horsepower is 550 pounds lifted 1 foot in 1 second. One horsepower is also equivalent to about 746 watts.

Power is dissipated in resistance as heat. Resistors that must radiate a lot of heat must be larger than those that only radiate a small amount of heat. To accommodate the need for various power requirements, color coded resistors are manufactured in a variety of sizes. Some resistors are wound with nichrome wire and can radiate far more heat than their smaller color coded cousins. These wire-wound resistors are often too hot to touch.

The power in a pulsating DC circuit can be deceiving. Sometimes engineers rely on a resistor's tolerance of taking short-term abuse and choose them based on their average power dissipation. Carbon film resistors are not as tolerant as wire-wound resistors. Also, circuit defects may cause the short-term power level to exceed the rating of the resistor while the average power remains within specifications.

Ohm's law and the power equation can be combined to form 12 very useful equations. There are many occasions when time can be saved by using the correct equation instead of performing several procedures in succession. The price for that convenience is having to deal with squares and square roots. If the problems are to be done mentally, the longest way may be faster. If a calculator is available, squares and square roots should not be a problem.

NEW VOCABULARY

energy	horsepower	kinetic energy	watt
ether	joule	potential energy	wire-wound
foot-pound	kilowatt-hour	power	work

Questions

1. What is the difference between potential energy and kinetic energy?
2. What is another name for watt-second?
3. What is another name for a volt-coulomb?
4. What is the basic power equation?
5. Is the kilowatt-hour a unit of power?
6. How much power would a 1-horsepower electric motor require if it were 100% efficient?
7. Why are wire-wound resistors made from nichrome?
8. What is the smallest available wattage in color coded resistors?

Problems

1. How much current would a 60-watt, 120-volt light bulb require?
2. What is the maximum voltage that could be placed across a 100-Ω 1/2-watt resistor using no safety factor?
3. How much power would be required by a 50% efficient 1-horsepower electric motor?

Critical Thinking

1. Why can't power be developed in a conductor with zero resistance?
2. Why would the power of an electric motor diminish after it had run for a while?
3. Is it safe to only consider the average power in a resistor?

Answers to Review Quizzes

6-1
1. work
2. joules
3. voltage; charge
4. joule

6-2
1. current
2. time
3. energy

6-3
1. 746
2. 550

6-4
1. size
2. heat
3. two
4. yes

6-5
1. V^2/R; $I^2 \times R$
2. twelve

Chapter 7

Electrical Measurement

ABOUT THIS CHAPTER

This chapter deals with the instruments and techniques used to measure voltage, current, and resistance. In so doing, it provides an overview of the purpose and features of the most commonly used pieces of test equipment. Voltage notation is emphasized because of the need to correctly identify electric polarities. The use of various

After completing this chapter you will be able to:

- List the applications of volt-ohm-meters and digital multimeters and show how they are different.
- Explain the use of the various ground symbols.

schematic notations involving ground symbols is explained with specific examples. Analog and digital meters are discussed and compared, and the oscilloscope is introduced as a means of analyzing rapidly changing voltage waveforms.

- Describe how resistance measurements differ from voltage and current measurements.
- Provide the purpose and general operation theory of oscilloscopes.

7-1 Measuring DC Voltages

Voltage is an electric difference of potential that is measured with an instrument called a **voltmeter.** Voltage measurements are always taken across a source, component, or device. To aid in making voltage measurements with the correct polarity, the leads on meters are often color coded red and black. Red is traditionally positive and black is traditionally negative. On dedicated DC voltmeters without wire leads, the terminals are marked plus (+) and minus (−).

Voltage Notation

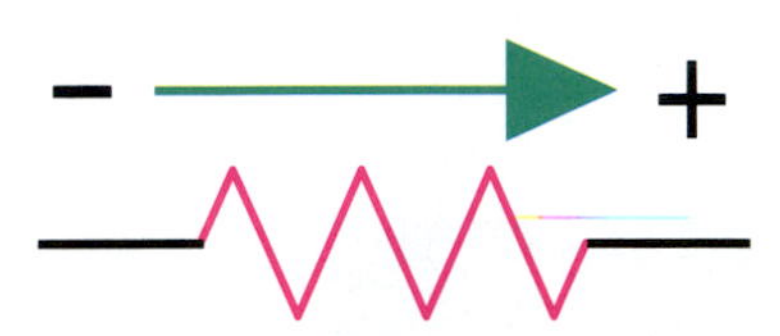

Figure 7-1 Resistors are polarized by the direction of current through them.

The polarity of a DC voltage is more important than its absolute value. The polarity of voltage sources is an integral part of their makeup and normally cannot be changed by circuit conditions. The voltage polarity across a resistor is a different matter. Resistors are polarized when current flows through them (Fig. 7-1). But before that polarity can be established, it is necessary to understand:

1. Electric current is a flow of electrons, unless otherwise indicated.
2. Electrons flow from negative to positive in circuits outside the source.
3. The end of the resistor that current enters is negative with respect to the end it leaves.

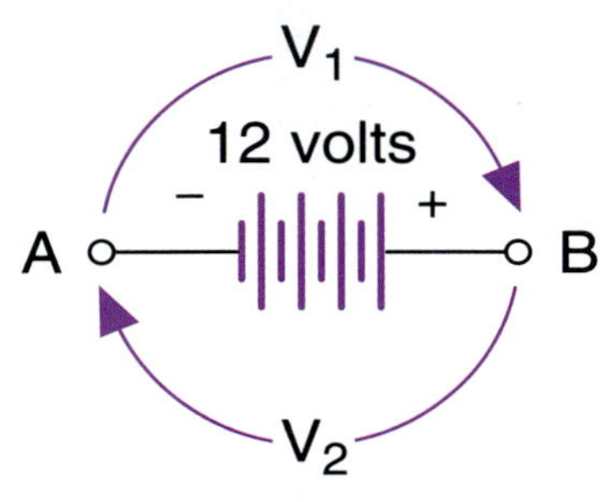

Figure 7-2 An example of voltage notation.

The voltage across two points in a circuit may be identified using an arrow with the symbol V drawn alongside it or separating it into two parts. Voltages are referenced from the tail of such an arrow and use the point to identify polarity. Figure 7-2 demonstrates the importance of a system for identifying polarities. Voltage V_1 is measured from the negative to the positive terminal of a battery. Because the arrow points to the positive terminal, the voltage is +12 volts. Voltage V_2 is measured from the positive terminal to the negative terminal of the same battery. This voltage is −12 volts because the arrow points to the negative battery terminal.

Another system is called **double subscript notation.** The test points in Fig. 7-2 are labeled A and B. The voltage across the battery can also be described as V_{AB} or V_{BA}. When double subscripts are used, the first subscript is the reference and the second subscript identifies the polarity. Therefore, V_{AB} is +12 volts and V_{BA} is −12 volts.

In written text the word *from* is the reference and the word *to* identifies the polarity. This form of notation is normally used when parts of a circuit have labeled **test points.** Test points are indicated on schematic diagrams as small circles that interrupt or terminate a wire. The test points are terminals in the physical circuit that extend above the circuit board. They may be a machined piece of metal or simply a short piece of wire that appears to have been soldered in by mistake. In some cases the test point may be formed by a resistor soldered on end into a circuit board. The lead stub at the top of the resistor becomes the actual test point. Although the arrangement may appear to be a mistake, there are frequently cases where the circuit under test must be isolated from the loading effects of the test equipment. Figure 7-3(*a*) shows these test points on a schematic diagram. Figure 7-3(*b*) shows how the points might look on an actual circuit board.

STUDENT to STUDENT

Understanding the various forms of voltage notation is very important. Without proper notation what you may see as positive voltage another person may see as negative voltage.

The terms *positive* and *negative* only apply to voltages that have a point of reference. Voltages without reference points should not have a polarity attached

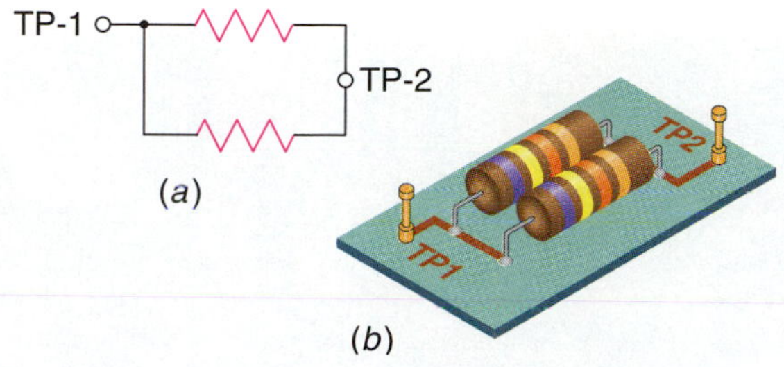

Figure 7-3 Test points are frequently available to aid technicians in taking measurements for troubleshooting and alignment purposes.

to them. For instance, an automobile normally has a 12-volt electrical system, and its headlights, radio, and other electrical devices operate on 12 volts. It would not be appropriate to say the automobile's radio works on +12 volts or −12 volts. It is appropriate to say that an automobile has a negative ground or positive ground. Ground is understood to be the automobile's metal frame, which is the reference point for voltage measurements within the automobile's electrical system. "Ground" becomes positive or negative by being connected to the positive or negative terminal of the automobile's battery.

Ground Symbols

All branches of electronics owe their existence to radio. When radio was in its infancy, the antenna systems of both the transmitter and receiver were connected to Earth with a copper-plated metal stake or wire mesh. The connection to Earth was necessary because the Earth became a part of the antenna system. The development of the vacuum tube made it possible to amplify radio signals and greatly improve the performance of the radio broadcast system. Vacuum tubes required electric power, which was supplied at first by batteries and later from power supplies connected to the public power distribution system. As a matter of convenience and necessity, the parts of the circuit that shared common power supply connections and were also common to the various signal paths were connected to the metal chassis used as a framework around which the equipment was built. Because the metal chassis was common to these circuit connections and connected to **Earth ground,** the chassis connections were called **common ground.**

The days of metal chassis are long past, having given way to the **printed circuit board.** Printed circuit boards are flat pieces of a copper-clad insulating material, such as fiberglass impregnated with epoxy resin. The unwanted portions of copper are etched away by chemical action to form the circuit conductors. The desired circuit image was originally silk-screened onto the board in the same way that posters are printed, thus the name *printed circuit board.* A sample circuit board is shown in Fig. 7-4 on the next page.

The terms *circuit (chassis) ground, common ground*, or simple *ground* remain. On many circuit boards, the ground path is larger than any of the other traces on the board. Circuit ground plays an important role in simplifying schematic diagrams. Components connected to ground are drawn using a special symbol. The symbol has several variations that distinguish Earth ground from circuit ground, or sometimes what is called **hot ground.** The three most popular symbols are shown in Fig. 7-5 on the next page.

Figure 7-4 Printed circuit boards are well suited to automated processes and allow a degree of repeatability not possible with hand wiring.

Component leads terminated with similar ground symbols are understood to be connected. The use of the ground symbol eliminates many lines from a schematic diagram, making the diagram easier to read. The ground symbol establishes a reference point for all voltages and signals within a circuit, as indicated by the following statement: *Unless otherwise specified, all voltages and waveforms are measured in respect to ground.* That statement is present on most schematic diagrams of commercial electronic equipment. The understanding imparted by the statement is that when any voltage appears on a schematic it is to be measured from circuit ground. Any exception to the rule will be clearly identified.

The official symbol for circuit ground, as shown in Fig. 7-5(*a*) resembles a small rake with three teeth. The symbol for Earth ground is that of a triangle formed from parallel horizontal lines decreasing in size from top to bottom, as shown in Fig. 7-5(*b*). For many years American and Japanese manufacturers of electronic equipment, particularly home entertainment products, reversed the official Earth ground and chassis ground symbols. This is not normally a problem

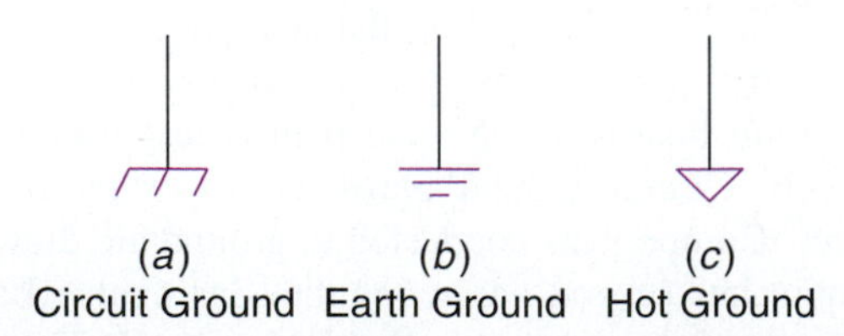

Figure 7-5 Ground symbols are used to identify (*a*) circuit (chassis) ground, (*b*) Earth ground, (*c*) hot ground. The symbols are not always used according to international standards, but their use should be self-evident on schematic diagrams.

when reading schematic diagrams because there is seldom more than one Earth ground symbol, but there will be many chassis ground symbols.

All electronic equipment requires one or more voltage sources in order to operate. The voltages are generally provided by **line-operated** power supplies. *Line-operated* refers to the use of the electrical wall outlet for power. In North America one side of the outlet is connected to Earth ground. The other side is often referred to as the "hot" side. Computers, televisions, video player/recorders, and computer monitors make use of special switching power supplies that have a considerable amount of circuitry closely connected to Earth ground. These grounds are identified with a symbol drawn as the outline of a triangle. This is the hot ground shown in Fig. 7-5(*c*).

Hot indicates that it is so closely connected to the voltage from the wall outlet that there is a strong potential for electric shock or damage to test equipment if certain types of contacts are made to this ground. The schematic diagrams of the sections of circuitry operated from line-operated supplies are normally enclosed in a shaded background and are clearly identified. All voltage measurements to these sections are made to hot ground. When a circuit contains sections using both hot grounds and chassis grounds, care must be taken that voltages are not measured from the wrong ground. This will almost certainly give incorrect readings and sometimes cause damage to sensitive low-voltage circuits. The right and wrong ways of testing circuits containing these two types of grounds are shown in Fig. 7-6.

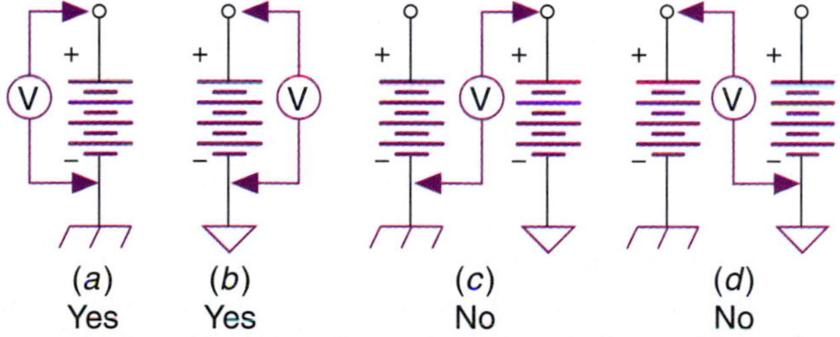

Figure 7-6 Care must be taken to use the correct ground when making voltage measurements in circuits containing both hot and circuit grounds.

Meter Loading

The ideal voltmeter would not require any current from the voltage being measured. However, in practical situations this is not the case. Every voltmeter requires some current to operate. The amount of current varies greatly with the type of meter used. A low-cost meter that is popular with electricians and automotive technicians is called a **volt-ohmmeter (VOM).** Although not implied by its name, a VOM can also measure current and resistance. Because VOMs make several kinds of measurements, they are commonly called **multimeters.** Figure 7-7 on the next page shows a typical VOM/multimeter.

The sensitivity of a VOM is rated in ohms per volt (Ω/V). For instance, if a VOM was rated at 1000 Ω/V and was set to a 1-volt scale, taking a voltage measurement would place a 1000 Ω resistance across the circuit under test (1000 Ω/volt × 1 volt = 1000 Ω). VOMs have multiple voltage ranges that can be manually selected. The selected range should always be above the expected voltage being measured or damage to the meter may result. The same meter set to a scale of 10 volts appears as a 10,000 Ω resistance to the voltage under test

Figure 7-7 VOMs are low-cost, multipurpose meters popular with electricians and automotive technicians.

(1000 Ω/volt × 10 volts = 10,000 Ω). The effect a meter has on a circuit because of any current it may take from the voltage being measured is called *loading*. Although less loading is obtained on higher scales, the voltage range is compressed because more voltage is indicated in the same space. This compression of the scale makes a low voltage reading less accurate than a reading on a lower scale.

Circuits that can deliver large amounts of current, such as public and automotive electrical systems, are not affected by the small amount of current required by even the least sensitive VOMs. Having to reduce loading by moving to a higher scale would not be required when measuring voltages in these circuits. A sensitive VOM may offer as much as 100,000 Ω/V sensitivity. Such a meter is adequate for testing many electronic circuits, but only those not severely affected by light loading and for which an error of from 2% to 5% is acceptable.

The Ideal Voltage Source

The **ideal voltage source** would produce a constant voltage under any load condition. Most power supplies are thought of as ideal sources. This point of view is especially appropriate if the source is regulated. Most voltage sources are nearly ideal if the current required of them is less than 1% of what it would be under short circuit conditions. But not all voltage sources are designed to be power supplies. Many voltage sources can only deliver small amounts of current. Even being connected to a modest load will cause their voltage to fall to a fraction of its unloaded value. Why this occurs is explained in detail in Chap. 8. For now it is important to know that the voltage of any source will fall when loaded—some far more than others. If a voltmeter with a small internal resistance is connected across a low current voltage source, the meter reading may contain an unacceptable error.

Electronic Voltmeters

The ideal voltmeter would not require any current whatsoever from the circuit being tested. That is not possible, but modern **digital multimeters (DMMs)** come very close. DMMs present the same resistance to a voltage source regardless of the selected scale. In fact, many DMMs automatically select the best scale for the selected function. Meters with this feature are said to be **autoranging.** A typical DMM has an input resistance of 10 million ohms, with some as high as 20 million ohms. These very high resistance values ensure that with today's low voltage circuits (compared to older vacuum tube circuits) only a minimum amount of current is required by the meter. The low current requirements ensure a more accurate voltage reading. In addition, the base accuracy of DMMs is far superior

Figure 7-8 Digital multimeters (DMMs) have replaced VOMs as the most frequently used piece of electronic equipment.

to any other meter. A typical DMM will accurately measure voltages with less than 0.25% error. Some of the more expensive meters designed for laboratory use have less than 0.005% error. A typical DMM is shown in Fig. 7-8.

REVIEW QUIZ 7-1

1. A voltmeter should be connected ______________ the voltage being measured.
2. Small circles that interrupt or terminate a line on a schematic diagram represent ______________.
3. A symbol that looks like a small rake at the end of a line represents ______________ ground.
4. The ground symbol drawn as the outline of a triangle represents ______________ ground.
5. Metal chassis and hand wiring have been replaced by ______________.

7-2 Measuring DC Current

Current is measured with an instrument called an **ammeter.** Because current is the rate of flow of electric charge, the ammeter must be placed in series with the device, circuit, or component through which the current is being measured. It is

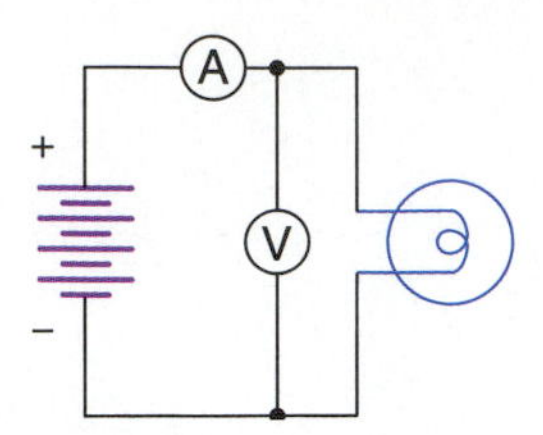

Figure 7-9 An ammeter that is a permanent part of an electric circuit is represented by a large circle that interrupts a wire. Normally a unit of electric current is placed inside the circle.

very important to remember that current always flows through something, never across it. In schematic diagrams, current meters are represented by large circles that interrupt a line. A capital letter A is usually placed inside the circle. Sometimes an M is used, and the type of meter is assumed from its placement in the circuit. The circle may also contain the symbol for a unit of current such as μA or mA. The "A" in μA and mA stands for amperes, which is the unit of measurement used for stating the amount of current. The symbol μA is read as "micro amp" and mA is "milliamp." How these are determined will be discussed later. The schematic diagram in Fig. 7-9 shows the use of the ammeter symbol and how its placement in a circuit differs from that of a voltmeter.

Just as a voltmeter consumes a small amount of current from the circuit being measured, some voltage will be dropped across an ammeter, albeit very small. Most analog ammeters are very sensitive and generally require less than 0.1 mA to produce full-scale deflection. Full-scale deflection means that the pointer on the analog ammeter will move all the way to the right. These meters may have an internal resistance of several hundred to several thousand ohms. To enable such meters to measure larger amounts of current, a **shunt** is connected across them. The shunt is basically a low-value resistor in the form of a metal strip that appears to short across the meter. The word *shunt* means to place across.

Sometimes current is measured using a special probe that detects the magnetic field around a wire. The output of the probe is a voltage that is supplied to a voltmeter whose scale is calibrated to read in amperes. All electric currents produce a magnetic field whose intensity is directly proportional to the strength of the current. These probes usually require that more than two amperes of current be present or they may not provide a reliable reading.

Direct current magnetic probes are based on the **Hall device.** The Hall device operates on a principle discovered in 1879 by E. H. Hall. Under normal conditions, the distribution of electric current is fairly even throughout a material. However, when a current-carrying material is placed in a magnetic field with the current and the field at a right angle, the current tends to move toward one side of the material. This difference in current density creates a very small voltage that is proportional to the strength of the current. The voltage is called the **Hall voltage** and is measured across the sides of the conductor perpendicular to both the current direction and the magnetic field. Small, flat, rectangular sections of certain semiconductor materials, such as indium arsenide (InAs), are normally employed in practical DC current probes. These slices of semiconductor material are between iron poles that channel the magnetic field through the material.

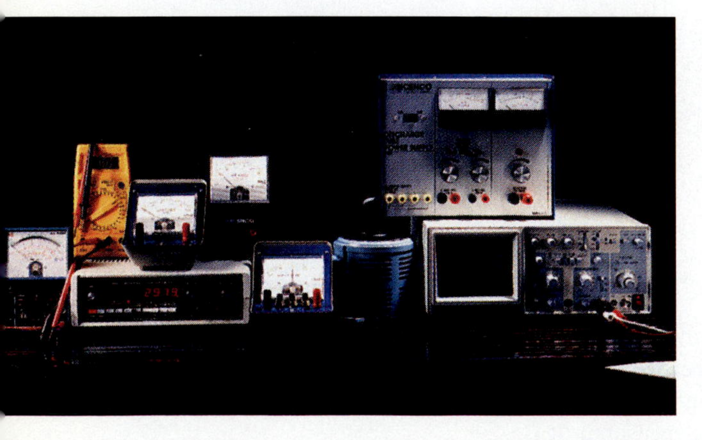

Digital Ammeters

The ammeter section of a DMM, or a dedicated digital ammeter, is used in the same way as an analog ammeter. The circuit under test must be broken and the meter placed in series with the path of current flow. Unlike analog meters, DMMs have a **fuse** in series with one of the test leads. These fuses are normally glass cylinders with metal end caps connected to the ends of a special metal conductor, as shown in Fig. 7-10(*a*). The conductors are made from a metal alloy that melts at a low temperature. If too much current is forced through a digital meter, the fuse element will melt and open the circuit. These fuses are usually located directly under one of the test jacks, as shown in Fig. 7-10(*b*). If a DMM seems defective because it will not measure current when properly connected to an operating circuit, the fuse is probably open.

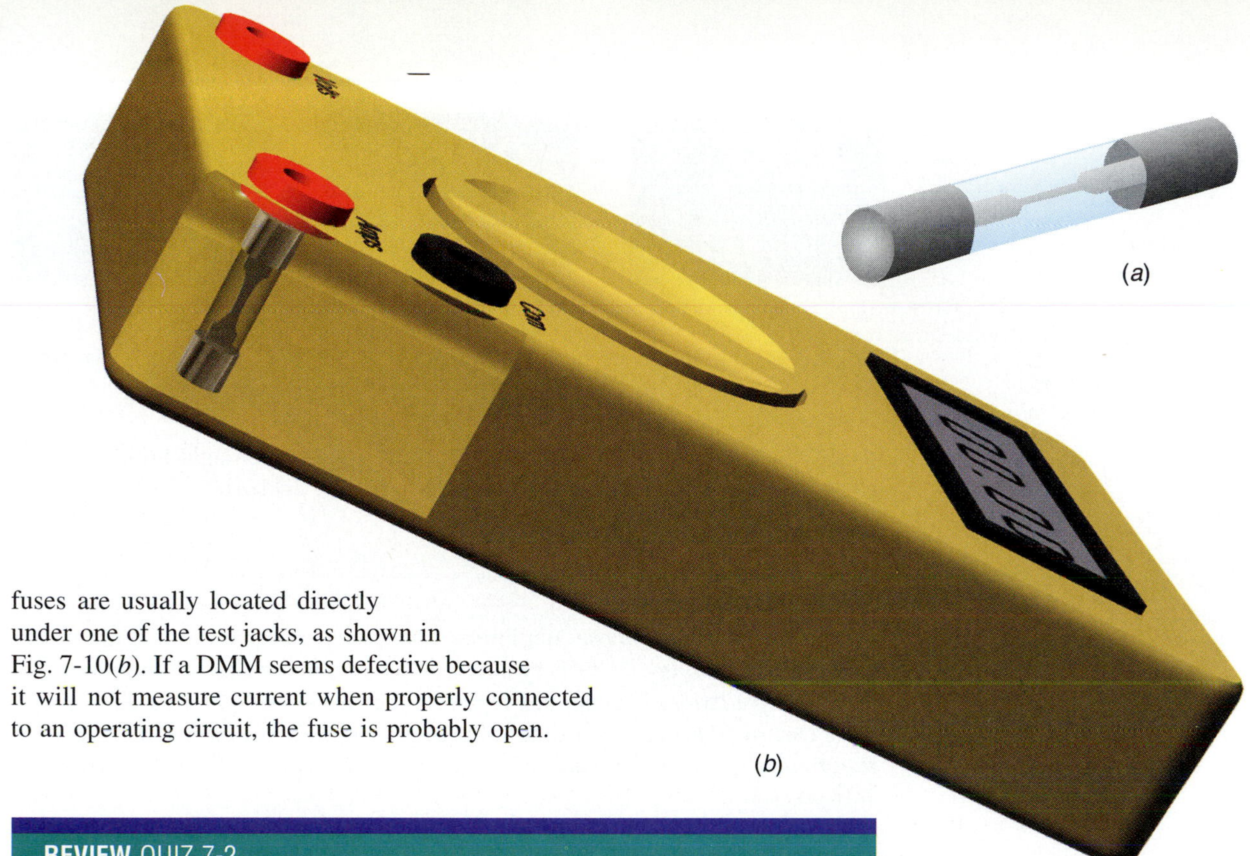

Figure 7-10 A fuse in the ammeter portion of a DMM will provide protection for the current shunt and some protection to a voltage source.

REVIEW QUIZ 7-2

1. Current is measured with an instrument called a(n) ______________.
2. A meter that measures current must be placed in ______________ with a load.
3. A solid state device that produces a voltage proportional to the strength of a magnetic field is called a(n) ______________ .
4. A DMM that fails to measure current when properly connected may have a defective ______________ .

7-3 Types of Direct Current

By definition, direct current flows only in one direction, which is determined by the polarity of the applied voltage. The use of the term *direct current* implies only that the direction of current does not change. It does not imply that the amplitude of a current or voltage must remain constant. Voltmeters and ammeters are designed to measure direct current as voltages and currents that do not change amplitude. Although this is normally the case with power supplies, it is not always the case within a circuit.

Pure DC

A steady current or voltage whose amplitude does not change is called **pure DC.** Pure DC is produced by chemical batteries and solar cells. A graph of a

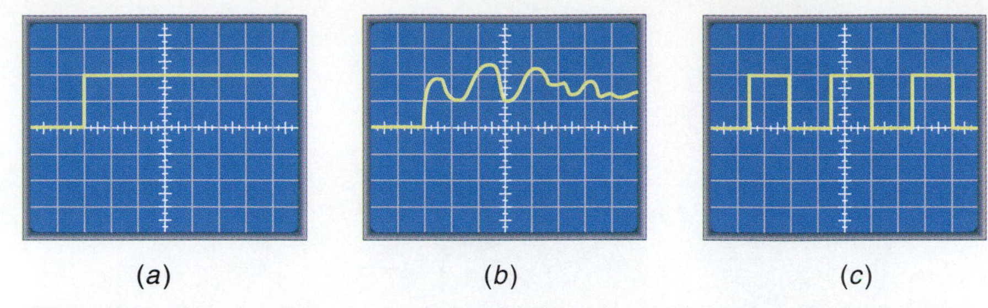

Figure 7-11 The type of direct current is determined by whether its amplitude changes, and if it does change, whether it periodically drops to zero.

pure DC current or voltage plotted over time produces a straight horizontal line above or below the zero reference, as shown in Fig. 7-11(*a*).

Varying DC

Currents or voltages whose amplitudes change but never drop to 0 are called **varying DC.** Varying DC is very common in circuits containing amplifying devices. The variations in amplitude represent the signals being processed. A graph of an electrical quantity over time is called a *waveform.* A varying DC waveform resulting from superimposing an audio signal onto pure DC is shown in Fig. 7-11(*b*).

Pulsating DC

A current or voltage that varies in amplitude and periodically drops to 0 is called **pulsating DC.** Pulsating DC is common in digital logic circuits. A graph of pulsating DC over time is shown in Fig. 7-11(*c*).

Most amplitude variations encountered in electronic circuits occur too frequently for a meter to follow. As a result, the meter will only indicate the average current or voltage. Regardless of the DC waveform, the measured DC value is an average of the area under the amplitude-versus-time waveform. For pure DC waveforms, the average area and the maximum amplitude are the same.

The procedure for calculating the average area under a waveform is dependent on the shape of the waveform. Consider the circuit of a battery-operated electric lamp, such as the one in Fig. 7-12. The lamp is activated by a momentary contact switch. While the switch is held in the on position, the lamp will

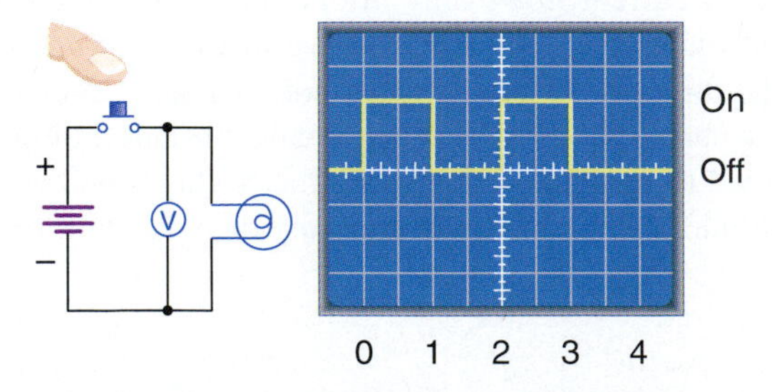

Figure 7-12 An electric lamp connected across a battery and activated by a momentary contact switch provides an example of pulsating DC. When the voltage across the lamp is graphed over time, the voltage waveform is revealed.

light; when the button is released, the lamp will go out. Assume the button is pressed for 1 second, released for 1 second, pushed for 1 second, then released for 1 second. The voltage across the lamp is pulsating DC because the voltage periodically dropped to 0. The average voltage over the 4-second interval would be one-half of the battery voltage. Because the button is being pushed at such a slow rate, the meter should have no trouble following the changes in voltage. However, if the changes were more frequent, say 100 times per second, the meter could not follow the changes. Here the meter would give a false reading and indicate only 50% of the applied voltage.

Voltage and Duty Cycle

The reason the average voltage in the previous example is 50% of the source is that the voltage was at maximum only 50% of the time. The waveform created by pressing the momentary contact switch is called **square-wave** because of its shape. The **duty cycle** of the waveform is the on time divided by the total time period, which is the sum of the on time and off time. This ratio is expressed as a percentage when it is multiplied by 100. A DC square-wave with a 100% duty cycle is pure DC. A DC square-wave voltage with a 10% duty cycle would only indicate 10% of the source voltage when measured with a standard meter.

REVIEW QUIZ 7-3

1. The type of DC produced by a battery is called ______________ .
2. Pulsating DC periodically drops to ______________ amplitude.
3. Direct current that changes in amplitude but never drops to 0 is called ______________ .
4. A DC square-wave voltage with a duty cycle of 75% will show an average meter reading of ______________ %.

7-4 Resistance Measurements

Resistance is measured with an instrument called an **ohmmeter.** All ohmmeters share the common trait of forcing a current through the component, circuit, or device being tested. An ohmmeter can be thought of as a voltage source in series with a current meter and a current limiting resistor. When the leads of an ohmmeter are shorted, the current in the circuit is only limited by the current limiting resistor, the meter deflects to its full-scale position. When the probes are not touching (open), the meter is in its zero position. The numbers on the scale used to indicate resistance read oppositely to the relative values of current through the meter. The maximum current through the meter produces a 0-ohm reading, and zero current produces a resistance reading of infinity. A simple ohmmeter circuit is shown in Fig. 7-13(*a*) on the next page. A simplified scale for a VOM is shown in Fig. 7-13(*b*). Notice that the ohm scale reads oppositely to that of voltage or current. The ohm scale is also very nonlinear, with scale markings at the higher resistance end of the scale very close together.

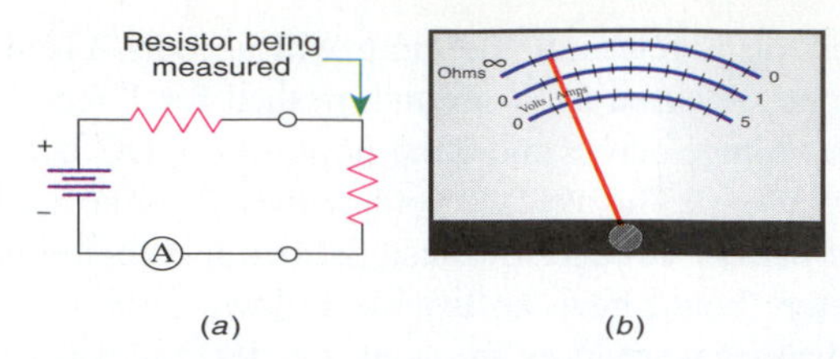

Figure 7-13 Ohmmeters must force a current through something to measure its resistance. The current supplied by an ohmmeter causes it to behave oppositely to voltage or current as indicated by the reversed ohms scale.

VOMs typically supply more current than DMMs during a resistance test. Some older VOMs can actually cause a small incandescent lamp to light. These older meters can damage modern solid state circuits and must be used with care. Modern DMMs have sensitive circuits that operate on far less current. These low ohmmeter currents are safe for all except the most sensitive circuits and provide better accuracy in the process. Some common meters are accurate to better than 0.01%.

REVIEW QUIZ 7-4

1. An ohmmeter must output a(n) ______________ to cause current to flow through the device under test.
2. The ohms scale on a VOM reads ______________ to the voltage and current scales.

STUDENT *to* **STUDENT**

Pay extra attention to this section on oscilloscopes. You'll be using "scopes" in a lot of your labs.

7-5 The Oscilloscope

An **oscilloscope** is an electronic instrument that graphically displays voltage over time on a screen that is typically 10 cm wide and 8 cm high. The screen is divided into a grid of 1 cm squares to aid the measurement process. During normal operation, a small point of light is moved horizontally across the center of the screen from left to right at a uniform rate. When the dot reaches the right side of the screen, it is turned off (blanked) and rapidly returned to the left side of the screen, where the process begins again. The time to traverse each centimeter is adjustable using a knob that indicates the dot's velocity in seconds, milliseconds, microseconds, and sometimes nanoseconds per centimeter.

Oscilloscopes have a vertical input that is sensitive to voltage. A positive voltage causes the dot to move upward and a negative voltage causes the dot to move downward. The sensitivity of the vertical input is adjustable with a knob labeled volts, millivolts, and sometimes microvolts per centimeter. Oscilloscopes are used more often than any other test instrument except the multimeter.

The most important uses for an oscilloscope are to measure the amplitude of voltages that change too quickly for a meter to follow and to measure the time interval between those changes. The changing of voltage over time creates a waveform. The oscilloscope provides a picture of that waveform as a graph of voltage versus time. The dot normally appears as a line because it is moving very fast. The human eye creates the illusion of a line because the retina stores the image of the dot at all its positions for about one-tenth of a second. Most oscilloscopes can analyze voltage waveforms that change millions of times per second.

Oscilloscopes are especially good at analyzing **periodic waveforms.** Periodic waveforms repeat at regular intervals. The time consumed by the waveform before it repeats is called the **period** of the wave. The period can be measured between any similar points on the waveforms using the graduations on the oscilloscope screen, as shown in Fig. 7-14.

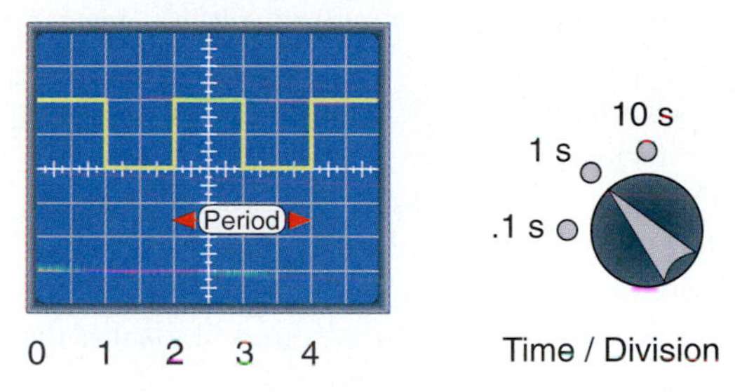

Figure 7-14 The time between like points on adjacent waveforms is called the period.

The **frequency** of a waveform is the number of times per second the waveform repeats itself. Frequency is measured in **hertz (Hz)** and calculated from the period by taking the reciprocal of the period in seconds.

Equation 7-1

$$f = \frac{1}{T} \Rightarrow \text{frequency} = \frac{1}{\text{time period}}$$

A wave form's period may be found easily when viewed on an oscilloscope. The period is the reciprocal of the frequency, as shown in Eq. 7-2. Period and frequency are reciprocals of each other. If the period is known, the frequency may be calculated.

Equation 7-2

$$T = \frac{1}{f} \Rightarrow \text{time period} = \frac{1}{\text{frequency}}$$

REVIEW QUIZ 7-5

1. The oscilloscope is an electronic instrument used to measure the ______________ and ______________ of a periodic waveform.
2. The oscilloscope provides a graph of ______________ versus ______________.
3. The vertical input of an oscilloscope is only sensitive to ______________.
4. The time between similar points on a waveform is called the ______________ of the wave.

SUMMARY

Voltages are measured with an instrument called a voltmeter and are always taken across the component, circuit, or device being measured. Proper voltage measurements depend on an understanding of voltage notation and an agreement on the direction of current and the type of charge carrier. Without these understandings, the polarity of a DC voltage may be read incorrectly. Polarity is more important than the absolute value of a voltage.

Schematic diagrams are easier to read when ground symbols are used. These symbols are the reference points from which all voltages and signals are measured. Having ground as an understood reference point simplifies the taking of voltage measurements. Three ground symbols are commonly employed to represent circuit ground, Earth ground, and hot ground.

Voltmeters and ammeters require a small amount of power from the circuits whose parameters are being measured. Digital meters are generally more sensitive than analog meters and have less effect on circuit operation. Digital ammeters usually have a fuse in series with one of the test leads to protect the meter. It is very easy to attempt to measure voltage when the meter is positioned for current measurement. This mistake will almost certainly cause the fuse to fail. The failure of the fuse will disable the current measuring function of the meter until the fuse is replaced.

Resistance measurements require that current be forced through the device under test. Therefore, all ohmmeters have an internal voltage source. When using older VOMs, the current supplied by the voltage source may damage sensitive components. DMMs seldom have this problem because they use much less current.

Direct current is divided into three categories called pure DC, varying DC, and pulsating DC. Pure DC is easily read with a DC meter, but varying DC and pulsating DC will not read correctly because they are in a constant state of change. Voltages that change rapidly are measured with an oscilloscope. The oscilloscope is also able to measure the time intervals between the repetitive patterns of voltage waveforms.

NEW VOCABULARY

ammeter
autoranging
circuit ground
common ground
digital multimeter (DMM)
double subscript notation
Earth ground
frequency
fuse
Hall device
hertz (Hz)
hot ground
ideal voltage source
multimeter
ohmmeter
oscilloscope
periodic waveform
printed circuit board
pulsating DC
pure DC
shunt
square-wave
test point
varying DC
voltmeter
volt-ohmmeter (VOM)

Questions

1. How is a voltmeter used in a circuit?
2. Which polarities are represented by the red and black leads on a meter?
3. Is the polarity of a voltage more important than its value?
4. How are test points identified on schematic diagrams?
5. Are circuit voltages normally measured from circuit ground?
6. Which would draw more current from a voltage under test, a VOM with 1000 Ω/V or one with 20,000 Ω/V?
7. Would the voltage produced by an ideal voltage source fall when a load is placed across it?
8. How are ammeters used in a circuit?
9. Why is a shunt placed across an ammeter?
10. What form of direct current changes amplitude by periodically dropping to 0 volts?
11. What piece of electronic equipment is used to measure and display voltages that vary rapidly with time?
12. What is meant by the term *periodic waveform*?

Problems

1. What is the period of a 100-kilohertz waveform?
2. What is the frequency of a waveform with a period of 100 microseconds?
3. If a 1-volt source had an internal resistance of 20,000 ohms and a 20,000 Ω/V VOM set on the 1-volt scale were connected across it, what voltage would the meter indicate?

Critical Thinking

1. A circuit diagram has dozens of Earth ground symbols and no other type of ground symbol is present. Why could you correctly assume that the Earth ground symbol is being used as the symbol for circuit ground?
2. Why must a circuit be broken before small currents can be measured?
3. Is the output of a current probe current or voltage?
4. Why can't an oscilloscope make unaided current measurements?
5. When can a VOM accurately measure pulsating DC?

Answers to Review Quizzes

7-1
1. across
2. test points
3. circuit
4. hot
5. printed circuit boards

7-2
1. ammeter
2. series
3. Hall device
4. fuse

7-3
1. pure DC
2. zero
3. varying DC
4. 75

7-4
1. voltage
2. opposite

7-5
1. voltage; period
2. voltage; time
3. voltage
4. period

Chapter 8

Series Circuits

ABOUT THIS CHAPTER

This chapter examines circuits made by connecting resistors in a series. The circuits may also contain multiple voltage sources. The series circuit is the second of five basic circuit types. When made using resistors, it is often called a voltage divider. The voltages across resistors in

After completing this chapter you will be able to:

- Recognize the series connection and understand its rules of operation.
- Describe the process of voltage division and calculate or approximate individual voltage values.

the divider can be calculated or estimated. The different procedures for doing so are presented with examples. Series circuits obey a specific set of rules that apply only to them. These rules are also presented and reinforced with examples.

- Explain how voltage sources function in series-aiding and series-opposing circuits.
- Understand internal voltage drops in voltage sources.

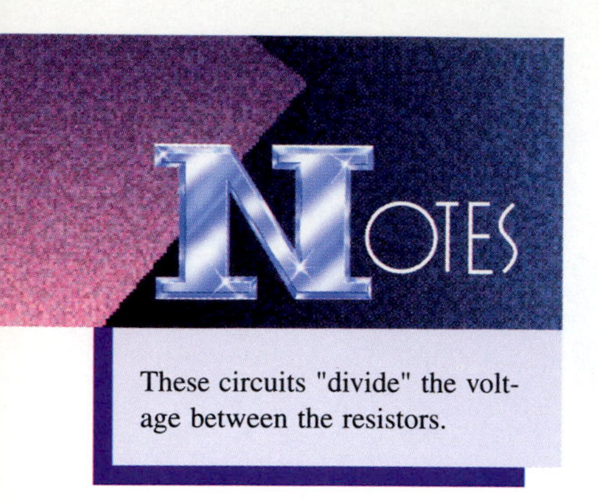

These circuits "divide" the voltage between the resistors.

8-1 Elements of Series Circuits

Components connected end-to-end form a **series circuit.** Each component in a series circuit has the same current flowing through it when connected across a voltage source. Series circuits made from resistors distribute the source voltage across each resistor. Such circuits are frequently called **voltage dividers.** An example of a series circuit is shown in Fig. 8-1.

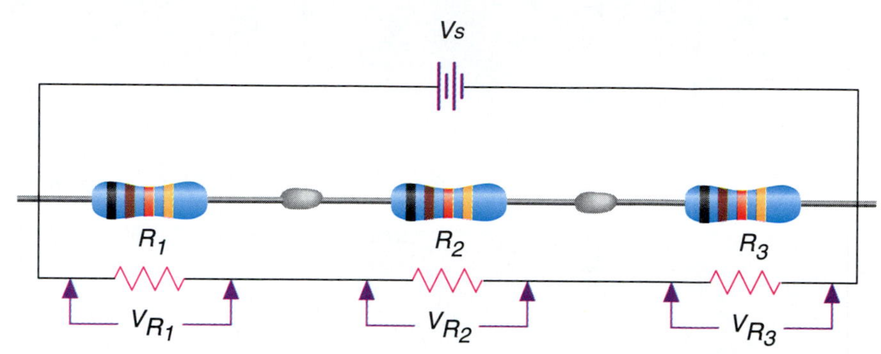

Figure 8-1 Series resistive circuits have the same current flowing through each resistor and divide the source voltage into several smaller voltages.

Voltage Distribution

Any resistor with current flowing through it has a voltage across it that can be found using Ohm's law. The distribution of that voltage is proportional to the resistance. For instance, wires are made from many different metals and are usually consistent in diameter and composition throughout their length. Therefore, the resistance of a wire is proportional to its length, meaning that if a wire is cut in half, each half would account for one-half of the wire's resistance. If that wire is placed across a voltage source, one-half of the voltage would be found between the center of the wire and either end (Fig. 8-2).

To avoid melting the wire, or damaging the source, the wire should be of small diameter and made from a high resistance material, such as nichrome. As the

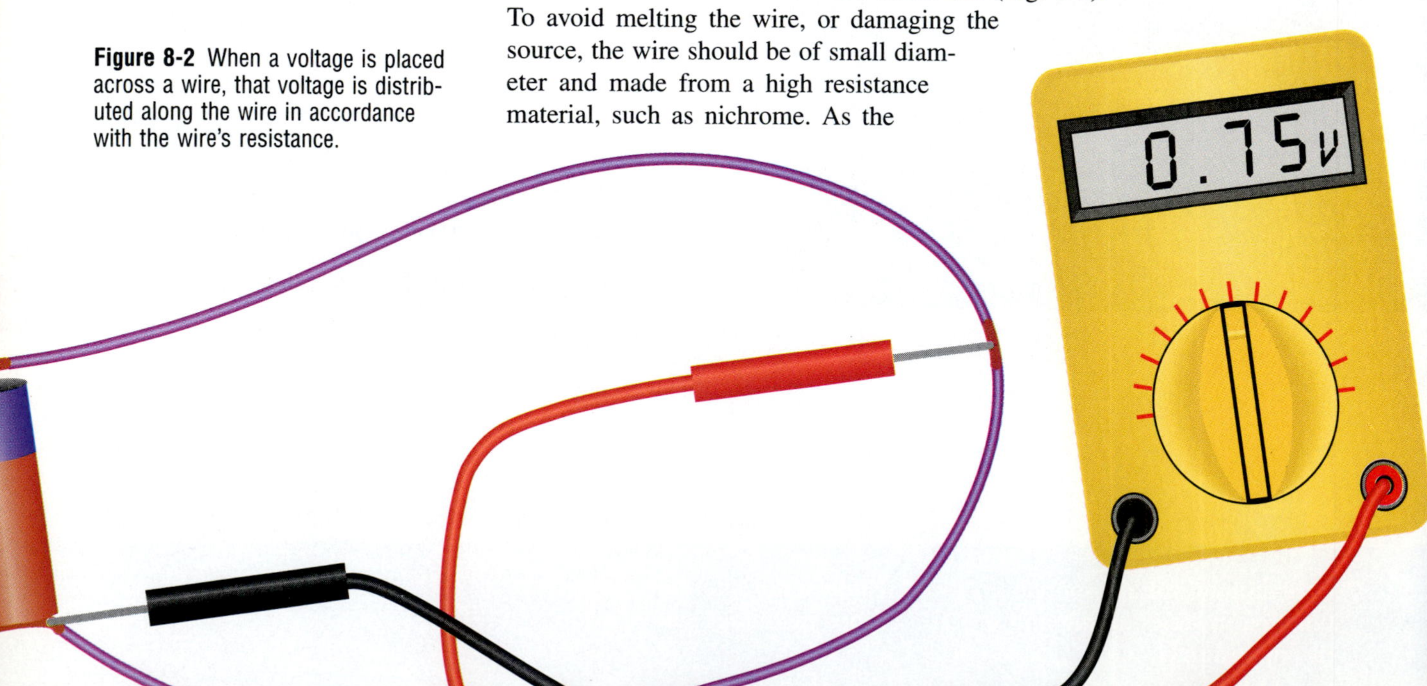

Figure 8-2 When a voltage is placed across a wire, that voltage is distributed along the wire in accordance with the wire's resistance.

meter probe is moved along the wire, the measurement between the probe and the end of the wire varies from zero to the source voltage.

Potentiometers

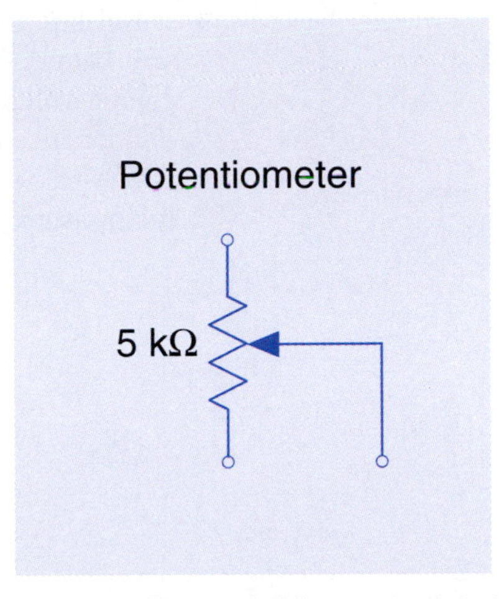

Figure 8-3 The symbol for a potentiometer is that of a resistor touched at the center by an arrowed line forming a three-terminal device.

A resistive component called a **potentiometer,** also known as a **pot,** distributes any applied voltage throughout its resistance. The potentiometer is a three-terminal component with two of the terminals connected to the ends of a resistance element. The elements are usually made from carbon, but when they are required to carry large currents they are wound with nichrome wire. The third terminal is connected to a moveable contact called the **wiper** that slides along the length of the element. More often than not, the element is arranged to form the better part of a circle. The wiper is moved by rotating an attached shaft or inserting a tool into a slot and turning when the shaft is not present. Potentiometers can also be linear with a sliding contact. Either way, potentiometers divide the resistance element into two sections, forming a variable voltage divider.

The circuit symbol for a potentiometer is that of a fixed resistor with an arrow pointing toward the symbol and touching it at midpoint. The arrow represents the moveable contact. The potentiometer's resistance is measured across the full element. Its resistance value is written alongside its symbol, as shown in Fig. 8-3. Two small potentiometers are shown in Fig. 8-4.

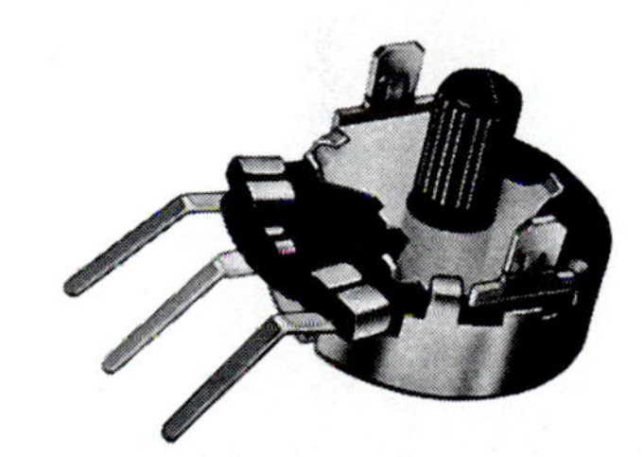

Figure 8-4 Small potentiometers are often used for variable controls in electronic circuits.

Rules for Series Circuits

Series circuits operate under a set of rules that, in conjunction with Ohm's law, provide a solution for determining the voltage at any point or across any element in the circuit. The total resistance of a series circuit and the relative current in a series circuit are defined within the rules themselves. The following set of six rules apply to resistive series circuits.

1. The current is the same at any point in a series circuit.
2. The total resistance in a series circuit is equal to the sum of its resistance values ($R_T = R_1 + R_2 + R_3 \ldots$).
3. If a series circuit is broken at any point, all current flow will stop.
4. The sum of the voltage drops around a series circuit is equal to the source voltage. This rule is known as **Kirchhoff's voltage law** and may also be expressed in the following form: The algebraic sum of the resistive voltage drops in a series circuit and the source voltage(s) is equal to zero.
5. The voltage across a resistor in a series circuit is to the source voltage as the resistor's value is to the total resistance. This is called the **voltage divider relationship.**

$$\frac{V_R}{V_S} = \frac{R_X}{R_T} \Rightarrow V_R \times R_T = V_S \times R_X \Rightarrow V_R = V_S\left(\frac{R_X}{R_T}\right)$$

6. The current in a series circuit is not affected by the arrangement of its components.

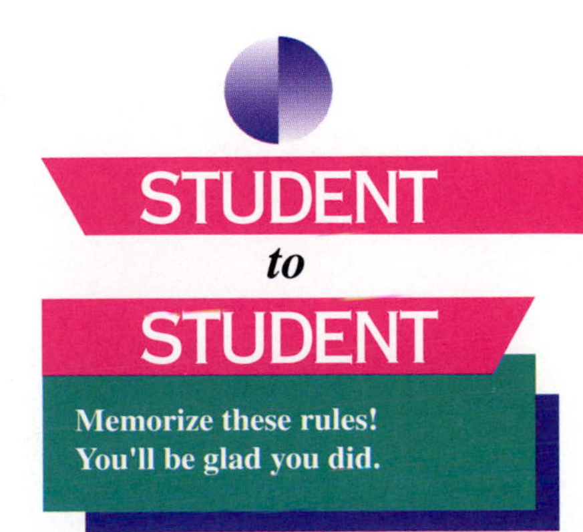

Voltage Dividers

Many divider ratios become intuitive with practice, as will the ability to estimate the voltage across a resistor in a series circuit by applying the voltage divider relationship. In Fig. 8-5, two 60-ohm resistors are connected in series across a 12-volt battery. The output voltage is identified as 6 volts, which is one-half of the source voltage. That should not be surprising, given that the resistances in the divider are equal values. The resistance that the output voltage is measured across is one-half of the total resistance, which, in accordance with the voltage divider relationship, has one-half of the source voltage across it.

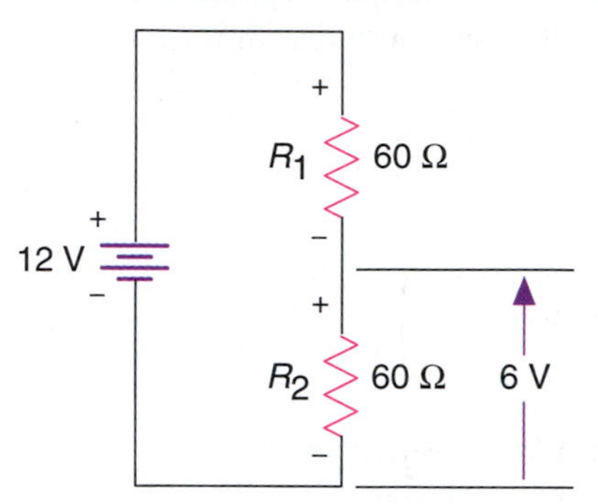

Figure 8-5 The voltage divider relationship is demonstrated in this circuit.

Referring back to Fig. 8-5, we see that the proof of the relationship between the voltage ratio and resistance ratio is provided by a combination of Ohm's law and the series circuit rules, as shown in Example 8-1.

Example 8-1

1. The goal is to prove that the voltage across R_2 is 6 volts. According to Ohm's law, the voltage across a resistor is found by multiplying the current through the resistor by the value of the resistor ($V_R = I \times R$). The formula cannot be implemented because the current through R_2 is not known.
2. The current through R_2 is equal to the total current in the circuit because of *Rule 1* for series circuits: *The current is the same at any point in a series circuit.*
3. The total current in a series circuit can be found by dividing the source voltage by the total resistance.

$$I_T = \frac{V_S}{R_T}$$

The total resistance of a series circuit can be found by applying *Rule 2* for series circuits: *The total resistance in a series circuit is equal to the sum of its resistor values* ($R_T = R_1 + R_2 + R_3 \ldots$). The total resistance R_T is equal to the sum of R_1 and R_2 ($R_T = R_1 + R_2 = 60\ \Omega + 60\ \Omega = 120\ \Omega$).

4. Step 2 can now be completed because the total resistance is known. The total circuit current is:

$$I_{R_2} = I_{R_1} = I_S = \frac{V_S}{R_T} = \frac{12 \text{ volts}}{120 \text{ ohms}} = 0.1 \text{ amp}$$

5. Return to Step 1 and plug in the values.

$$V_{R_2} = I_{R_2} \times R_2 = 0.1 \text{ amp} \times 60 \text{ ohms} = 6 \text{ volts}$$

The combination of Ohm's law and series circuit rules can be tedious because so many steps are involved. A simpler solution is to use the voltage divider relationship mentioned in Rule 5: *The voltage across a resistor in a series circuit is the same percentage of the source voltage that the resistor is of the total resistance.* Once the voltage across the resistor in question has been identified, its percentage of the overall resistance can be found by dividing its value by the total circuit resistance. The voltage across the resistor is determined by multiplying that percentage by the source voltage.

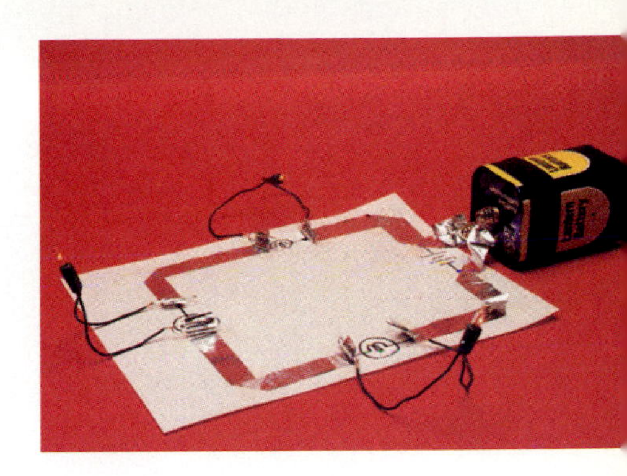

Equation 8-1

$$V_R = \left(\frac{R}{R_T}\right)V_S = \left(\frac{60 \text{ ohms}}{120 \text{ ohms}}\right)12 \text{ volts} = 6 \text{ volts}$$

In this equation V_R is the voltage across the resistor in question, R_T is the total resistance including the resistor in question, and V_S is the voltage of the source.

Voltage Divider Estimates

The voltage divider relationship provides a procedure for estimating the voltage across a resistor in a series circuit that is accurate, fast, and can be done mentally without a calculator. Consider the circuit in Fig. 8-6, containing three resistors connected in series, and estimate the voltage value across R_1.

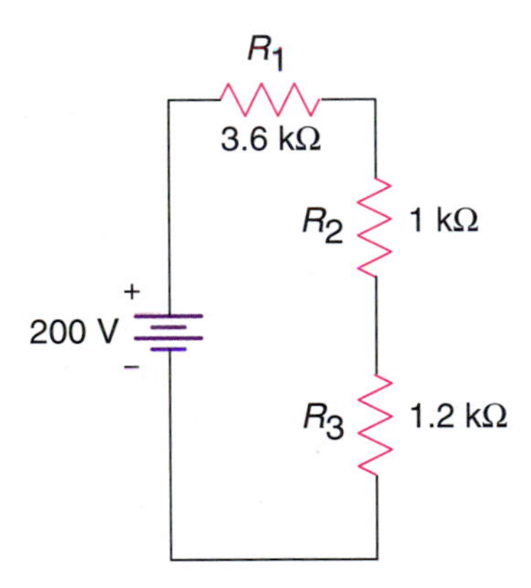

Figure 8-6 Series circuit with three resistors.

The total resistance in the circuit is found by adding the resistor values which is equal to 5.8 kΩ (kilohms), or 5800 ohms. The next step is to estimate what percentage 3600 is of 5800. Here is a simplified way of approaching the estimation problem. Clearly, 3600 is more than half of 5800; therefore, more than half of the source voltage will be across it. How much more than half is not very intuitive with the values given. One-tenth of the total resistance is 580 ohms, and would be a lot more convenient if it were 600 ohms, so we'll say that it is. The 3600 ohms of the resistor in question is evenly divisible by 600, which will go into 3600 six times. That means that about six-tenths of the total resistance is contained in the 3600-ohm resistor. Actually, there will be a little more than six-tenths because 580 is smaller than 600 and would go into 3600 more times. That will be dealt with later. One-tenth of the source voltage is 20 volts. Six-tenths is 20 times 6, or 120 volts. We know that this value is a little low, because we raised 580 to 600, so add another 5 volts. The estimate is now 125 volts. Applying the actual voltage divider formula, as shown in Eq. 8-2, produces an answer of 124.138 volts, an error of less than 1%. Using approximation methods the voltage anywhere in a series circuit can usually be estimated to within ±5%. Even without the 5-volt correction the estimate has only a 4% error.

Equation 8-2

$$V_{R_1} = V_S\left(\frac{R_1}{R_1 + R_2 + R_3}\right) = 200\text{ V}\left(\frac{3.6\text{k}\Omega}{3.6\text{k}\Omega + 1\text{k}\Omega + 1.2\text{k}\Omega}\right)$$

$$= 200\text{ V}\left(\frac{3.6\text{k}\Omega}{5.8\text{k}\Omega}\right)$$

$$= 124.138\text{ V}$$

The Rheostat

A component that is often confused with a potentiometer by calling it a pot is the **rheostat.** The difference is that rheostats are two-terminal adjustable resistors with connections made only to one end of the resistance element and the wiper contact. Rheostats are physically similar to the circular types of potentiometers. Electrically they are used as current-limiting devices and must be connected to another resistor to provide voltage divider action. The potentiometer is a self-contained voltage divider and can be used as a rheostat by making a connection only to one end of the wiper.

Two symbols are used to represent a rheostat. Both use the standard resistor symbol as a base. One is a simple resistor symbol with an arrow drawn through it at an angle, the other is identical to a potentiometer, except one end of the resistance element is not connected to anything. Occasionally the open end of the element is terminated by a short line to confirm that no access is provided to that part of the element. When a potentiometer is used as a rheostat, the unused end of the resistance element is often connected to the wiper. That way if the wiper should lose contact with the element, a portion of the resistance would remain in the circuit. Figure 8-7 shows schematic representations of rheostats.

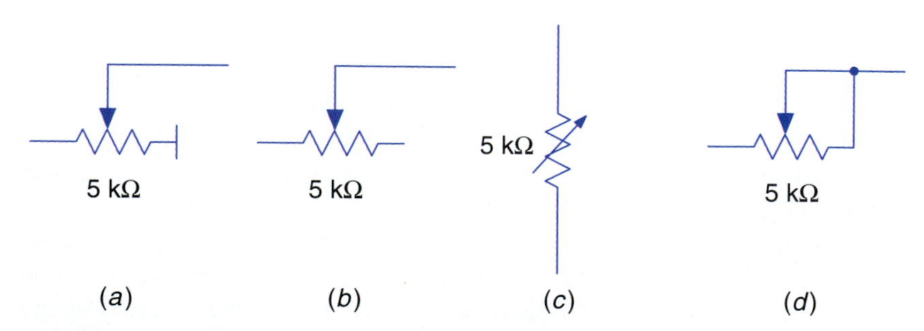

Figure 8-7 (*a*)–(*c*) Rheostats are two terminal adjustable resistors that may use several different symbols. (*d*) A potentiometer may also be used as a rheostat.

Special Rheostats

Rheostats normally have a resistance range that is adjustable from zero to the maximum resistance of the element, but some have a mechanical stop that prevents the resistance from reaching a 0-ohm value. The schematic symbol for these special rheostats have a line drawn through their elements to indicate the stop. The value of the rheostats as shown on the schematic will indicate its maximum and minimum resistance. A rheostat with a stop cannot safely be replaced by one without a stop. If such a replacement should become necessary, and the original part is not available, a standard rheostat in series with a fixed resistor can be used. The standard rheostat should have a value equal to the original minus the stop value. The fixed resistor should be the same as the stop value. Figure 8-8 shows the schematic representations of this special rheostat and its equivalent using standard parts.

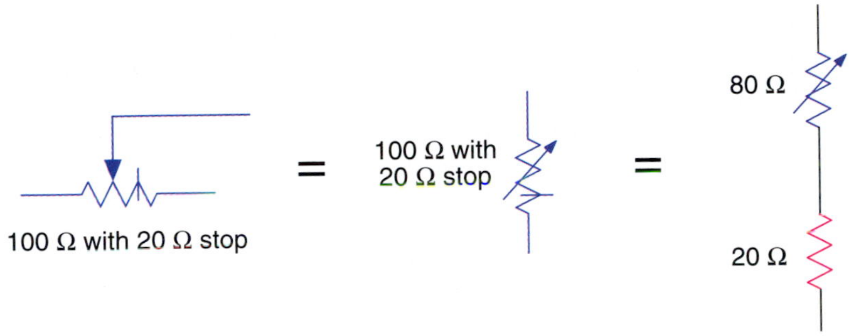

Figure 8-8 The line through the resistor symbol represents a mechanical stop that limits the range of resistance.

REVIEW QUIZ 8-1

1. In a series circuit, what electrical quantity remains constant?
2. The name of a three-terminal component used as a variable voltage divider is ______________.
3. The name of a two-terminal component used to vary the current in a circuit is ______________.
4. ______________ law states that the sum of the voltage drops around a series circuit is equal to the source voltage.

8-2 Voltage Sources in Series

The source voltage of a circuit often consists of several individual voltage sources connected in series. This is sometimes necessary because of the need for a higher voltage than any single source can provide, which is often the case in battery-operated equipment. Multiple sources are also the consequence of circuit interactions and are not always obvious. Electronic circuitry can be quite involved and any voltage buried within a network may have to be considered as a source for diagnostic purposes. For that reason, voltages connected in series-aiding and series-opposing must be considered.

Series-Aiding

In the **series-aiding** connection, the positive terminal of one source is connected to the negative terminal of another. The resultant voltage is the numeric sum of the individual sources. A 12-volt automobile battery is composed of six lead-acid cells each producing a nominal 2 volts. Because the cells are connected in series-aiding, the six cells produce the required 12 volts. A popular rectangular 9-volt battery, used in portable radios, toys, and remote control hand units, also contains six cells. However, these cells are usually carbon-zinc or alkaline and produce 1.5 volts each, which is the reason for the overall voltage being less than the voltage of the automobile battery.

The schematic representation for a group of cells manufactured as a single unit is different from cells connected by wires or contained in a holder. Individual cells are represented schematically by two parallel lines, one long and one short. The long line is the positive terminal and the short line is the negative terminal. Groups of cells combined into a single unit are represented by any number of cell symbols, with the number of symbols having nothing to do with the actual voltage. The voltage is indicated by writing it alongside the battery symbol. The distinguishing factor is that the cells are not joined by lines representing wires. Schematic representations of power sources made from individual cells, or groups of batteries, use lines to connect the symbols. The lines imply that the cells, or batteries, are connected by wires. Figure 8-9 demonstrates batteries of various voltages connected in series-aiding.

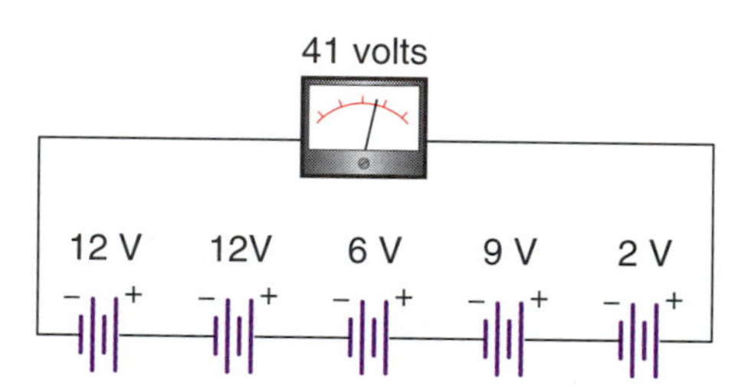

Figure 8-9 Voltage sources are connected in series-aiding when the positive terminal of one is connected to the negative terminal of another.

Series-Opposing

Sources connected in series with positive to positive or negative to negative are **series-opposing.** This configuration delivers current to a load by forcing it through the weaker source against the pressure of its voltage. The effective voltage is the difference of the two and retains the polarity of the stronger source. If the voltages are equal and opposite, they cancel. Deliberately placing sources in series-opposing to obtain a lesser voltage is not a normal design practice, but the series-opposing configuration appears often because of circuit interactions. Batteries are used to explain these interactions for the sake of simplicity.

When more than two sources are placed in series-opposing, some sources work together to supply current to the load and others work together in opposition to it. There are several techniques for determining the effective voltage. In series-opposing circuits, the source with the highest voltage always supplies the current. The pressure directions of the individual sources are identified and those that work together are added. The smaller of the two voltages is then subtracted

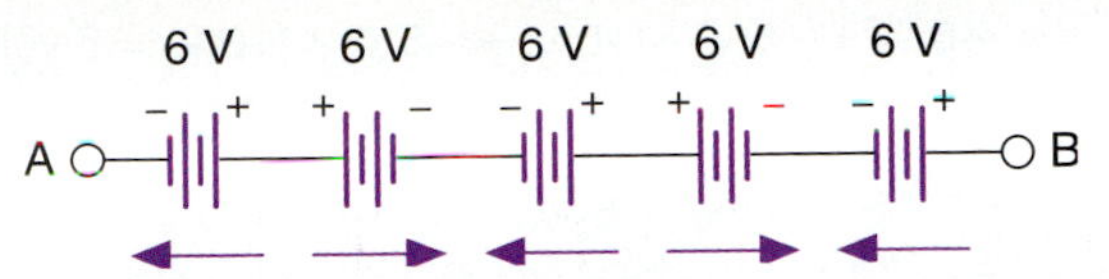

Figure 8-10 In some electric circuits, situations often occur where circuits behave as batteries connected in series-opposing.

from the larger to obtain the effective voltage of the aggregate source. Consider the circuit of Fig. 8-10, which contains five 6-volt batteries. The arrows on the schematic indicate the direction of pressure placed on the free electrons in the circuit wiring. It is the opposing pressures that account for the overall voltage reduction when sources are connected in series-opposing.

Three of the batteries in Fig. 8-10 would push current to the left; the other two would push it to the right. Those pushing to the left create a pressure of 18 volts, and those pushing to the right create a pressure of 12 volts. The effective voltage of this system is 6 volts, with point A negative and point B positive, as shown in Fig. 8-11.

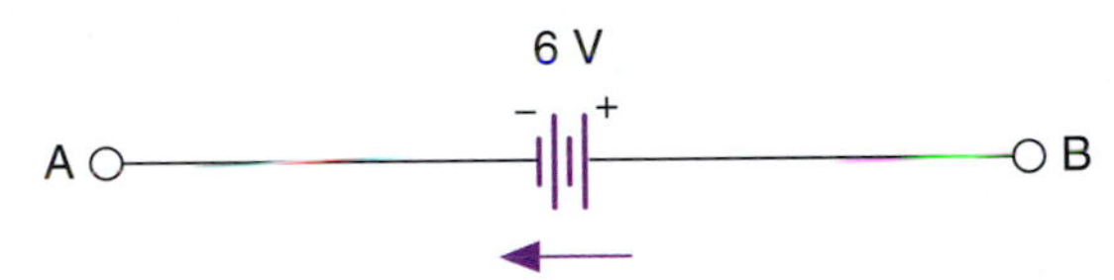

Figure 8-11 The five batteries connected in series-aiding and series-opposing behave as a single 6-volt battery.

Rule 6 for series circuits states: *The current in a series circuit is not affected by the arrangement of its components.* Since voltage is the pressure that creates current flow, rearranging the individual sources will not change the effective source voltage. If it did, the current would change in defiance of Rule 6. Figure 8-12 shows a rearrangement of the five batteries shown in Fig. 8-10. The batteries have been rearranged to group those pushing in the same direction. However, the result is still the same: three push to the left and two push to the right, creating an effective voltage of 6 volts.

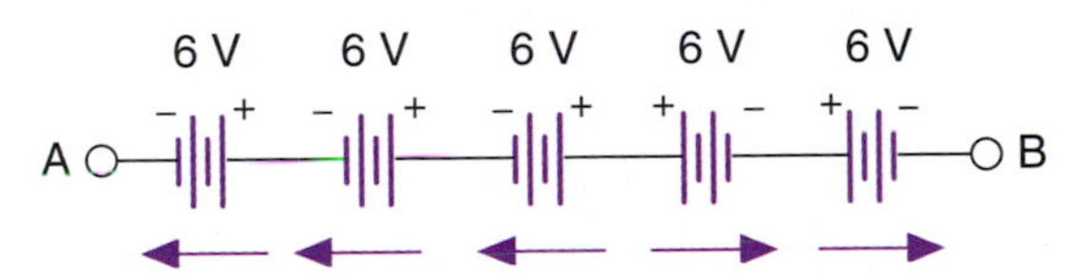

Figure 8-12 The arrangement of series-connected voltage sources does not affect the combined voltage of the group.

Algebraic Addition

Voltages must often be added algebraically because they may be connected in series-opposing. Algebraic addition is less difficult when voltages are measured

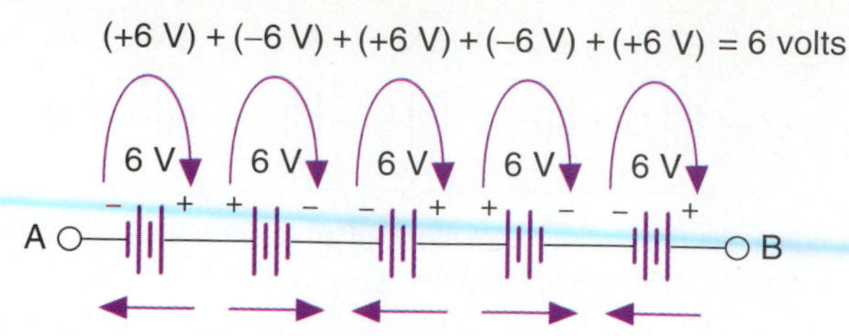

Figure 8-13 The effective voltage of sources in series-opposing is found easily and accurately using algebraic addition.

from a reference point. Consider the circuit of Fig. 8-13, which contains five 6-volt batteries with two end points labeled A and B. The voltage from A to B is 6 volts, and the voltage from B to A is minus 6 volts. Each of these voltages is the algebraic sum of the voltages between A and B. Start at A and move across each component. Write the polarity and voltage in parentheses. Continue the process until B is reached. A plus sign is placed between successive quantities to indicate they will be added.

Imagine that instead of volts you are dealing with money and you started with nothing. For each positive quantity you receive money and for each negative quantity you give it up. Assuming the denomination is in dollars, you first receive \$6, then give it up, then you get another \$6, then give it up; finally, you get yet another \$6 and keep it. In all, you have handled \$30, but only \$6 remain. Now let's return to the circuit in Fig. 8-13. Moving from B to A will produce a negative result. Starting at point B and moving across the first source discovers a minus sign. As before, the voltage value and its polarity are placed between parentheses (−6 V). Continuing the process produces (−6 V) + (+6 V) + (−6 V) + (+6 V) + (−6 V) = −6 volts. If the money analogy were applied to this example, you would owe \$6.

The effective voltage of a source made from several voltages of mixed polarity is often determined by subtraction. Technically, the subtractive process is the result of removing the parentheses. To remove the parentheses, each quantity within the parentheses, including the sign, is multiplied by +1. Applying this procedure to the previous example results in the following: + 1(+6 V) = 6 volts; + 1(−6 V) = −6 volts; + 1(+6 V) = 6 volts; + 1(−6 V) = −6 volts; and + 1(+ 6 V) = 6 volts. The procedure results in the following chain calculation:

$$6\text{ V} - 6\text{ V} = 0$$
$$0 + 6\text{ V} = 6\text{ V}$$
$$6\text{ V} - 6\text{ V} = 0$$
$$0 + 6\text{ V} = 6\text{ V}$$

With very little practice, the subtractive process becomes intuitive and the effective voltage of a source can be identified at a glance.

Split Sources

Determining the voltages within an electric network is usually more involved than in power supplies. Any current-carrying resistor can be thought of as a voltage source for the sake of circuit analysis. This postulate can be verified by creating a circuit containing several sources separated by resistors, as in Fig. 8-14, and applying Kirchhoff's voltage law.

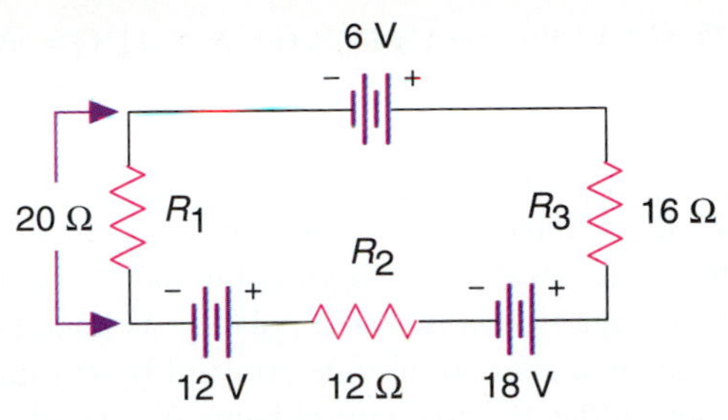

Figure 8-14 The treatment of circuits containing sources separated by resistors is different from circuits where the sources are not separated. It must be remembered that the arrangement of the components in a series circuit does not alter the current.

In Fig. 8-13, the voltage sources were wired directly together; here they are separated by resistors. This situation can sometimes generate confusion because there is a tendency to be concerned about how the resistive voltage drops might affect the overall source voltage. Rule 6 is the key to solving this dilemma. *The current in a series circuit is not affected by the arrangement of its components.* The voltage across the resistors is a direct result of the current being pushed through them by the source. That current is identified by applying Ohm's law, but first the effective source voltage and the total circuit resistance must be determined.

The 12-volt and 18-volt batteries are connected in series-aiding, as identified by their polarities, producing 30 volts. The 6-volt battery is connected in series-opposing. The voltage pushing current through the circuit is found by subtracting 6 volts from 30 volts for a total of 24 volts. Finding the total resistance in the circuit is even more straightforward because the values are added, in accordance with Rule 2, for a total resistance of 48 ohms. The circuit current is found by dividing 24 volts by 48 ohms, resulting in 0.5 ampere. The components in the circuit of Fig. 8-15 are rearranged to clarify the circuit's operation.

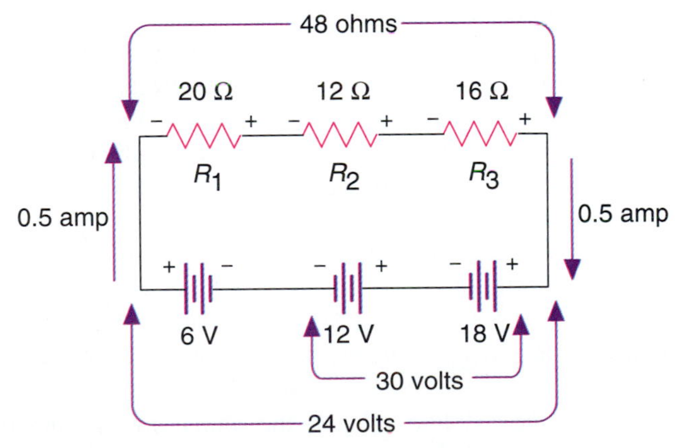

Figure 8-15 Grouping like components and recalculating the current proves that rearranging the components does not alter the circuit current.

The voltage across a resistor is found by multiplying its value by the current flowing through it. In each case that current will be 0.5 ampere because the current is the same at any point in a series circuit. Therefore, the voltage across

R_1 is 0.5 A × 20 Ω = 10 volts; across R_2 is 0.5 A × 12 Ω = 6 volts; and across R_3 is 0.5 A × 16 Ω = 8 volts. The total of the voltage drops across the resistors should equal the 24 volts of the source according to Kirchhoff's voltage law, and it does: 10 V + 6 V + 8 V = 24 volts.

The polarity of the voltage across a component is usually far more important than the actual voltage. It is not uncommon to have a circuit failure that results in a plausible voltage with a reversed polarity. Voltage polarity, excluding voltage-producing elements, is normally determined by the direction of current through a component. In the example circuit being discussed, current is flowing in a clockwise direction, as dictated by the polarity of the source. To identify the polarity of the voltage developed across the resistors, place a minus sign (−) where the current enters and a plus sign (+) where it leaves.

Ample information has now been placed on the example schematic to verify the statement that any resistive voltage drop could, for the sake of circuit analysis, be considered as a source conforming to Kirchhoff's voltage law. Assume the 10 volts across R_1 in the schematic shown in Fig. 8-16 is the source

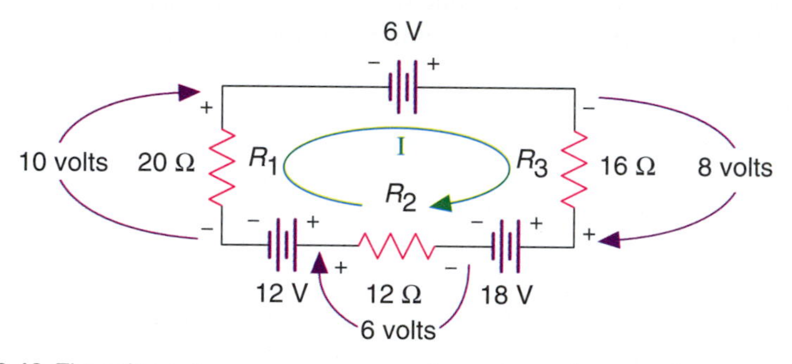

Figure 8-16 The voltage drop across any current-carrying resistor can be thought of as a source for the sake of circuit analysis.

voltage. Then, moving clockwise around the circuit, note the voltage and polarity of each component and add them algebraically.

The first component encountered is the 6-volt battery. Moving across it produces a positive polarity. The first signed value is therefore: (+ 6 V). The next component is R_3. The signed value for the voltage across R_3 is: (+8 V). Moving across the 18-volt battery the polarity is negative, producing (−18 V). The voltage across R_2 is 6 volts, and continuing in a clockwise direction discovers a positive polarity. The next value is (+6 V). The last value needed is that of the 12-volt battery. Moving across it produces a negative polarity. The last value is therefore (−12 V). The sum of the voltage drops around the circuit is (+6 V) + (+8 V) + (−18 V) + (+6 V) + (−12 V) = −10 volts. Thus, if we measured from the original starting point to where we are now, the voltage would read −10 volts.

Considering only the absolute value of the source voltage, the 10 volts of our hypothetical source equaled the absolute value of the algebraic sum of the voltage drops around the circuit. The term *absolute value* had to be added because several batteries of opposing polarities were involved and viewing a resistor as a source is somewhat unconventional. If we had continued around the circuit to our original starting point, the second version of Kirchhoff's voltage law would come into play: *The algebraic sum of the resistive voltage drops in a series circuit and the source voltage(s) are equal to zero.* Adding all of the drops around the circuit algebraically proves this: (+6 V) + (+8 V) + (−18 V) + (+6 V) + (−12 V) + (−10 V) = 0. Kirchhoff's voltage law simply states that all the voltages in a circuit are accounted for; no voltage mysteriously appears or disappears.

REVIEW QUIZ 8-2

1. When two voltage sources are connected in series-opposing, the one with the ______________ voltage will supply current to the load.
2. If voltage sources connected in series are separated by resistors, would the current be different if the sources were connected directly together?
3. For diagnostic purposes, can a current-carrying resistor be considered a source?
4. The voltages across polarized components connected in series should be added ______________ to obtain the correct value.
5. Could a battery connected in series-opposing with current forced through it against its pressure direction be replaced by a fixed resistor?

8-3 Internal Voltage Drops in Sources

There are no perfect conductors in nature. All materials, even the best natural conductors, have a measurable resistance. This is especially important to realize when dealing with electrical sources. They have an internal resistance that is in series with whatever they are operating. Any load that is placed across a source will cause a reduction in output voltage because of the voltage divider action between the source's internal resistance and the load. The voltage produced by a source does not actually change but appears to drop because direct access to the voltage-producing elements is not available. In the circuit of Fig. 8-17, the dashed

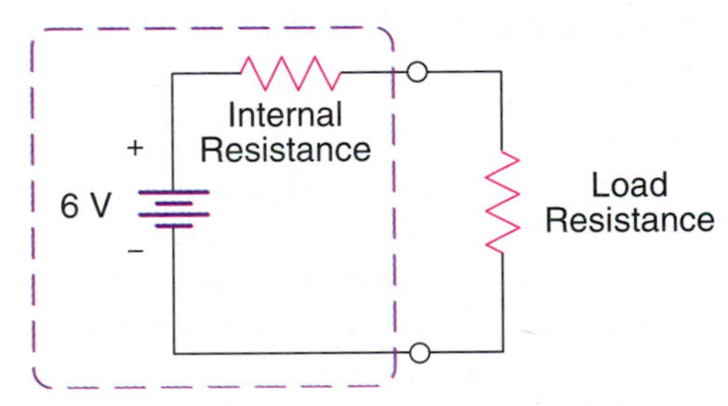

Figure 8-17 All voltage sources have an internal resistance that causes the voltage at their terminals to lessen when the source is required to deliver current.

lines indicate how the voltage source and the internal resistance are isolated from the load resistance.

Any current delivered by the source flows through both the internal resistance and the load dividing the source voltage between them. A low-output voltage from a source may be caused by an increase in the internal resistance or a decrease in the load resistance. In power supplies, the excessive current caused by a decrease in load resistance is normally accompanied by physical symptoms, such as heat, unusual odors, and smoke.

If a 10-volt battery with an internal resistance of 0.1 ohm is connected to a load resistance of 10 ohms, the total resistance seen by the source is 10.1 ohm. The current delivered by the source, 0.9901 ampere, is found by dividing 10 volts by 10.1 ohms. The actual output voltage of 9.901 volts is found by multiplying 0.99 ampere by 10 ohms. The remaining 0.099 volt is dropped across the internal resistance as shown in Fig. 8-18.

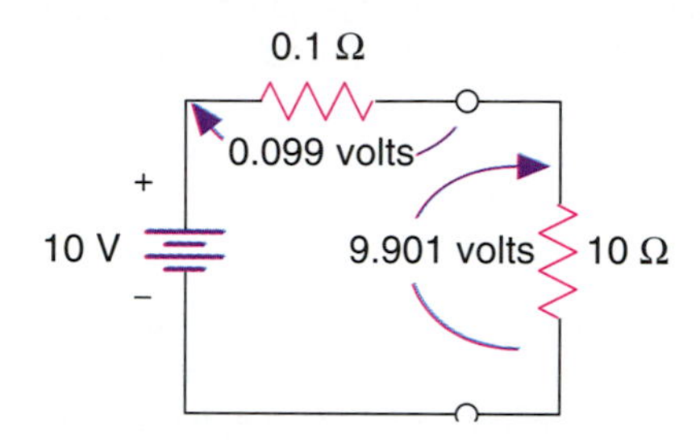

Figure 8-18 The difference between the loaded and unloaded voltage of a source is dropped across its internal resistance.

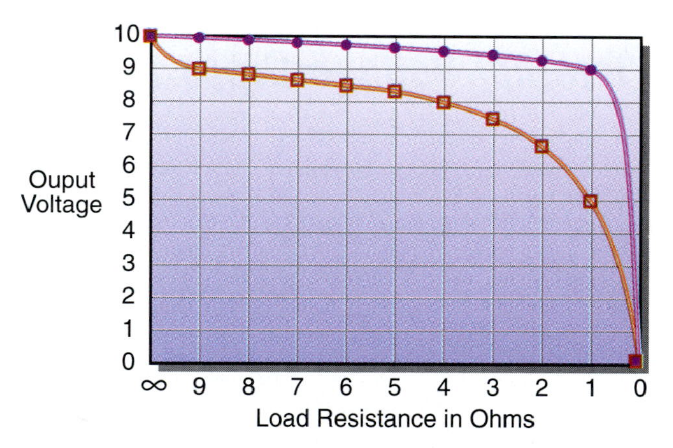

Figure 8-19 A load should not drop the voltage of a source to less than 90% of its unloaded value.

The purple line on the graph in Fig. 8-19 plots the output voltage of the previous circuit over a load range from infinity to 0. When the load resistance has fallen to 20 times the internal resistance, the output voltage has dropped about 5%. Under normal circumstances, the load should not drag the steady-state output voltage of a power source to less than 90% of its unloaded value. The brown line with the square markers plots the output voltage under the same load conditions but with a source resistance of 1 ohm. The higher internal resistance of the source represents a faulty condition that causes the output voltage to fall much faster than normal.

Fig. 8-20 can be compared with Fig. 8-19. From the bulb's point of view, it is being connected across a 24-volt source with an internal resistance of 60 ohms. When the 60-ohm resistor is considered as the source resistance, the circuit conforms to the maximum power transform. The maximum power transform states that maximum power will be delivered to a load when the resistance of the load

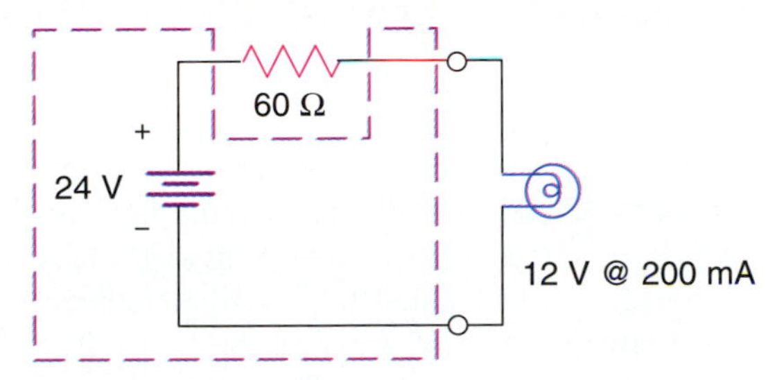

Figure 8-20 A nonlinear resistance, such as that of an incandescent lamp, may never reach its expected value when connected in series with a linear resistance.

matches the internal resistance of the source. The bulb's resistance is the same as the source resistance when it is lit to full brilliance. However, because the resistance of the bulb cannot match the resistance of the source until it has reached the proper temperature, and it cannot reach the proper temperature until it has reached the proper resistance, it is almost a certainty that the bulb's filament will stabilize at something less than the ideal voltage.

REVIEW QUIZ 8-3

1. The difference between the loaded and unloaded voltage of a source is dropped across its ______________.
2. The loaded voltage of a source should not drop below ______________% of its unloaded value.

SUMMARY

The series circuit represents the second of five circuit types. The current is the same at any point in a series circuit, but the voltages across the individual circuit elements may be different. The major function of a series circuit is as a voltage divider. Many electronic devices require that multiple voltages be developed from a single source. The resistive voltage divider is a very cost-effective way of accomplishing this. The series circuit is governed by a set of six rules that apply only to it. These rules, in conjunction with Ohm's law, make solving for individual currents and voltages a simple task.

Voltage sources are placed in series to obtain higher voltages than a single source provides. When used to provide higher voltages, the sources are connected in series-aiding. Voltage sources may also be placed in series-opposing. In this configuration, the voltage of the smaller source is subtracted from that of the larger. This is not an efficient or recommended way to reduce a voltage, but it is a consequence of circuit interactions. Sources in series-opposing are generally used as an example. Voltages in series-opposing must be added algebraically.

All sources have an internal resistance that cannot be separated from the voltage-producing elements. This internal resistance forms a voltage divider with any load connected across the source and causes its effective voltage to fall. This internal resistance limits the amount of current a source can deliver.

NEW VOCABULARY

Kirchhoff's voltage law
pot
potentiometer
rheostat
series-aiding
series circuit
series-opposing
voltage divider relationship
voltage dividers
wiper

Questions

1. What is the same at any point in a series circuit?
2. Will the algebraic sum of the voltage drops around a series circuit always equal the source voltage in resistive circuits?
3. What component behaves as a variable voltage divider?
4. How is the total resistance in a series circuit determined?
5. What happens to the current in a series circuit if the circuit is broken at any point?

6. Does the arrangement of the components in a series circuit affect the current in that circuit?
7. Would the highest voltage always be found across the resistor with the highest value in a series circuit?
8. How many electrical contacts are there on a rheostat?
9. What is indicated by a short perpendicular line drawn through the schematic symbol for a rheostat?
10. What are two ways batteries may be connected in series?

Problems

1. What is the voltage across a 20-ohm resistor in series with a 30-ohm resistor when the series combination is placed across a 100-volt source?
2. Three resistors are connected in series across a 150-volt source. The first resistor has 75 volts across it and the second resistor has 50 volts across it. The current in the circuit is 0.5 amp. What is the value of the third resistor?
3. A 50-ohm rheostat has a stop at 5 ohms. The rheostat is connected in series with a 50-ohm fixed resistor. What range of voltage is possible across the rheostat when a 25-volt source is connected across the series combination?
4. A 50-ohm resistor placed across a 12-volt battery causes the battery's voltage to decrease to 6 volts. What is the internal resistance of the battery?
5. A 40-ohm and 10-ohm resistor are connected in series. What voltage must be applied across the series combination to produce 5 volts across the 10-ohm resistor?

Critical Thinking

1. A 200-ohm resistor is 25% of the resistance in a series circuit. If 33 volts is measured across this resistor, what is the voltage applied across the series circuit?
2. If a source did not have any internal resistance, would its voltage fall under load?
3. When two voltage sources are connected in series-opposing, will the weaker source dissipate power?

Answers to Review Quizzes

8-1

1. current
2. potentiometer
3. rheostat
4. Kirchhoff's voltage

8-2

1. highest
2. no
3. yes
4. algebraically
5. yes

8-3

1. internal resistance
2. 90

Chapter 9

Parallel Circuits

ABOUT THIS CHAPTER

Components connected across each other form parallel circuits. The parallel circuit is the third of five circuit types. This chapter discusses resistors and conductors connected in parallel. Unlike series circuits, where voltage sources could be intermixed freely, voltage sources with different voltages or opposite polarities should never be placed in parallel. Even when the sources seem to be

After completing this chapter you will be able to:

- Recognize parallel connections and calculate their equivalent resistance.
- Make accurate mental approximations of parallel resistance values.

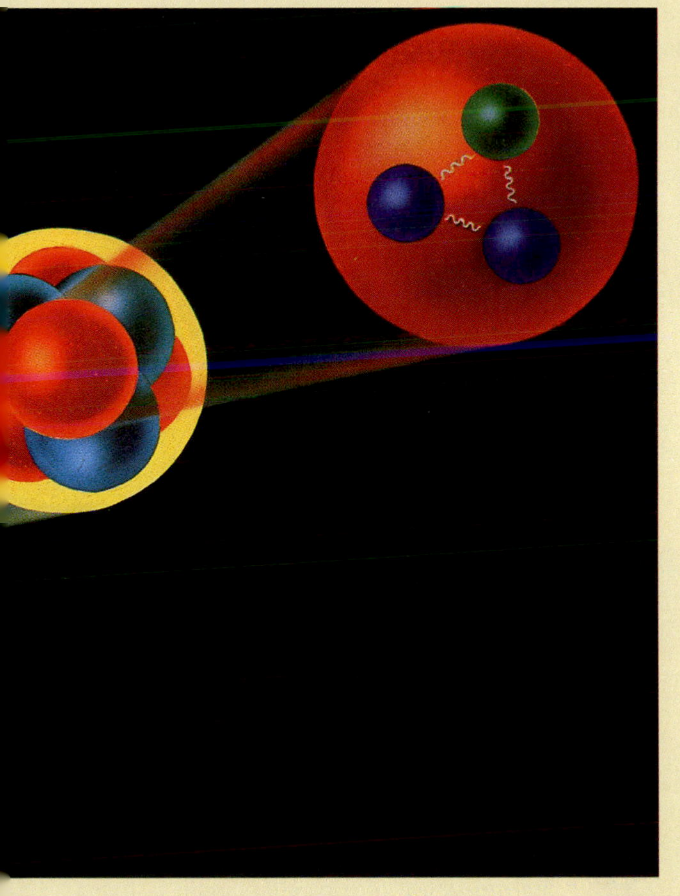

properly matched, there is the possibility that one source will force large amounts of current through the other.

It is more difficult to determine the equivalent resistance of a parallel circuit than to determine the resistance of a series circuit. To make troubleshooting easier, parallel resistance values are approximated using a variety of techniques. These approximations are frequently better than 95%.

- Calculate the resistance of conductors too large to be measured with an ohmmeter.
- Explain the rules of operation that apply only to parallel circuits.

9-1 The Parallel Connection

Resistors are in **parallel** when connected directly across each other, as shown in Fig. 9-1. The parallel circuit is the third of five basic circuit types. The most distinguishing characteristic of a parallel circuit is that the same voltage is applied across every component. As a result, the current through each resistor is determined solely by its value. These currents can be found easily using Ohm's law. The equivalent resistance of a parallel circuit may be known by dividing the applied voltage by the sum of these currents.

A voltage source sees any circuit as a single load. The source has no way of knowing the number of paths current follows in its journey around a circuit. It is only aware of how much voltage it is producing, how much current it is delivering, and if there is a difference in time between the production of the maximum voltage and maximum current.

When resistors are connected in parallel across a voltage source, their individual currents are added to obtain the source current ($I_S = I_1 + I_2 + I_3 \ldots$). Dividing the source voltage by the source current determines the equivalent circuit resistance ($R_{eq} = V_S / I_T$). Adding additional components increases the number of current paths making it easier for current to flow by reducing the circuit resistance. Because increasing the number of resistors in parallel reduces the circuit resistance, the equivalent resistance cannot be determined from the sum of the resistor values.

Parallel Calculations

In a parallel resistive circuit, the total conductance is found by adding the individual conductance values ($G_T = G_1 + G_2 + G_3 \ldots$). Refer back to Chap. 5 for the conversion between resistance and conductance. Remember that conductance is the reciprocal of resistance ($G = 1/R$), and resistance is the reciprocal of conductance ($R = 1/G$).

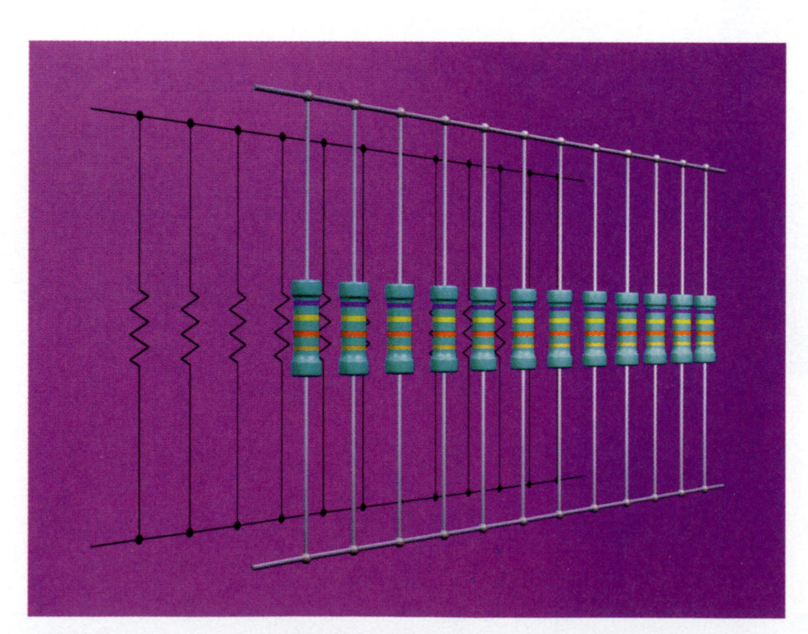

Figure 9-1 Components connected directly across each other are in parallel.

The basic equation for finding the equivalent resistance of a parallel circuit is called the **reciprocal equation.** In the reciprocal equation, each resistance is first converted into a conductance:

Equation 9-1

$$G_T = \frac{1}{R_1} + \frac{1}{R_2} + \frac{1}{R_3} \ldots$$

Then, the individual conductances are added to obtain the total conductance. The equivalent resistance of a parallel circuit is found by taking the reciprocal of the total conductance:

Equation 9-2

$$R_{eq} = \frac{1}{G_T}$$

The reciprocal equation for total conductance is not as simple as indicated by Eq.

9-2. Instead, the individual reciprocal equations of Eq. 9-1, are substituted for G_T:

Equation 9-3

$$R_{eq} = \frac{1}{\frac{1}{R_1} + \frac{1}{R_2} + \frac{1}{R_3} \ldots}$$

The reciprocal solution is very reliable and works for any number of resistors. It is particularly useful in conjunction with scientific calculators, which usually have a reciprocal key ($1/X$) as one of the standard functions. On virtually any calculator, X represents the number in the display. Pressing the $1/X$ button divides the number in the display into 1, which is the reciprocal of the original number.

When only two resistors are connected in parallel, a mathematical derivative of the reciprocal equation called **prodivisum** is often employed. One of the resistor values is multiplied by the other, then the resultant product is divided by the sum of the two values, as shown in Eq. 9-4. The name *prodivisum* means product divided by sum.

Equation 9-4

$$R_{eq} = \frac{R_1 R_2}{R_1 + R_2}$$

The prodivisum equation only works with a specified number of resistors—usually two. It is possible to develop prodivisum equations for any number of resistors. However, equations for more than two resistors become very cumbersome, thus decreasing the advantages of using prodivisum. Equation 9-5 shows the prodivisum equation for three resistors, and Eq. 9-6 shows the prodivisum equation for four resistors. With three resistors the efficiency of the equation is questionable and with four resistors it is too labor-intensive to be practical.

Equation 9-5

$$R_{eq} = \frac{R_1 R_2 R_3}{R_1 R_2 + R_2 R_3 + R_1 R_3}$$

Equation 9-6

$$R_{eq} = \frac{R_1 R_2 R_3 R_4}{R_1 R_2 R_3 + R_1 R_2 R_4 + R_1 R_3 R_4 + R_2 R_3 R_4}$$

The preceding examples of prodivisum equations demonstrate the impracticality of the equations when using more than two resistors. However, the two-resistor prodivisum equation is very useful when calculations must be performed manually. The reciprocal equation is more prone to round-off error because the reciprocal fraction often results in a long series that must be rounded off. Because the prodivisum equation requires that the division process be performed only once, the equation is easier to do mentally and less subject to error when done manually. But the high degree of precision inherent in modern calculators, in conjunction with a reciprocal key, establishes the reciprocal equation as the equation of choice when a calculator is available.

The fact that resistors are in parallel is often indicated by two vertical parallel lines placed between resistor values (12 kΩ||4 kΩ). These lines only indicate that the resistors are in parallel; they do not dictate or even suggest which parallel equation should be used to determine the equivalent resistance. They are only a shorthand method of saying ***in parallel with.***

Memorize these equations. They are important.

There are occasions when a certain value resistor is needed but not available. The solution to the problem may be to determine what value resistor to place in parallel with another in order to obtain the desired value. Determining the value of that resistor lies in a manipulation of the basic prodivisum equation. Two versions of the equation are shown in Eq. 9-7 and Eq. 9-8.

Equation 9-7

$$R_1 = \frac{R_2 R_{eq}}{R_2 - R_{eq}}$$

Equation 9-8

$$R_2 = \frac{R_1 R_{eq}}{R_1 - R_{eq}}$$

REVIEW QUIZ 9-1

1. The _______________ is the same across the elements of a parallel circuit.
2. The _______________ equation is based on the fact that conductances add in parallel.
3. The _______________ equation gets its name from the fact that the equivalent resistance of two parallel resistors is the product of the resistor values divided by the sum of the resistor values.

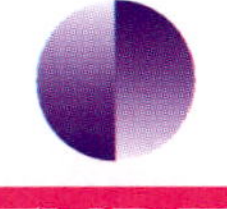

STUDENT

to

STUDENT

Pay extra attention to any troubleshooting hints. If you are good at troubleshooting, you'll make more money as a technician.

9-2 Mental Approximations

It is seldom necessary to use a calculator when troubleshooting electronic circuits. Instead, the equivalent resistance of parallel circuits may be approximated mentally. The simplest of these **mental approximations** is applied to a parallel circuit where all the resistor values are identical. The value of any one of the resistors is divided by the number of resistors in the circuit to obtain the equivalent resistance. This relationship is shown in Eq. 9-9.

Equation 9-9

$$R_{eq} = \frac{R}{N}$$

Where R_{eq} = the equivalent resistance of a parallel circuit
R = the value of any of the resistors
N = the number of resistors

STUDENT

to

STUDENT

This is a useful equation and is really easy to learn.

Although the equation is often thought of as just an approximation, it is actually a precise determination of the equivalent resistance.

According to Ohm's law, the current through a resistor is determined by its value and the voltage across it. In a parallel circuit, that voltage is the same across each resistor. If the values of the resistors are also the same, then equal currents must flow in each resistor. Adding additional resistors in parallel increases the number of current paths. These paths divide the source current in inverse proportion to the resistor values, thus lowering the overall resistance of the circuit. Inverse proportion means that the lowest value resistor will carry the largest cur-

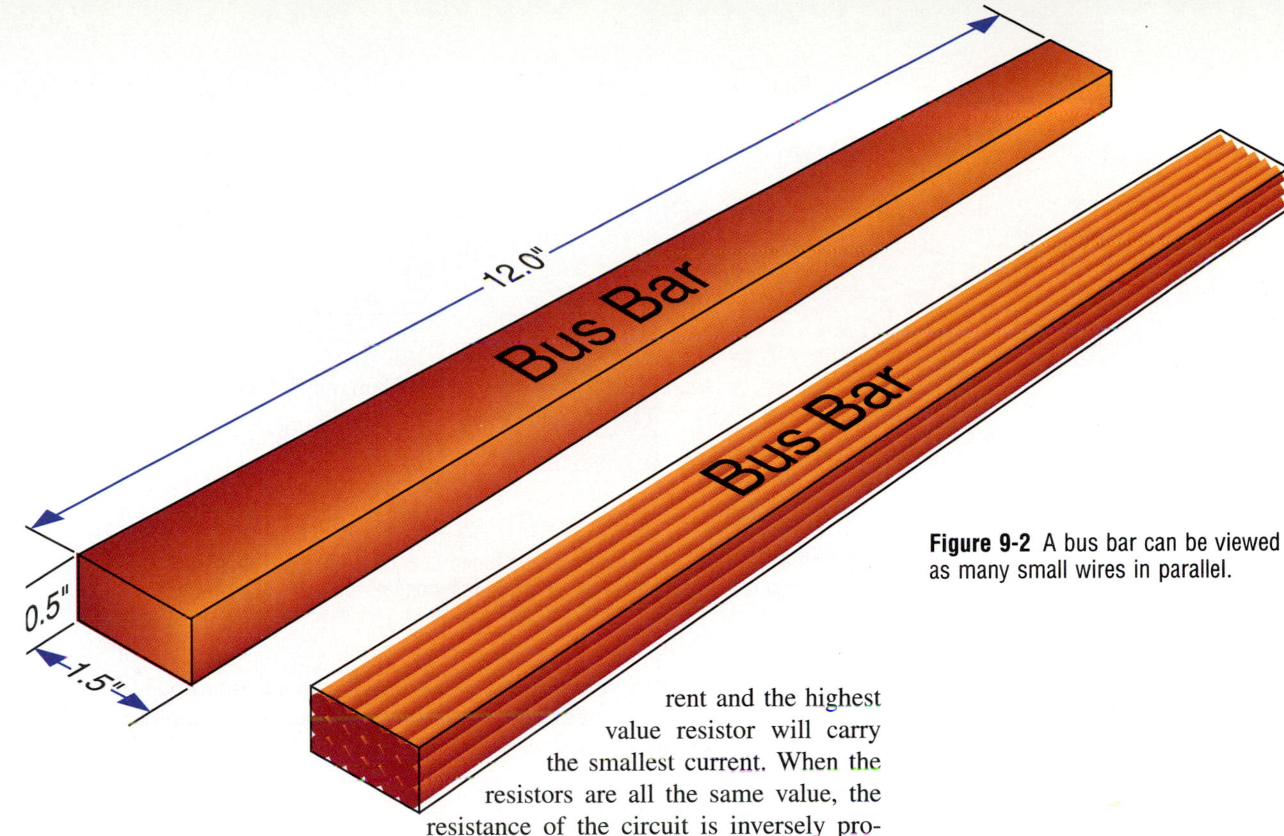

Figure 9-2 A bus bar can be viewed as many small wires in parallel.

rent and the highest value resistor will carry the smallest current. When the resistors are all the same value, the resistance of the circuit is inversely proportional to the number of resistors. For instance, two similar resistors in parallel cut the resistance in half. Three equal value resistors in parallel decrease the resistance to one-third. Four similar resistors in parallel drop the resistance to one-fourth and so on for any number of similar resistors as described in the ***R*-over-*N* equation**.

The *R*-over-*N* equation is extremely useful in calculating the resistance of conductors too large to be accurately measured with an ohmmeter. The specific resistance of a metal is standardized around a unit called the circular-mil-foot, as described in Chap. 4.[1] The resistance of the wire is measured at room temperature, which is usually considered to be 20°C (68°F). When comparing the resistance of one metal to another, the temperature is very important because the resistance of some metals changes greatly with changes in temperature.

The cross-sectional area of a large metal conductor can be viewed as many small wires connected in parallel. If each of these hypothetical wires is thought of as the resistance standard for the material, from which the wire is made, the resistance of the conductor can be determined using the *R*-over-*N* equation. Dividing the resistance of the standard by the number of circular mils in the cross-sectional area of the conductor yields its resistance for a 1-foot length. Assume you want to know the resistance of a copper bus bar 1.5 inches wide, 0.5 inches thick, and 1 foot long, as shown in Fig. 9-2. A bus bar is a rectangular conductor that is too large to be called a wire. The procedure described in Example 9-1 will yield the result.

[1]As a reminder, a circular-mil-foot consists of a round wire 1 mil (0.001 inch) in diameter and 1 foot long.

Example 9-1

1. Find the specific resistance of the conductor from a table giving the specific resistance of common metals. Copper has a specific resistance of 10.37 Ω per circular-mil-foot at 20°C.

2. Find the area of the bus bar in the specified units: 1.5 inches × 0.5 inch = 0.75 square inches.
3. Determine the number of square mils in 1 square inch, as shown in Fig. 9-3.

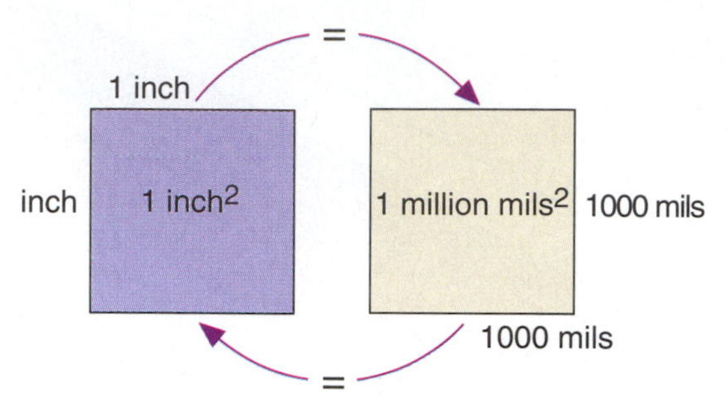

Figure 9-3 The area of a square can be found using the same technique to find the area of a rectangle: Multiply any side by one of the sides adjacent to it.

4. Convert the cross-sectional area of the bus bar to square mils. Multiply the cross-sectional area of the bus bar in square inches by the number of square mils in 1 square inch, as shown in the following equation.

$$0.75 \text{ in}^2 \left(\frac{1{,}000{,}000 \text{ mil}^2}{1 \text{ in}^2}\right) = 750{,}000 \text{ mil}^2$$

5. Convert the cross-sectional area from square mils to circular mils by dividing the number of square mils by 0.7854, as shown in the next equation.

$$\frac{750{,}000 \text{ square mils}}{0.7854} = 954{,}927.4 \text{ circular mils}$$

Dividing the true area of a circle by the area of a square having the same diameter will always yield the constant of 0.7854, as shown in Fig. 9-4.

6. The resistance of a 1-foot length of the bus bar is found using the R-over-N equation as shown:

$$R_{eq} = \frac{R}{N} = \frac{10.37 \ \Omega/\text{ft}}{954{,}927.4} = 1.086 \times 10^{-5} \ \Omega/\text{ft}$$

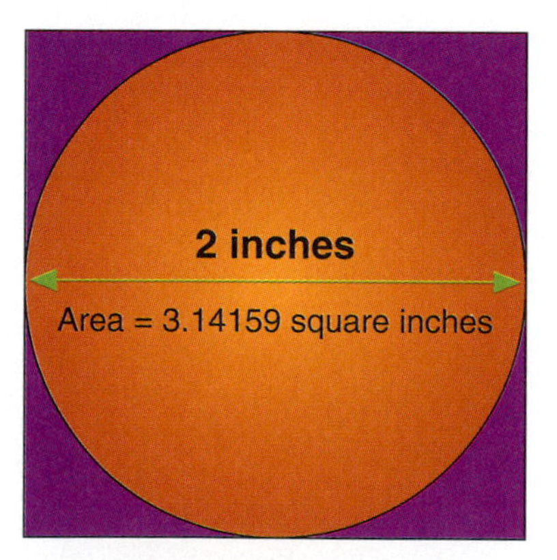

Area = 4 square inches

Figure 9-4 The ratio of the area of a circle to the area of a square having the same diameter.

The Ratio Approximation

Another approximation technique that can give the exact value of two parallel resistors relies on one of the resistors being an exact multiple of the other. The smaller of the two resistors is first divided into itself, which always equals 1. The smaller resistor is then divided into the larger resistor; the result is added to 1. This yields the total number of the smaller resistor values present in both resistors. The value of the larger resistor is divided by this number. The result is the exact resistance of the parallel circuit.

Equation 9-10

$$R_{eq} = \frac{R_1}{\dfrac{R_2}{R_2} + \dfrac{R_1}{R_2}} = \frac{R_1}{1 + \dfrac{R_1}{R_2}}$$

In actual practice, the assignment of R_1 and R_2 to the larger or smaller value is unimportant. The equation will work either way, and the resistors may be of any value. However, for mental approximations, there is a need for a close multiple relationship. A typical situation is shown in Example 9-2.

Example 9-2

Assume that a 400-ohm and a 100-ohm resistor are connected in parallel. There is one 100-ohm resistor in 100-ohms, and four 100-ohm resistors in 400 ohms, for a total of five 100-ohm resistors. If the larger of the two resistors (400 ohms) is divided by 5, the result is 80 ohms, the exact resistance of the circuit.

This method can be reduced further by memorizing the ratio results.

1. If two resistors are equal in value, the circuit resistance is one-half or 50% of either one.
2. If one resistor is twice the value of the other, the circuit resistance is two-thirds or 66.7% of the smaller.
3. If one resistor is three times the value of the other, the circuit resistance is three-fourths or 75% of the smaller.
4. If one resistor is four times the value of the other, the circuit resistance is four-fifths or 80% of the smaller.
5. If one resistor is five times the value of the other, the circuit resistance is five-sixths or 83.3% of the smaller.

Most of these ratios are easily remembered. Some may require a bit of work. However, the purpose of these approximations is to approximate. In actual circuits, there are usually many variables, and exact values may not be needed. The standard tolerance for resistors is ±5%. Power supply voltages often vary from ±1% to ±10%. A close approximation will quickly reveal if a problem exists in a circuit, thus demonstrating that exact calculations are usually unnecessary.

Assume two resistors are in parallel and that one is 10 times the value of the other. Between the two resistors there would be 11 units equal to the value of the smaller resistor. In the approximation process, the larger resistor could be divided by 11, or, one-eleventh could be subtracted from the smaller resistor. One-eleventh of something does not lend itself to mental computation, but one-tenth of something is easy. Dividing by 10, or subtracting one-tenth is simple yet the error is only 1%. The graph in Fig. 9-5 will help you learn the approximation method and show you which ratios are more critical than others.

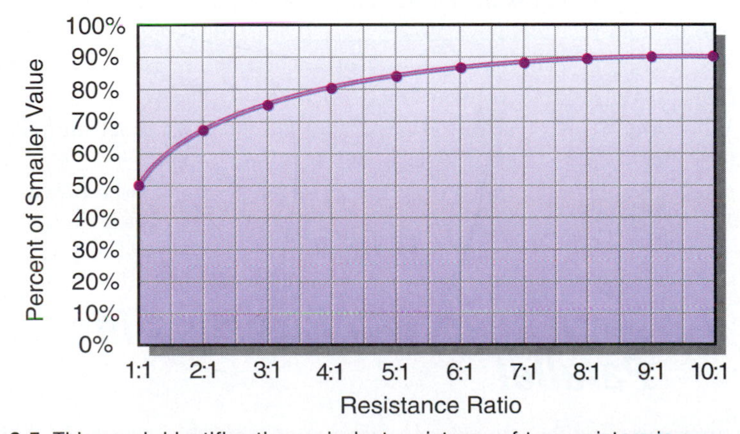

Figure 9-5 This graph identifies the equivalent resistance of two resistors in a parallel circuit as a percentage of the lowest value resistor for resistance ratios from 1:1 to 10:1.

The Averaging Technique

Another approximation method relies on finding the average resistance of two parallel resistors whose values are within 50% of each other. If the difference in the values is more than 50%, the error will exceed 10%. This method has its roots in the *R*-over-*N* equation. First, a range of values is obtained by assuming that both resistors are equal to the larger value. Using the *R*-over-*N* method, the upper limit would then be one-half of the larger value. Next, both resistors are assumed to be equal to the smaller value. Again, using the *R*-over-*N* method, the lower limit would be one-half of the smaller value. The equivalent resistance cannot be more than one-half of the larger nor less than one-half of the smaller. The actual value is somewhere in between. Averaging the two values will give a close approximation. Assume that a 100-ohm and 82-ohm resistor are connected in parallel and that the averaging technique is to be used to find the equivalent resistance. Follow the procedure shown in Example 9-3.

Example 9-3

1. Assume that the resistance of both resistors is 100-ohms. Using the *R*-over-*N* procedure will make the equivalent resistance 50 ohms.
2. Now assume that the resistance of both resistors is 82 ohms. The equivalent resistance using the *R*-over-*N* method is 41 ohms. You may want to mentally alter this value to 40 ohms. This will make the approximation much easier.
3. The first two steps produce the maximum and minimum limits of resistance. The resistance cannot be more than 50 ohms or less than 40 ohms. The actual value lies somewhere between. The average of the two values can be found by adding them and taking half the value: $50\ \Omega + 40\ \Omega = 90\ \Omega$ and $90\ \Omega/2 = 45\ \Omega$. The actual calculated value is 45.05 ohms. The error in this procedure was only 0.1%. If the 41-ohm value had not been changed to 40 ohms, the average would have been 45.5 ohms and the error would have been 1.1%.

REVIEW QUIZ 9-2

1. The *R*-over-*N* approximation is actually an exact calculation when the two parallel resistors have ______________ values.
2. The averaging technique is only useful when one resistor is more than ______________ % of the other.
3. When one of two resistors in a parallel circuit is an exact multiple of the other, the resistance is determined by dividing the ______________ resistor by the total number of smaller resistor values in both resistors.

9-3 More Than Two Resistors in Parallel

When approximating the equivalent resistance of more than two resistors in parallel, only two resistors are considered at a time. The order in which the resistors are combined may affect the accuracy of the result when approximating. It is usually advantageous to combine the larger resistances first and then work the resultant against one of the smaller values. Then, this resultant is combined with another resistor to obtain a second resultant. The process is repeated until only one resistance remains. Once two resistors are combined, only the resultant is considered. The original resistances are hypothetically discarded.

Assume that a 620-ohm, 390-ohm, 1200-ohm, and 180-ohm resistor are all connected in parallel. The steps in Example 9-4 are used to approximate the equivalent resistance of this circuit.

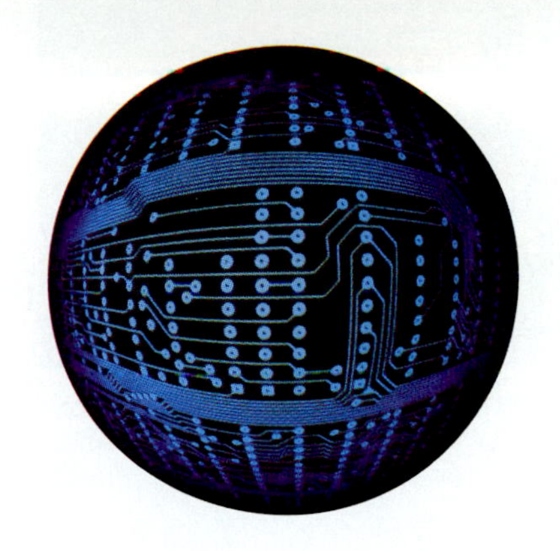

Example 9-4

1. Select the two largest resistors and determine if their values conform to one of the approximation keys. It is possible that after the other resistors are reduced to one equivalent, the largest resistor will make little or no difference.
2. The 1200-ohm and 620-ohm resistors are the largest. The 1200-ohm resistor is about twice the value of the 620-ohm resistor. The ratio method tells us that when a resistor is twice the value of another in a parallel circuit, the equivalent is two-thirds of the smaller value. Therefore, the equivalent resistance of the two resistors is about 400 ohms.
3. Using the *R*-over-*N* method, the 400-ohm resultant is about the same value as the 390-ohm resistor. The result of these two would be approximately 200 ohms.
4. The last step is to combine this 200-ohm equivalent with the remaining 180-ohm resistor. These resistors are almost the same value, but the averaging method will yield a closer result than assuming they are identical. If both resistors were 200 ohms, the resultant would be 100 ohms. If both resistors were 180 ohms, the resultant would be 90 ohms. The difference between the two is 10 ohms. Splitting the difference between the two values results in 5 ohms. This 5-ohm resistance can be added to the lower resultant of 90 ohms, or subtracted from the higher resultant of 100 ohms. Either way, the equivalent resistant is about 95 ohms. An actual calculation yields 94.64 ohms as the circuit's equivalent resistance. The approximation produces an error of less than 1.5%.

REVIEW QUIZ 9-3

1. When approximating the equivalent resistance of a parallel circuit containing three or more resistors, only ______________ are worked together at any one time.
2. When working with three or more parallel resistors, approximations are more accurate if the resistor values are worked from the ______________ value to the ______________ value.

9-4 Rules for Parallel Circuits

Like the series circuit, the operation of the parallel circuit is governed by its own set of rules. These rules, in conjunction with Ohm's law, allow the parameters of the parallel circuit to be understood and predicted. The rules are as follows:

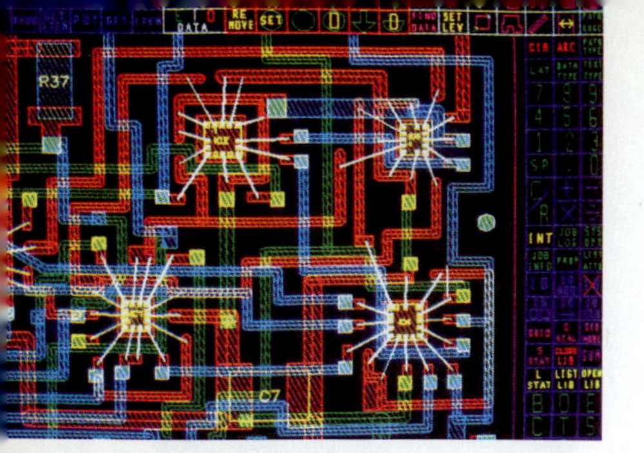

1. The same voltage is applied across all of the elements in a parallel circuit.
2. The total conductance in a parallel circuit is equal to the sum of the conductance values ($G_T = G_1 + G_2 + G_3 \ldots$).
3. The sum of the currents in the branch elements of a parallel resistive circuit is equal to the main line current **[Kirchhoff's current law (KCL)]**.
4. The equivalent resistance of a parallel circuit is always less than the smallest resistor in that circuit.
5. The loads in a parallel circuit are independent of each other. If one load should open, the current through the other loads will not be affected.
6. The currents through two parallel resistors are inversely proportional to their values. That is, if one resistor is twice the value of the other, the current through it will be one-half that in the other. The relationship forms the basis for what is known as the **current divider relationship.** The current divider relationship is expressed in Eq. 9-11 and Eq. 9-12. These equations are useful for finding the branch currents if the resistor values and source current are known.

Equation 9-11

$$I_2 = \frac{R_1}{R_1 + R_2} I_T$$

Equation 9-12

$$I_1 = \frac{R_2}{R_1 + R_2} I_T$$

REVIEW QUIZ 9-4

1. ______________ current law states that the sum of the branch currents in a parallel circuit is equal to the main line current.
2. The equivalent resistance of a parallel circuit is always less than the ______________ resistor in the circuit.

9-5 Kirchhoff's Current Law

Kirchhoff's current law describes a conservation of current in parallel circuits. Current does not magically appear or disappear. All the current in a circuit is accounted for at all times. Kirchhoff's current law views this conservation of current from three perspectives:

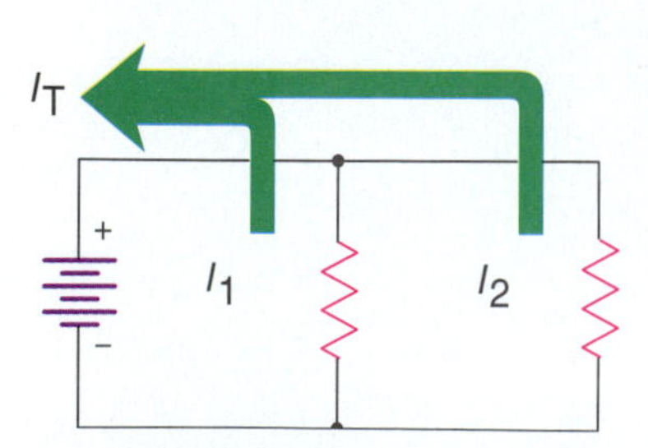

Figure 9-6 The sum of the currents leaving a point is equal to the current entering that point.

1. When a current path divides into two or more branches, *the sum of the currents leaving the point of division are equal to the current entering that point.* Sometimes this relationship is written from a more mathematical perspective: *The algebraic sum of the currents leaving a point and the current entering that point is equal to zero.* The relationship is depicted in Fig. 9-6.
2. The relationship of the branch currents in a parallel circuit to the main-line current can be stated as: *The sum of the currents in the branch circuits of a parallel circuit is equal to the main-line current.* This relationship is also

sometimes written from a mathematical perspective: *The algebraic sum of the currents in the branch circuits of a parallel circuit and the source current is equal to zero.* This relationship is shown in Fig. 9-7.

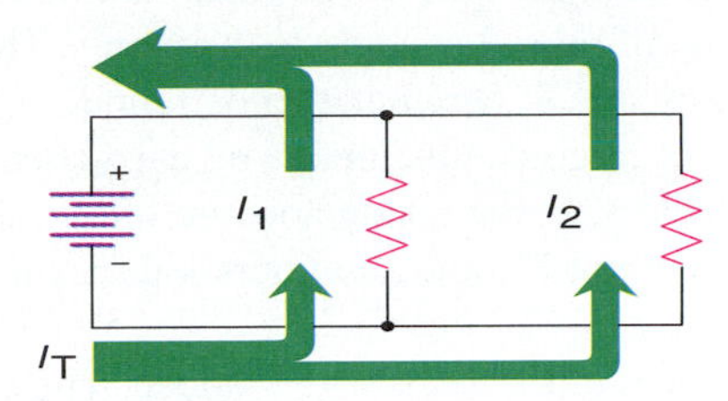

Figure 9-7 The sum of the branch currents in a parallel circuit is equal to the source current.

3. From the perspective of a point that collects the currents from two or more branches and combines them into a single current: *The current leaving a point is equal to the sum of the currents entering that point.* Also: *The algebraic sum of the current leaving a point and the currents entering that point is equal to zero.* This relationship is shown in Fig. 9-8.

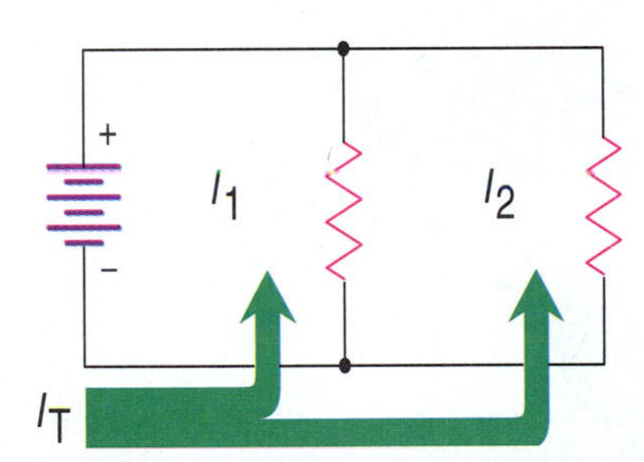

Figure 9-8 The sum of the currents entering a point is equal to the current leaving that point.

When adding the currents algebraically, the current entering a point is considered positive, and the current leaving a point is considered negative. Because the current entering a point and the current leaving a point are always the same, the two equal and oppositely signed values must always equal zero.

REVIEW QUIZ 9-5

1. The algebraic sum of the current entering a point and the current leaving that point equals ______________.
2. The currents in the branch circuits of a parallel circuit are equal to the ______________ current.
3. Kirchhoff's current law states that there is a conservation of ______________ in a parallel circuit.

SUMMARY

The parallel circuit is the third of the five basic circuit types. Its major characteristic is that the same voltage appears across all the circuit elements. The currents in the circuit may be different, but the sum of those currents is equal to the main line current. Like the series circuit, parallel circuits have a set of rules that apply to them only. These rules, in conjunction with Ohm's law, allow any circuit parameter to be found.

Parallel calculations allow the resistance of large conductors to be calculated by treating them as many identical parallel conductors. Large conductors are often called bus bars and are used to carry extremely large currents. In the aluminum industry these bus bars often carry as much as 300,000 amperes.

In a parallel resistive circuit, the conductances of the resistors are added. The conductance of a resistor is determined by taking the reciprocal of its resistance value. An indirect method for calculating the equivalent resistance of a parallel circuit is to find the current through each of the resistors, add them together, and divide the total current into the voltage across the circuit. The procedure uses a combination of Ohm's law and Kirchhoff's current law. Such a procedure does not yield as direct a result as the parallel equations, but it is reliable.

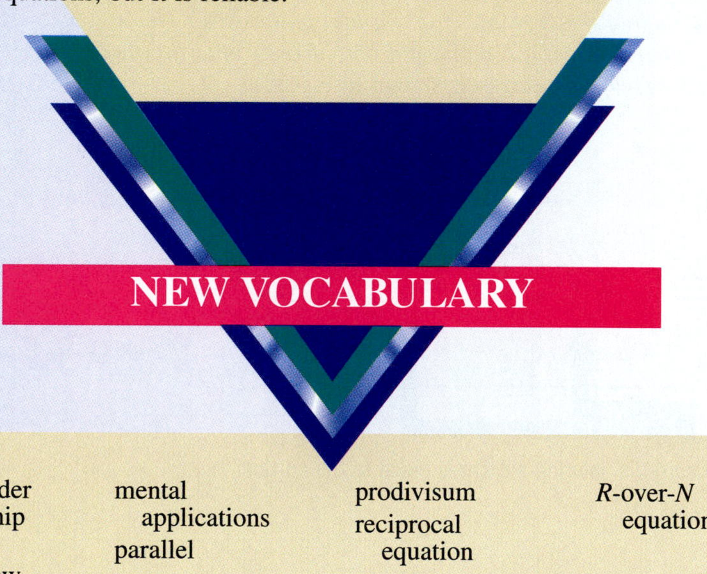

NEW VOCABULARY

- current divider relationship
- Kirchhoff's current law
- mental applications
- parallel
- prodivisum
- reciprocal equation
- *R*-over-*N* equation

Questions

1. What is the same across all the elements of a parallel circuit?
2. Do resistor values add when they are connected in parallel?
3. When is the prodivisum equation most useful?
4. What is a simple way to determine the equivalent resistance of a parallel circuit when all the resistors have the same value?
5. Is the equivalent resistance of a parallel circuit always less than the lowest value resistor in the circuit?
6. If one of the resistors in a parallel circuit were to open, would the currents through the other resistors be affected?

7. What law states that the sum of the currents in the branch circuits of a parallel circuit is equal to the main-line current?
8. Are the currents through the resistors in a parallel circuit directly or inversely proportional to their values?
9. Does the word *shunt* mean *in parallel with?*

Problems

1. What is the equivalent resistance of two 50-ohm resistors in parallel?
2. A 1200-ohm, 600-ohm, and 400-ohm resistor are connected in parallel across a 100-volt source. What is the source current?
3. A 10,000-ohm, 560-ohm, and 390-ohm resistor are connected in parallel. What is the difference in circuit resistance when the 10,000-ohm resistor is treated as though it did not exist?
4. What is the resistance at room temperature of a copper strip 1 inch wide, 6 inches long, and 10 mils thick?
5. What value resistor must be placed in parallel with a 100-ohm resistor to achieve a circuit resistance of 80 ohms?
6. The current through one of two resistors in a parallel circuit is 10 mA and the second resistor has twice the value. What is the source current?

Critical Thinking

1. Which resistor in a parallel circuit would dissipate the most power?
2. If one of the resistors in a parallel circuit is shorted, are the other resistors also shorted?
3. Are the currents entering a point in a circuit considered to be positive or negative?
4. What is the algebraic sum of the currents entering and leaving a point in a circuit?
5. What is the algebraic sum of the currents in the branch circuits of a parallel circuit and the main-line current?
6. What is the algebraic sum of the currents leaving a point and the current entering that point?

Answer to Review Quizzes

9-1
1. voltage
2. reciprocals
3. prodivisum

9-2
1. identical
2. 50
3. larger

9-3
1. two
2. larger; smaller

9-4
1. Kirchhoff's
2. smallest

9-5
1. zero
2. main-line
3. current

Chapter 10

Series-Parallel Circuits

ABOUT THIS CHAPTER

This chapter introduces a new type of circuit called series-parallel. These circuits are composed of series circuits connected in parallel, parallel circuits connected in series, or a combination of both. They represent an increased level of difficulty in determining their circuit parameters because more steps are usually required to arrive at the desired solution. In addition, the paths to those solutions are not always obvious and may require some

After completing this chapter you will be able to:

- Recognize the types of series-parallel circuits.
- Identify the processes used to determine the voltages and currents associated with each component.

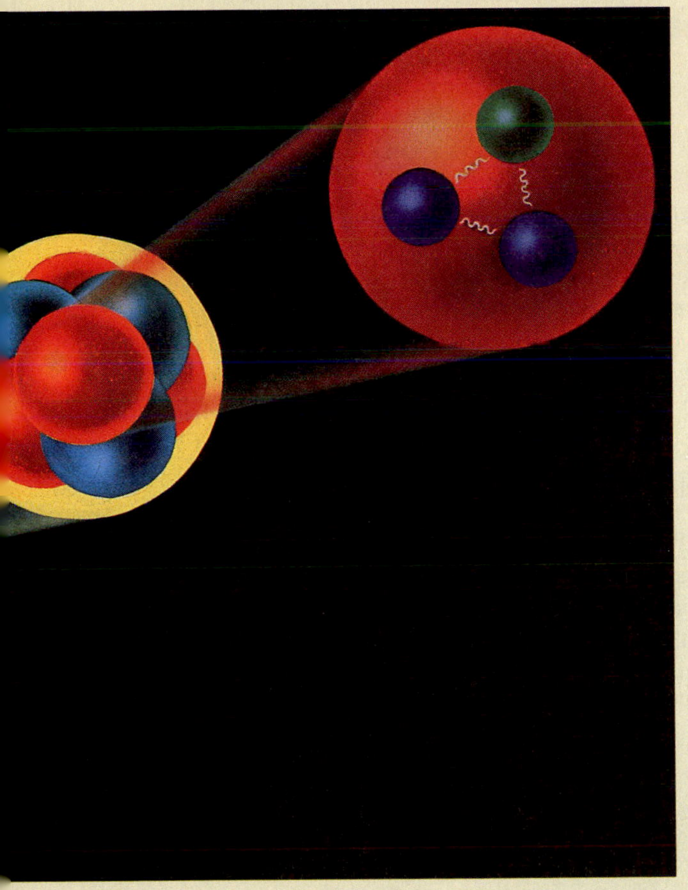

ingenuity. On the brighter side, no new rules or formulas are necessary for evaluation of these circuits, but series-parallel circuits test one's knowledge of those rules and formulas already presented.

One of the most common occurrences of series-parallel circuits is a loaded voltage divider, which allows circuits to operate at different voltages developed from a single source. This chapter gives considerable attention to this topic. It also presents a number of worked examples that are designed to reveal some of the techniques used to determine circuit parameters as well as make approximations when possible.

- Explain the effects a load has on a voltage divider.
- Describe how series and parallel rules work hand-in-hand with Ohm's law.

10-1 Series and Parallel Circuit Combinations

A fourth circuit classification called **series-parallel** is created when series and parallel circuits are combined. These circuits may consist of series circuits connected in parallel or parallel circuits connected in series. There are no new rules governing series-parallel circuits. Any series-parallel circuit may be analyzed using the previously established rules of circuit operation and Ohm's law. Nevertheless, these circuits represent a new level of complexity because the procedures used to determine the various circuit parameters must be performed in a specific order. A key factor in analyzing series-parallel circuits is the ability to recognize and isolate the series and parallel sections. However, circuit identification can often become blurred when series and parallel circuits are mixed.

As the Source Sees It

General rules of circuit operation are considered to be common knowledge among those individuals with even a minimal amount of training in electronics. The application of these rules is essential in determining the various voltages and currents within any circuit.

One of the most useful truisms of electronics is that any circuit composed entirely of resistors appears as a single resistor to a source connected across it. A source is unaware of the many paths current may follow in its journey around a circuit. To the source, any resistive circuit is always a single resistor. Knowing the value of that resistor is often the key to determining the currents and voltages associated with individual components. When the circuit resistance and source voltage are known, the source current can be determined using Ohm's law.

An Example of Parallel Circuits in Series

Example 10-1 traces the steps involved in determining the voltage across and the current through each resistor in a basic series-parallel circuit. The circuit is converted to a single resistor by finding the equivalent resistance of each parallel section and then adding those resistances to find the total resistance. The original circuit and the stages of reduction are shown in Fig. 10-1. The example demonstrates the importance of being able to find the source current.

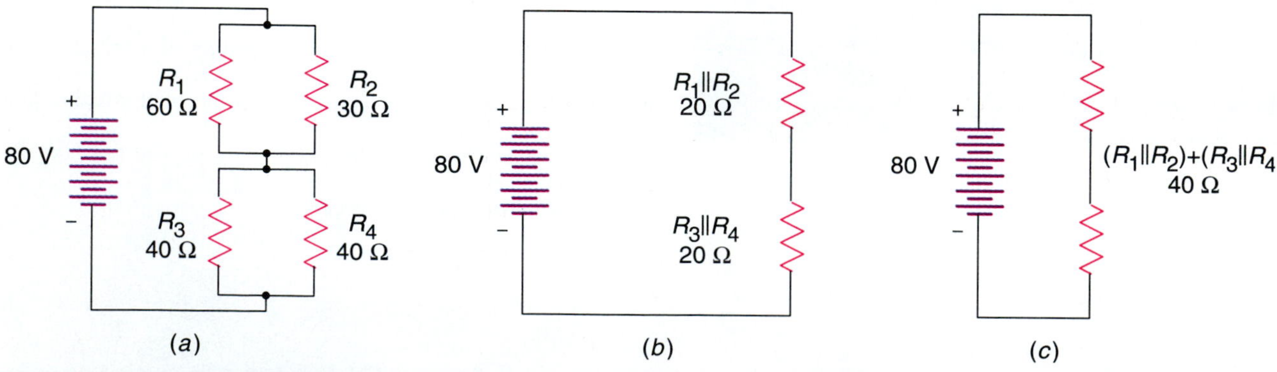

Figure 10-1 Series-parallel circuits are often equated to a single resistor to determine the source current. The original circuit is shown in (*a*) and the stages of reduction shown in (*b*) and (*c*).

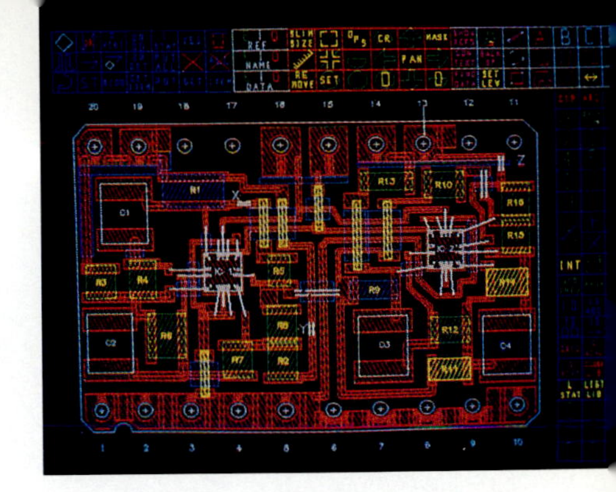

Example 10-1

Consider the series-parallel circuit shown in Fig. 10-1(*a*). It consists of a 60-Ω and 30-Ω resistor in parallel connected in series with two 40-Ω resistors in parallel. The circuit is connected across an 80-V source. The procedures for determining the circuit parameters are outlined in the following steps.

1. *Determine the equivalent resistance of each parallel circuit.* The upper parallel circuit consisting of R_1 and R_2 has an equivalent resistance of 20 Ω, as determined in Eq. 10-1.

Equation 10-1

$$R_{eq} = \frac{R_1 R_2}{R_1 + R_2} = \frac{60 \times 30}{60 + 30} = \frac{1800}{90} = 20\ \Omega$$

The lower parallel circuit also has an equivalent resistance of 20 Ω, as determined in Eq. 10-2.

Equation 10-2

$$R_{eq} = \frac{R}{N} = \frac{40}{2} = 20\ \Omega$$

After implementing Eq. 10-1 and Eq. 10-2, the circuit becomes two 20-Ω resistors connected in series as shown in Fig. 10-1(*b*).

2. *Add the equivalent resistances from Eq. 10-1 and Eq. 10-2 to obtain the total resistance of the circuit.* The two 20-Ω resistors form a total resistance of 40 Ω, as shown in Fig. 10-1(*c*) and verified in Eq. 10-3.

Equation 10-3

$$\begin{aligned} R_T &= (R_1 \parallel R_2) + (R_3 \parallel R_4) \\ &= (60 \parallel 30) + (40 \parallel 40) = 20\ \Omega + 20\ \Omega = 40\ \Omega \end{aligned}$$

3. *Determine the voltage across each parallel section.* This can be done in several ways. The simplest is to apply the voltage divider relationship. Each of the 20-Ω equivalent resistances represents one-half of the circuit resistance. Therefore, according to the voltage divider relationship, one-half of the source voltage, 40 V, will appear across each resistor. An alternative method would be to find the total current in the circuit and multiply the equivalent resistance of each parallel branch by it. The total circuit current I_T is found by dividing the voltage of the source by the total circuit resistance, as shown in Eq. 10-4.

Equation 10-4

$$I_T = \frac{V_S}{R_T} = \frac{80}{40} = 2\ \text{A}$$

Multiplying each 20-Ω equivalent resistance by 2 A, in accordance with Ohm's law, produces two voltage drops of 40 V.

4. *Find the current in each resistor.* This is now a matter of applying Ohm's law because the voltage across each resistor and the value of each resistor are known. The current through R_1 is found by dividing 40 V by 60 Ω. The result is 0.667 A. The current through R_2 is found by dividing 40 V by 30 Ω. The current through R_2 is 1.333 A. If these values are correct, their sum will be the source current of 2 A. The sum of 0.667 A and 1.333 A is 2 A. If it had been otherwise, an error in one of the calculations would be indicated.

   The current through R_3 is found by dividing 40 V by 40 Ω. The result is 1 A. Because R_3 and R_4 are the same value and in parallel, the current through R_4 will also be 1 A. The sum of the currents in the lower parallel circuit must equal the source current of 2 A, which it does. The current in one element of a series circuit may never be different than the current in any other element. Kirchhoff's voltage and current laws may be applied to any circuit to ensure that there is a conservation of voltage and current. The circuit is shown again in Fig. 10-2 with all voltages and currents.

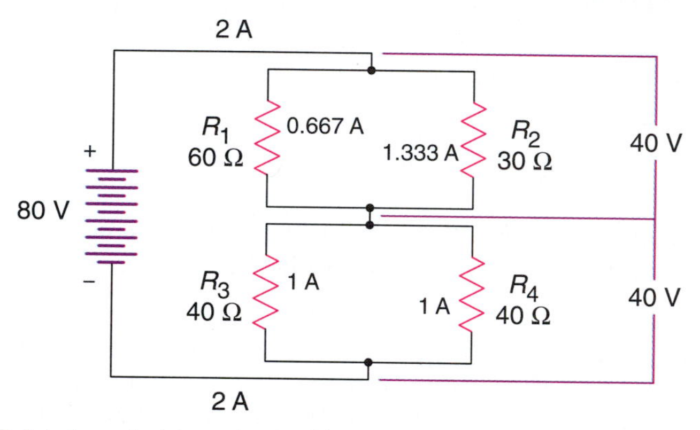

Figure 10-2 It is important to understand how the voltages and currents are distributed within a circuit.

Series Circuits in Parallel

When series circuits are placed in parallel, the individual currents and voltages can be found without knowing the source current. Each branch of the circuit is independent of the other; the only thing they share is being connected across a common source. Such is the basic trait of all parallel circuits.

Example 10-2 traces the steps involved in determining the equivalent resistance of series circuits connected in parallel. This circuit has the same values as the circuit used in Example 10-1, and it is also connected across an 80-V source. The basic circuit and its stages of reduction are shown in Fig. 10-3, although the stages of reduction are shown only to demonstrate how the equivalent resistance may be found. When series circuits are connected across a common source, their equivalent resistance is unimportant. However, if even one resistor is connected in series with paralleled series circuits, knowing the equivalent resistance of the circuit is essential.

Example 10-2

1. *Add the resistor values in each branch.* In this case, R_1 and R_3 are added to obtain 100 Ω. Adding R_2 and R_4 produces 70 Ω.

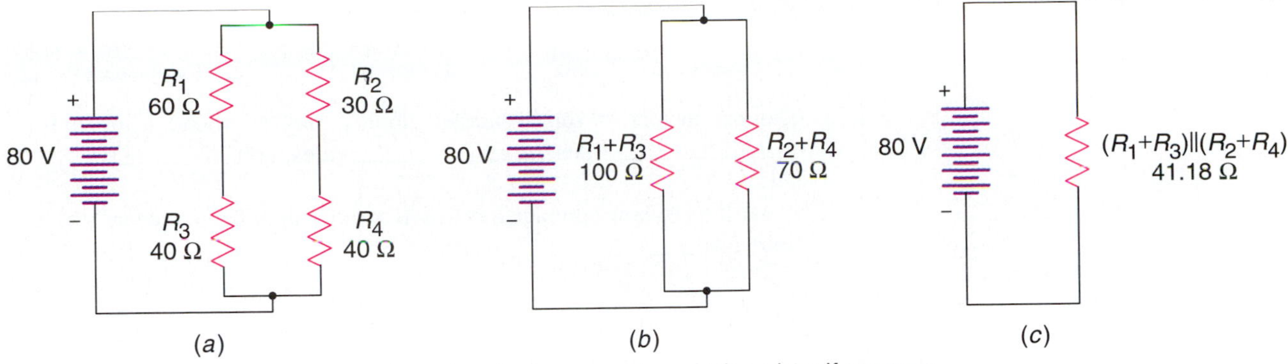

Figure 10-3 Parallel-series circuits can be systematically reduced to a single resistor if necessary to aid in circuit evaluation. A parallel-series circuit is shown in (*a*). The stages of reduction are shown in (*b*) and (*c*).

2. *Find the current in each of the branch circuits.* The current in the left branch is found by dividing 80 V by 100 Ω. The current in this branch is 0.8 A. As is typical with parallel circuits, one branch does not have an effect on the other. The current in the right branch is 80 V divided by 70 Ω. This current is 1.14286 A.
3. *Find the voltage across each resistor.* Ohm's law defines voltage as current multiplied by resistance. The currents through R_1 and R_3 are the same: 0.8 A. The voltage across R_1 is found by multiplying 60 Ω by 0.8 A. The result is 48 V. The voltage across R_3 is found by multiplying 40 Ω by 0.8 A. The result is 32 V. If these voltages are correct, the sum of each pair of voltages should equal 80 V, which they do.

   The voltage calculations on the other branch will be prone to round-off error because the number of decimal places exceeds the limit of the calculator's display and must be rounded off. The error can be greatly reduced by using a calculator to hold the result of 80 divided by 70. The result may be stored in the calculator's addressable memory. If the calculator does not have an addressable memory, the 80 ÷ 70 calculation may be performed for each resistor in the branch. With only two resistors, this should not be a problem. The voltage across R_2 is found by multiplying 30 Ω by 1.14 A. The result is 34.3 V. The voltage across R_4 is found by multiplying 40 Ω by 1.14 A. The result is 45.7 V. The sum of these voltages should also equal 80 V, which it does. The circuit is shown with all values in Fig. 10-4.

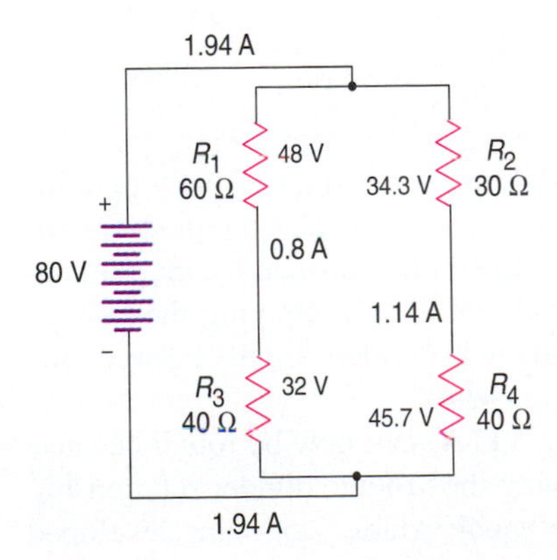

Figure 10-4 Series circuits in parallel are independent of each other when connected across a common source.

REVIEW QUIZ 10-1

1. The parameters of series-parallel circuits may be found using only ______________ rules, ______________ rules, and ______________ law.
2. Regardless of how complex a resistive circuit might be, it appears to the source as ______________ resistor.

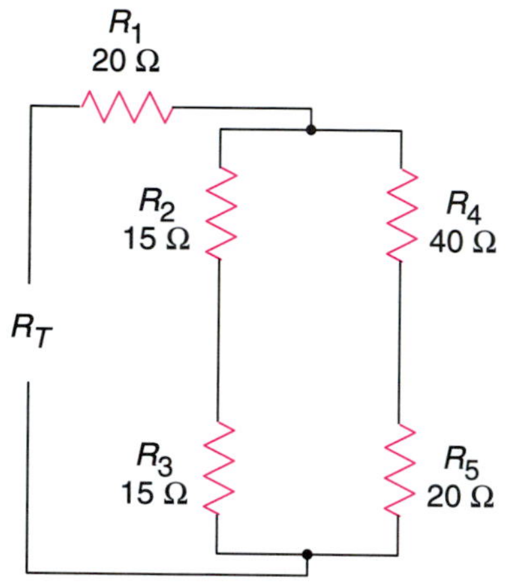

Figure 10-5 The parameters of some apparently complicated circuits can be solved mentally.

10-2 A More Complex Example

The normal procedure used for solving resistive series-parallel circuits is to first identify any parallel circuits and reduce them to a single resistor. These parallel circuits may contain branches with resistors connected in series, as shown in Fig. 10-5. The series resistors must be added before the equivalent resistance of the parallel group can be found.

The circuit in Fig. 10-5 appears to be dominated by the three series circuits, but the solution to finding the total circuit resistance lies in determining the resistance of the parallel group formed by $R_2 + R_3$ and $R_4 + R_5$. The values of these resistors must be added before a parallel solution can be found. Resistor R_1 is in series with the parallel group and is the last thing to be considered when solving the total resistance.

The resistance of the circuit can be approximated mentally using the techniques presented in Chap. 9. The procedures are presented in Example 10-3.

Example 10-3

1. *Find the resistance of each series branch.* Resistors in series are simply added: $R_2 + R_3 = 15\ \Omega + 15\ \Omega = 30\ \Omega$, and $R_4 + R_5 = 40\ \Omega + 20\ \Omega = 60\ \Omega$.
2. *Find the equivalent resistance of 30 Ω in parallel with 60 Ω.* In this case, the value can be accurately estimated because the 60-Ω resistor is twice the value of the 30-Ω resistor. That ratio equates to an equivalent resistance that is two-thirds the value of the smaller resistor, or 20 Ω.
3. *Add the 20-Ω resistance of the series resistor R_1 to the 20-Ω equivalent resistance of the circuit formed by R_2, R_3, R_4, and R_5.* The total resistance of the circuit is 40 Ω.

Circuit Perspective

It is often necessary to view a circuit from different perspectives, usually in terms of voltage, current, or resistance. The last example contained two potential voltage drops, one across R_1 and the other across the circuit formed by the remaining resistors. Placing a 30-V source across the circuit and applying the voltage divider relationship shows that the applied voltage is divided equally between the two voltage drops. The circuit is shown in Fig. 10-6.

The individual voltages across R_2, R_3, R_4, and R_5 can now be found because the voltage across the group is known. Applying the voltage divider relationship to R_2 and R_3 reveals that, because they have equal values, 7.5 V are developed

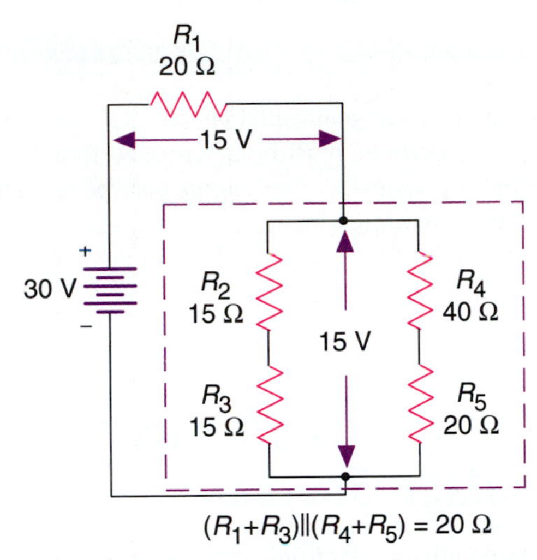

Figure 10-6 The applied voltage is split equally across the two voltage drops.

across each resistor. Focusing our attention on R_4 and R_5, we see that the voltage across R_4 is twice that across R_5 because the two resistors form a series circuit where R_4 has twice the resistance of R_5. This necessitates dividing 15 V into three parts of 5 V each. Two of the three parts, 10 V, is across R_4 and the remaining 5 V is across R_5. The proportion of voltage across each resistor in series is proportional to the resistor's ohmic value.

The Accuracy of Approximation

When circuits values are approximated using the prescribed procedures, the results are usually accurate to within ±5%, which coincides with standard resistance tolerances. Most circuit parameters do not vary outside that limit in modern equipment. In older equipment, components with tolerances of ±10% or even as high as ±20% were common. The 5% accuracy achievable with mental approximation techniques is often more accurate than necessary, but that potential accuracy testifies to the value of the approximation procedures.

Circuit malfunctions severe enough to cause improper operation will generally lead to voltage errors greater than the ±5% window provided by the approximations. Many circuit failures are caused by undesired changes in the conduction levels of various components. These resistance or conductance changes produce changes in circuit voltages because the currents through the affected components have altered. This means that approximately 80% of circuit failures can be found with a voltmeter. Therefore, understanding the procedures for approximating the voltages in a circuit is of paramount importance in circuit analysis.

Schematic diagrams usually indicate the voltages of power sources and the terminals of key components. Voltages are almost never supplied at points within component networks. The given voltages serve as starting points for determining the other voltages within a circuit. Unfortunately, the voltages printed on schematic diagrams are not always correct. The ability to approximate what a voltage should be is very beneficial when these infrequent errors do occur. Applying Kirchhoff's voltage law to the voltage drops around a circuit will usually reveal any problems. Remember that Kirchhoff's voltage law implies a conservation of voltage in that no voltage will be lost or appear without explanation.

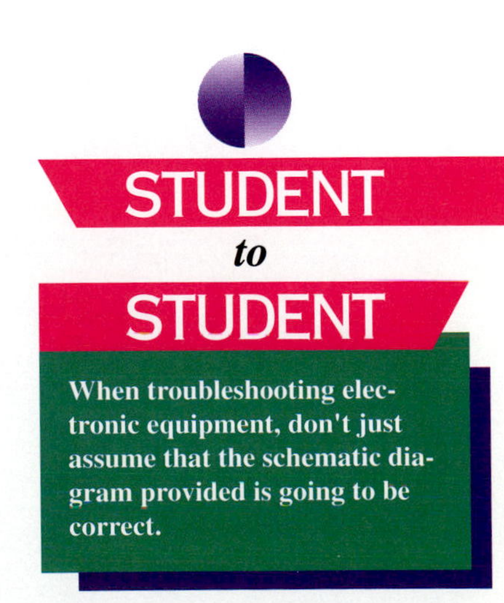

REVIEW QUIZ 10-2

1. When series circuits are connected in parallel, the resistance of the ______________ portions must be determined first.
2. A voltage drop in a series-parallel circuit may be a single resistor or a group of resistors connected in ______________.
3. Voltages or currents within series-parallel circuits may be approximated with an accuracy of ± ______________ %.

10-3 Voltage Dividers as Multiple Voltage Sources

Voltage dividers are popular in electronic circuitry because they provide an economical means of obtaining several voltages from a single source. They are also a natural consequence of other goals of circuit design. Chapter 8 discussed the voltage division characteristics of series circuits, but the voltages obtained from those voltage dividers were not required to supply any current. When the output of a voltage divider supplies current to a load, the circuit becomes series-parallel.

A simple voltage divider consisting of two resistors, R_1 and R_2, has only a single path for current to flow when connected across a source. As with any voltage divider, the voltage division is dependent on the ratio of the resistors. Assuming the output voltage is to be taken across R_2, that voltage will conform to the voltage divider relationship. However, when the divider must supply current to a load, the load represents an additional current path placed in parallel with R_2, effectively reducing its resistance. The equivalent resistance of the load often remains obscure because the load may be a very complex circuit or combination of circuits. Figure 10-7(*a*) is a diagram of a voltage divider without an external load. The same divider is shown in Fig. 10-7(*b*) with a load.

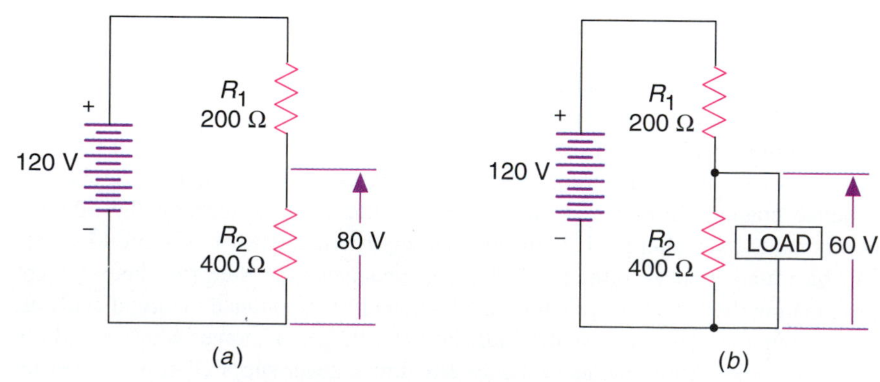

Figure 10-7 Placing a load across the output of a voltage divider causes its output voltage to fall. (*a*) Voltage divided without external load. (*b*) The same voltage divider with an external load.

The load reduces the output voltage of the divider by changing the divider ratio. In this case, the output voltage has dropped from 80 V to 60 V. Although the resistance of the load is not known, the parallel equivalent of R_2 and the load can be determined using the voltage divider relationship. In a voltage divider, the voltage ratio and resistance ratio are the same. Since the load has caused the output voltage to drop to one-half of the source voltage, the parallel equivalent of R_2 and the load must be one-half of the total resistance. The other half is contained in R_1, which is given as 200 Ω. Therefore, the addition of the load has reduced the effective value of R_2 from 400 Ω to 200 Ω.

Approximating what value resistor to place in parallel with another resistor to create a lower value can be difficult. In this case, the solution is simplified because the resistance of the lower half of the circuit has been cut in half from 400 Ω to 200 Ω. One of the postulates for approximating the value of parallel resistors states that: *If two resistors are equal in value, the circuit resistance is one-half, or 50%, of either one.* Applying a reverse view of that postulate to R_2 and the load establishes the load's resistance at 400 Ω.

Voltage Divider Current

The circuit in the previous example could be simplified by eliminating R_2 and increasing the value of R_1 to 400 Ω, but not without some disadvantages. When a resistor is placed in series with a load to drop a portion of the source voltage, the voltage across the load may vary excessively with changes in load current. If the load is a linear resistance there is no need to use a separate divider resistor, but more often than not the conductance of the load is not constant. The variations in load voltage caused by changes in load current are greatest when all of the source current flows through the load, as illustrated in Fig. 10-8.

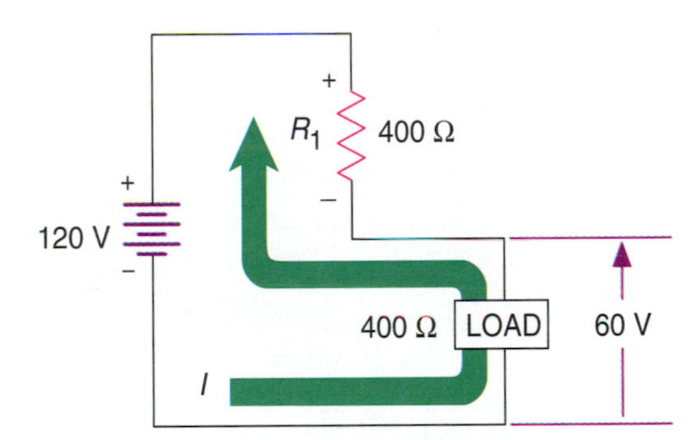

Figure 10-8 Variations in load voltage are greatest when the load forms a complete section of a voltage divider.

The original circuit as shown in Fig. 10-7(*b*) allowed only a portion of the source current (I_S) to flow through the load. For obvious reasons, that portion is called the **load current** (I_L). The remaining current flows through R_2 and is called the **divider current** (I_{DIV}). Large divider currents produce a more stable output, but the ratio of divider current to load current must exist within practical limits. The division of the source current between the divider and the load is illustrated in Fig. 10-9 on the next page.

A general rule in circuits using voltage dividers is that loads requiring significant amounts of power carry a much larger percentage of the source current than loads consuming negligible amounts of power. Voltage dividers are often referred to as being **firm** or **stiff.** In a firm voltage divider, the divider current is about 10 times greater than the load current. In a stiff voltage divider, the divider

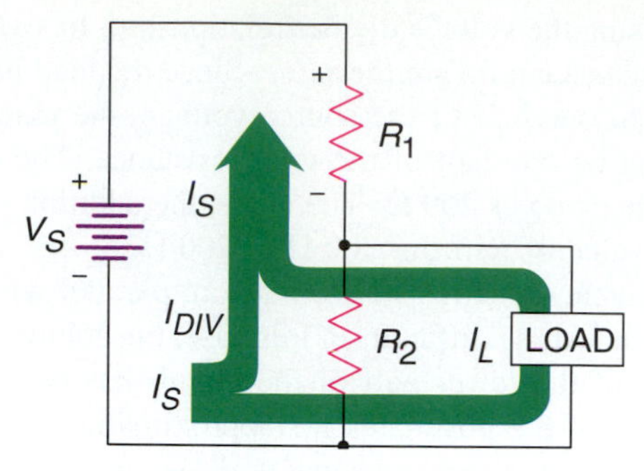

Figure 10-9 The source current consists of the divider current and load current.

current may be as much as 100 times greater than the load current. Obviously, this is a situation that can only be justified in low power circuits. The larger divider current is made possible by the divider resistor in parallel with the load. This resistor is sometimes called a **bleeder** resistor because one of its purposes is to "bleed" or drain current in order to reduce the shock hazard. The current through this resistor may be called either the divider current or **bleeder current.** The two terms are synonymous.

The output voltage of firm and stiff voltage dividers may be closely approximated by applying the voltage divider relationship to the divider resistors and ignoring any current that the load may draw. In any event, the output voltage of a divider circuit cannot be greater than that predicted by the voltage divider relationship and could conceivably drop to zero if the load were shorted.

Example 10-4: Designing a Simple Voltage Divider

Assume that a load requiring 20 V at 50 mA is to be operated from a 30-V source. The expected variations in load current require the use of a firm divider, as shown in Fig. 10-10.

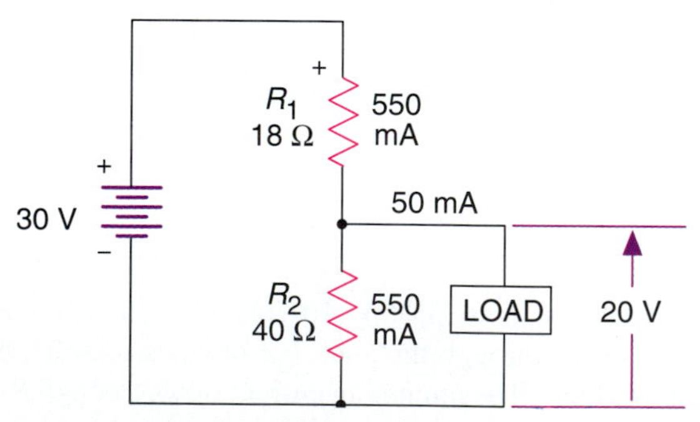

Figure 10-10 A firm voltage divider requires that at least 10 times more current flow in the divider than in the load.

The following steps will produce the indicated values for R_1 and R_2.

1. Because a firm divider requires the divider current to be 10 times greater than the load current, the 50 mA load current must be multiplied by 10:

$$I_{DIV} = 10(50\text{ mA}) = 500\text{ mA} = 0.5\text{ A}$$

2. The voltage across the divider resistor R_2 is 20 V because it is in parallel with the load. The current in the resistor is 500 mA (from Step 1). The value of the resistor can now be found using Ohm's law.

$$R_2 = \frac{V_{R_2}}{I_{R_2}} = \frac{20\text{ V}}{0.5\text{ A}} = 40\ \Omega$$

3. The source current flows through the upper resistor R_1 and is the sum of the divider current and the load.

$$500\text{ mA} + 50\text{ mA} = 550\text{ mA} = 0.55\text{ A}$$

4. The voltage across R_1 is the difference between the source voltage and the load voltage.

$$30\text{ V} - 20\text{ V} = 10\text{ V}$$

5. The value of R_1 can now be found using Ohm's law.

$$R_1 = \frac{V_{R_1}}{I_{R_1}} = \frac{10\text{ V}}{0.55\text{ A}} = 18\ \Omega^{1}$$

[1]The value of the resistor is actually 18.1818 Ω, but unless the application is so critical that using a custom value resistor is justified, the EIA standard value of 18 Ω would be used.

The unloaded voltage of the divider, as determined by the voltage divider relationship, is 20.960 V. That is only 0.960 V more than the full-load voltage. Had the load been placed in series with a dropping resistor, the output voltage would vary from 30 V to 20 V when going from no-load to full-load conditions. Clearly, the use of the series-parallel circuit employing a firm-to-stiff voltage divider produces a secondary supply voltage with a reasonable amount of voltage regulation. For that reason, it is commonly encountered in electronic devices.

Example 10-5: A Complex Voltage Divider

Electronic circuitry often requires the development of several voltages from a single source. Once again, this may be accomplished with a voltage divider. Assume that a single 100-V source is to produce three additional sources of 80 V at 20 mA, 50 V at 75 mA, and 35 V at 10 mA, as shown in Fig. 10-11.

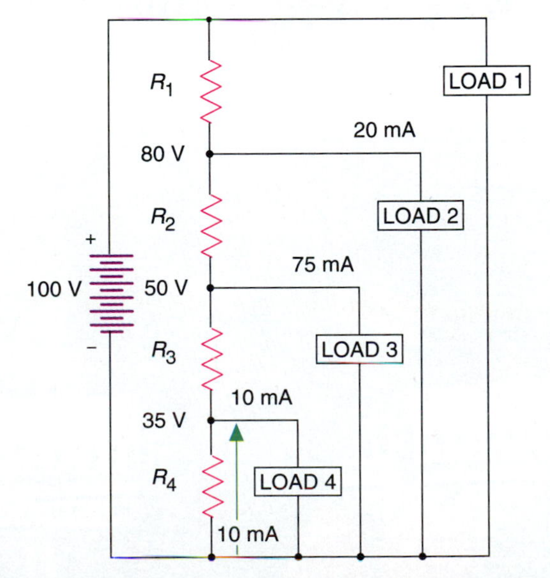

Figure 10-11 Voltage dividers provide an economical method of creating multiple voltage sources from a single source if the loads are fairly constant.

The diagram shows a fairly complex, but typical, series-parallel circuit. The major advantage of the circuit is that it provides several voltages at a minimum cost. A disadvantage is that the output voltages, other than the one taken directly from the source, are dependent on the loads drawing the correct current. Should any load that is supplied by the divider fail, the other divided voltages, and therefore the circuits they operate, will be adversely affected.

There are several methods for finding the resistor values, but the most straightforward is to begin at R_4. The voltage across R_4 and the current through R_4 are given on the schematic. Using R_4 as the starting point, we can obtain the values of the resistors through the following procedures.

1. *Find the value of R_4 using Ohm's law.* The voltage across R_4 is given as 35 V, and the current through R_4 is 10 mA. According to Ohm's law, resistance is voltage divided by current. Therefore, the resistance is 35 V divided by 10 mA, or 3.5 kΩ.

$$R_4 = \frac{V_{R_4}}{I_{R_4}} = \frac{35\text{ V}}{10\text{ mA}} = 3.5\text{ k}\Omega$$

2. *Find the value of R_3 using Ohm's law.* The voltage at the upper end of R_3 is 50 V, and the voltage at the lower end of R_3 is 35 V. The voltage across R_3 is the difference of these two voltages, or 15 V.

$$V_{R_3} = 50\text{ V} - 35\text{ V} = 15\text{ V}$$

The current through R_3 is the sum of the currents through R_4 and LOAD 4. The current through R_4 is given as 10 mA, and the current through LOAD 4 is also 10 mA. The current through R_3 is the sum of these currents, or 20 mA.

$$I_{R_3} = I_{R_4} + I_{L_4} = 10\text{ mA} + 10\text{ mA} = 20\text{ mA}$$

The value of R_3 is found by dividing the voltage across R_3, 15 V, by the current through R_3, 20 mA. Therefore, R_3 is 500 Ω.

$$R_3 = \frac{V_{R_3}}{I_{R_3}} = 15\frac{\text{V}}{20\text{ mA}} = 0.5\text{ k}\Omega = 500\ \Omega$$

3. *Find the value of* R_2. The value of R_2 is found using Ohm's law. The voltage at the upper end of R_2 is given as 80 V and the voltage at the lower end of R_2 is given as 50 V. The voltage across R_2 is the difference of these two voltages, or 30 V.

$$V_{R_3} = 80\text{ V} - 50\text{ V} = 30\text{ V}$$

The current through R_2 is the sum of the currents through R_3 and LOAD 3. The current through R_3, as determined in Step 2, is 20 mA, and the current through LOAD 3 is given as 75 mA. Therefore, the current through R_2 is 95 mA.

$$I_{R_2} = I_{R_3} + I_{L_3} = 75\text{ mA} + 20\text{ mA} = 95\text{ mA}$$

The value of R_2 is found using Ohm's law. The voltage across it, 30 V, is divided by the current through it, 95 mA. Therefore, R_2 is 316 Ω.

$$R_2 = \frac{V_{R_2}}{I_{R_2}} = \frac{30\text{ V}}{95\text{ mA}} = 0.316\text{ k}\Omega = 316\ \Omega$$

4. *Find the value of* R_1 *using Ohm's law.* The voltage at the upper end of R_1 is given as 100 V, and the voltage at the lower end of R_1 is given as 80 V. The voltage across R_1 is the difference between these two voltages, or 20 V.

$$V_{R_1} = 100\text{ V} - 80\text{ V} = 20\text{ V}$$

The current through R_1 is the sum of the currents through R_2 and LOAD 2. The current through R_2, as determined in Step 3, is 95 mA, and the current through LOAD 2 is 20 mA. Therefore, the current through R_1 is 115 mA.

$$I_{R_1} = I_{R_2} = 95\text{ mA} + 20\text{ mA} = 115\text{ mA}$$

Applying Ohm's law, the value of R_1 is found by dividing the voltage across R_1, 30 V, by the current through it, 115 mA. The result is that R_1 equals 261 Ω.

$$R_1 = \frac{V_{R_1}}{I_{R_1}} = \frac{30\text{ V}}{115\text{ mA}} = 0.261\text{ k}\Omega = 261\ \Omega$$

The voltage divider in Example 10-5 represents a commonly encountered circuit. Understanding its operation requires a firm understanding of series rules, parallel rules, and Ohm's law.

REVIEW QUIZ 10-3

1. When a load is placed across one of the resistors in a voltage divider, the voltage across that resistor will ______________.
2. A resistor in a voltage divider with a load connected across it that does not supply load current is called a(n) ______________ resistor.
3. If the divider current is at least 10 times greater than the load current, the divider is said to be ______________ .
4. If the divider current is at least 100 times greater than the load current, the divider is said to be ______________ .
5. Voltage divider circuits are not practical when ______________ amounts of power are required by the loads.

SUMMARY

When series and parallel circuits are combined, a new type of circuit called series-parallel is created. Series-parallel circuits may consist of series circuits connected in parallel or parallel circuits connected in series. These circuits may be very complex, but their circuit parameters can always be solved using basic circuit rules and Ohm's law.

Series-parallel circuits are often the result of using a voltage divider to develop one or more voltages from a single source. Such voltage sources are economical, but they rely heavily on the load currents' being stable. One solution to the problems encountered when using voltage dividers to provide power to fluctuating loads is to have the divider current much larger than the load current. When the divider current is about 10 times the load current, the divider is said to be firm. When the divider current is about 100 times the load current, the divider is said to be stiff. Firm and stiff voltage dividers are only practical when supplying small amounts of current. Series-parallel circuits often go unrecognized because one or more circuit elements may be a complete circuit that requires a fixed current at some fixed voltage. These unidentified circuit elements are generally identified as box-labeled load for purposes of circuit analysis.

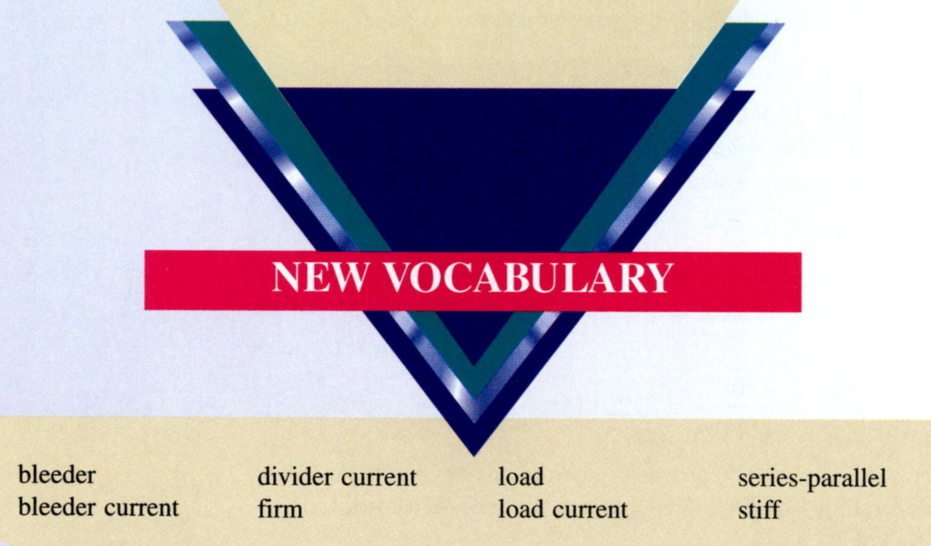

NEW VOCABULARY

bleeder
bleeder current
divider current
firm
load
load current
series-parallel
stiff

Questions

1. Are there new rules associated with series-parallel circuits?
2. Does adding a load across a voltage divider create a series-parallel circuit?
3. What is the ratio of divider current to load current in a firm voltage divider?
4. Would a stiff voltage divider be a practical solution when a load demands large amounts of power?
5. Should the voltage output of a voltage divider decrease when a load is placed across it?

Problems

1. A load requiring 12 V at 5 mA, as shown in Fig. 10-12, is to be operated from a 24-V source using a stiff voltage divider. What are the values of R_1 and the bleeder resistor R_2?

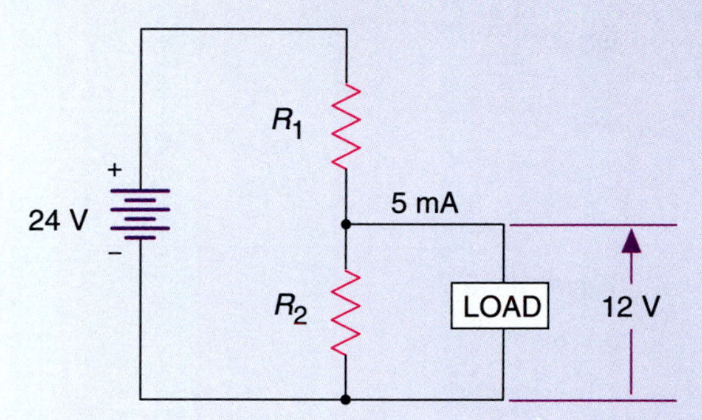

Figure 10-12

2. What is the total resistance of the series-parallel circuit in Fig. 10-13?

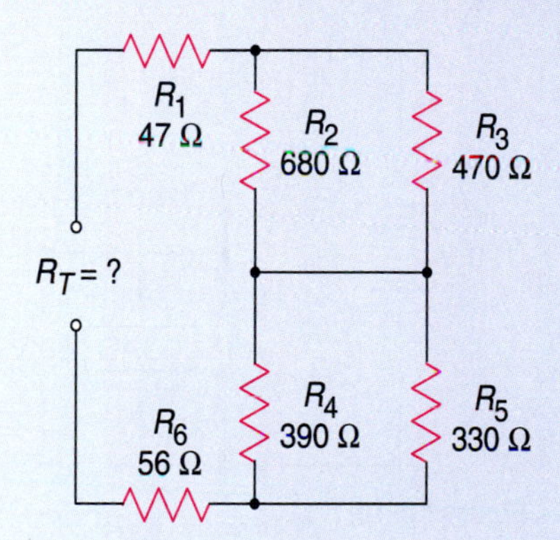

Figure 10-13

3. What is the voltage across R_2 in Fig. 10-14?

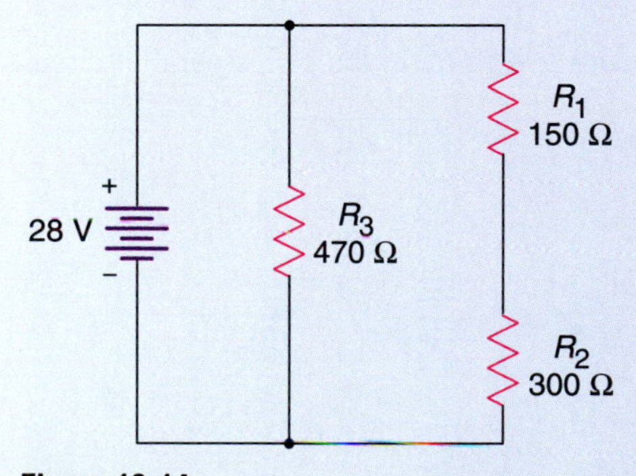

Figure 10-14

4. Approximate the voltage between points A and B in the series-parallel circuit of Fig. 10-15.

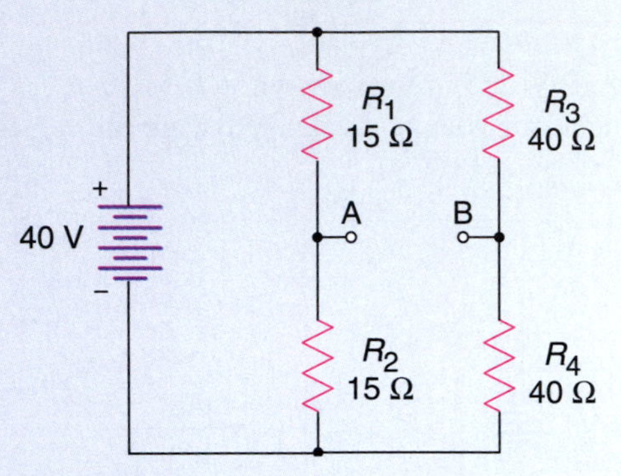

Figure 10-15

5. What value of load resistance would be required to reduce the output voltage to 60 V in the circuit of Fig. 10-16?

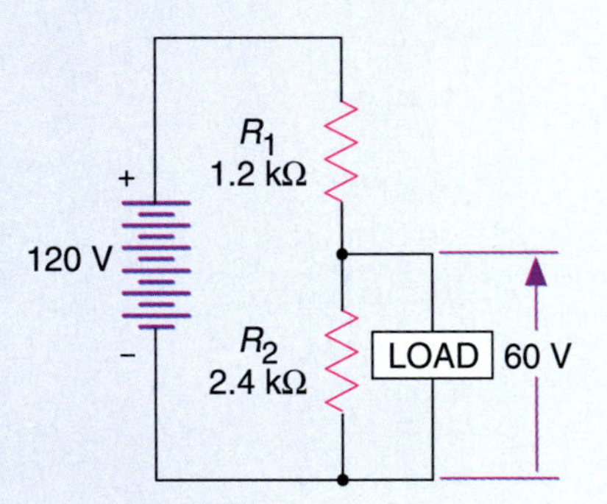

Figure 10-16

6. What is the current through R_1 in the series-parallel circuit of Fig. 10-17?

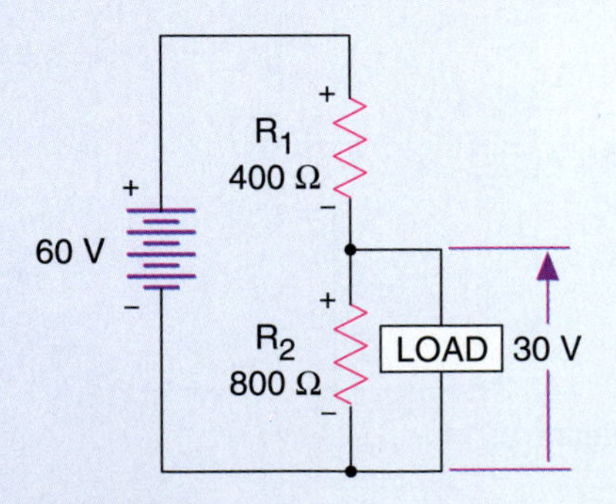

Figure 10-17

7. What is the resistance of each of the four loads in the complex voltage divider of Fig. 10-18?

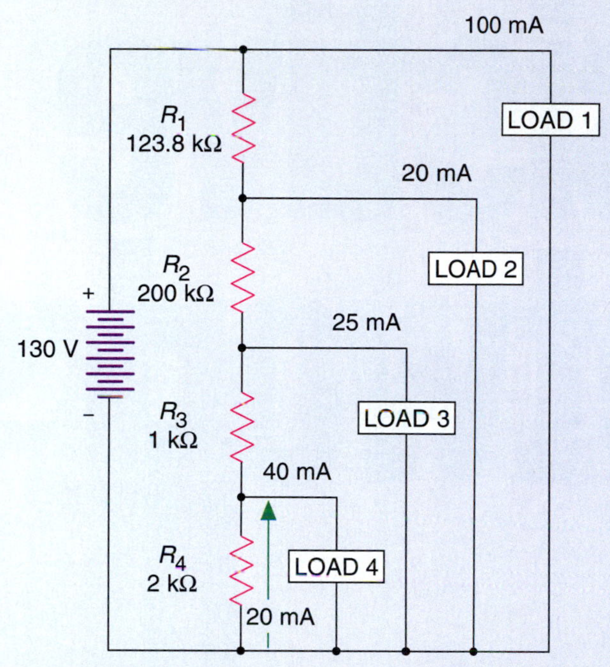

Figure 10-18

Critical Thinking

1. Why can't the resistance of a series-parallel circuit be calculated by first combining the series resistor with only one of the parallel resistors?
2. What is the advantage of using a stiff voltage divider over a simple dropping resistor?
3. If the load on a resistive voltage divider should open, why will the voltage of the source increase?
4. If the bleeder resistor in a voltage divider were to open, why would the output voltages increase?

Answers to Review Quizzes

10-1
1. series; parallel; Ohm's
2. one

10-2
1. series
2. parallel
3. 5

10-3
1. decrease
2. bleeder
3. firm
4. stiff
5. large

Chapter 11

Resistive Networks

ABOUT THIS CHAPTER

Resistive networks represent a very wide range of circuits that are generally more difficult than the series-parallel circuits discussed in Chapter 10. Resistive networks have the common trait of revealing only one circuit parameter per component, making an Ohm's law solution very rare. Several theorems and circuit transformations

After completing this chapter you will be able to:

- Determine the type of network by the arrangement of its components on schematic diagrams.
- Describe the possible paths for current in some common networks.

have been developed to provide any desired parameter, but these solutions are often based on algebraic techniques. It is not the intent of this chapter to provide a complete treatment of resistive networks. Instead, several common procedures are presented using worked-out examples as a building foundation.

- ▽ Apply network theorems and transformations to resistive networks to determine their circuit parameters.
- ▽ Understand how mathematics is used to determine a circuit's parameters and current flow.

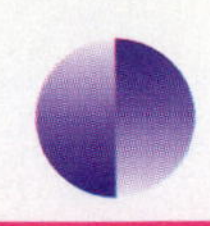

STUDENT *to* STUDENT

Focus on the general concepts of networks and the nonalgebraic theorems.

11-1 Network Configurations

A fifth class of electric circuit is called a **network**. The word *network* is used because the arrangement of the components on a schematic diagram looks similar to a netted material. Almost any circuit could be called a network using that definition, but the name is usually reserved for those circuits whose parameters cannot be found using Ohm's law and basic circuit rules. The circuit parameters of networks are determined using **network theorems** that frequently employ algebraic techniques.

Circuits are often given names that refer to the arrangement of their components on schematic diagrams. The most common names are ***T* delta**, **wye**, **pi**, **bridge**, and **ladder**. Circuits with special names are usually called networks, even if their parameters can be determined without the use of network theorems. Schematic representations of some basic networks are shown in Fig. 11-1.

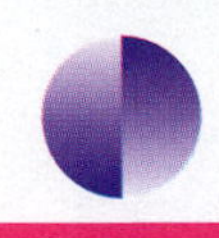

STUDENT *to* STUDENT

The bridge circuit takes many forms. You will see a lot of them.

The Bridge Circuit

A **bridge circuit** is a very common network that is sometimes called a **double delta**. The name *delta* is derived from the triangular arrangement of the components in the common schematic representation. Delta is the fourth letter of the Greek alphabet (Δ-delta). However, the name *bridge* is not derived from the network's shape; it refers to a component spanning two independent circuits.

Resistive bridge circuits are frequently drawn as a diamond with a resistor along each of the four edges and a fifth resistor connected across the two side points as shown in Fig. 11-2(*a*). This fifth resistor is called the **bridge resistor**.

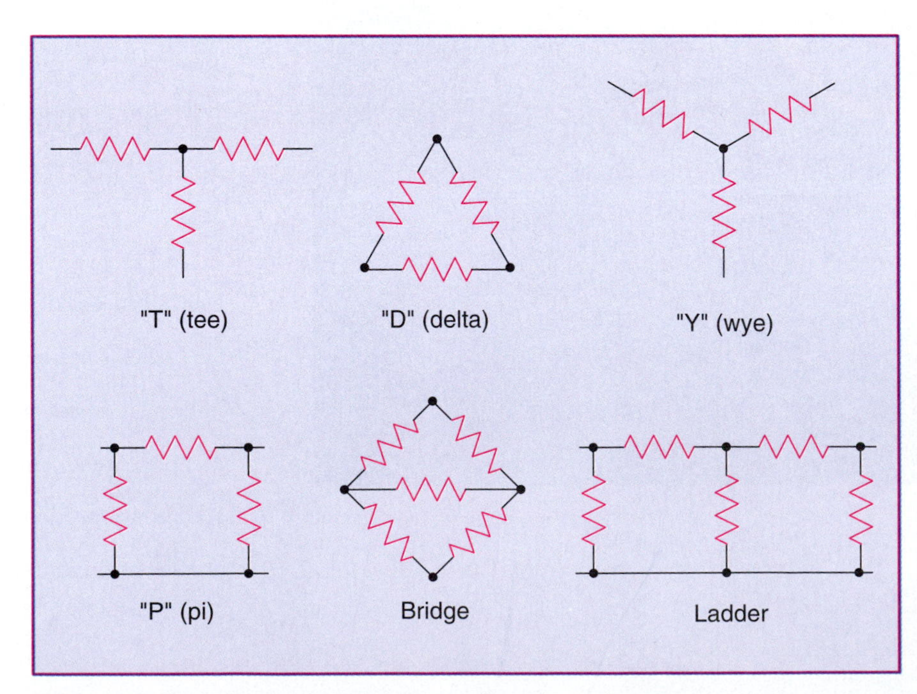

Figure 11-1 Networks are often given names that refer to the arrangement of their components on a schematic diagram.

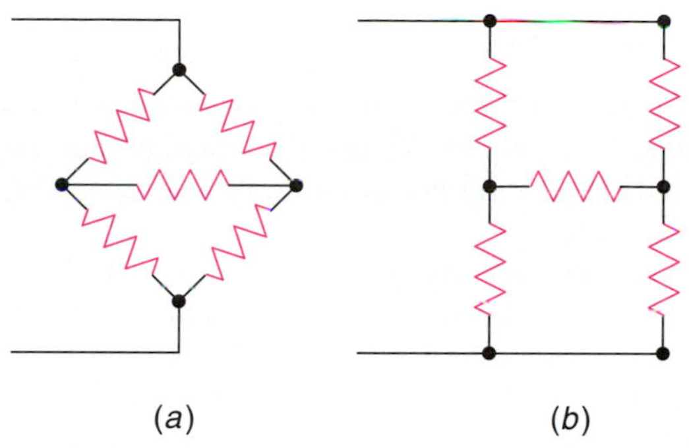

Figure 11-2 The bridge is usually drawn as (*a*) a divided diamond or (*b*) in an *H* configuration.

Networks are often named after letters of the English or Greek alphabets (t, wye, delta, pi).

It is the addition of the bridge resistor that divides the diamond into two triangles, giving it the name double delta.

The bridge is also drawn in an *H* configuration as shown in Fig. 11-2(*b*). The *H* configuration reveals the true nature of the bridge circuit more clearly than the diamond configuration.

The bridge circuit should be viewed as two voltage dividers connected in parallel. When placed across a voltage source, the circuit parameters of the dividers may be determined using Ohm's law and the basic rules of circuit operation. The addition of the bridge resistor converts the two voltage dividers into a bridge network by "bridging" the outputs of the dividers. With the addition of the bridge resistor the circuit parameters may no longer be determined using Ohm's law because no series or parallel portions of the circuit can be isolated.

Current flows through the bridge resistor any time the voltage dividers produce different voltages. The direction of current through the bridge resistor is dependent on the relative polarities created by the dividers. The flow may be from left to right, as shown in Fig. 11-3(*a*), or right to left, as shown in Fig. 11-3(*b*). It is also possible for the voltages to be equal. This would result in zero current through the bridge resistor as shown in Fig. 11-3(*c*). When the current through the bridge resistor is zero, the bridge is said to be **balanced**. The value of a balanced bridge resistor is unimportant and may be ignored for the purpose of calculating the current through the other resistors. The circuit may once again be viewed as two parallel voltage dividers whose parameters may be solved using Ohm's law. In the balanced condition the ratios of the dividers are identical.

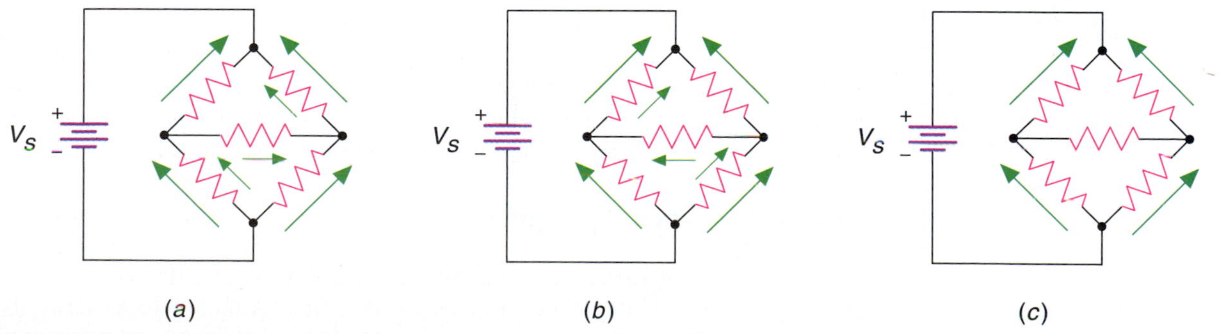

Figure 11-3 The current through the bridge resistor may be (*a*) from left to right, (*b*) from right to left, or (*c*) zero.

The Wheatstone Bridge

The principles just explained find a practical application in a circuit called a **Wheatstone bridge**. Although the Wheatstone bridge is named after Sir Charles Wheatstone (1802–1875), he did not invent it, but he did modify it into its better known form.

The bridge resistor in a Wheatstone bridge is replaced with a zero-center microammeter, also known as a Galvanometer. The voltage divider ratio on one side of the meter is fixed and used to establish a reference voltage. The voltage divider on the other side is made from a rheostat equipped with a calibrated knob and the resistor whose value is to be determined. The resistor being measured is placed across a pair of external terminals. The rheostat is then adjusted to bring the needle of the meter to zero center. At that point the value of the meter can be read from a scale attached to the rheostat's knob as shown in Fig. 11-4(*a*). The schematic of a Wheatstone bridge is shown in Fig. 11-4(*b*).

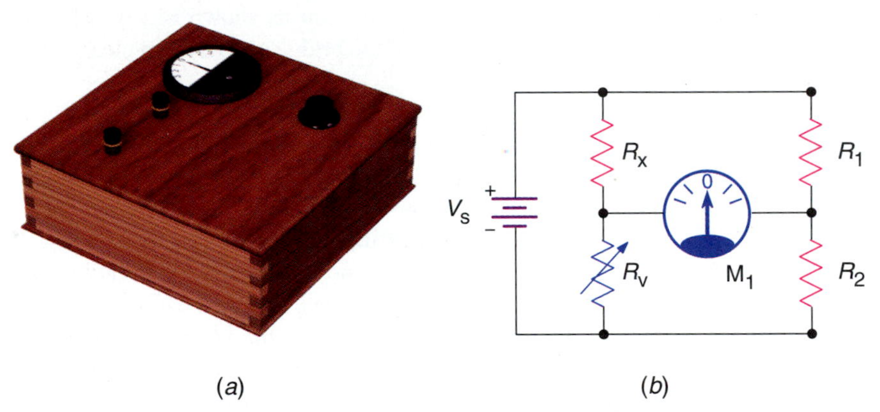

Figure 11-4 (*a*)The Wheatstone bridge is used to measure resistance or to compare one resistor to another. (*b*) The schematic of a Wheatstone bridge.

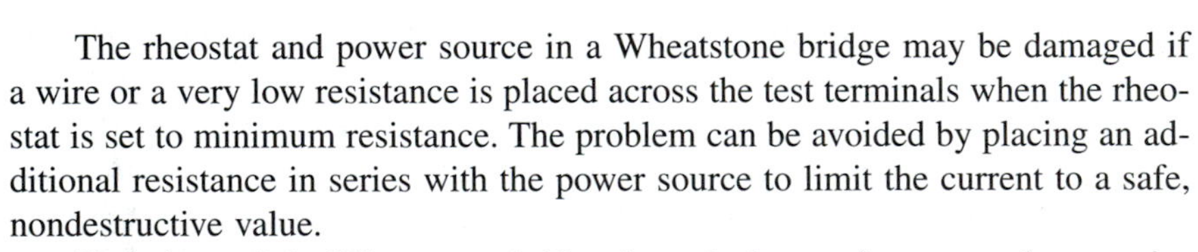

Figure 11-5 A modified Wheatstone bridge can quickly and easily establish the tolerance of a resistor when compared to a standard.

The rheostat and power source in a Wheatstone bridge may be damaged if a wire or a very low resistance is placed across the test terminals when the rheostat is set to minimum resistance. The problem can be avoided by placing an additional resistance in series with the power source to limit the current to a safe, nondestructive value.

Variations of the Wheatstone bridge form the heart of many modern test instruments. An example is a device used to sort resistors into various tolerance groupings. The device is shown in Fig. 11-5. The normal scale on the meter is replaced with zones of color on either side of center to give a relative comparison to a highly accurate standard. For instance, green might represent a zone equal to ±2%, yellow ±5%, and red ±10%. This system requires only minimal operator skill and no knowledge of the resistor value being tested.

The *T* And Wye Networks

The ***T*** and **wye networks** are electrically identical. The only difference between them is the way they are drawn on schematic diagrams. Although these networks appear to be simple voltage dividers, placing a voltage source at each end eliminates the possibility of determining the circuit parameters using an Ohm's law solution. The current paths in such networks may be difficult to determine from

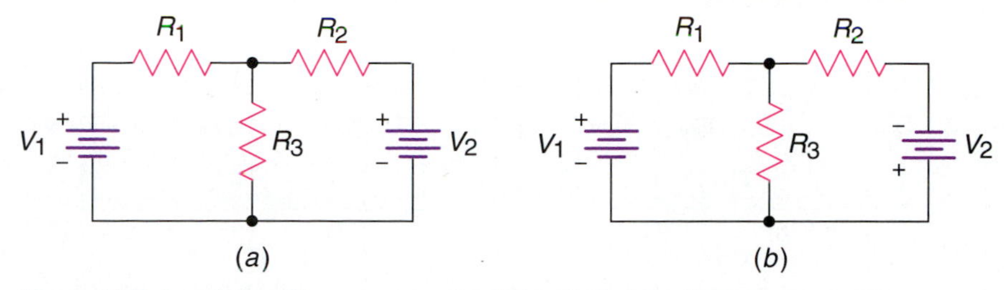

Figure 11-6 A resistive *T* becomes a network by placing a voltage source between the base of the stem and the end of each arm. The sources may be (*a*) series-aiding or (*b*) series-opposing.

looking at the schematic, but only a limited number of possibilities exist. The two voltage sources may be connected in series-aiding or series-opposing, with the base of the *T* connected between the two sources as shown in Fig. 11-6.

The *T* Network With Series-Opposing Sources

Three closed loops are formed by the addition of the two sources. Each loop represents a possible current path, but only two loops are active at the same time. The active loops are determined by the voltage and resistor values. There are five possible configurations for current flow when the sources are connected in series-opposing as shown in Fig. 11-7.

When the sources of the *T* network are connected in series-opposing, current will always flow through R_3. However, this is not the case for R_1 and R_2. Under certain conditions there will be no current through one or the other of these resistors. That situation occurs when one voltage source is higher than the other and the voltage across R_3 is equal to the smaller source. The resistor between the smaller source and the junction of the three resistors can have no current flowing through it because there is no difference of potential across it. The current paths in that situation are shown in Fig. 11-7(*b*) and (*d*).

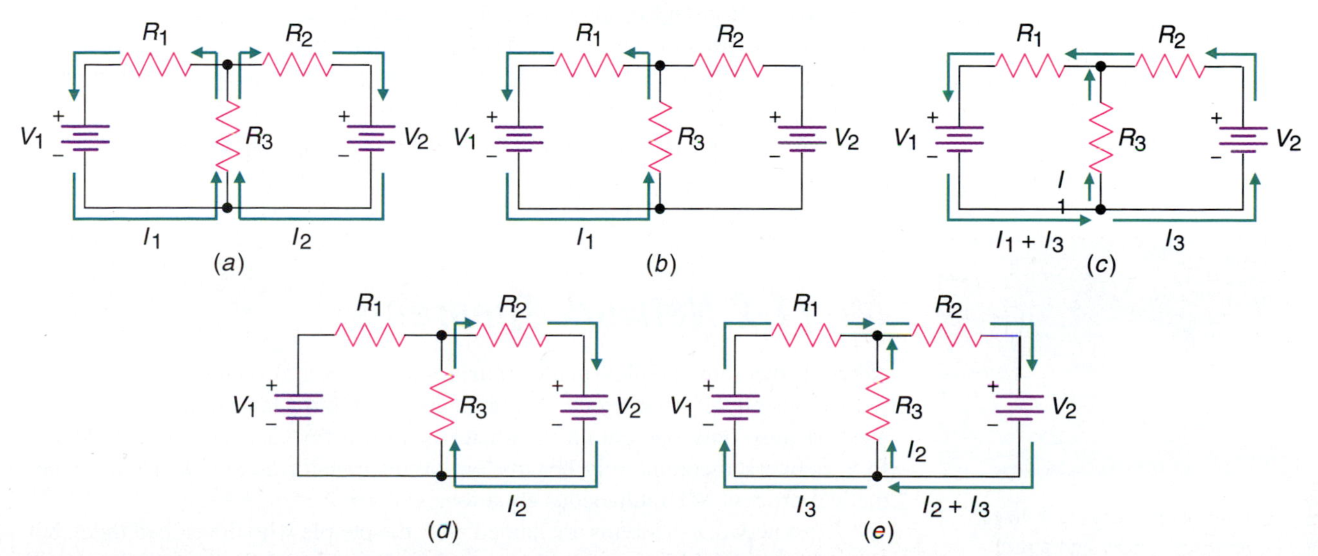

Figure 11-7 There are five possible current loops in a *T* network when the sources are connected in series-opposing.

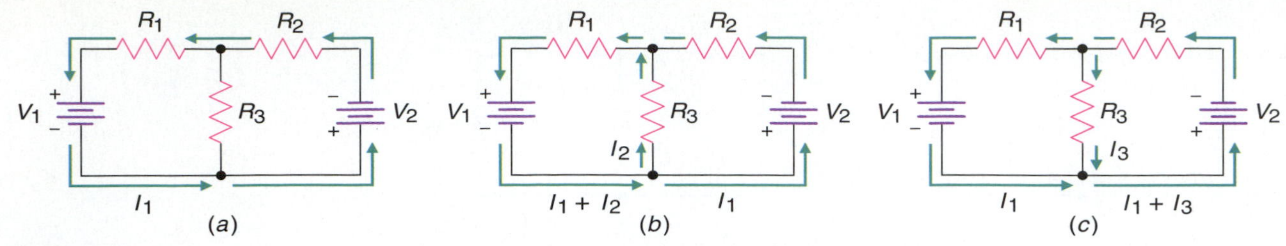

Figure 11-8 There are only three possible current paths in a *T* network when the sources are connected in series-aiding.

The *T* Network With Series-Aiding Sources

The *T* network with series-aiding sources appears very similar to the *T* network with series-opposing sources, but the current paths through the two configurations are different. Again, there are three closed loops through which current can flow. The main loop is along the outside perimeter through V_1, R_1, R_2, and V_2 as shown in Fig. 11-8(*a*). Current will flow through this loop at all times regardless of the voltage or resistor values.

Two additional current loops are created by the presence of R_3. The loop on the left is formed by V_1, R_1, and R_3. The loop on the right is formed by R_3, R_2, and V_2. Current does not flow through the two loops at the same time. Current will flow in the left loop when V_1 is higher than V_2; it will flow in the right loop when V_2 is higher than V_1. The direction of current through R_3 is upward if V_1 is higher than V_2 as shown in Fig. 11-8(*b*). The current through R_3 is downward if V_2 is higher than V_1 as shown in Fig. 11-8(*c*).

REVIEW QUIZ 11-1

1. Generally speaking, the circuit parameters of a network cannot be solved using ______________ .
2. A bridge circuit is sometimes called a(n) ______________ .
3. The Wheatstone bridge is a device that measures ______________ .
4. When the current in the bridge resistor is zero, the bridge is said to be ______________
5. The *T* and wye networks are electrically ______________ .

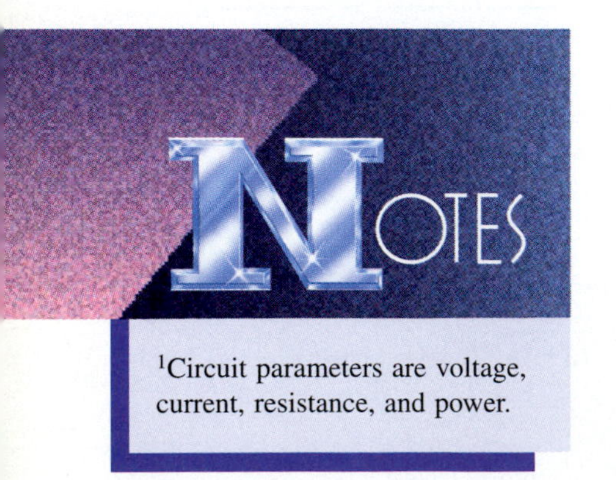

[1]Circuit parameters are voltage, current, resistance, and power.

11-2 Network Theorems

Network theorems employ mathematical and procedural methods to determine the various circuit parameters[1] within a network. Ohm's law solutions cannot be used because only one circuit parameter is known for each component. Therefore, network theorems rely heavily on circuit transformations and the mathematical process of simultaneous equations.

Most network theorems are named after the people who developed them, but several are named for the procedures used to determine their circuit parameters. Network theorems should be thought of as a set of tools for solving specific prob-

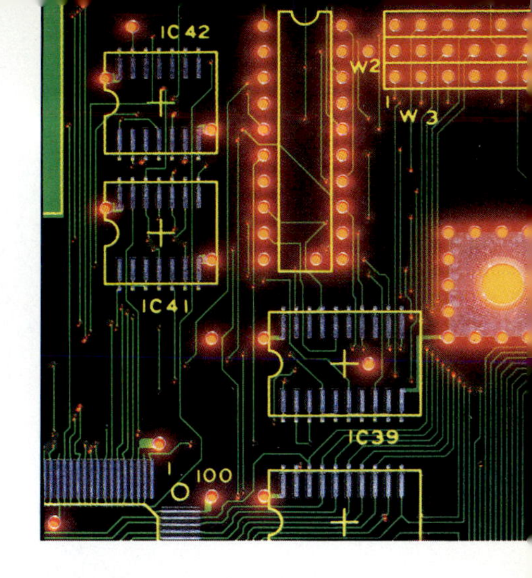

lems. Like tools, some theorems are better suited to certain jobs than others and none is universally applicable to all types of networks.

If the electrical parameters of a network are to be found, one or more network theorems or procedures must be employed. It is very important to understand the strengths and weaknesses of a theorem before attempting to use it. The major theorems are **mesh, superposition**, and **Thévenin's**. Several other network theorems exist but they are used less frequently and can be very math intensive.

Kirchhoff's laws represent a conservation of voltage and current; you will not lose any voltage or current through some unexplained process or gain any voltage or current from some mysterious source. Kirchhoff's laws are used as logical statements of truth to formulate algebraic equations that allow unknown circuit quantities to be calculated. The difficulty in using Kirchhoff's laws lies in setting up and implementing the algebraic equation. The process is made even more confusing because there are often several approaches to a problem.

Mesh Analysis

A very reliable method of calculating the loop currents in a simple two-dimensional network is called **mesh analysis**. The method requires a moderate level of algebraic manipulation. The circuit elements must be linear and the sources should have very little internal resistance. The *T* network with two sources is a good circuit to work with when first learning the process. Example 11-1 describes the step-by-step procedure for determining the circuit parameters of the circuit shown in Fig. 11-9.

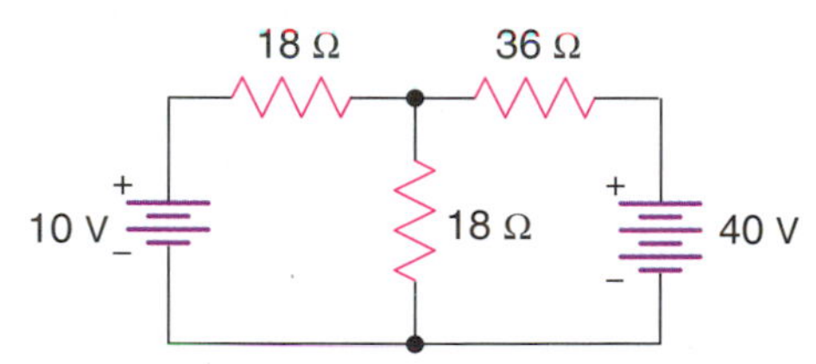

Figure 11-9 Mesh analysis may be used to determine the loop currents in a two-dimensional network with linear elements.

Example 11-1 Loop 1 is formed by V_1, R_1, and R_3, and loop 2 is formed by V_2, R_2, and R_3. Note that R_3 is common to both loops.

1. Starting at the positive terminal of V_1, draw a line in a clockwise direction that follows loop 1 back to the negative terminal of V_1 in Fig. 11-10.

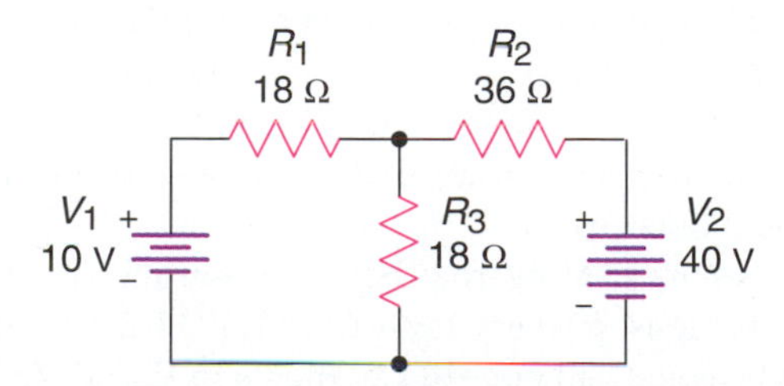

Figure 11-10 When using mesh analysis, the currents in each loop are assumed to all circulate in the same direction.

2. Starting at the negative terminal of V_2, draw a line in a clockwise direction that follows loop 2 back to the positive terminal of V_2 in Fig. 11-10. It is very important that both lines are drawn in the same direction. Clockwise was suggested, but both could have been drawn counter-clockwise. The polarities of the source voltages are identified by the arrowheads on the lines tracing the current loops.

   The current in loop 1 is called I_1, and the current in loop 2 is called I_2. If either I_1 or I_2 is known, then any other circuit parameter can be found using conventional methods.

Mesh analysis relies on the version of Kirchhoff's voltage law which states that the algebraic sum of the voltage drops around a series circuit is equal to the algebraic sum of the source voltages.

Remember that according to Ohm's law, voltage is current multiplied by resistance. Therefore, the voltage across R_1 is $I_1(R_1)$ and the voltage across R_3 is $I_1(R_3)$, but $I_1(R_3)$ does not take into account that current I_2 also flows through R_3. By initially assuming that I_1 and I_2 flow clockwise, the assumption is also made that those currents flow in opposite directions through R_3 at the same time, which is an impossible situation. The name *mesh* refers to this apparent mixing of currents.

For the purpose of calculation it is assumed that the voltage across R_3 is $I_1(R_3) - I_2(R_3)$.

3. Set up the Kirchhoff's law equation for loop 1.

   Equation 11-1

$$18\,I_1 + 18\,I_1 - 18\,I_2 = -10\text{ V} \Rightarrow 36\,I_1 - 18\,I_2 = -10\text{ V}$$

   Note that the source polarity pointed to by the I_1 arrow is used when identifying the source voltage.

4. Set up the equation for loop 2.

   Equation 11-2

$$18\,I_2 - 18\,I_1 + 36\,I_2 = 40\text{ V} \Rightarrow 54\,I_2 - 18\,I_1 = 40\text{ V}$$

5. Place Eq. 11-2 directly under Eq. 11-1 and rearrange the terms so that I_1 and I_2 line up in the same column.

   Equation 11-3

$$\begin{aligned} 36\,I_1 - 18\,I_2 &= -10\text{ V} \\ -18\,I_1 + 54\,I_2 &= 40\text{ V} \end{aligned}$$

   Loop 1 and loop 2 are now expressed in terms of Kirchhoff's voltage law and are set up as simultaneous equations. One or both of the equations must be scaled in such a way that either the I_1 terms or the I_2 terms are equal and opposite. This is easier in some situations than others.

6. Determine how to cancel either the I1 terms or the I2 terms by making them equal and opposite.

   In this case the I_1 term from Eq. 11-1 and the I_1 term from Eq. 11-2 are of opposite signs. If every term from Eq. 11-2 is multiplied by 2, the I_1 terms will be equal and opposite as shown in Eq. 11-4. The I_1 terms may then be canceled leaving only one unknown, I_2.

Equation 11-4

$$2(-18\,I_1 + 54\,I_2 = 40\text{ V}) = -36\,I_1 + 108\,I_2 = 80\text{ V}$$

$$\begin{aligned} \cancel{36}\,I_1 - 18\,I_2 &= -10\text{ V} \\ -\cancel{36}\,I_1 + 108\,I_2 &= 80\text{ V} \end{aligned}$$

7. Add the remaining columns including their signs in the two simultaneous equations.

Equation 11-5

$$\begin{aligned} -18\,I_2 &= -10\text{ V} \\ 108\,I_2 &= 80\text{ V} \\ \hline 90\,I_2 &= 70\text{ V} \end{aligned}$$

8. Divide both sides of the equation by 90 to determine I_2.

Equation 11-6

$$\frac{\cancel{90}I_2}{\cancel{90}} = \frac{70\text{ V}}{90} = 0.7778\text{ A}$$

The clockwise direction of I_2 was assumed correctly because the result is positive. It will be positive any time the direction of the assumed mesh current matches the direction of electron flow. A negative sign would have indicated that I_2 flowed oppositely to the assumed direction. The value of I_2 is actually $0.777\overline{7}$ amperes[2].

9. Find the voltage across R2 using Ohm's law.

Equation 11-7

$$\begin{aligned} V_{R_2} &= R_2 I_{R_2} = 36\ \Omega\ (0.7778\text{ A}) \\ V_{R_2} &= 28\text{ V} \end{aligned}$$

The polarity of the voltage across R_2 is in series-opposing with V_2, making the voltage across R_3 equal to V_2 minus the voltage across R_2.

Equation 11-8

$$\begin{aligned} V_{R_3} &= V_2 - V_{R_2} \\ V_{R_3} &= 40\text{ V} - 28\text{ V} \\ V_{R_3} &= 12\text{ V} \end{aligned}$$

10. Find the voltage across R_1.

Equation 11-9

$$\begin{aligned} V_{R_1} &= V_{R_3} - V_1 \\ V_{R_1} &= 12\text{ V} - 10\text{ V} \\ V_{R_1} &= 2\text{ V} \end{aligned}$$

The voltage across R_1 is the difference between the 12 volts across R_3 and the 10 volts of V_1. The fact that the voltage across R_3 is higher than V_1 indicates that current is being forced through V_1 against its pressure direction by V_2.

This example has shown how Kirchhoff's voltage law was used to establish a pair of simultaneous equations that provided an algebraic solution for determining the circuit parameters.

ELECTRONIC FACTS

[2]An overbar in a repeating decimal (example $0.777\overline{7}$) indicates that the series continues indefinitely.

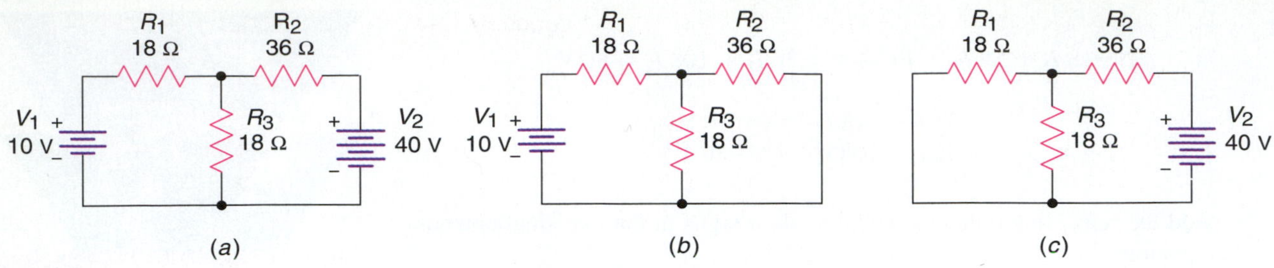

Figure 11-11 The superposition theorem can be used to find the common currents of two loops using basic math and circuit transformations.

Superposition

The superposition procedure calculates the voltage across R_3 from two different perspectives and then adds the voltages for the correct result. The same circuit used in the mesh example is used for comparison as shown in Fig. 11-11(*a*). The procedure for calculating the voltage across R_3 is described in Example 11-2.

Example 11-2

1. Remove V_2 and replace it with a wire. This places R_2 in parallel with R_3 and transforms the circuit to series-parallel as shown in Fig. 11-11(*b*).
2. Calculate the parallel resistance of R_3 and R_2.

Equation 11-10

$$R_3||R_2 = \frac{R_3\,R_2}{R_3+R_2} = \frac{36 \times 18}{36 + 18} = \frac{648}{54}$$

$$R_3||R_2 = 12\ \Omega$$

3. Using the voltage divider relationship, calculate the voltage across the parallel equivalent of R_3 and R_2.

Equation 11-11

$$V_{R_3}||R_2 = V_1\left(\frac{R_3||R_2}{(R_3||R_2) + R_1}\right)$$

$$V_{R_3}||R_2 = 10\left(\frac{12}{12+18}\right) = 10\left(\frac{12}{30}\right) = 10(0.4)$$

$$V_{R_3}||R_2 = 4\ \text{V}$$

4. Restore the circuit to its original form as shown in Fig. 11-11(*a*).
5. Remove V_1 and replace it with a wire. This places R_1 in parallel with R_3 as shown in Fig. 11-11(*c*).
6. Calculate the parallel resistance of R_1 and R_3.

Equation 11-12

$$R_1||R_3 = \frac{R}{N} = \frac{18}{2}$$

$$R_1||R_3 = 9\ \Omega$$

STUDENT *to* **STUDENT**

You'll need to use basic math and Ohm's law here.

7. Using the voltage divider relationship, calculate the voltage across the parallel equivalent of R_1 and R_3.

Equation 11-13

$$V_{R_1} \| R_3 = V_2\left(\frac{R_1 \| R_3}{(R_1 \| R_3) + R_2}\right)$$

$$V_{R_1} \| R_3 = 40\left(\frac{9}{9 + 36}\right) = 40\left(\frac{9}{45}\right) = 40(0.2)$$

$$V_{R_1} \| R_3 = 8 \text{ V}$$

8. Add the 4 volts from Eq. 11-11 to the 8 volts from Eq. 11-12 to obtain 12 volts. Twelve volts is the actual voltage across R_3 in the original circuit, which agrees with the value obtained in the mesh example.

The superposition procedure provides a relatively simple solution to certain types of network problems. It will work only if the components in the circuit are linear and **bilateral**. A bilateral component is one that conducts equally in both directions.

Thévenin's Theorem

The most commonly used network theorem is **Thévenin's theorem**, named after M. L. Thévenin, a French engineer. The primary function of Thévenin's theorem is to determine the current through a selected resistor somewhere in a network. Once that current is known, the voltage across the resistor may be calculated using Ohm's law. Thévenin's theorem states that any circuit, regardless of its complexity, can be reduced to a series circuit containing a single voltage source called **Thévenin's voltage** (V_{TH}), the internal resistance of the source called **Thévenin's resistance** (R_{TH}), and the resistor you wish to know, the current through which acts as a load. **Thévenin's circuit** is shown in Fig. 11-12. The remaining resistors are combined to form a single resistor that represents the internal resistance of the Thévenin voltage source.

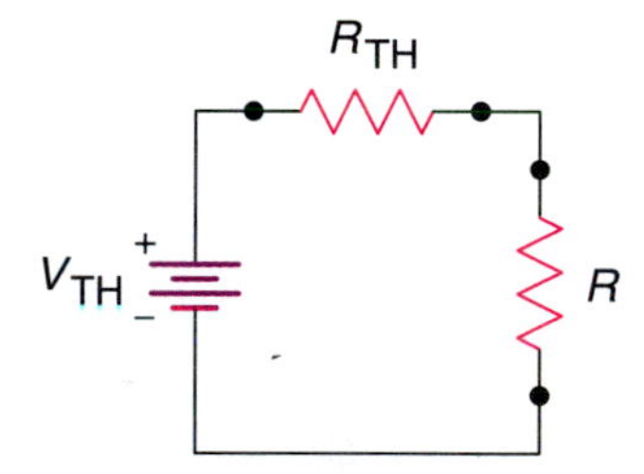

Figure 11-12 Thévenin's theorem reduces any circuit to a single source called Thévenin's voltage (V_{TH}), the internal resistance of that source called Thévenin's resistance (R_{TH}), and the resistor (R) through which you are attempting to find the current.

Converting a circuit to its Thévenin equivalent involves the following six steps:

1. Select the resistor you wish to know the current through and remove it from the circuit. This is done physically if dealing with an actual circuit and mentally if dealing with a schematic diagram.
2. Measure or calculate the voltage across the points in the circuit where the resistor was connected. This is Thévenin's voltage (V_{TH}).
3. Remove all voltage sources and replace them with a short circuit, taking care to never short an active source in an actual working circuit.
4. Measure or calculate the resistance at the points in the circuit where the resistor was removed. This is Thévenin's resistance (R_{TH}).
5. Create a series circuit with Thévenin's voltage as the source and Thévenin's resistance placed in series with the resistor that was removed. The polarity of V_{TH} must be defined in the Thévenin equivalent circuit to produce the same voltage polarity across R_{TH} as in the original circuit. This becomes Thévenin's circuit (see Fig. 11-12).
6. Now, calculate the current in Thévenin's circuit using Ohm's law. Although the circuit is different, the current through the chosen resistor is precisely the same as in the original circuit.

Although the procedure just described works well when the desired values can be measured, it may not work when the values must be calculated. The circuit transformations made while applying Thévenin's theorem must produce a circuit that can be solved using Ohm's law or the advantage of the theorem is lost. When dealing with a complex network, applying Thévenin's theorem sometimes produces yet another network. If that proves to be the case, a different resistor should be chosen as the key to the solution.

Thévenin's theorem is perhaps the most useful and common network theorem. It is so versatile that it may be used with any type of circuit, even those with only a single source and load. Using Thévenin's theorem to determine the current in such a simple circuit would be impractical because Ohm's law could establish the current in a single step. But being able to apply Thévenin's theorem to basic circuits provides an opportunity to use it in a known situation and become familiar with it.

Example 11-3 traces the Ohm's law procedure for determining the current through one of the resistors in the parallel branch of a basic series-parallel circuit. Then Example 11-4 applies Thévenin's theorem to the same problem and demonstrates that equivalent results are achieved.

Example 11-3 Determine the current through R_3 in Fig. 11-13. The steps for finding it using Ohm's law are as follows:

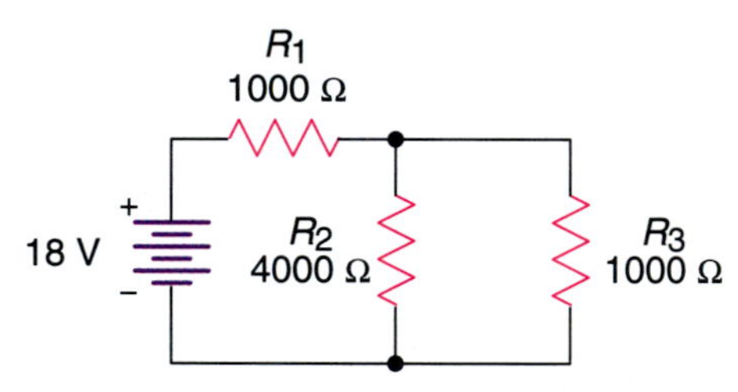

Figure 11-13 Thévenin's theorem can be as easy to use as Ohm's law in even a basic series-parallel circuit.

1. Find the parallel resistance of R_2 and R_3.

Equation 11-14

$$R_2 \| R_3 = \frac{R_2 R_3}{R_2 + R_3}$$

$$R_2 \| R_3 = \frac{4000 \times 1000}{4000 + 1000}$$

$$R_2 \| R_3 = \frac{4{,}000{,}000}{5000}$$

$$R_2 \| R_3 = 800\ \Omega$$

2. Find the total circuit resistance.

Equation 11-15

$$R_T = R_1 + (R_2 \| R_3)$$

$$R_T = 1000 + 800$$

$$R_T = 1800\ \Omega$$

3. Calculate the source current using Ohm's law.

Equation 11-16

$$I_S = \frac{V_S}{R_T}$$

$$I_S = \frac{18\ \text{V}}{1800\ \Omega}$$

$$I_S = 0.01\ \text{A}$$

$$I_S = 10\ \text{mA}$$

4. Find the voltage across R_1 using Ohm's law.

Equation 11-17

$$V_{R_1} = I_{R_1}R_1$$

$$V_{R_1} = 0.01\ \text{A} \times 1000\ \Omega$$

$$V_{R_1} = 10\ \text{V}$$

5. Using Kirchhoff's voltage law, determine the voltage across R_1.

Equation 11-18

$$V_{R_2} = V_{R_3} = V_S - V_{R_1} = 18 - 10 = 8\ \text{V}$$

6. Ohm's law may now be applied to determine the current through R_3.

Equation 11-19

$$I_{R_3} = \frac{V_{R_3}}{R_3} = \frac{8\ V}{1000\ \Omega} = 0.008\ \text{A} = 8\ \text{mA}$$

The Ohm's law solution required six steps to determine the current through R_3. Each step involved a simple equation using basic math. Thévenin's theorem takes this type of simplicity to network solutions. Example 11-4 shows how Thévenin's theorem can be applied to the problem in Example 11-3.

Example 11-4

1. Remove R_3 from the circuit as shown in Fig. 11-14(a) on the next page and determine the voltage across the open terminals.

   The voltage across the open terminals left by the removal of R_3 is found using the voltage divider relationship. This voltage is Thévenin's voltage.

Equation 11-20

$$V_{\text{TH}} = V_s\frac{R_2}{R_1 + R_2}$$

$$V_{\text{TH}} = 18\left(\frac{4000}{4000 + 1000}\right)$$

$$V_{\text{TH}} = 18\left(\frac{4000}{5000}\right)$$

$$V_{\text{TH}} = 18(0.8)$$

$$V_{\text{TH}} = 14.4\ \text{V}$$

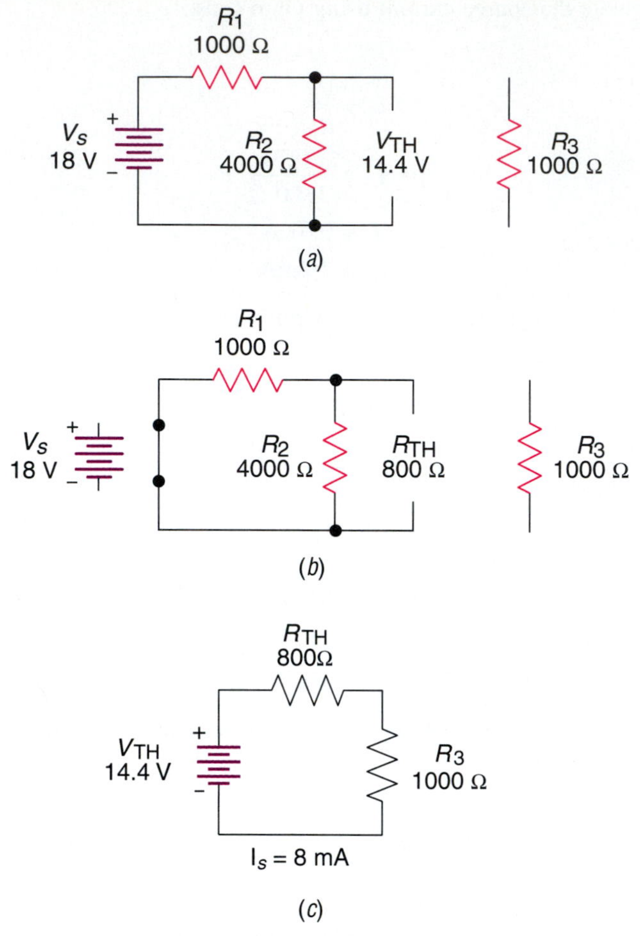

Figure 11-14 Applying Thévenin's theorem. (*a*) Remove R_3 and determine Thévenin's voltage. (*b*) Determine Thévenin's resistance. (*c*) Once Thévenin's circuit has been established, the last step is to use Ohm's law to determine the current through the resistor in question.

2. Remove the 18-V source and replace it with a wire as shown in Fig. 11-14(*b*). This places R_1 in parallel with R_2. The result is Thévenin's resistance as shown in Equation 11-21.

Equation 11-21

$$R_{TH} = R_1 \| R_2$$

$$R_{TH} = \frac{R_1 R_2}{R_1 + R_2}$$

$$R_{TH} = \frac{1000 \times 4000}{1000 + 4000}$$

$$R_{TH} = \frac{4{,}000{,}000}{5000}$$

$$R_{TH} = 800\ \Omega$$

3. Thévenin's voltage, Thévenin's resistance, and R_3 are now used to form Thévenin's circuit as shown in Fig. 11-14(*c*). The current in this circuit is found using Ohm's law.

Equation 11-22

$$I_{R_3} = \frac{V_{\text{TH}}}{R_{\text{TH}} + R_3}$$

$$I_{R_3} = \frac{14.4 \text{ V}}{800 + 1000}$$

$$I_{R_3} = \frac{14.4 \text{ V}}{1800\ \Omega}$$

$$I_{R_3} = 0.008 \text{ A or } 8 \text{ mA}$$

The current found using Thévenin's procedure is the exact current flowing through R_3 in the original circuit.

REVIEW QUIZ 11-2

1. Mesh analysis requires the use of ______________ equations that must be solved algebraically.
2. Superposition allows the parameters of networks with two sources to be determined using ______________ .
3. Thévenin's theorem is generally used to find the ______________ through a resistor somewhere in a network.

SUMMARY

Networks represent the most complicated type of electric circuit. They are mainly identified by their inability to be broken down into series and parallel groups, which makes it difficult to determine their circuit parameters using Ohm's law and the basic rules of circuit operation. In most cases only one electrical parameter is identified for each component in the network. This lack of information spawned the development of the network theorems and algebraic techniques to accomplish what Ohm's law can accomplish in simpler circuits.

Networks are frequently given names that resemble the arrangement of their components on schematic diagrams. These names are often letters of the English or Greek alphabets, such as tee, wye, delta, or pi. The names may also resemble objects, such as a ladder or bridge. Many networks are networks in name only until more than one voltage source is present. Several voltage sources can turn a relatively simple series-parallel circuit into a very complex network.

Determining the circuit parameters in a complicated network often requires the use of Kirchhoff's laws to establish the guidelines for an algebraic solution. Fortunately, some network theorems are based on rearranging the circuit, which results in an Ohm's law solution.

NEW VOCABULARY

balanced
bilateral
bridge
bridge circuit
bridge resistor
delta
double delta
ladder
mesh
mesh analysis
network
network theorems
pi
superposition
T network
Thévenin's circuit
Thévenin's resistance
Thévenin's theorem
Thévenin's voltage
Wheatstone bridge
wye network

Questions

1. What distinguishes a network from a series-parallel circuit?
2. How important is the value of the bridge resistor in a bridge network if there is no current flowing through it?

3. Under what condition is a bridge circuit balanced?
4. In a T network with series-opposing sources, is it possible for one of the sources to have zero current flowing through it? Explain your answer.
5. What is the purpose of a Wheatstone bridge?
6. Why is the mesh analysis theorem more difficult than the superposition theorem?

Problems

1. Determine the current through R_3 in Fig. 11-15 using Thévenin's theorem.
2. What value should resistor R_4 be to bring the bridge circuit of Fig. 11-16 into balance?

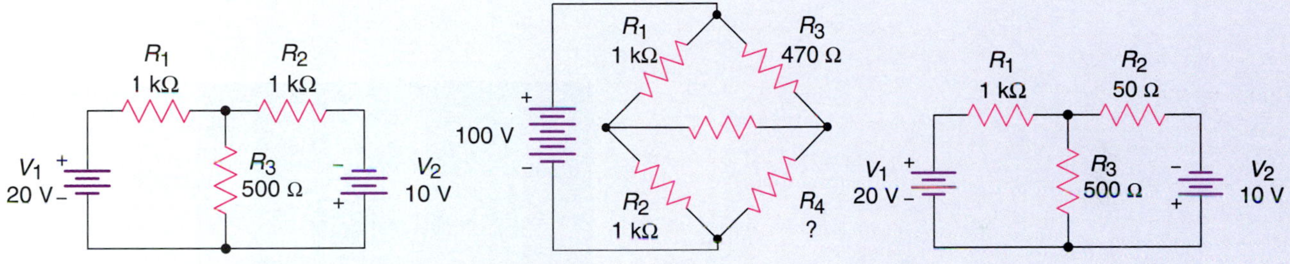

Figure 11-15 Circuit for Problem 1. **Figure 11-16** Circuit for Problem 2. **Figure 11-17** Circuit for Problem 4.

3. What is the current through the bridge resistor in Fig. 11-16 if R_4 is open?
4. What value resistor will cause zero current to flow through R_3 in Fig. 11-17?
5. Using the superposition theorem, determine the current through R_3 in Fig. 11-17 if R_2 is 50 Ω.

Critical Thinking

1. Assume R_2 in Fig. 11-17 to be 100 Ω. Can the superposition theorem be used to find the voltage across R_3 if R_3 is open? Explain your answer.
2. Why is the value of the bridge resistor unimportant when the bridge is balanced?
3. In the T network of Fig. 11-17, is there any combination of voltage and standard resistance values that will cause the current through either source to drop to zero? Explain your answer.
4. If Thévenin's theorem is applied to a resistor in a network and the simplified circuit remains a network, could the process work by Thévenizing the circuit using another resistor? Explain your answer.

Answers to Review Quizzes

11-1

1. Ohm's law
2. double delta
3. resistance
4. balanced
5. identical

11-2

1. simultaneous
2. Ohm's law
3. current

Chapter 12

Magnetism

ABOUT THIS CHAPTER

The observable characteristics of magnets as they are found in nature were all that was known about magnetism for almost 3000 years. That changed with Oersted's discovery that electricity and magnetism were directly related. His observations and the work of others, inspired by his accidental discovery, represent some of the most important contributions to electrical knowledge.

After completing this chapter you will be able to:

- Determine why some materials are magnetic and others are not.
- Explain how magnetism is created from electricity.

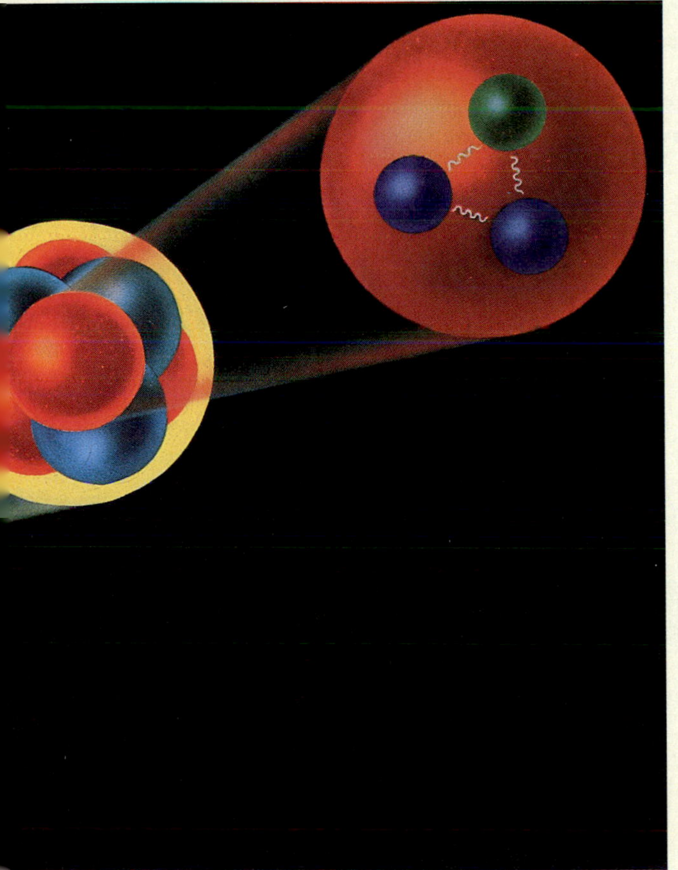

This chapter discusses the characteristics of magnetic fields, along with their units of measure. Permanent and temporary magnets are introduced and the differences between them are examined. Moreover, electromagnetism is viewed from the perspective of wires and coils of wire without the benefit of magnetic cores. Finally, the elements are divided into five magnetic classifications, revealing that materials are not simply magnetic or nonmagnetic.

- ▽ Describe units of magnetic flux and flux density.
- ▽ Identify five classifications of magnetic materials.

ELECTRONIC FACTS

In Europe magnetite became known as **lodestone**. *Lode* is an Anglo-Saxon word that means "lead." Lodestone literally means lead stone. It was so named because it was used to lead the way.

12-1 Natural Magnetism

An iron oxide called **magnetite** (Fe_3O_4) is the only **magnet** that occurs naturally on Earth. Magnetite was discovered more than 3000 years ago near Magnesia, an ancient city in Asia Minor, as a black stone that could attract iron. The stone was called *magnes lapis*, which means Magnesian stone. The word *magnet* is derived from *Magnesian*.

When suspended from a thread, or otherwise allowed to move freely, magnetite will normally align itself in a north-south direction. The direction-seeking property of magnetite was discovered in China as early as A.D. 85. Chinese experimenters found that iron needles stroked with magnetite inherited its magnetic properties. These needles were used by the Chinese to make reliable compasses. Knowledge of their discovery soon spread throughout the known world.

Although numerous improvements in compass design were achieved, few attempts were made to discover the nature of magnetism until the thirteenth century. A crusader named Petrus Peregrinus de Maricourt performed numerous experiments with magnetite. He discovered that small bits of iron sprinkled onto pieces of magnetite were attracted to a pair of points that he called **poles**. He named the magnetic poles *north* and *south* after the poles of the celestial sphere. He also discovered the **law of magnetic poles**, which states that *like poles repel* and *opposite poles attract*.

Interest in magnetism remained dormant until 1600 when Sir William Gilbert (1540–1603), an English physician, published a book detailing his experiments with a spherically shaped lodestone. He called the book *De Magnete* (*About Magnets*), and in it he described the Earth as a giant magnet. According to Gilbert, it was the Earth's natural magnetism interacting with the magnetism of the compass needles that gave the needles their directional properties. He concluded that the Earth's geographic north pole was a **south magnetic pole** and that its geographic south pole was a **north magnetic pole**. Gilbert's book assigned the first physical properties to the Earth since it was decided that the Earth was round. The study of magnetism would go no further than terrestrial magnetism, lodestone, and compass needles for another 219 years.

REVIEW QUIZ 12-1

1. The two common names applied to the Earth's only natural magnetic material are ______________ and ______________ .
2. The names of the two magnetic poles are ______________ and ______________.
3. The first practical use for magnets was a(n) ______________ .
4. The law of magnetic poles states that ______________ poles attract and ______________ poles repel.

12-2 Electromagnetism

One of the most important electrical discoveries was made accidentally toward the end of 1819. While lecturing to his class, Hans Christian Oersted (1777–1851), a Danish professor of physics, placed a wire over the top of a magnetic compass and connected the ends of the wire to a voltaic battery as shown in Fig. 12-1. He was attempting to demonstrate that an electric current had no effect on the compass. Oersted was amazed to see the compass needle swing perpendicular to the wire. He then reversed the battery connections, causing the current to flow through the wire in the opposite direction. The needle swung around and completely reversed its direction.

Oersted had tried the experiment many times previously, but always started by placing the wire perpendicular to the compass needle. He must have also kept the battery polarities the same each time he performed the experiment because, had he reversed the polarity of the applied voltage, he would have observed a reversal of the compass needle.

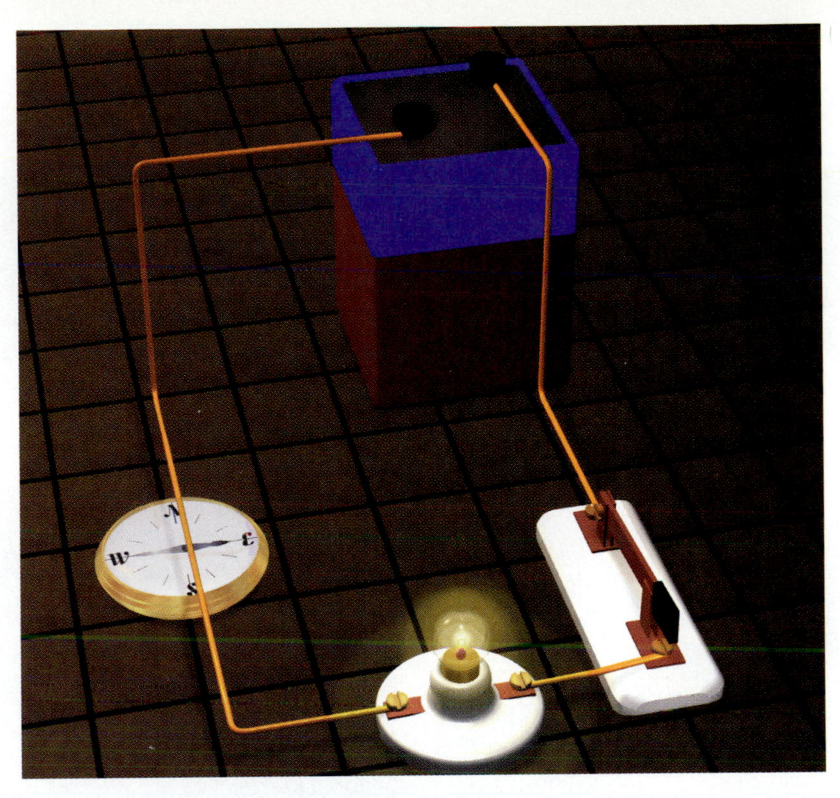

Figure 12-1 Oersted discovered that a current-carrying wire will deflect a compass needle at a right angle to the wire.

About three months after his discovery of electromagnetism, Oersted used a stronger battery and expanded on his original experiment. He placed various materials between the wire and the compass to see if the magnetic strength was affected. He found that a magnetic field passes through virtually any material without loss. He also found that when the wire was held below the compass, the needle deflected oppositely to what it did when the wire was above the compass.

Oersted published the results of his experiments on July 21, 1820. His brief four-page account was read enthusiastically by the French experimental physicist André Ampère, who guessed that two parallel wires carrying current would affect each other. Ampère developed several experiments to verify his assumptions. From those experiments, he concluded that two parallel wires attract each other when the currents through them flow in the same direction and repel each other when they flow in opposite directions.

Between 1822 and 1823, Ampère published a complete set of formulas by which the forces between two parallel current-carrying wires could be calculated. The measured forces between conductors were accurately predicted by Ampère's formulas, thus erasing any doubt that electricity and magnetism were inseparable. The *ampere* unit of electrical current is based on Ampère's experiments and remains a lasting tribute to his contributions to electrical science.

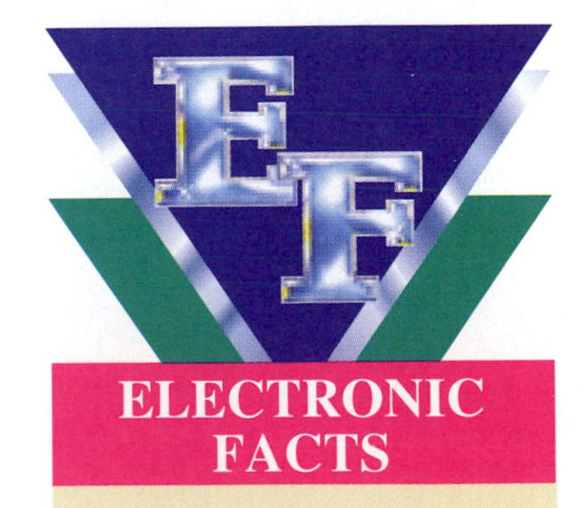

Oersted's results were first published in Latin. A French translation was published in August 1820 and an English translation in October 1820.

REVIEW QUIZ 12-2

1. The fact that a current-carrying wire produces a magnetic field was discovered by ______________ .
2. ______________ was the first to calculate the force between two current-carrying wires.

12-3 Natural and Electromagnetic Parities

Figure 12-2 De la Rive discovered that a current-carrying coil behaved like a compass needle when it was allowed to move freely.

Oersted's discovery of a link between electricity and magnetism, coupled with Ampère's formulas, caused many to wonder whether electromagnetism and natural magnetism were somehow the same. Auguste Arthur de la Rive (1801–1873), a Swiss physicist who lived in Paris, devised a clever experiment to show that electromagnetism behaved in the same way as the natural magnetism of lodestone. He pierced a cork disk with a piece of copper and a piece of zinc, being careful that the two metals did not touch. He then connected a helical coil of wire between the copper and zinc. The coil would align itself in a north-south direction when the assembly was floated in a bath of dilute sulfuric acid as shown in Fig. 12-2. The copper, zinc, and sulfuric acid formed a simple battery, creating a current through the coil that caused it to become a magnet with distinct poles at each end. The behavior of the energized coil was identical to lodestone and magnetized needles.

In an article[1] dated September 11, 1821, the English physicist Michael Faraday added his opinion to the growing consensus that electromagnetism and natural magnetism are one in the same. Faraday wrote, "Thus the phenomena of a helix, or a solid cylinder of spiral silked wire, are reduced to the simple revolution of the magnetic pole round the connecting wire of the battery, and its resemblance to a magnet is so great that the strongest presumption arises in the mind that they both owe their powers, as M. Ampère has stated, to the same cause."

NOTES

[1]Published in the *Quarterly Journal of Science,* Vol. XII, 1822, p. 74.

REVIEW QUIZ 12-3

1. Oersted's major contribution to electrical science was that an electric current creates a(n) ______________ .
2. The magnetic field produced by a natural magnet and an electromagnet are ______________.

12-4 Modeling the Magnetic Field

Faraday conducted many experiments with magnetism and is responsible for both magnetic and electric fields being described as lines of force. On October 28, 1831, Faraday referred for the first time to "magnetic curves." In his own words: "By curves I mean lines of magnetic forces which would be depicted by iron filings." Faraday had placed a magnet under a piece of stiff paper and sprinkled iron filings over it. The filings formed a pattern around the magnet that resembled lines. Gently tapping the paper as the filings were sprinkled clarified the iron filing signature of the normally invisible magnetic field. Under the influence of the magnet, the individual iron filings became magnetized. These miniature magnets attracted each other end to end and repelled each other side to side producing the appearance of lines.

Faraday detested mathematics, with its complex symbolism and intangible ideas. He was even uncomfortable with the concept of atoms and proposed that belief in the existence of matter was founded in the forces which it seemed to exert. He contended that it was these forces that should be studied, not matter itself. Paradoxically, Faraday's lines of force gave mathematicians the tool they needed to model and mathematically describe electric and magnetic fields.

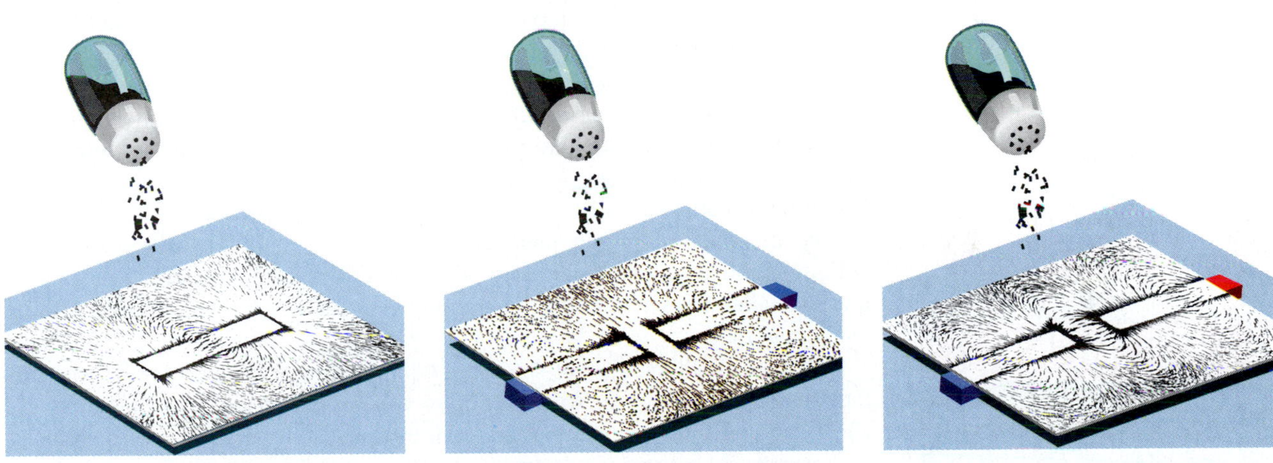

Figure 12-3 Faraday discovered that iron filings sprinkled onto a piece of paper placed over a magnet form patterns that look like lines.

Figure 12-3 demonstrates an expanded version of Faraday's iron filing experiment. When two magnets have their poles facing each other, the interaction of like poles and opposite poles can be seen in the iron filings. The filings confirm what can be sensed by holding two magnets and allowing them to interact; magnetic fields are very elastic and readily change their shape in response to another magnetic field.

REVIEW QUIZ 12-4

1. Faraday referred to a magnetic field as being composed of _______________ after performing his iron filing experiment.
2. Faraday believed that the important area of matter to study was not atoms but the _______________ associated with matter.

12-5 Permanent and Temporary Magnets

The Latin word for iron is ***ferrum***. Materials that contain iron, or have properties similar to iron, are called **ferrous**. The terms *ferrous* and *nonferrous* are frequently used in electronics to mean magnetic and nonmagnetic because they use the magnetic properties of iron as a standard by which other materials are judged. This is a restricted interpretation of a terminology that is widely used. Metal trades have a much broader usage of ferrous, and in reality many ferrous metals are nonmagnetic.

Temporary Magnets

Iron, nickel, and cobalt are ferrous metals that become magnetized after being stroked with a magnet or placed inside an electrified coil of wire. Magnets so created are called **temporary magnets**. Once magnetized, the magnetic forces within the metal dominate the internal electrical forces. If these temporary magnets are dropped to a hard surface or receive a strong mechanical shock, the magnetism may be lost as the molecules return to their normal alignments. Conversely, if any of these magnetic metals are struck with a hard object, such as a hammer, their molecules may align in the earth's magnetic field, causing them to become magnetized. The ability of a temporary magnet to remain magnetized can be improved by using hard metals. Some very soft irons will not retain any magnetism. Temporary magnets are also susceptible to external magnetic fields. These fields can weaken or redefine their magnetic properties.

No magnet is permanent under all environmental conditions because all magnetic materials become nonmagnetic at some critical temperature. That temperature is called the **curie point**, which may be well below room temperature for some materials.

A major manufacturer of soldering equipment uses the curie point to regulate the temperature of the soldering tip in a line of soldering irons they produce. Electrical contacts used to supply current to the heating element are held closed by a permanent magnet that is attracted to the base of the tip. The magnetic force acts against a spring that would open the contacts if the magnetism were not present. At a predetermined temperature, usually 600°, 700°, or 800° Fahrenheit—the tip base becomes nonmagnetic and the spring opens the contacts. This causes the tip to cool off and again become magnetic, closing the contacts and repeating the process.

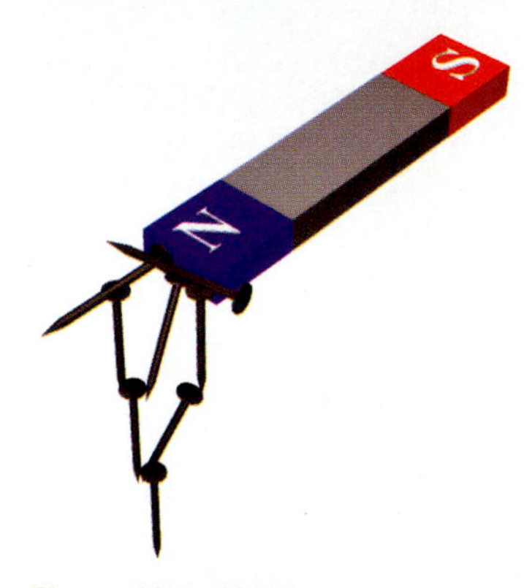

Figure 12-4 Objects, such as nails, can become magnetized through a process called magnetic induction.

Magnetic Induction

The attraction between a magnet and a ferrous material is mutual. Ferrous materials become magnets while under the influence of an external magnetic field. This can be demonstrated by dipping a magnet into a pile of nails. When the magnet is withdrawn, it will not only have nails attached to it, but those nails will have other nails attached to them. The process may repeat through several nails, as shown in Fig. 12-4, but eventually the magnetism in the nails will become too weak to support any additional weight. The fact that nails are attracted to nails implies that some or all of them have become magnets. They do so through a process called **magnetic induction**.

Magnetic induction is the process by which a material, in the presence of an external magnetic field, is magnetized. The atoms or molecules of magnetic materials behave as individual magnets and undergo a physical realignment in response to the external magnetism. This realignment sets up magnetic polarities in the material that are opposite to the poles of the influencing magnet. It is the opposite poles' facing each other that accounts for the mutual magnetic attraction.

Permanent Magnets

For magnets to find commercial usefulness they have to produce strong magnetic fields and be stable over a wide range of environmental conditions. Magnets with those properties are called **permanent magnets**. Nearly permanent magnets can be made by heating steel and allowing it to cool under the influence of a strong external magnetic field. The process was improved dramatically by using hard metal alloys, such as **alnico**, which was developed by General Electric in 1933. Alnico is an acronym for *al*uminum, *ni*ckel, and *co*balt. Permanent magnets retain their magnetism longer when a steel bar called a **keeper** is placed across their poles when the magnet is not in use.

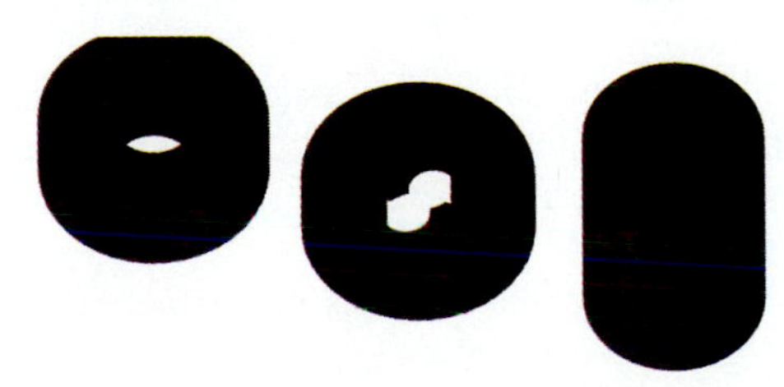

Figure 12-5 Ferrite magnets are available in a wide variety of shapes and sizes.

Permanent magnets are also made from certain **ferrites**. These ferrites are ferrous compounds ground into a fine powder. The powder is then fused into a ceramic base by heating it to a very high temperature while under the influence of a strong magnetic field. The use of magnetic ferrites allows magnets to be manufactured with multiple poles and in almost any shape. An example of the special shapes possible with ferrites is shown in Fig. 12-5. Ferrite magnets have replaced alnico magnets in most applications.

Magnetic Observations

Experimenting with magnets and objects made from steel quickly raises the question of whether magnets are attracted to steel or steel is attracted to magnets. The sensation of attraction depends greatly on the size and weight of the objects used in the experiment. A large magnet clearly seems to attract small pieces of steel, but the reverse can also seem true. A large piece of steel will seem to attract small magnets. If a magnet and a steel bar look alike, it may be difficult to identify which is the magnet using only those two objects and the sense of attraction. There is a tendency to assume magnetic interactions occur only at the ends of the bars. While that is true for the magnet, it is not true for the steel. The poles of a

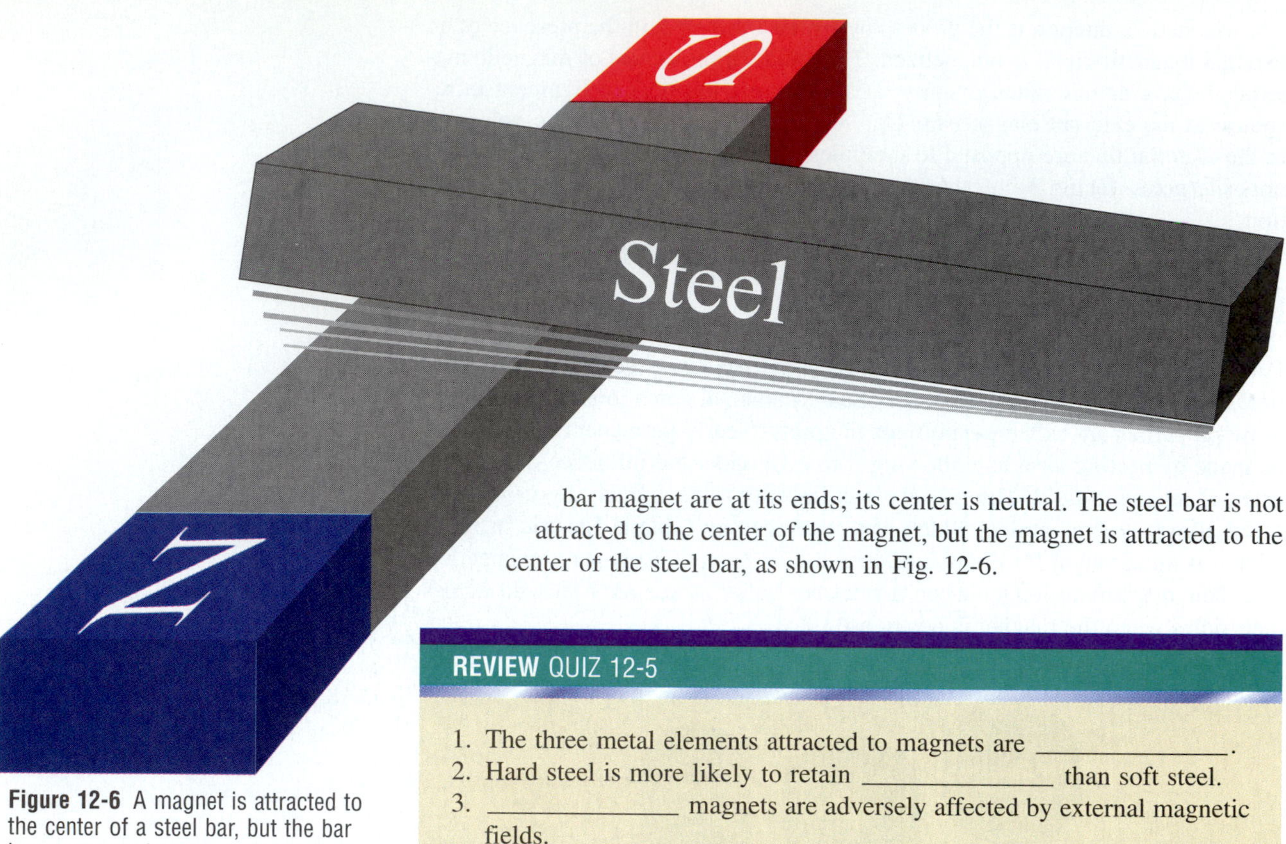

bar magnet are at its ends; its center is neutral. The steel bar is not attracted to the center of the magnet, but the magnet is attracted to the center of the steel bar, as shown in Fig. 12-6.

Figure 12-6 A magnet is attracted to the center of a steel bar, but the bar is not attracted to the center of the magnet.

REVIEW QUIZ 12-5

1. The three metal elements attracted to magnets are ______________.
2. Hard steel is more likely to retain ______________ than soft steel.
3. ______________ magnets are adversely affected by external magnetic fields.
4. Objects such as nails become magnets under the influence of a magnetic field through a process called ______________ .

12-6 Magnetic Flux (ϕ)

Magnetic flux is a lines-of-force representation of a magnetic field. The lines are assigned arrowheads that point away from the north magnetic pole and toward the south magnetic pole. The Greek letter ϕ (phi) identifies magnetic flux in mathematical equations and is often used to represent the word *flux*. The terms *magnetic lines of force*, *flux loops*, and *flux lines* are used interchangeably. Loops of magnetic flux are usually referred to as lines because a portion of the loop is always contained within a coil, in the body of a magnet, or inside the surface of a spinning electron, causing the loops to appear as lines.

Magnetic fields are created by the movement of electric charge. That movement may be along a straight path, a circular path, or the simple spin of an electron. Magnetic flux is directly proportional to the velocity of the charge, that is, how fast it is spinning.

The magnetic flux around a straight current-carrying wire or any charge moving along a straight path does not have identifiable north and south poles. The flux loops exist as concentric circles centered over the wire and are distributed evenly along its length. The directivity of the loops can be determined when the curled fingers of the left hand grasp the wire and the thumb points in the direction of the current. The fingers curled around the wire will point the same way as the assigned arrowheads as shown in Fig. 12-7.

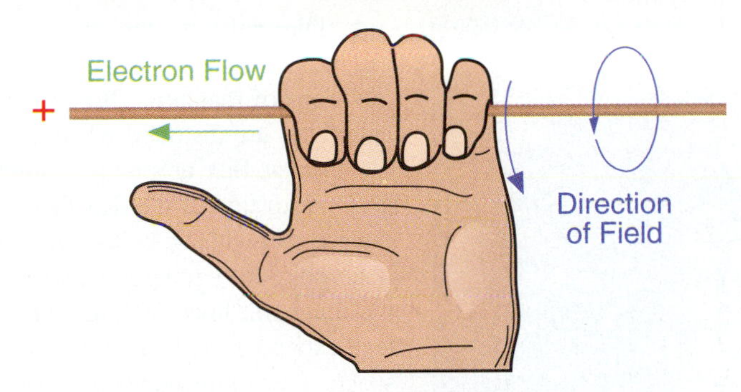

Figure 12-7 When your left hand is placed around a current-carrying wire, with your thumb pointing in the direction of the current flow, your fingers indicate the north-south directivity of the magnetic lines of force.

The closed loop nature of magnetic lines of force, collectively called magnetic flux, gives them a property not shared by electric or gravitational fields. Each individual loop of magnetic flux has an elasticity that causes it to contract inward toward its point of origin. In fact, if the electric charge creating the magnetic field should stop moving, the magnetic flux collapses to the point of vanishing. That is not to say that something has been converted into nothing. When magnetic lines of force are created, energy is consumed from the source that causes the charge to move. That energy is returned to the source when the charge stops moving. The magnetic field, like electric and gravitational fields, cannot be consumed.

Magnetic flux expands away from a current-carrying wire, regardless of its shape, because new flux lines are formed within old ones—like the layers of an onion. The creation of the new lines produces an outward pressure on the old ones, which results from the mutual repulsion between them. With a steady-state current, or uniform charge velocity, the magnetic field will achieve a balanced condition where it is neither expanding nor contracting. The field only expands when the current increases and only contracts when the current decreases. The magnetic expansion, as reflected in the increased flux value, is determined solely by the amount of current that can be made to flow. Therefore, there is no theoretical limit to the strength a magnetic field may achieve, although there are certainly some practical ones.

The shape of the magnetic field around an electrified coil of wire is very different from its shape when the wire is straight. The loops of magnetic flux expand to encompass some or all of the coil's turns. The magnetic flux that "leaks" from between the turns of the coil depends on the turns' spacing and the length of the coil. Increasing either one will increase the flux leakage. The length of a coil is determined by the number of turns, the size of the wire, the thickness of the insulation, and the spacing between the turns. Two coils with their turns wound as close together as possible may have different lengths and therefore different amounts of magnetic leakage.

The magnetic lines of force created by an individual turn of an electrified coil are confined within the turn because they must surround the moving charge that creates them. If the turns of a coil are wound tightly together and the coil is short, all of the magnetic lines of force are confined to the interior of the coil. Consequently, the lines of flux inside the coil are far more concentrated than those outside the coil, which causes them to become distorted from their normally circular shape.

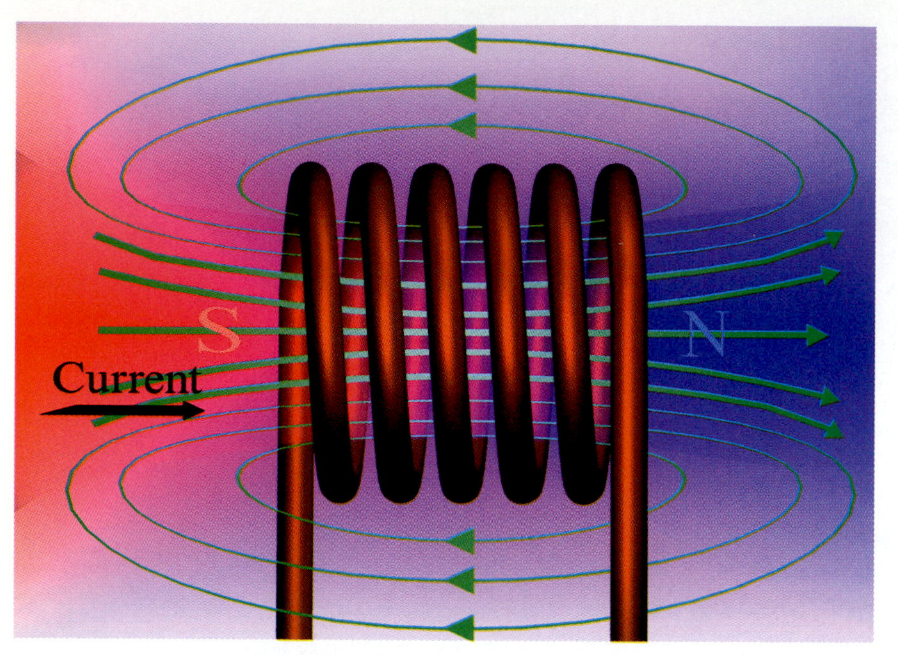

Figure 12-8 Magnetic poles occur where the magnetic flux undergoes the greatest changes in density.

Diagraming the magnetic field around an energized coil confirms that the loops of magnetic flux have remained intact, but are distorted when compared to the circular flux around a straight wire. It is the distortion of the flux that creates the illusion of magnetic poles. The north magnetic pole is identified at the point where the magnetic lines of force leave the coil and expand rapidly into the surrounding space. The south magnetic pole is identified at the point where the magnetic lines of force merge to enter the coil as shown in Fig. 12-8.

Left-Hand Polarity Rule

As already noted, the left hand may be used to determine the magnetic polarity of an electrified coil. When the coil is grasped by the left hand with the curled fingers pointing in the direction of electron flow, the extended thumb points toward the north magnetic pole as shown in Fig. 12-9.

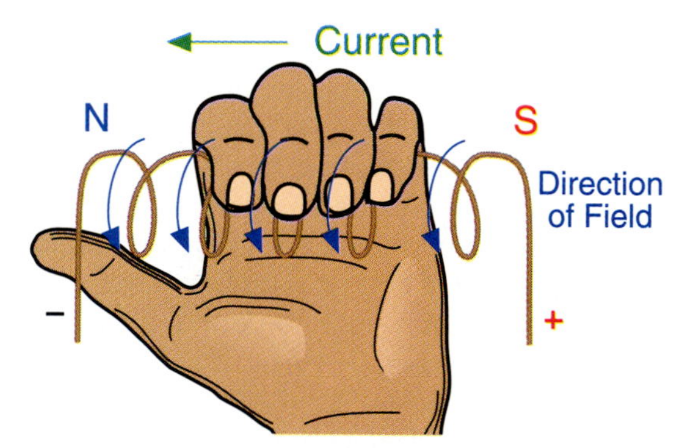

Figure12-9 The thumb of the left hand can be used to identify the north magnetic pole when the fingers are held around the coil pointing in the direction of electron flow.

Magnetic Polarities

Magnetic polarities do not exist in the same sense as electric polarities. The repulsion of like magnetic poles, and attraction of opposite magnetic poles, is explained using the directions assigned to the interacting loops of magnetic flux. Loops of magnetic flux with a similar directivity repel each other and those with an opposite directivity attract each other. This was the conclusion Ampère came to when he calculated the forces between two current-carrying wires.

Ampère observed that when the current through two parallel wires is in the same direction the wires attract each other. When the current through one of the wires is reversed, so that the currents flow in opposite directions, they repel each other. The attraction and repulsion cannot be explained using opposite and like magnetic poles because the circular flux around a current-carrying wire does not have identifiable poles.

The magnetic flux created by a permanent magnet is similar to that created by an electromagnet. In a permanent magnet, or any material that can be magnetized, the orbital paths of the valence electrons represent small current loops. The cumulative effect of countless millions of these circulating currents within a material can create magnetic fields of substantial strength.

The overall magnetic field surrounding a bar magnet is identical to that created by a current-carrying coil because, in effect, they are the same. The similarity becomes even greater when an iron core is inserted into the coil.

When two bar magnets are placed end to end with opposite magnetic poles touching each other, the flux loops of each magnet expand to encompass both. The number of magnetic lines of force has not changed. It is equal to the numeric sum of the lines from both magnets. The flux loops do not open and join with other loops to form larger ones. Each flux loop is created as a separate entity and once created will remain as such. The increased length of the magnet pair encourages a considerable increase in flux leakage through the sides of the magnets, making the sensation of magnetic strength less than expected.

REVIEW QUIZ 12-6

1. Magnetic flux is a(n) ______________ representation of a magnetic field.
2. Magnetic fields are created by the movement of ______________ .
3. Magnetic flux always exists as closed ______________ .

12-7 Magnetic Units

The units of magnetic flux are defined by three systems of measurement. The first is the English system, the second is the metric system, and the third is the Système International, abbreviated SI.

Maxwell

The unit for magnetic flux in the English system is the **maxwell**. One maxwell (Mx) is one line of force. The centimeter-gram-second (cgs) units of magnetic flux are the same as in the English system, the maxwell.

Weber

The meter-kilogram-second (mks) unit of magnetic flux is the **weber**. One weber (Wb) is equal to 100 million lines of force. The SI unit of magnetic flux is also the weber, as the SI system is based on mks dimensions.

Some manufacturers of magnetic materials continue to use cgs units. They do so because SI units are often too large to be practical. The SI system recognizes and accommodates the need to represent small values by prefixing its di-

ELECTRONIC FACTS

The maxwell (Mx) is named for the Scottish mathematical physicist James Clerk Maxwell. The weber is named for Wilhelm Weber, a German physicist.

mensions with a submultiple unit. Small quantities of magnetic flux are described using the microweber (μWb), which is one-millionth of a weber. One microweber is also equal to 100 maxwells (1 μWb = 100 Mx).

REVIEW QUIZ 12-7

1. A unit of magnetic flux that represents 100 million lines of force is the ____________.
2. The maxwell represents ____________ line(s) of magnetic flux.
3. The unit of magnetic flux in the English system is the ____________ .
4. Flux is another way of stating the quantity of ____________ .

12-8 Flux Density (B)

ELECTRONIC FACTS

The gauss is named for the German mathematician Karl F. Gauss. The tesla is named for the Yugoslav inventor Nikola Tesla.

The strength of a magnetic field is called **flux density**. It is gauged by the number of lines of force in a specified area. The symbol for flux density is a capital letter *B*.

Lines Per Square Inch

In the English, or American Standard system flux density is measured in *lines per square inch* and has no unit, as shown in Fig. 12-10(*a*).

Gauss

In the cgs system flux density is measured in **gauss** (G). One gauss is equal to one maxwell per square centimeter as shown in Fig. 12-10(*b*).

Tesla

The **tesla** (T) is the unit of flux density in both the mks and SI systems. One tesla is equal to one weber per square meter, or 10,000 gauss, as shown in Fig. 12-10(*c*).

The standard unit of magnetic flux is the tesla. However, the tesla is very large when compared to the flux density of commonly encountered magnetic

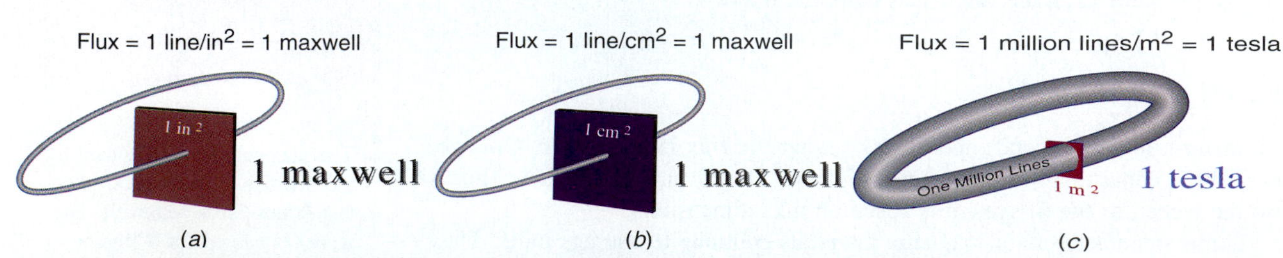

Figure 12-10 Flux density is a measure of the number of lines of force in a given area. (*a*) American Standard system, (*b*) cgs system, and (*c*) tesla used in both the mks and SI system.

fields. The magnetic field of the Earth is only about 5×10^{-5} T. The most powerful superconducting magnets have achieved 20 T. Manufacturers of magnetic materials often opt to specify flux density in gauss. For example, the flux density required to saturate the soft steels used in making magnetic cores for motors and transformers seldom exceeds 16 kilogauss. When the tesla is used, flux density is most likely to be expressed in scientific notation.

REVIEW QUIZ 12-8

1. The ______________ system has no unit for magnetic flux density.
2. The cgs unit of flux density is the ______________ .
3. The unit of flux density for 100 million lines of force contained in one square meter is the ______________ .

12-9 Classification of Magnetic Materials

Materials are classified magnetically into five general categories:

- Diamagnetic
- Ferromagnetic
- Paramagnetic
- Antiferromagnetic
- Ferrimagnetic

The **diamagnetic** elements, shown in Fig. 12-11, are beryllium (Be), copper (Cu), silver (Ag), gold (Au), germanium (Ge), and bismuth (Bi). These materials do

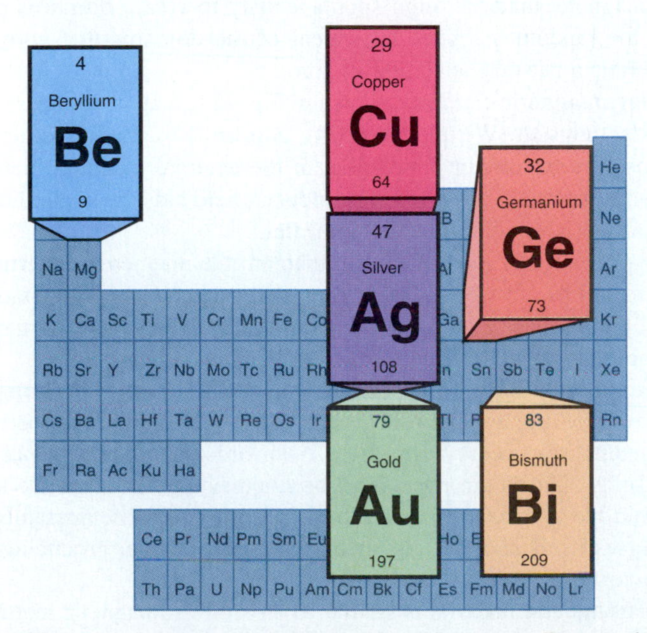

Figure 12-11 There are six elements on the periodic table that are diamagnetic.

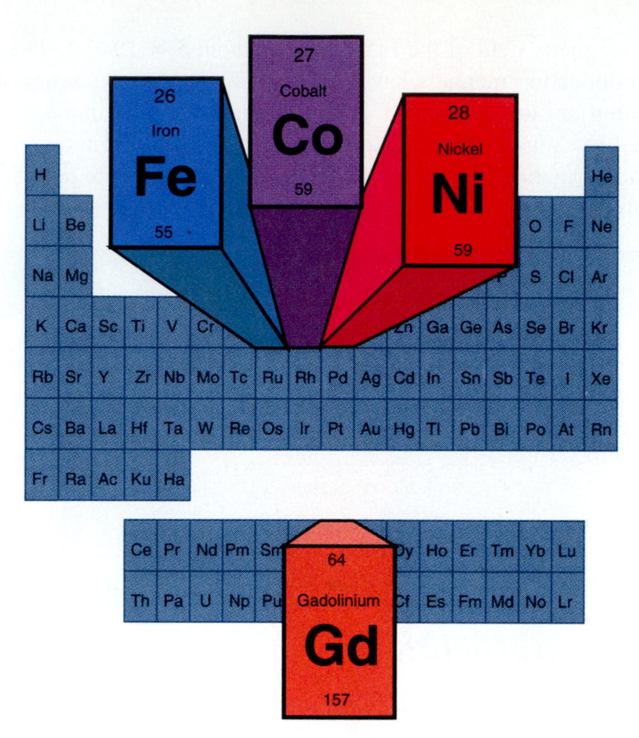

Figure 12-12 There are four elements on the periodic table that are ferromagnetic.

not normally produce any magnetism, but they are slightly repelled by a magnet because their magnetic properties change when under the influence of an external field. If these materials are allowed to rotate freely between the poles of a very strong magnet, they will turn at a right angle to the lines of force.

The **ferromagnetic** elements, shown in Fig. 12-12, are iron (Fe), nickel (Ni), cobalt (Co), and gadolinium (Gd). We normally think of these metals as being magnetic because magnetic effects are easily observable in them. Magnetic spins in a ferromagnetic material align spontaneously to create **domains** of magnetism. It is the random magnetic alignment of the domains that normally gives these materials a net magnetization of zero.

The **paramagnetic** elements, shown in Fig. 12-13, are manganese (Mn), aluminum (Al), tungsten (W), platinum (Pt), and tin (Sn). Unpaired electrons can align themselves parallel or antiparallel to the external magnetic field. The antiparallel alignments are preferred. The induced field aids the applied field so that the material is susceptible to the external field.

Being only slightly more magnetic than air, the magnetic properties of paramagnetic materials are hard to identify without careful experimentation. When suspended freely, these materials will align themselves in an external magnetic field, as the ferromagnetic materials do, but not as enthusiastically.

Chromium (Cr) is the only element that exhibits **antiferromagnetism** at room temperature. Figure 12-14 shows chromium's position on the periodic table. Antiferromagnetism occurs when there is an ordered nonparallel alignment of magnetic spins. Chromium appears to be nonmagnetic because the spin alignments within the atom cancel each other. At some critical temperature this condition reverses itself and the alignments become additive, producing paramagnetic behavior.

A **ferrimagnetic** material is similar to an antiferromagnetic material in that internal magnetic fields cancel each other. However, the magnetism produced by

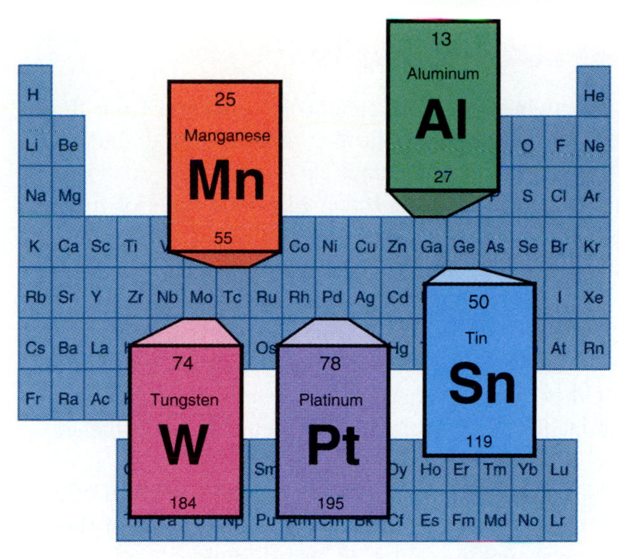

Figure 12-13 There are five elements on the periodic table that are paramagnetic.

the atoms is not in balance. This tends to give the material a slight paramagnetic behavior. A majority of ferrites fall into this category.

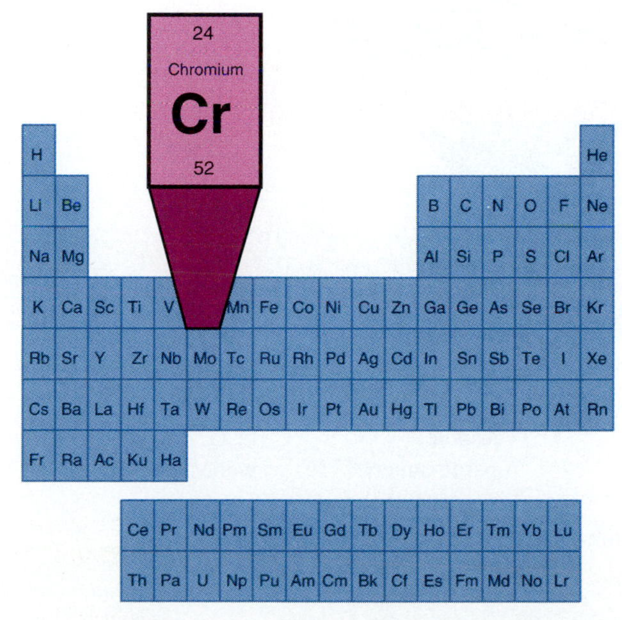

Figure 12-14 Only one element on the periodic table is antiferromagnetic.

REVIEW QUIZ 12-9

1. ________________ elements are easily attracted to magnets.
2. ________________ materials are slightly repelled by magnets.
3. ________________ materials are only slightly attracted to magnets.

SUMMARY

Magnetism and electricity are so intertwined that one cannot be separated from the other. Like electric fields, magnetic fields have two polarities that both repel and attract. Unlike the electric field, the poles of a magnet cannot exist separately. Also like electric fields, magnetic fields are modeled as lines of force. Magnetic lines of force always exist as closed loops, where electric lines of force are open ended.

The magnetic properties of materials are determined by the orbital paths of their valence electrons. These orbital paths fit the charge-in-motion definition of electric current, so in a sense all magnetism is electromagnetism. Permanent magnets have molecular alignments that support electron orbits that circulate in the same direction and have the same general orientation. Other materials that are considered to be magnetic can only support a temporary realignment of their molecules while under the influence of an external magnetic field.

The amount of magnetism is measured in lines of flux. The perceived strength of a magnet is determined by the number of flux lines in a given area. The concepts of flux and flux density are similar to those presented in relation to electric fields.

NEW VOCABULARY

alnico
antiferromagnetism
curie point
diamagnetic
domains
ferrimagnetic
ferrites
ferromagnetic
ferrous
ferrum
flux density
gauss
keeper
law of magnetic poles
lodestone
magnet
magnetic flux
magnetic induction
magnetite
maxwell
north magnetic pole
paramagnetic
permanent magnets
poles
south magnetic pole
temporary magnets
tesla
weber

Questions

1. In the field of electronics, what distinguishes a ferrous material from a nonferrous one?
2. What should be placed across a magnet to preserve its magnetism?

3. Into what five magnetic classifications are elements divided?
4. Do the terms *magnetic lines of force*, *flux loops*, and *flux lines* have the same meaning?
5. What is the relationship between the current through a conductor and the magnetic field surrounding it?
6. Is the magnetic field created by a permanent magnet any different from that created by an electromagnet? Explain your answer.
7. How do loops of magnetic flux with the same directivity affect each other?
8. What is the SI unit for flux density?
9. How many lines of force are represented by one maxwell?
10. Will soft iron retain magnetism very well?

Critical Thinking

1. Are there any magnetic monopoles?
2. Why can't north and south magnetic poles be identified around a straight wire?
3. Is magnetism always present in iron, nickel, and cobalt? If so, why?
4. How is the magnetic polarity of a current-carrying coil determined using the left hand?
5. What is the difference between flux and flux density?

Answers to Review Quizzes

12-1
1. magnetite, lodestone
2. north, south
3. compass
4. opposite, like

12-2
1. Oersted
2. Ampère

12-3
1. magnetic field
2. identical

12-4
1. lines of force
2. forces

12-5
1. iron, nickel, and cobalt
2. magnetism
3. temporary
4. magnetic induction

12-6
1. lines of force
2. electric charge
3. loops

12-7
1. weber
2. one
3. maxwell
4. magnetism

12-8
1. English
2. gauss
3. tesla

12-9
1. ferromagnetic
2. diamagnetic
3. paramagnetic

Chapter 13

Magnetic Circuits and Devices

ABOUT THIS CHAPTER

A discussion of electromagnetic units begins this chapter; these units are used to describe the magnetic circuits created when magnetic fields are channeled through ferrous materials. Magnetic circuits have properties similar to Ohm's law. These properties are often referred to as *Ohm's law for magnetism.*

The interaction of magnetic fields is often used to create motion. This magnetic motor force is discussed us-

After completing this chapter you will be able to:

- Identify and describe the magnetic units associated with ferrous materials.
- Describe the basics of magnetic motor force and identify some of its common applications.

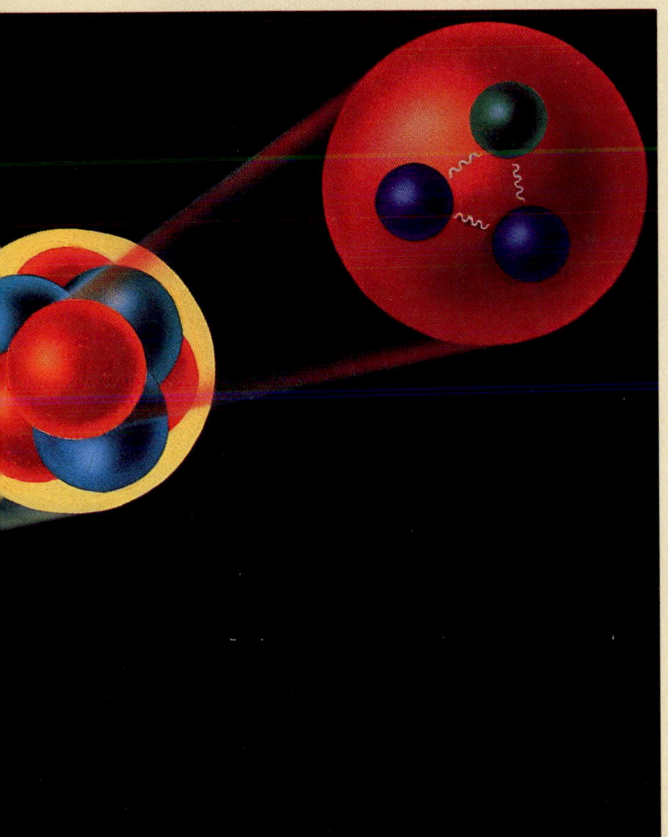

ing common electrical devices as examples. Magnetic fields also create motion at the molecular level in ferrous materials. This molecular motion can produce heat because of friction or change the length of a ferrous material as the molecules alternately align in one direction and then the other.

- ▽ Discuss the effects magnetic fields have upon the molecules of a ferrous material.
- ▽ State the operating principles of electromechanical devices.

13-1 Electromagnetic Units

The magnetic fields created by electric currents flowing through coils of wire have their own set of electromagnetic units. These units are based on the physical dimensions of the coils and the amount of current flowing through them.

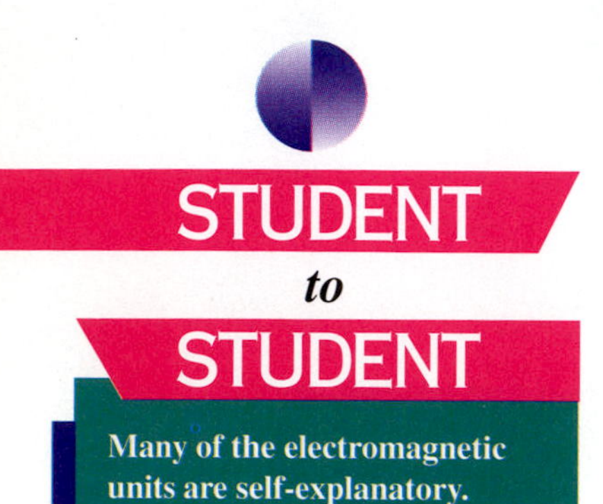

Magnetomotive Force ($\mathscr{F}$)

Current-carrying coils are sometimes called **electromagnets**. The magnetizing force created by an electromagnet is called **magnetomotive force** and is abbreviated *mmf*. Magnetomotive force is often represented by a stylized capital F ($\mathscr{F}$). It is determined by multiplying the current (I) through a coil by its number of turns of wire (N) as shown in Eq. 13-1.

Equation 13-1

$$\mathscr{F} = N \times I = \text{magnetomotive force}$$

The formula for magnetomotive force may seem incomplete. It does not include the diameter of the coil, the length of the coil, or the number of layers of wire. However, those things do not alter the relative flux produced by a coil's individual turns. The magnetic flux created by a single turn is dependent on the current through the coil. Since each turn carries the same current, the overall flux is proportional to the number of turns in the coil; two turns will produce twice as much flux as one turn. Equation 13-1 is based on the magnetic potential of the coil, not the practical restrictions imposed by its physical dimensions.

Ampere-turn

The mks and SI systems use the compound unit of **ampere-turn** (A·t) to represent magnetomotive force. In the SI system ampere-turn is abbreviated with a capital *A*. The use of that abbreviation can be confusing because it is more commonly used to abbreviate the word "ampere."

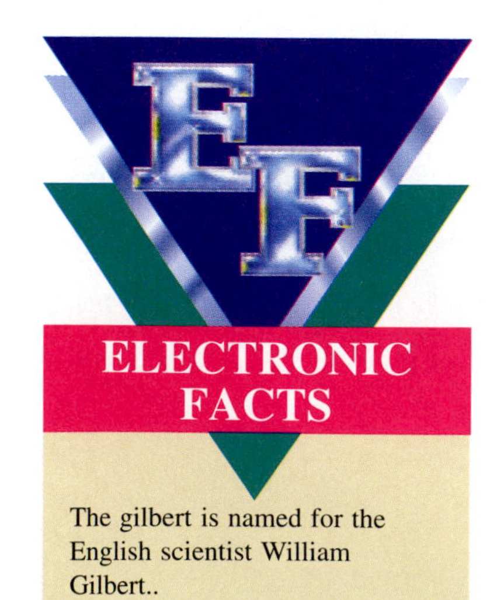

Gilbert

The cgs unit for magnetomotive force is the **gilbert** (Gb). One gilbert is equal to 0.796 ($10/4\pi$) ampere-turns.

Field Intensity (H)

The magnetic **field intensity** created by a coil is measured at its center, where the flux density is greatest. Field intensity depends on the coil's length. Long coils do not produce as much field intensity as short coils that have the same number of turns. Consequently, the length of a coil is an important part of the field intensity equation. That equation is an extension of the one used to determine magnetomotive force. In the SI system field intensity is equal to magnetomotive force divided by the length of the coil in meters, as shown in Eq. 13-2. There is no distinct unit for field intensity; instead, it is measured in ampere-turns/meter and is represented by the capital letter H.

Equation 13-2

$$H = \frac{\mathcal{F}}{l}$$

$$= \frac{\textit{magnetomotive force}}{\textit{length}} = \frac{N \times I}{l} = \frac{\textit{ampere-turns}}{\textit{length}\ \text{(meters)}} = \textit{field intensity}\ \text{(SI)}$$

If a coil has an iron core, l represents the length of the core instead of the length of the coil because l is actually the distance between the magnetic poles. The use of a magnetic core can extend those poles well beyond the limits of the coil as shown in Fig. 13-1.

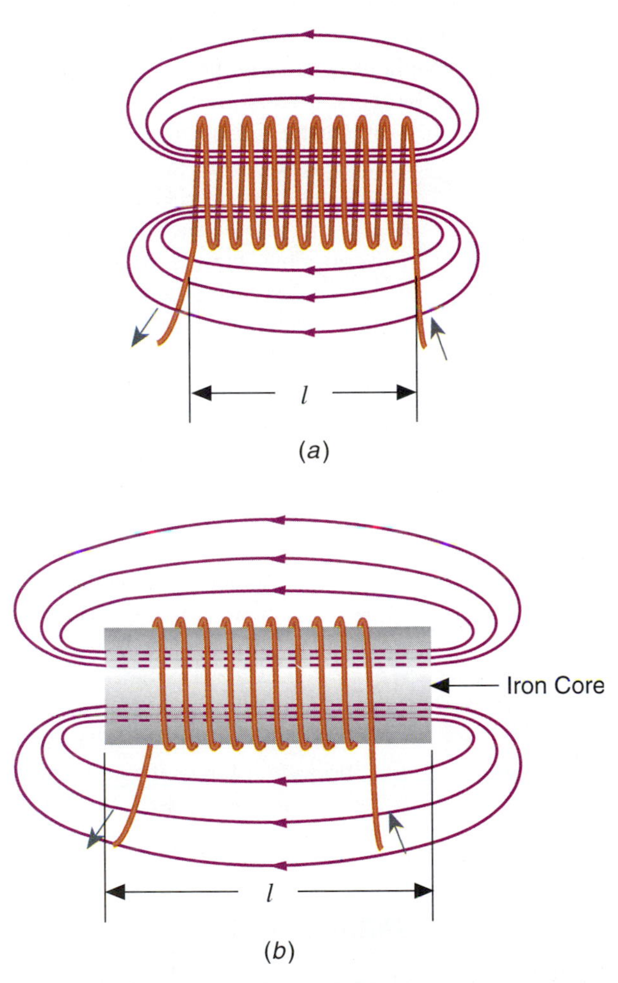

Figure 13-1 If a magnetic core is present inside a coil, the magnetic length of the coil is equal to the length of the core. Note the difference in length in (*a*) and (*b*).

Oersted

The cgs unit of field intensity is the **oersted**, abbreviated *Oe*. The oersted is one gilbert per centimeter. Equation 13-3 shows the basic mathematical relationships used to calculate the field intensity in oersteds.

Equation 13-3

$$H \text{ (oersteds)} = \frac{\textit{magnetomotive force} \text{ (gilberts)}}{\textit{length} \text{ (centimeters)}} = \frac{0.796 \textit{ ampere-turns}}{l\text{(cm)}} = \textit{field intensity} \text{ (cgs)}$$

The Effects of a Magnetic Core

Adding an iron core to a coil can increase its magnetic flux thousands of times without increasing the coil's current. Oersted's compass experiment has conclusively proved that electric current creates magnetic fields. Since magnetic lines of force always exist in loops that surround a moving charge, an electrified coil must confine a portion of each loop to the center of the coil. Thus, the overall increase in a coil's magnetic flux owing to the addition of an iron core cannot be accounted for by concentrating the flux produced by the coil because it was already within the area being sampled, as shown in Fig. 13-2.

The additional magnetic flux is always present in the iron. It is not normally noticed because it is created by individual atoms whose random alignments cancel their magnetic effects. The magnetomotive force created by a coil brings the magnetic polarities of the atoms into a common alignment. That magnetomotive force is sometimes called a **coercive force** because it forces the atoms to align in opposition to the patterns they would normally form.

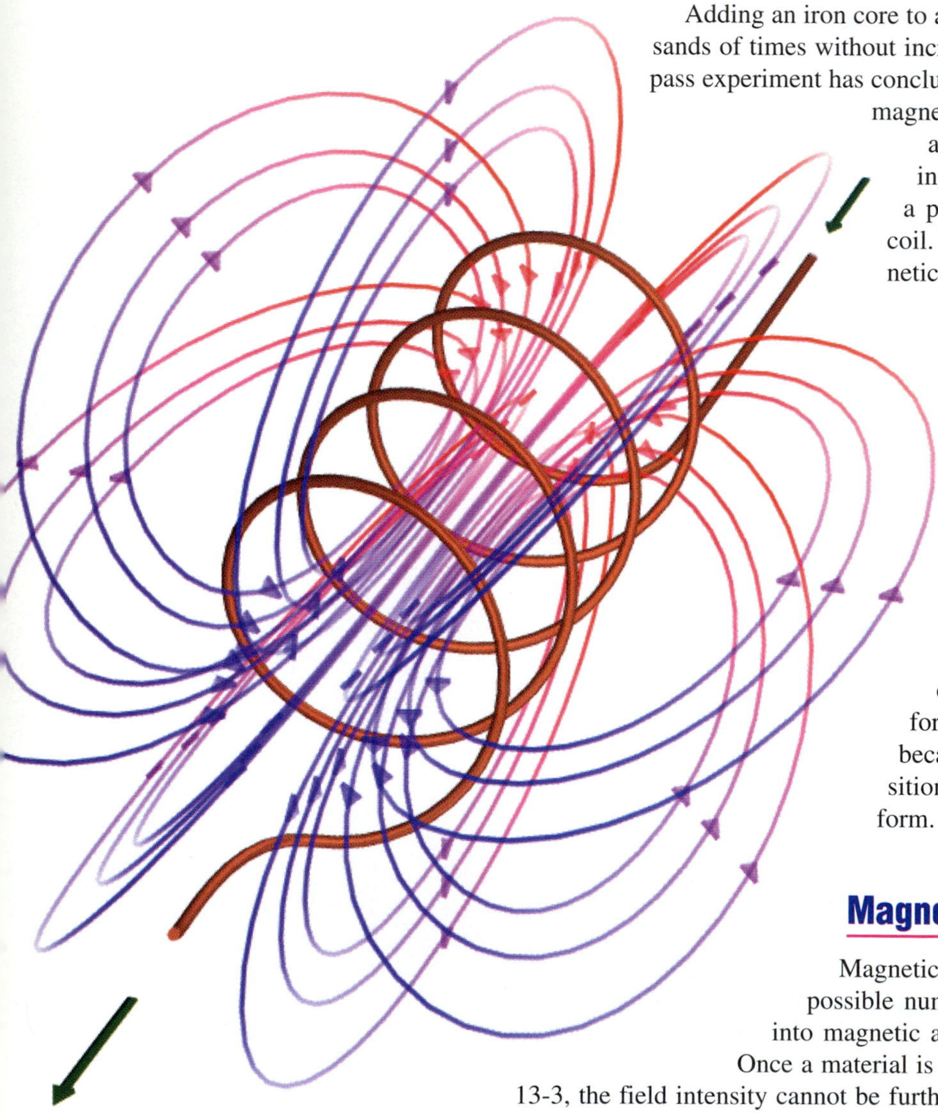

Magnetic Saturation

Magnetic **saturation** occurs when the maximum possible number of atoms or molecules are brought into magnetic alignment by an external magnetic field. Once a material is magnetically saturated, as shown in Fig. 13-3, the field intensity cannot be further increased.

Figure 13-2 All of the lines of force created by an electrified coil are confined to the center of the coil.

Permeability (μ)

Materials that concentrate magnetic fields have a high **permeability**. To understand the effect of permeability, think about what happens when a towel is dipped

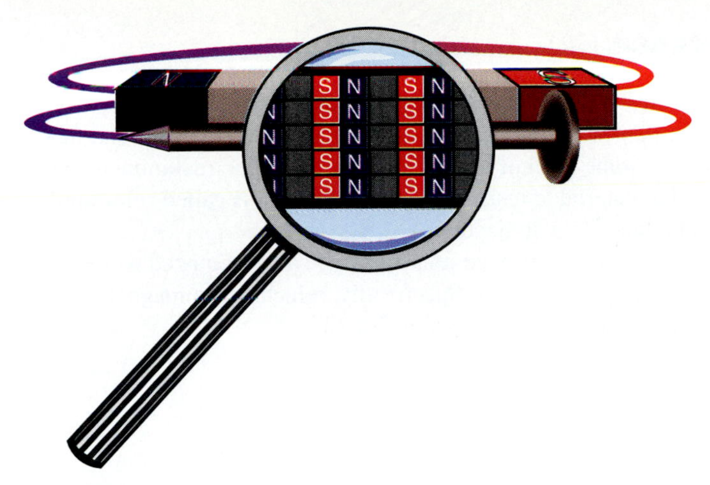

Figure 13-3 When all the molecules of a magnetic material are magnetically aligned, the material is magnetically saturated.

into water. Towels are very permeable; water can easily exist in them. However, there is a limit to how much water towels can hold. That limit is defined by their saturation point.

Permeability is represented by the Greek letter μ (mu) and can be viewed in the following three ways:

1. The magnetic alignments of atoms or molecules in a permeable material result in an increase in magnetic flux. Therefore, permeable materials act as flux multipliers when they are used as magnetic cores for electrified coils.
2. Magnetic lines of force exist more easily in a permeable material than in a nonpermeable one. Permeable materials behave as magnetic lenses that concentrate magnetic lines of force into a smaller area.
3. The magnetic alignment created in a permeable core by an external magnetic field produces flux loops with an opposite directivity. Because magnetic flux loops with opposite directivities attract, the flux loops of the external field are drawn into the permeable material, where they can exist more easily. Permeable materials behave as magnetic conductors to external magnetic fields.

Permeability is similar to electrical conductance. Magnetic fields exist more easily in materials with a high permeability, just as electric currents flow more easily through materials with a high conductance. Permeability does not have a unit of measurement. It is the increase in the number of lines of force produced by a current-carrying coil after a magnetic core is inserted. Air is generally used as a reference and has a permeability of 1. For example, if a piece of steel that just matches the area of an air-core coil is inserted into the coil and its magnetic flux increases 500 times, the steel has a permeability of 500. Permeability may be calculated by dividing flux density (B) by field intensity (H) as shown in Eq. 13-4.

Equation 13-4

$$\mu = \frac{B}{H} = \frac{\textit{flux density}}{\textit{field intensity}} = \frac{Wb/m^2}{A \cdot t/m} = \frac{T}{A \cdot t/m} = \frac{G}{Oe} = \textit{permeability}$$

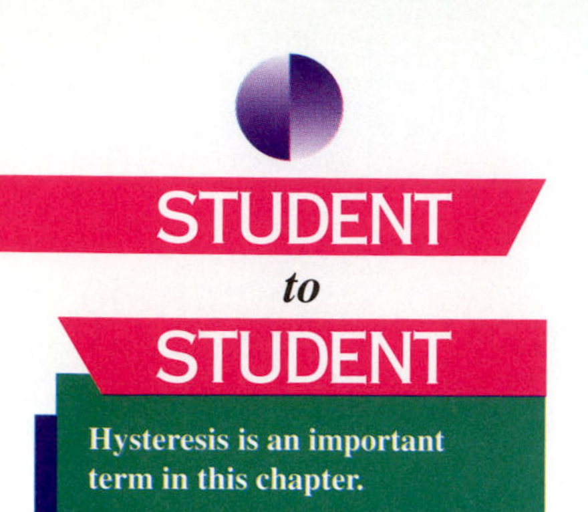

Reluctance ($\mathcal{R}$)

The molecules of a magnetic material are often reluctant to conform to an ordered alignment when exposed to an external magnetic field. The field must usually be above some level of intensity before magnetic realignment can occur. The property of a material to resist magnetic alignment is called **reluctance**. The symbol for reluctance is a stylized $\mathcal{R}$.

Reluctance does not have a unit, but it can be equated to the ratios of several other magnetic quantities. Specifically, reluctance is magnetomotive force divided by magnetic flux as shown in Eq. 13-5.

Equation 13-5

$$\mathcal{R} = \frac{\mathcal{F}}{\Phi} = \frac{\textit{magnetomotive force}}{\textit{magnetic flux}} = \frac{A \cdot t}{Wb} = \textit{reluctance}$$

Retentivity

When exposed to a magnetic field, a material may retain some magnetism after the magnetizing force is removed. There is residual magnetism because of a property called **retentivity**, which is the ability of a material to retain magnetism after being exposed to an external magnetic field. It is retentivity that allows materials, such as hard steel, to become temporary magnets. Retentivity is identified in a material by stating that it is **magnetically hard** or **magnetically soft**. The terms *soft* and *hard* can usually be taken literally with iron or steel. Soft steel does not hold residual magnetism when removed from the influence of an external magnetic field. Hard steels, such as those used to make screwdriver shafts, can retain magnetism for long periods of time. The terms *magnetically soft* and *magnetically hard* usually conform to the physical hardness of permeable metals, such as steel, but not necessarily with other magnetic materials. Ferrites are usually very hard, but many of them are magnetically soft and have virtually no retentivity.

Hysteresis ($\mathcal{H}$)

The word *hysteresis* means "to lag behind." In magnetic terms **hysteresis** refers to the hesitation of a material's atoms or molecules to follow the changes in an external magnetic field. The situation is similar to spraying water from a garden hose. If the nozzle of the hose is moved back and forth at even a moderate rate, the stream of water will lag behind the nozzle's movement. The inability of the water to track the movement of the nozzle, as shown in Fig. 13-4, fits the definition of hysteresis.

Figure 13-4 Hysteresis is the inability to follow in step with an action.

Hysteresis occurs in magnetic materials when they are exposed to a changing magnetic field because of reluctance and retentivity. Those properties are inherent in magnetically hard materials. The molecules of such materials are reluctant to align in a magnetic field until the strength of that field exceeds a certain level. That reluctance decreases as the magnetomotive force increases. The molecules conform to a common magnetic alignment as an increasing magnetomotive force moves them toward saturation.

Once saturation is reached, additional increases in magnetomotive force will not change the flux density. In magnetically hard materials the magnetic alignments achieved through saturation are not easily altered. Decreasing the magnetomotive force may have little or no effect on the magnetic state of the material.

The return to a random molecular alignment depends on the natural internal forces being stronger than those induced by the magnetomotive force. If the internal forces are weaker, then the retentivity of the material will delay the return to magnetic normalcy, if it occurs at all.

It is very common for the current through a coil to alternate directions at regular intervals. The reversal of the current means that the polarities of the magnetizing force created by the coil have reversed. This is a factor that even a magnetically hard material cannot ignore. Reluctantly, the molecules conform to the coercion of the reversing magnetomotive force. As the frequency of the current reversals is increased the molecules of the magnetic material will reach a point when they simply cannot respond.

Hysteresis can also occur in magnetically soft materials, but in these materials it does not become a factor until the magnetomotive force changes very rapidly. The molecules of any magnetic material will eventually fail to respond to a rapidly changing magnetomotive force. The critical rate of change for iron or steel is usually less than 25,000 times per second. The molecules of some members of the ferrite family can follow magnetic changes that occur many millions of times per second.

The effects of hysteresis are presented graphically in Fig. 13-5. The flux density (B), which was introduced in Chapter 12, is plotted against the field intensity (H) for a coil wound around an iron core. Flux density and field intensity are shown using both positive and negative values because the current through the coil may alternate at regular intervals. The negative values of B and H only serve to indicate that the magnetic polarities were reversed; they do not indicate that either one has a negative value. The graph is called a **B-H curve** and traces out a traditional **hysteresis loop**. The shape of the loop (⎎) is commonly employed to denote the hysteresis function in circuits and components.

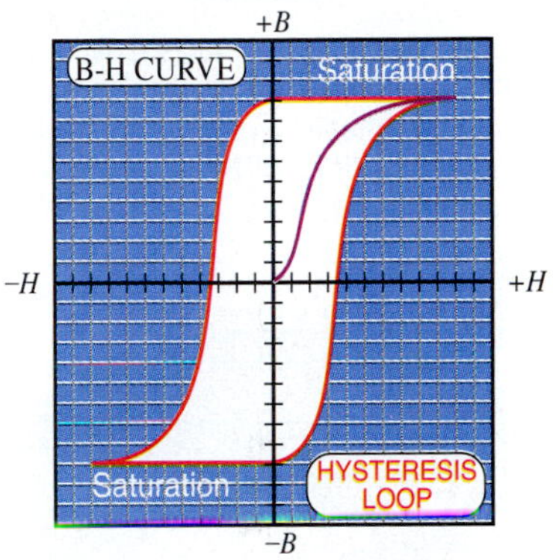

Figure 13-5 The hysteresis curve of a magnetic core identifies how easily the magnetization of the core can follow an external magnetic field. The flux density peaks as the core becomes saturated, thus creating the overall S shape.

REVIEW QUIZ 13-1

1. The number of turns in a coil multiplied by the current through the coil is called _______________ .
2. The field intensity created by a current-carrying coil is found by dividing magnetomotive force by the _______________ of the coil.
3. The cgs unit of magnetic force is the _______________ .
4. The hesitancy of the molecules in a magnetic material to align in a magnetic field is called _______________.

13-2 Magnetic Circuits

Magnetically permeable materials, such as iron, behave as magnetic conductors. This allows magnetic lines of force to be channeled through magnetic materials forming **magnetic circuits.**

A Closed Magnetic Circuit

If a wire is wound into a coil around a **toroidal** (doughnut-shaped) iron core and then a current is made to flow through that coil, the magnetic flux is completely

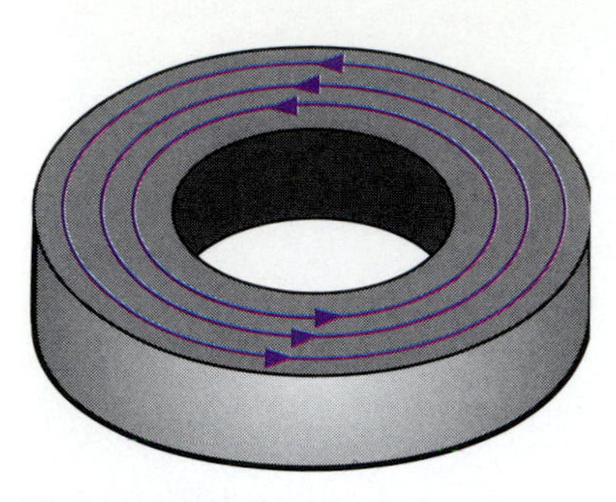

Figure 13-6 A doughnut-shaped core called a toroid is one of the most efficient core shapes available. Virtually all magnetic lines of force are contained within it.

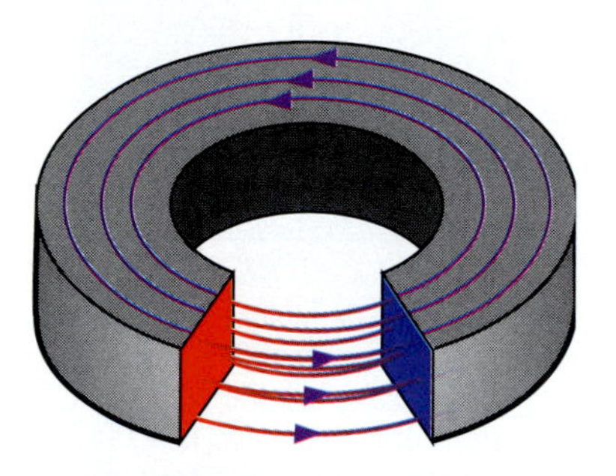

Figure 13-7 Placing an air gap into a core greatly reduces the magnetic flux by introducing the equivalent of magnetic resistance.

confined to the core and adopts its circular shape, as shown in Fig. 13-6. Like the flux loops created around a straight current-carrying wire, circular loops of magnetic flux do not exhibit magnetic polarities.

The reluctance of the iron toroid is very low, and virtually all of the magnetic flux created by the coil is confined within it. Reluctance is the equivalent of electrical resistance. The iron toroid forms a closed magnetic circuit with very little opposition.

An Open Magnetic Circuit

If a section of the iron toroid is removed, an **air gap** is created, as shown in Fig. 13-7. The air gap has a very high reluctance; consequently, the magnetomotive force is greatest at the gap. Magnetomotive force is the magnetic equivalent of voltage. Cutting the ring greatly reduces the magnetic flux by introducing a large magnetic resistance. Magnetic flux is the equivalent of current.

Magnetic Equivalent of Ohm's Law

Magnetic circuits have their own form of Ohm's law. Magnetic lines of force pass through permeable materials in the same way that current flows through a conductor. Magnetic flux can be equated to current, magnetomotive force equated to voltage, and reluctance equated to resistance. The three quantities are defined in terms of each other in the following equations.

Equation 13-6

$$\mathcal{R} = \frac{\mathcal{F}}{\phi} = \frac{\textit{magnetomotive force}}{\textit{magnetic flux}} = \textit{reluctance}$$

Equation 13-7

$$\phi = \frac{\mathcal{F}}{\mathcal{R}} = \frac{\textit{magnetomotive force}}{\textit{reluctance}} = \textit{magnetic flux}$$

Equation 13-8

$$\mathcal{F} = \phi \times \mathcal{R} = \textit{magnetic flux} \times \textit{reluctance} = \textit{magnetomotive force}$$

REVIEW QUIZ 13-3

1. Magnetically permeable materials, such as iron, behave as magnetic ______________ .
2. The magnetic field inside a doughnut-shaped core does not have identifiable ______________ .
3. Reluctance is the magnetic equivalent of electrical ______________ .

13-3 Magnetic Motor Force

Oersted's compass experiment was an inspiration to others and prompted them to investigate the magnetic fields created by electric current. They soon discovered a direct relationship between the amplitude of the current and the strength of the magnetic field it produced. These investigations led to the development of several instruments to measure electric current that were based on the movement created by magnetic interactions.

The First Current Meters

Less than two months after Oersted published the results of his compass experiment, Johann Schweigger (1779–1857), a German chemist, developed the first current meter. Schweigger reasoned that if a single wire held above a compass needle would deflect the needle to the right and the same wire held below it would deflect it to the left, one turn of wire would exert twice the deflecting force of a single wire. Continued experiments demonstrated that the deflecting force is proportional to the number of turns of wire wound around the needle. The additional turns multiplied the sensitivity of the instrument. Schweigger's current meter is shown in Fig. 13-8.

Figure 13-8 Current meters relied on magnetic motor force until the development of digital meters.

In 1881, French scientist Jacques Arsène d'Arsonval took the opposite approach to metering current by using a moving coil instead of a moving magnetic needle. In d'Arsonval's meter, the coil was wound over an iron core and suspended vertically between the poles of a permanent magnet by spring-loaded wires. A small mirror attached to the coil reflected light from a built-in source onto a scale. The scale had zero (0) at its center, with increasing numbers on each side. With no current flowing through the coil the light would usually illuminate the zero. The coil would rotate one way or the other in response to the direction of the current through it. The meter had many disadvantages: it was large, it had to be used in a dimly lit area, it required a power source for the light, and it was definitely not portable.

Edward Weston, a young inventor and businessman from England, made significant improvements to d'Arsonval's meter. He replaced the mirror with a pointer that moved along a calibrated scale. Weston's modified meter did not require a light source and was truly portable. He formed the Weston Electric Company and began manufacturing his meters in 1882.

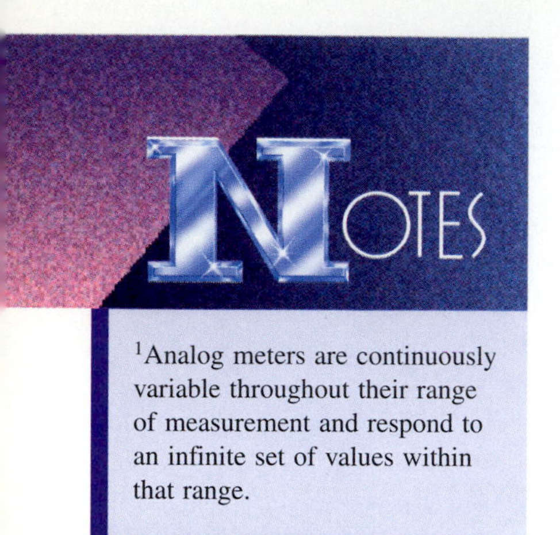

[1]Analog meters are continuously variable throughout their range of measurement and respond to an infinite set of values within that range.

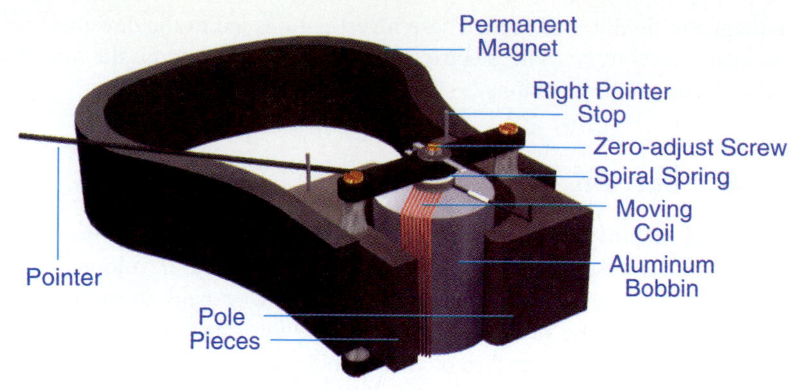

Figure 13-9 The d'Arsonval/Weston meter movement is the most popular analog mechanism ever developed.

The **d'Arsonval/Weston meter** movement, shown in Fig. 13-9, was unsurpassed in popularity by any other **analog**[1] design and has only recently yielded supremacy to its more accurate **digital**[2] counterpart. The d'Arsonval/Weston meter movement continues to have the advantage of a shorter response time to changes in voltage. The faster response provides more tactile feedback when making metered adjustments.

[2]Digital meters measure and display a discrete quantity.

Motor Action of Magnetic Fields

Oersted's compass experiment showed that a current-carrying wire exerts a force on a magnet. By applying Newton's third law of motion[3] it follows that the magnet must also exert a force on the wire. That force can be demonstrated by causing current to flow through a loosely held wire placed between the poles of a strong horseshoe magnet and observing the wire's motion. The force that causes the wire to move is perpendicular to the lines of force and also perpendicular to the length of the wire. The force is strongest when the wire is at a right angle to the magnetic lines of force. A force will diminish in strength as that angle is decreased. There is no motor action when the wire and the lines of force are parallel to each other.

Assuming that the magnetic polarities and direction of electron flow are known, the direction of force can be predicted by the **left hand motor rule**, which is as follows: When the thumb, first finger, and second finger of the left hand are held at right angles to each other, the thumb will indicate the direction of motion when the first finger is pointing along the current path and the second finger is pointing in the north-south direction of the magnetic field. The left-hand motor rule is shown in Fig. 13-10.

The effect of a magnetic field on electric current is very different from the interactions of electric charges or the effect of gravity on mass. Electric and gravitational fields produce motion along their lines of force. The interactions between a fixed magnetic field and moving electric charges produces motion at a right angle to the magnetic lines of force.

While it appears that a magnet is acting directly on the moving charges placed between its poles, to do so would be a violation of the **rule of fields,** which states: Fields interact only with other fields and only with other fields of their own type. The perpendicular motion of the wire occurs because the magnetic lines of force developed around the wire are at a right angle to the movement of the current

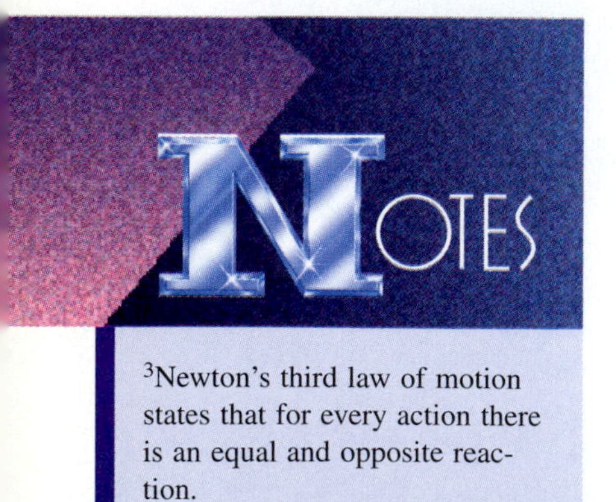

[3]Newton's third law of motion states that for every action there is an equal and opposite reaction.

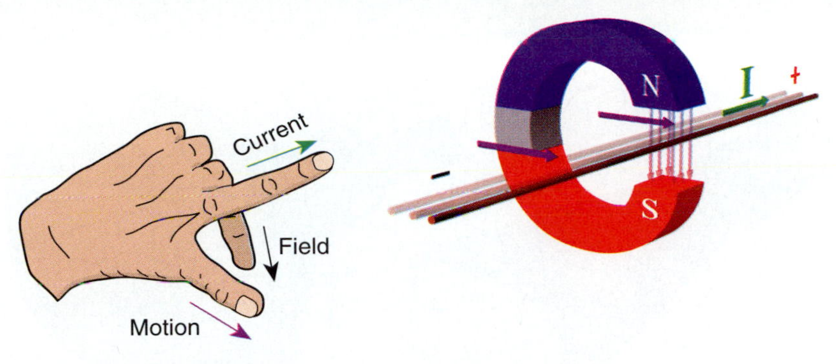

Figure 13-10 The thumb of the left hand shows the motion of a current-carrying wire using the left-hand motor rule.

within it. The interaction is between the two magnetic fields; it is never directly between magnetic and electric fields.

An Electric and Magnetic Motor Comparison

The difference between electric and magnetic interactions is clearly demonstrated by comparing the deflection mechanism used with the CRT (cathode-ray tube) in an oscilloscope with the deflection system used with a television picture tube. Technically, the picture tube is also a CRT and operates on the same principle as the one in the oscilloscope. CRTs are vacuum tubes that have an electron gun on one end and a phosphor coated screen on the other. The electron gun projects a tightly focused beam of electrons at the screen, making a bright dot. The dot is moved rapidly around the screen to produce an image.

Although the applications of oscilloscopes and televisions are very different, each requires the electron beams in their CRTs—which can be thought of as electric currents without wires—to be moved over the surface of their phosphor screens. In the oscilloscope's CRT the beam is moved using the interaction of electric fields, a process called **electrostatic deflection**. But in the television picture tube, the beam is moved using the interaction of magnetic fields, which is called **magnetic deflection**. The electron beam in the oscilloscope's CRT and the electron beam in the picture tube require both vertical and horizontal deflection.

The deflection mechanism in the oscilloscope's CRT consists of four metal plates arranged like the sides of a square box, but not touching each other. Inside the CRT, two of the plates are positioned vertically; the other two are positioned horizontally at the desired deflection point, as shown in Fig. 13-11(*a*) on the next page. The plates are named *vertical deflection plates* and *horizontal deflection plates* according to their position. The electron beam is projected through the center of the plate assembly.

The beam's position on the screen is changed by applying a voltage to the deflection plates. The negative electron beam is attracted to a positive plate and repelled away from a negative plate, following the law of charges. The beam is moved horizontally by changing the voltage on the horizontal deflection plates. It is moved vertically by changing the voltage on the vertical deflection plates. The magnetic field created by the electron beam is not affected by the deflection plate voltages.

The deflection mechanism for a picture tube lies outside the tube and is magnetic. It consists of two pairs of wire coils placed perpendicular to each other in

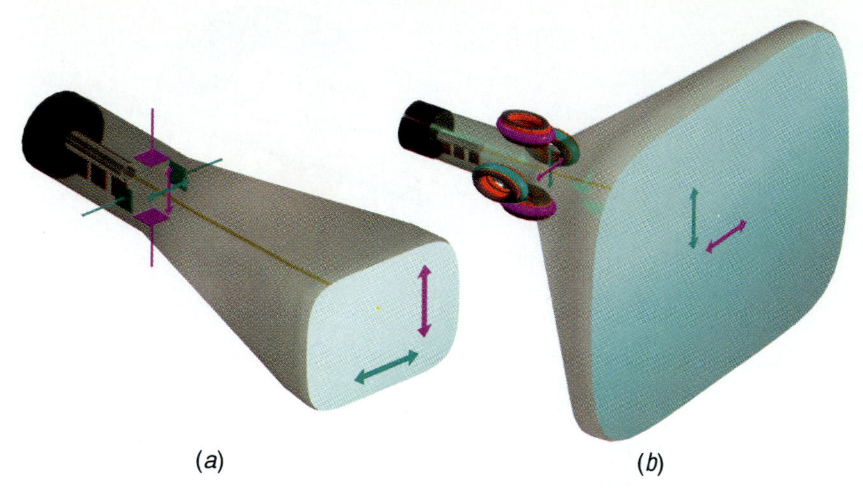

Figure 13-11 The electron beams in CRTs can be deflected (*a*) electrostatically or (*b*) magnetically.

an assembly called a *deflection yoke*. The deflection yoke is placed around the neck of the picture tube at the desired point of deflection as shown in Fig. 13-11(*b*.) The coils that deflect the electron beam horizontally are mounted vertically. The coils that deflect the beam vertically are mounted horizontally. This physical arrangement is reflected in the schematic diagrams of televisions and computer monitors. The coils move the electron beam perpendicular to their position, which is also perpendicular to the magnetic fields they produce. The magnetic fields produced by the deflection coils only interact with the magnetic field created by the moving electrons. They do not interact with the charge of those electrons.

REVIEW QUIZ 13-3

1. The d'Arsonval meter movement was based on a moving ______________ attached to a pointer.
2. A current-carrying wire is deflected at a(n) ______________ to the magnetic lines of force when the wire is immersed in a magnetic field.
3. The electron beam in a cathode ray tube may be deflected ______________ or ______________ .

13-4 Magnetically Induced Friction

Molecular movement resulting from a rapidly changing magnetic field can cause several side effects, which may or may not be desirable. Molecules in motion create friction and friction creates heat. The heat represents a loss of energy that must be supplied by the magnetic field and ultimately by the power source delivering current to the magnetic coils. In most cases the heat is undesirable, so great efforts are made to keep it at a minimum. However, there are times when magnetically induced friction is the ultimate goal.

Modern foundries often use electric furnaces that melt metal using rapidly oscillating, intensely powerful magnetic fields. These furnaces can bring a ton or more of iron to pouring temperature (≈3000°F) in about 20 minutes. The internal friction created by the magnetic interactions between the external field and those of the metal's atoms or molecules produce the heat.

The same form of magnetic vibration is used by microwave ovens to heat food. The frequency of the vibrations closely matches the natural magnetic rhythms within water molecules, causing them to vibrate and produce heat from friction. The wood products industry also uses magnetically induced friction to dry the glue in plywood and laminated beams. The magnetic fields used for that purpose reverse polarity approximately 40 million times every second.

Magnetostriction

Magnetically induced molecular motion can produce effects other than heat. In some metals, such as nickel, magnetic alignment brings about a change in the physical length of the material. If the current in a wire coil wrapped around a bar of nickel is made to change directions at regular intervals, the bar will grow longer and shorter each time the magnetic field reverses polarity. The change in the length of a material with a change in magnetic condition is called **magnetostriction.**

The magnetostrictive properties of magnetic materials are employed primarily by manufacturers of specialized cutting tools and ultrasonic transducers. Hard, brittle materials such as ceramics and thin glass are difficult to cut with any precision. A tungsten-carbide or diamond cutting tip placed on the tapered end of nickel solenoid core is an amazing cutting tool.

An electrical current that alternates at 40,000 to 60,000 times per second will cause the cutting tip to vibrate at twice that rate. The mechanical frequency is double the electrical frequency because the bar is longest when it is magnetically polarized—which is any time the current in the coil is maximum, regardless of the current's direction. The mechanical pressure per unit area is amplified at the tip because of the tapered core. The pressure created on the cross-sectional area of the bar is applied to the much smaller area at the tip. Since only the area and not the pressure has changed, the pressure per unit area is increased by the ratio of the area of the body divided by the area of the tip. This cutting device is so effective that a glass Christmas tree ornament can be cut into a spiral without breaking as shown in Fig. 13-12—a feat that is almost impossible by any other means.

Figure 13-12 Cutting tools based on magnetostrictive principles are used to cut hard, brittle materials.

Although the human ear can at best respond to sound vibrations of 20,000 times per second, there are numerous applications where the production sound at higher frequencies is necessary and desirable. One of the most popular applications is sonar. Powerful ultrasonic waves are sent out by a transducer[4] to produce echoes from objects in the area. The echoes are then assembled into images, allowing the sonar operator to "see" with sound. The higher the frequency of the sound, the better the resolution of the image. The process works far better in water than in air because of the difference in densities. It is the magnetostrictive properties of metals like nickel that make high-frequency, high-powered transducers possible.

[4]A transducer is any device that converts energy from one form to another. The term is used most often in electronics to describe a device that converts electrical energy into mechanical energy.

REVIEW QUIZ 13-4

1. Magnetically induced friction produces ______________ .
2. The process by which a material changes length when its molecules are magnetically aligned is called ______________ .

13-5 Electromechanical Devices

One of the most important uses of electricity is the conversion of electrical energy into mechanical energy. With few exceptions that process involves **electromagnetism**. The ability of a magnet to attract iron and other similar materials is used to convert electricity into physical movement. That movement is either linear or rotational. Most electromechanical devices use coils that are wound with **magnet wire**. Magnet wire is solid copper wire that is coated with a tough, thin layer of varnish or enamel. The thin insulation allows the maximum number of turns to be wound in the smallest possible area.

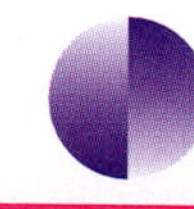

STUDENT to STUDENT

Magnetic flux loops are like stretched rubber bands that pull whatever they can with them as they "shrink."

Relays

One of the earliest electromagnetic devices to find a practical application was the **relay**. A relay can be thought of as an electrically operated switch. The heart of a relay is an electromagnet that is made by winding many turns of wire on a bobbin made from an insulating material. The bobbin is placed over an iron core that is fastened to an iron U-shaped bracket. The coil is fastened to the bottom of the bracket and enclosed by its sides. The arrangement forms two parallel magnetic circuits with the gaps across the top of the coil. A moveable iron plate called an **armature** is fastened to one of the upper ends of the bracket on a pivot mechanism that allows it to move toward the coil. A spring is positioned alongside the bracket and fastened between the armature and the bracket's lower end. The spring holds the armature in the desired position. The other side of the bracket supports at least one switch contact, which is mounted using an insulating material. The other contact is fastened to the armature.

When the coil is energized, the magnetic flux follows the core in the coil, the iron bracket, and the armature as shown in Fig. 13-13. The maximum magnetomotive force is developed between the top of the coil and the raised armature. The armature moves toward the coil, completing the magnetic circuit and closing the switch contacts. The magnetic flux is greatly increased as the arma-

Figure 13-13 The core, frame, and armature of a relay form a magnetic circuit. The magnetic resistance is least when the relay is energized.

Figure 13-14 As the armature of a solenoid is pulled into the coil, a magnetic circuit is completed.

ture moves toward the coil, which in turn increases the pressure holding the contacts together and ensures a good electrical connection.

When voltage is removed from the coil, the spring returns the armature to its upright position. It is important that the core and bracket have a very low retentivity or the relay might remain closed after the power to the coil is removed. Sometimes there are two contacts supported by the frame: one above and one below the armature contact. That arrangement creates a *single-pole double-throw* (SPDT) switch. The upper contact provides the normally closed circuit and limits the travel of the armature.

Modern relays are often placed in a sealed housing that hides the mechanism from view. Such relays generally have a legend near their electrical terminals that may be in the form of a full or partial schematic. Simple relays with SPDT contacts are labeled "NO," "C," and "NC." The abbreviations stand for normally open, common, and normally closed, respectively.

Solenoids

A solenoid is a hollow coil of wire with a moveable core called an *armature*, or *plunger*. When current flows through the coil, the armature is drawn into it, creating a linear motion. There is a basic similarity between a solenoid and a relay in terms of construction. Like the relay, the coil of a solenoid is also mounted to an iron U-shaped bracket that forms part of a magnetic circuit. The armature completes the circuit when it is pulled inside the coil, as shown in Fig. 13-14.

The armature of a solenoid must be fairly hard to keep from being deformed as it slams into the base of the U-shaped bracket when it is pulled into the coil. The hardness of the armature makes it prone to accumulating some residual magnetism. However, retentivity is not a significant factor in most solenoids because any residual magnetism would normally be insufficient to overcome the pull of the return spring.

The pull of a solenoid is very nonlinear. When the armature is fully extended, the magnetic flux—and therefore the magnetomotive force—is minimum. As the armature is pulled into the solenoid the magnetic flux increases dramatically. The armature of a solenoid is usually returned to its nonactivated position by an external spring, but sometimes the return spring is located inside the solenoid.

[4]The commutator switches the current through the armature coils so that the magnetic fields are of the proper polarity to maintain the armature's rotation.

Electric Motors

Electric motors are available in almost limitless variety, but they all have one thing in common: They create a rotational force. As with the relay and solenoid, the force of an electric motor is created during the process of completing a magnetic circuit. A motor is divided into two major parts called an *armature* and a *field*. The armature, which is the part that moves, is usually on the inside of the motor and the stationary field surrounds it. This configuration is so common that should the armature be on the outside it would most likely be called a *rotating field*.

A simple DC motor, as shown in Fig. 13-15, has a permanent magnet field with a single north-south pole on opposite sides of the circular opening. The armature will have a minimum of three poles to ensure that the motor is self-starting. The poles are made from laminated disks of iron that are deeply slotted every 120° to allow them to be wound with wire. Only two of the poles are active at any one time. One will be a north pole, one will be a south pole. The third pole will be neutral. As the armature rotates the magnetic properties of the poles are constantly being changed by a **commutator.**[4]

The commutator consists of a copper tube that has been split along its length to form three separate segments. The segments are insulated from each other by a plastic cylinder fitted over the armature's shaft. The three segments of the commutator are positioned so that the slits between them are in line with the three pole pieces of the armature, thus ensuring that an armature pole is neutral when it is directly in line with one of the field poles (see Fig. 13-15). Each armature coil is soldered to terminals on the commutator segments located at its base. A DC voltage is supplied to the commutator using graphite brushes that glide over its surface as it rotates.

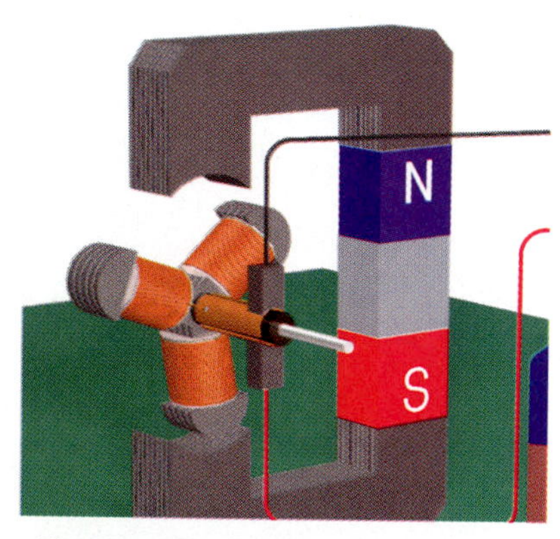

Figure 13-15 A simple DC motor has a permanent magnet field and a three-pole armature to ensure that it is self-starting.

REVIEW QUIZ 10-2

1. Electromechanical devices create ______________ from the interaction of magnetic fields.
2. The motion created by electric motors is usually ______________ .
3. A solenoid is a(n) ______________ coil of wire with a moveable core.
4. A relay is used to open or close a set of ______________ .

SUMMARY

Electromagnetism is described using a set of units related to the physical dimensions of wire coils, the currents through them, and whether or not the coil has a permeable core. These units are a part of every technician's vocabulary. Some of the units address changes or restrictions in electromagnetic fields caused by the permeable core.

Magnetic circuits can be made from permeable materials. These circuits are subject to a form of magnetic Ohm's law where magnetomotive force is equivalent to voltage, reluctance is equivalent to resistance, and magnetic flux is equivalent to current.

The magnetic motor force created between the magnetic field surrounding electric current and the field surrounding a permanent magnet is often used to advantage. The first electric meters relied on this interaction, which has been expanded to include numerous other applications.

Changing magnetic fields often produce heat in permeable materials because of friction between their molecules. The heat is used to advantage in devices like microwave ovens and induction furnaces. The magnetic alignment of a material's molecules can also change its length through a process called magnetostriction.

In their attempts to shrink to a minimum size, magnetic flux loops can develop a mechanical attraction between portions of magnetic circuits. This attraction is used to advantage in several electrical components and devices, such as motors, relays, and solenoids.

NEW VOCABULARY

air gap
ampere-turn (A·t)
analog
armature
B-H curve
coercive force
commutator
d'Arsonval/Weston meter
digital
electromagnetism
electromagnets
electrostatic deflection
field intensity
gilbert (Gb)
hysteresis
hysteresis loop
left-hand motor rule
magnet wire
magnetic circuits
magnetic deflection
magnetically hard
magnetically soft
magnetomotive force
magnetostriction
oersted (Oe)
permeability
relay
reluctance
retentivity
rule of fields
saturation
toroidal

Questions

1. Is the magnetomotive force developed by a current-carrying coil independent of the physical dimensions of the coil?
2. What is the SI unit of magnetomotive force?
3. What is a gilbert?
4. Where is the magnetic field intensity of a current-carrying coil measured?
5. Can adding an iron core to a current-carrying coil change the distance between its magnetic poles?
6. Is iron more permeable than air?
7. What property of an iron core causes its molecules to lag behind a magnetizing force?
8. Can iron act as a magnetic conductor?
9. What name is applied to the force created between a current-carrying coil and a magnetic field?
10. What property causes the length of a magnetic material to change when it becomes magnetized?

Problems

1. What is the permeability of a material with a flux density of 500 Wb/m^2 and a field intensity of 4 ampere-turns per meter?
2. What is the reluctance of a magnetic material with a magnetomotive force of 100 ampere-turns and a magnetic flux of 0.4 Wb?
3. What is the field intensity of a 50 turn coil 0.1 meter long with a current of 2 amperes?
4. What is the magnetomotive force of a current-carrying coil with a magnetic flux of 0.005 Wb and a reluctance of 200?
5. How many turns of wire are needed to produce a magnetomotive force of 50 ampere-turns with a current of 1 ampere?

Critical Thinking

1. How is magnetic motor force produced from loops of magnetic flux?
2. What physical interaction occurs between adjacent turns of wire in a current-carrying coil?
3. Describe why the magnetic field around the electron beam in a CRT does not have identifiable polarities?
4. Why should the core of a relay be magnetically soft?
5. How is the direction of current reversed through the armature coils of a DC electric motor?

Answers to Review Quizzes

13-1
1. magnetomotive force
2. length
3. gilbert
4. reluctance

13-2
1. conductors
2. poles
3. resistance

13-3
1. coil
2. right angle
3. magnetically, electrostatically

13-4
1. heat
2. magnetostriction

13-5
1. motion
2. circular (rotational)
3. hollow
4. contacts

Chapter 14

Alternating Current

ABOUT THIS CHAPTER

This chapter begins by describing the differences between the various forms of direct current and alternating current. The discussion of alternating current is limited primarily to sinusoidal wave forms. How the different amplitude measurements are derived and interrelated is

After completing this chapter you will be able to:

- State the difference between alternating current and direct current.
- Describe the relationships between the units of amplitude associated with sinusoidal AC waveforms.
- Show how sine wave generation is related to the trigonometry of a right triangle.

discussed. The key formulas for converting one unit to another are also included. Moreover, concepts of frequency and angular measurement are presented as the amplitude of AC waveforms is normally plotted against angular units instead of against time. However, these angular measurements are viewed as time when using an oscilloscope to observe AC waveforms. The fundamentals of AC generators are explored using simple generators as examples.

- ▽ Describe the basic principles of an AC generator.
- ▽ Explain how an AC generator can be made to produce pulsating or varying direct current.

14-1 Direct Current and Alternating Current: An Overview

A movement of charge in only one direction is called **direct current** (DC), which may be **pure DC**, **varying DC**, or **pulsating DC**. Pure DC always maintains a constant amplitude and is the type of current produced by batteries. The amplitude of varying DC periodically changes without dropping to zero. Pulsating DC regularly drops to zero, as in the case of a circuit being turned on and off repeatedly. Graphs of pure DC, varying DC, and pulsating DC are shown in Fig. 14-1.

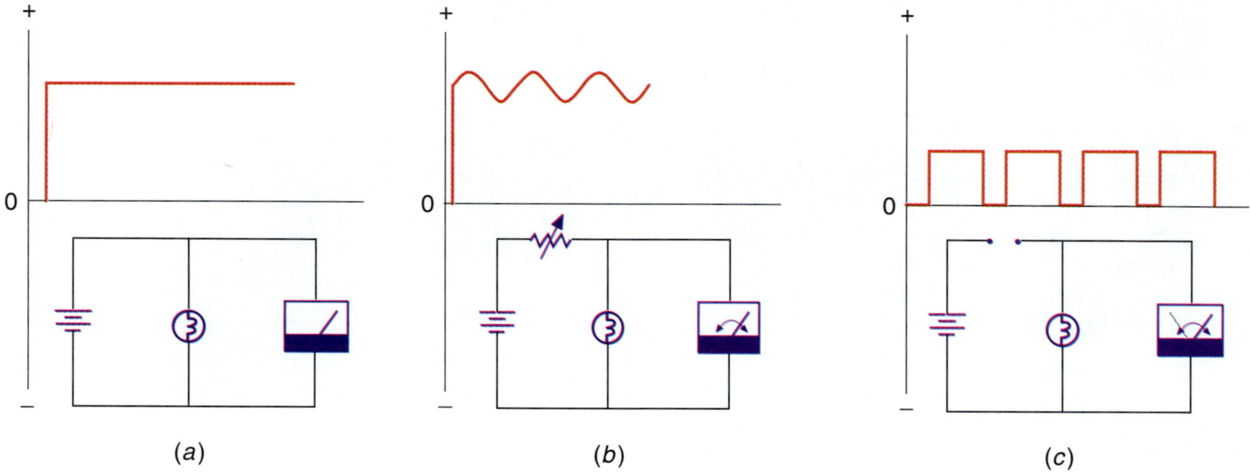

Figure 14-1 Graphs of (*a*) pure DC, (*b*) varying DC, and (*c*) pulsating DC voltages or currents may indicate a steady or changing amplitude, but the voltage polarities and direction of current never change.

A movement of charge that reverses direction at regularly recurring intervals is called **alternating current** (AC). The periodic reversals are in response to the constantly transposing polarities of the voltage source. As with any movement that reverses direction, the current must stop before the reversal can begin. During that brief instant no work is performed. A graph of a typical AC voltage or current is shown in Fig. 14-2.

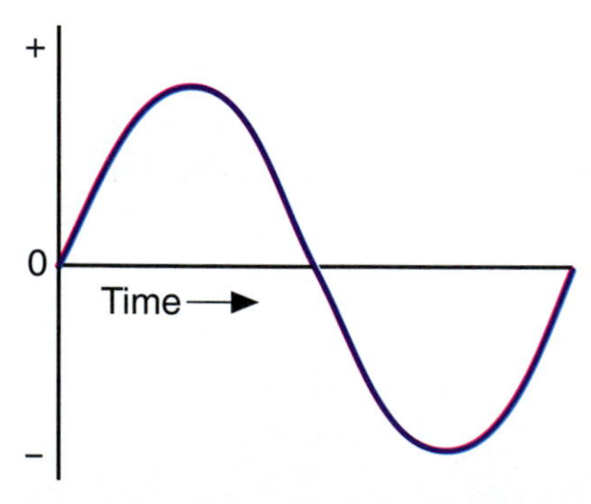

Figure 14-2 Alternating current is distinguished from direct current in that the voltage polarities and direction of current are constantly changing.

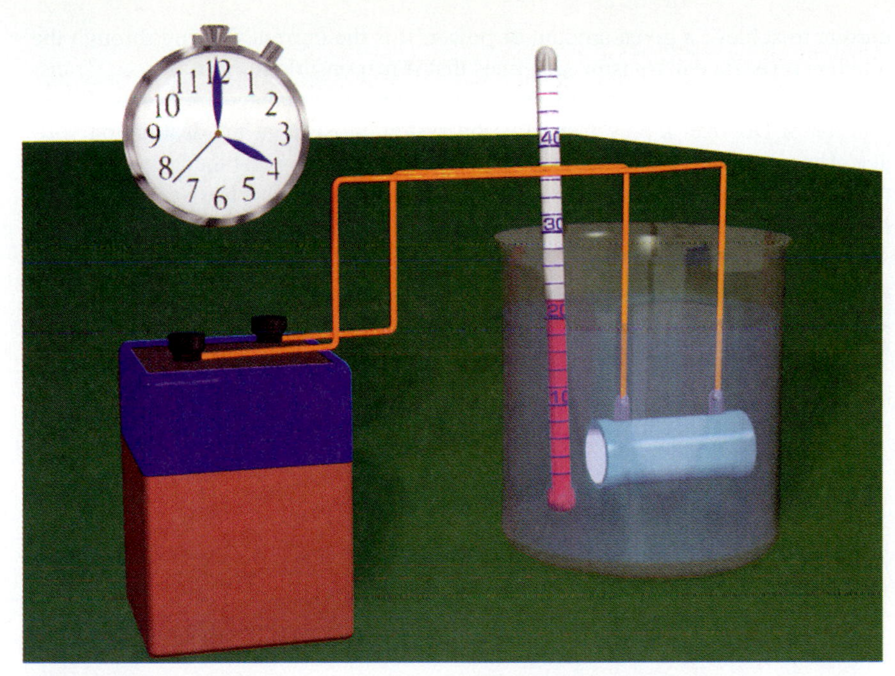

Figure 14-3 Determining how quickly a powered resistor heats a known volume of water provides a mechanism for comparing AC and DC voltages for their work potential.

Factors of AC Measurement

The term *alternating current* is often confusing because it applies equally to voltage and current. An alternating current source may not actually be supplying any current, but its changing voltage polarities identify it as AC. The fact that alternating current is in a constant state of change generates several methods of measurement or evaluation. Its amplitude may be graphed against time or angular measure, calculated or measured at some specified instant of time, or compared with direct current for its work potential. The methods of measurement apply equally to voltage and current.

When used as a source of power, AC voltages and currents are equated to a DC value that will produce an equivalent amount of power. The power equivalency can be verified by placing a resistor in a premeasured container of water at some specified temperature as shown in Fig. 14-3. Applying power to the resistor heats the water. Monitoring how fast the temperature of the water rises allows a precise calculation of the work performed by either DC or AC and establishes an equality between the two. Alternating current is continuously building toward or falling away from its maximum amplitude, where it only remains for a brief instant of time. When compared to DC, AC must work harder part of the time to produce equivalent amounts of power overall.

Power Distribution Using AC

The advantage of AC over DC in power distribution systems is that AC is in a constant state of change. This continuous variation in voltage and current allows the transformation of AC voltages to very high levels. High voltages require less current to achieve a given amount of power. It is the current flowing through the inherent resistance in transmission lines that is responsible for most losses. Transformers at the destination points reduce the voltages to manageable levels. Transmission of DC over a power distribution system must be at the destination voltage. The wires used as transmission lines in a DC system must therefore be larger to accommodate the necessarily higher currents. The larger wires keep resistance losses to a minimum, but they are far more costly.

REVIEW QUIZ 14-1

1. May the amplitude of direct current change? ______________
2. Will the direction of direct current periodically reverse? ______________
3. Is the term *alternating current* applied to voltage sources with transposing polarities? ______________
4. Must AC work harder part of the time to do the same amount of work as DC? ______________

14-2 Waveforms

A mathematical plot of how the amplitude of an electrical quantity changes with time is called a **waveform**. Electrical waveforms normally occupy the right two quadrants of the Cartesian coordinate system. The horizontal *x-axis* is used as a reference and represents zero amplitude. Positive values are drawn above it and negative values are drawn below it. The left two quadrants of the Cartesian coordinates are not used when plotting amplitude variations over time because time does not have a negative value. Electrical waveforms may be drawn on graph paper or, if they can be represented by voltage, displayed on an **oscilloscope**. An oscilloscope is an electronic instrument that displays a graph of voltage versus time on a ruled screen.

All electronic technologies developed out of experimentation with radio transmission and reception. Early radio transmitters and receivers were always connected (grounded) to the Earth. The incorporation of vacuum tubes into radio equipment also meant that batteries or other DC sources were required to supply power to them. The negative side of the main DC power source was connected to a point called **common ground**. It was called common ground because it was common to the signal information and any other voltages necessary for circuit operation. Common ground was usually connected to the Earth by a wire attached to a rod driven into the ground.

The positive side of the power source was said to be "above ground." Voltages negative with respect to ground were said to be "below ground." As a result, positive voltages were higher than negative voltages—not in voltage, because any voltage may be just as negative as another is positive, but in a physical frame of reference relating to up and down. To this day, voltages and currents that are displayed upward are positive and voltages or currents that are displayed downward are negative. This positive and negative relationship is shown in Fig. 14-4.

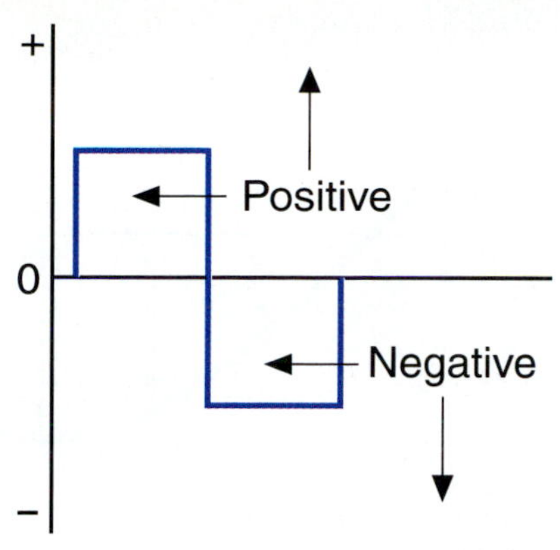

Figure 14-4 All portions of a waveform drawn above the x-axis are positive, and all portions of a waveform drawn below the x-axis are negative.

Parameters of AC Waveforms

There are four important parameters associated with alternating current. They are **amplitude**, **cycle**, **period**, and **frequency**. Amplitude refers to the **peak** values of voltage, current, or power. Cycle relates to the repetitive actions of AC, no matter how complex they may be. A cycle is complete when its actions begin to repeat themselves. It is very common for AC voltages or currents to repeat the pattern of their amplitude variations at regular intervals.

Each AC cycle is divided into two parts: a positive half-cycle and a negative half-cycle. The half-cycles have amplitude, shape, and duration, however, they are not necessarily identical. When these three elements are identical in both half-cycles, the waveform represents **pure AC**. If the three elements are not identical, the waveform is **impure AC**. A term used to represent each half-cycle, especially with pure AC waveforms, is **alternation**. There are two alternations per cycle, one positive and one negative.

The period is the amount of time required to complete the cycle's actions. Frequency is the number of periods occurring in a given time frame, usually seconds. The period is the reciprocal of frequency, and frequency is the reciprocal of period. One may be found from the other by dividing the known quantity into 1, as shown in Eqs. 14-1 and 14-2.

Equation 14-1

$$\textit{period} = \frac{1}{\textit{frequency}} \rightarrow p = \frac{1}{f}$$

Equation 14-2

$$\textit{frequency} = \frac{1}{\textit{period}} \rightarrow f = \frac{1}{p}$$

For many years frequency was measured in **cycles per second** (cps). Nowadays **hertz** (Hz) is used instead.[1] Cycle is still correct when describing the events occurring during the period of a wave, but frequencies should always be expressed in hertz.

[1]The hertz is named for the German physicist Heinrich R. Hertz. He discovered electromagnetic waves and built the first spark-gas radio transmitter.

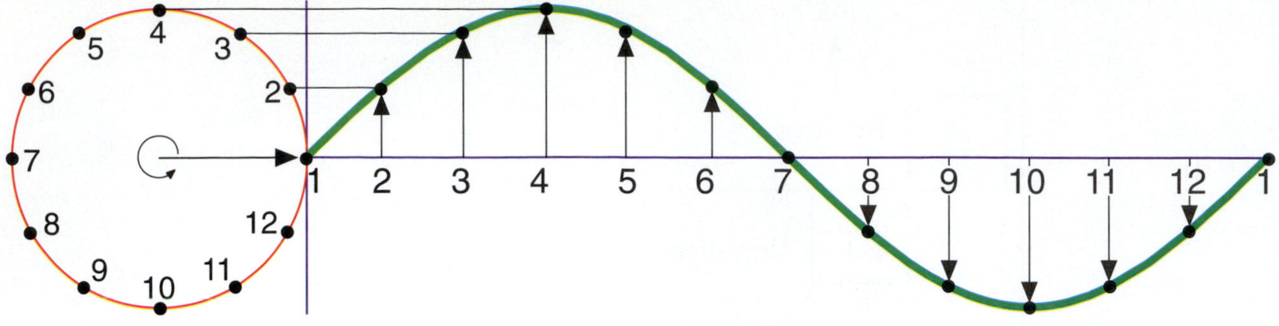

Figure 14-5 The sine wave is developed by plotting the height of a point on the circumference of a circle as the circle is rotated counterclockwise about its origin. The starting position of the point is to the right of the origin, midpoint between its maximum and minimum heights.

Sine Waves

The most common AC waveform is the sine wave, which is also called **sinusoidal**. The word *sinusoidal* describes a curving snakelike motion similar to that of a sidewinder rattlesnake. A sinusoidal waveform can be derived from circular motion. To do so, a point is placed on the circumference of a circle midway between its top and bottom. Plotting the vertical height of the point against the time it takes to complete a trip around the circumference produces the sinusoidal waveform. The normal starting position for the point is on the right of the circle. The point then follows the circumference in a counterclockwise direction until it returns to its original starting position. The point reaches its maximum height when it is directly above or below the center of the circle. When the point is above the center its height is positive, and when the point is below the center its height is negative. The height of the point can only be as far above or below the center of the circle as the length of the radius. The development of the sine wave from circular motion is shown in Fig. 14-5.

Sine Waves and Trigonometry

Sinusoidal waveforms are called **sine waves** because the amplitude at any point along the waveform is determined using the trigonometric sine function. All AC voltages and currents are assumed to be sine waves unless otherwise indicated.

Trigonometry is the study of triangles. There are six trigonometric functions: **sine**, **cosine**, **tangent**, **cotangent**, **secant**, and **cosecant**. The six functions represent the six possible ratios that can be formed using the three sides of a **right triangle**. Trigonometric functions are independent of the units of length assigned to the sides of a triangle because they are derived from ratios. An understanding of the sine function is an important aid to understanding sinusoidal alternating current. The word *sine* is usually abbreviated to *sin* on calculator keys and in written formulas.

The distinguishing feature of a right triangle is that the sides adjacent to one of the angles are always perpendicular to each other, forming an enclosed right angle. The side opposite the right angle is called the **hypotenuse**. The two *acute* angles are named using the Greek characters **theta** (Θ) and **phi** (Φ). Theta is the key reference point used to identify the sides of a right triangle, called *opposite* and *adjacent*. The opposite side is the *height* of the triangle and the adjacent side

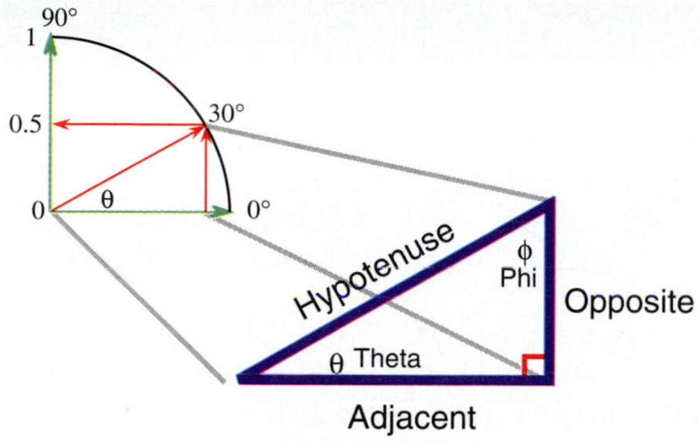

Figure 14-6 The angle theta (Θ) becomes the major reference point when naming the sides of a right triangle. The sides are called opposite and adjacent. The remaining side, the one opposite the right angle, is called the hypotenuse.

is its *base*. The symbol Θ and the word *theta* are used interchangeably. An angle designator (∠) often precedes theta to emphasize its angular meaning (∠Θ), but usually it is omitted (Θ). A typical right triangle with named sides and angles is shown in Fig. 14-6. The sine ratio is the result of dividing the side opposite ∠Θ by the hypotenuse.

The Sine Function and Circular Motion

There is a direct relationship between the sine wave generated by a point moving around the circumference of a circle and the trigonometric functions of a right triangle. A line from the center of the circle to the point, in addition to being the radius, forms the hypotenuse of a right triangle. The height of the point forms the opposite side and a line from the intersection of the height and the x-axis to the circle's center forms the adjacent side. The height divided by the hypotenuse is the sine ratio. The sine is 0 when the radius lies along the x-axis because the height is 0. The sine is 1 when the radius is completely vertical above the circle's center because the height and radius are equal. When the radius is again vertical, but pointing down, the sine is −1 because the height is negative. In this case, the amplitude of the sine wave is plotted directly from the sine values as shown in Fig. 14-7.

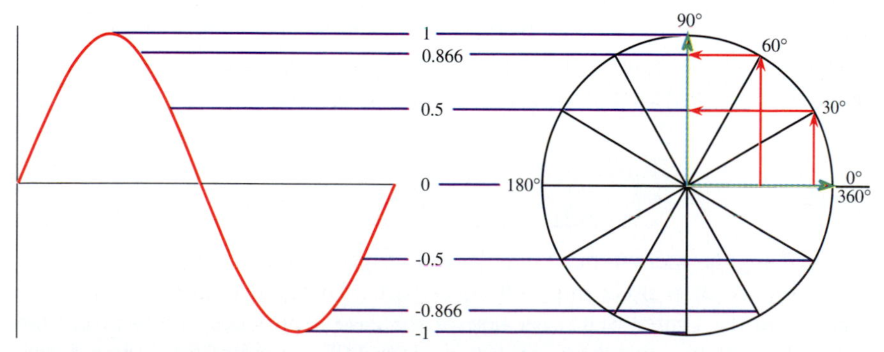

Figure 14-7 The amplitude at any point on a sine wave is equal to the sine of angle theta (∠Θ) multiplied by the peak amplitude. The sine of ∠Θ is found by dividing the side opposite Θ (height) by the hypotenuse.

REVIEW QUIZ 14-2

1. A mathematical plot of how the amplitude of a voltage or current changes with time is called a(n) ______________.
2. The frequency of an AC waveform is measured in ______________ .
3. A(n) ______________ is complete when its actions begin to repeat themselves.
4. Frequency is the reciprocal of ______________ .
5. The sine wave is created by plotting the______________ of a point on the circumference of a circle as the circle is rotated.

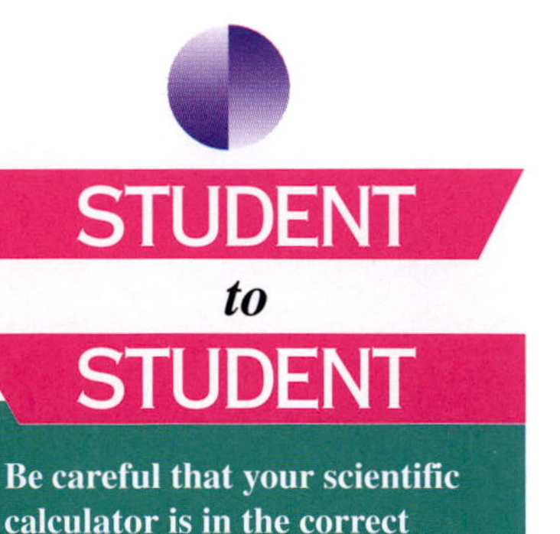

14-3 Angular Measurement

Although plotting the sine ratio over time does produce a sine wave, it is more convenient to divide the circumference of the circle into specific divisions and plot the sine against them. Each of the divisions mark the angular measure between the rotating radius and the starting point along the x-axis. The common divisions are called **degrees**, **radians**, and **grads**. Tables of trigonometric functions have been developed for each of the angular units, but values for the degree unit are the most common. Scientific calculators allow the user to easily work in any of the three angular modes.

Degrees

The *degree* unit divides the circumference of a circle into 360 equal parts. The symbol (°) is usually used in place of the word degree (for example, 360°). Degrees are very popular when describing mechanical systems and spatial orientation. The degree is the unit of choice in electronics outside of engineering.

Radians

A *radian* is a line the length of a circle's radius that is curved to fit its circumference. There are 2π ($\approx$6.28) radians in the circumference of a circle. Remember that π is the ratio of the circumference of any circle to its diameter, and the diameter is twice the length of the radius. There are approximately 57.3° in one radian. Radians are used in engineering and mathematics.

STUDENT to STUDENT

Make sure that you are not in the grad mode when you really should be in the degrees mode. It's a common error.

Grads

A unit of angular measurement popular in Europe is the *grad*. A grad divides the circumference of a circle into 400 equal parts. The grad is used in road construction in the United States because of the ease it provides in expressing the slope of a hill. Because there are 100 grads in 90°, a grad of 5 means a 5 percent slope in a hill's angle of rise from horizontal to vertical.

Angular Divisions of Sine Waves

The horizontal scale of a sine wave graph is marked in units of angular measure. At a minimum, it is marked at each of the three zero crossings and each of the two peaks. Graphs using the radian measurement seldom exceed the five mark minimum and are labeled as follows: 0, $\pi/2$, π, $3\pi/2$, and 2π. The ratios are used because π is an **irrational number**. In irrational numbers the decimal fractions go on forever without repeating. The fraction $\pi/2$ is one-half π and $3\pi/2$ is one and one-half times π. The values, in relation to π, are then: 0, 0.5, 1, 1.5, and 2.

Graphs using the degree measurement are often marked every 30° along the horizontal scale. At a minimum, they should be marked every 90°. Graphs using the grad measurement would follow the five mark minimum by placing a mark along the horizontal scale every 100 grads. Fig. 14-8 plots a sine wave using degrees and a sine wave using radians.

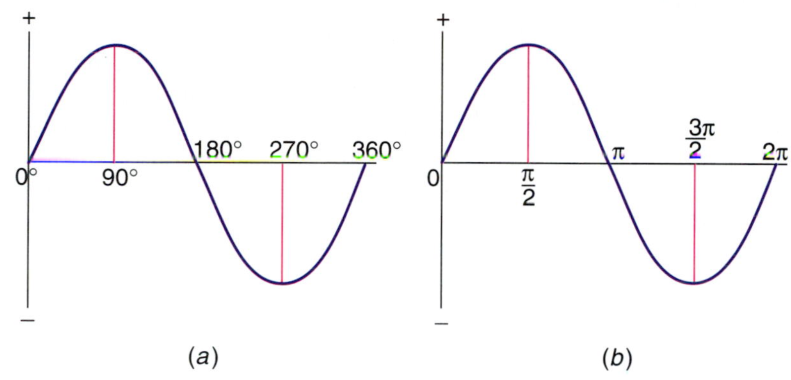

Figure 14-8 A sine wave may be plotted using (*a*) degrees or (*b*) radians as the angular measurement.

REVIEW QUIZ 14-3

1. Three angular measurements used to graph AC waveforms are ______________ .
2. ______________ are used in engineering and mathematics.

14-4 Magnetism and AC

Faraday's discovery that inserting a magnet into a coil of wire produces a brief electric impulse ultimately led to the development of alternating current. His simple experiment showed that magnetism acted as a catalyst in converting mechanical energy into electrical energy.

Voltage is proportional to the strength of the magnet, the speed of insertion or removal, and the number of turns in the coil. Experimentation reveals that relative motion between magnetic lines of force and a conductor produces the maximum voltage when the two cross each other at a right angle, because the angle affects the effective magnetic strength.

The relative motion between the magnetic field and the coil determines the polarity of the induced voltage. Moving the magnet into the coil produces a voltage of one polarity. Pulling it out produces a voltage of the opposite polarity. Turning the magnet around so that the opposite pole penetrates the coil also reverses the polarity of the induced voltage. And reversing the direction of the coil's windings reverses the polarity of the induced voltage.

Electric Generators

Generators convert mechanical energy into electrical energy. The mechanical energy usually rotates a shaft, but it could also be a reciprocating motion. Almost all generators operate on the principle of magnetic induction discovered by Faraday.

The Generating Process

Figure 14-9 The output voltage of a generator is very dependent on the angle between the moving conductor and the magnetic lines of force.

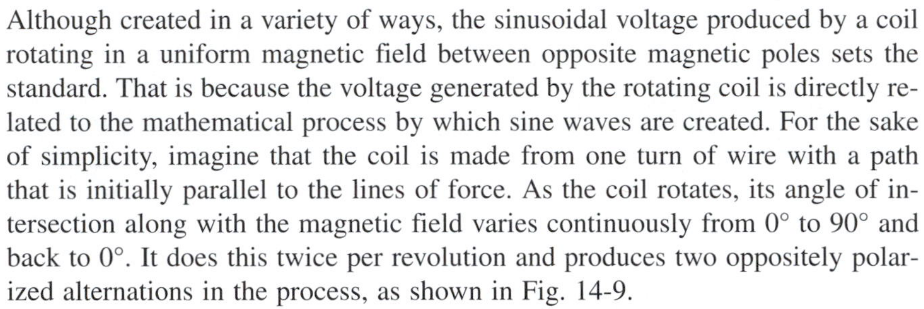

Although created in a variety of ways, the sinusoidal voltage produced by a coil rotating in a uniform magnetic field between opposite magnetic poles sets the standard. That is because the voltage generated by the rotating coil is directly related to the mathematical process by which sine waves are created. For the sake of simplicity, imagine that the coil is made from one turn of wire with a path that is initially parallel to the lines of force. As the coil rotates, its angle of intersection along with the magnetic field varies continuously from 0° to 90° and back to 0°. It does this twice per revolution and produces two oppositely polarized alternations in the process, as shown in Fig. 14-9.

The output voltage of a generator varies with the angular relationship between the moving conductors and the magnetic field. The peak voltage occurs when the conductor's motion is at a right angle to the magnetic lines of force. That voltage becomes a reference by which all instantaneous voltages along the waveform are calculated. Instantaneous voltage and current values are represented by a lower case v and i respectively. The trigonometric sine of the incident angle between the coil and the magnetic lines of force multiplied by the peak voltage accurately defines the instantaneous voltage at any angle of rotation, as indicated in Eq. 14-3.

Equation 14-3

$$v = V_{\text{max}} \times \sin \Theta$$

The radius of the coil's circular path has little to do with the generated voltage. However, the peak voltage is represented by the radius in the mathematical model. The electrical radius is called a **phasor**. Phasors are similar to vectors; they are also represented as arrows. Like a vector, the length of the phasor represents **magnitude**. But unlike the vector, the direction of the phasor arrow does not represent the direction of force.

The intersecting angle between the coil and the magnetic field is the same as the coil's angle of rotation for only the first 90°. The rotational angle continues increasing to 360° through four 90° segments called **quadrants**. The quadrants are numbered from one to four in a counterclockwise direction. The voltage phasor forms a separate right triangle in each of those quadrants; the right triangle is used to calculate the respective sine ratio. If sine calculations are done manually, the correct quadrant will have to be identified in relation to the coil's

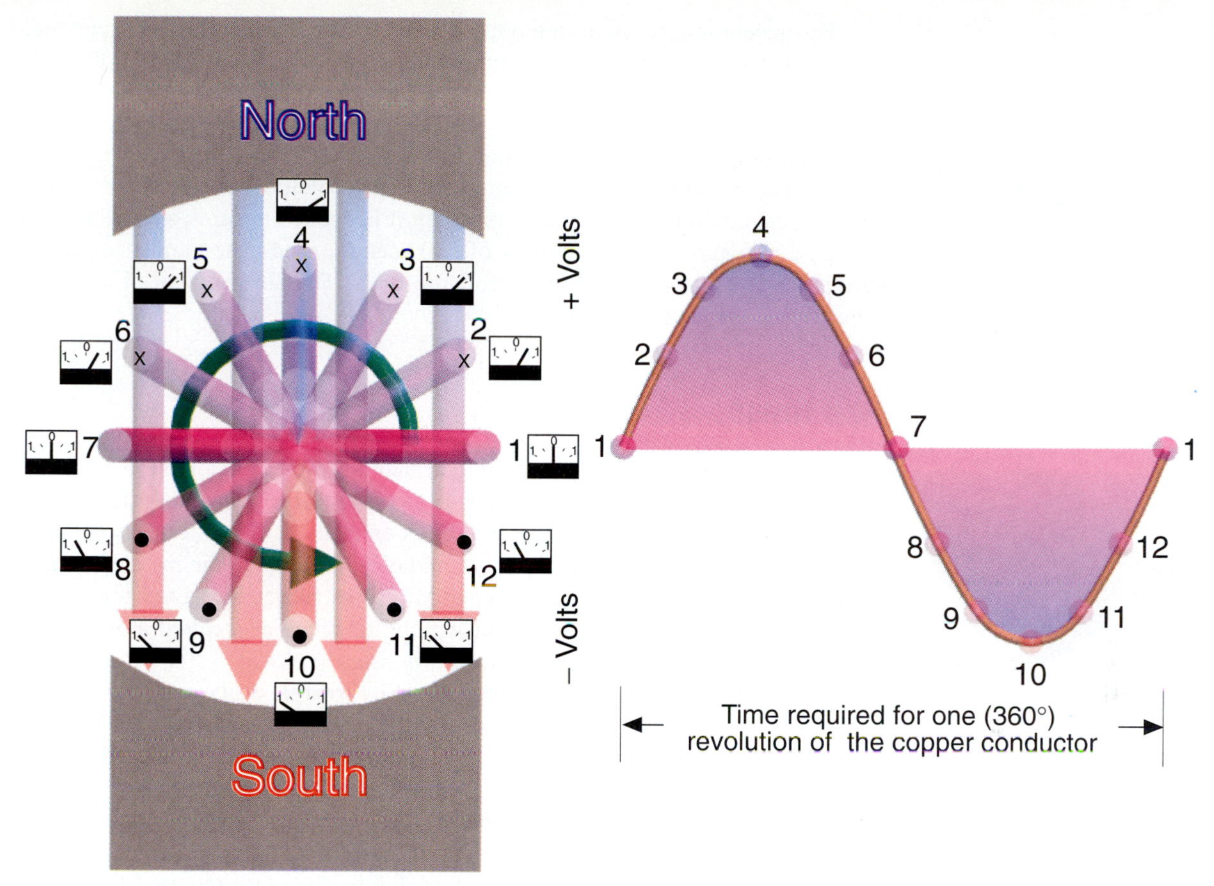

Figure 14-10 The 360° rotation of a conductor in a magnetic field is divided into four 90° quadrants numbered 1 to 4 in a counterclockwise direction.

rotational angle. Scientific calculators do this automatically and take the process even further by calculating the correct sine value after many rotations. Figure 14-10 illustrates the arrangement of the quadrants.

Angular Velocity

The point of a rotating phasor covers a distance of 2π radians per revolution. The number of revolutions in one second corresponds to the frequency of the generated sine wave. A term that is sometimes used instead of frequency is **angular velocity**. The angular velocity is measured in radians per second. The symbol for angular velocity is the lowercase Greek omega (ω). The angular velocity in radians per second may be calculated using the formula $2\pi f$.

Basic Generator Construction

Faraday's magnet and coil was transformed into an electric generator in the hands of such geniuses as Edison and Tesla. In its simplest form, a generator consists of a coil of wire rotating between the poles of a magnet. The magnetic field is concentrated and formed into the desired shape by pieces of soft iron that create a magnetic circuit. The magnet and iron assembly becomes the generator **field**.

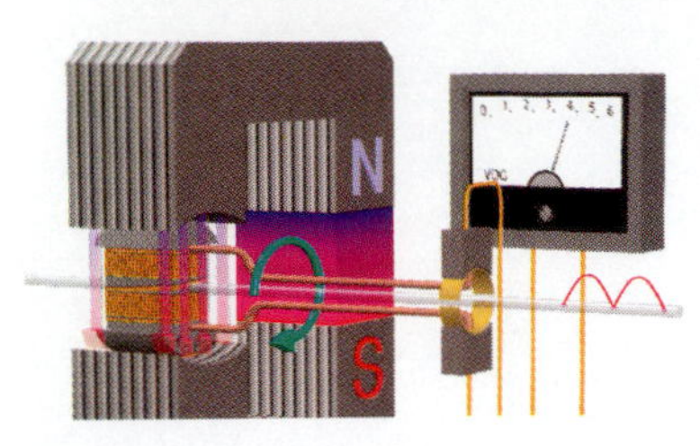

Figure 14-11 A simple generator consists of a coil rotating between the poles of a magnet.

Permanent magnet fields are only suitable for small generators. A generator field is normally made from an electromagnet. That way the output voltage is less dependent on the rotational speed of the generator because the magnetic strength of the field can be varied by controlling the current in the coil.

The complete coil assembly rotating in the magnetic field is called an armature. The placement of an iron core within the armature coil ensures the maximum concentration of magnetic flux where it is needed most. A single coil is a two-pole armature, but any number of coils can be used. The number of armature poles is twice the number of armature coils. The construction of a simple generator is shown in Fig. 14-11.

Slip Rings

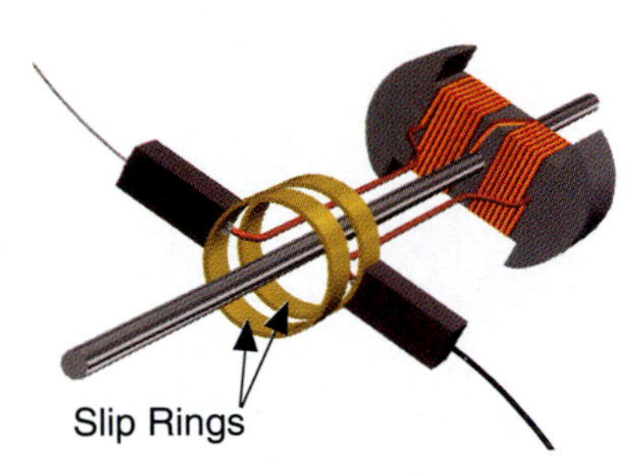

Figure 14-12 Slip rings allow the electricity produced by a rotating coil to be used by an external circuit. The output of a generator using slip rings is alternating current.

The electricity produced by the rotating armature becomes available outside the generator through the use of **slip rings** or a **commutator**. Slip rings are circular bands of copper worn by the shaft-like rings on a finger. They serve as contacts for the moving coil. An insulating material holds the rings to the metal shaft and prevents them from shorting. The use of slip rings causes the generator to produce alternating current. Generators using slip rings are also called **alternators**. Figure 14-12 illustrates a typical pair of slip rings.

Alternating current can also be produced by rotating a magnet inside a coil. For higher power applications, a coil energized by a DC current supplied through slip rings replaces the magnet. The spinning coil is sometimes referred to as a rotary field. However, the name *field* is usually applied to the stationary element.

In the early development of electric power generation, Edison and many others considered alternating current useless. Edison thought that the work used to push the current in one direction was canceled by the work required to bring it back again, making the effort wasted. To maintain a unidirectional current, the power developed by the rotating coils was coupled to the outside world through a commutator that converted the alternating current to direct current.

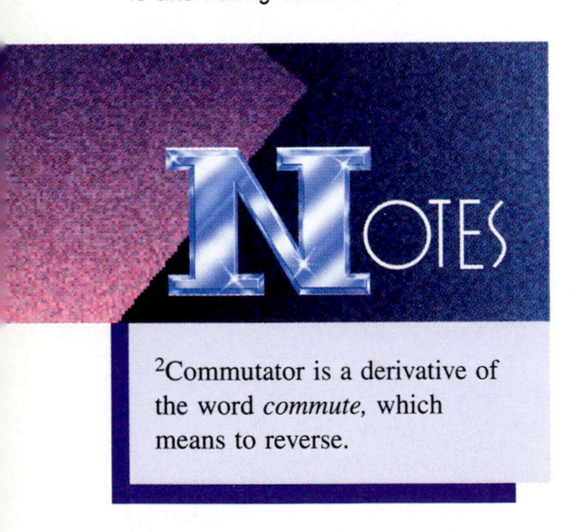

[2]Commutator is a derivative of the word *commute,* which means to reverse.

Commutators

A commutator[2] makes the conversion from AC to DC by constantly switching the coil connections coincident with the voltage alternations. The simplest form of commutator consists of a single slip ring split into two semicircles and insulated from each other. Each of the two halves of the commutator is connected to one end of the coil. Many electrical pioneers considered the production of alternating current to be an incorrect situation—to be **rectified** by the commutator as it converts AC to DC. The term *rectify* identifies any process that is used to that end. Figure 14-13 illustrates the use of a commutator.

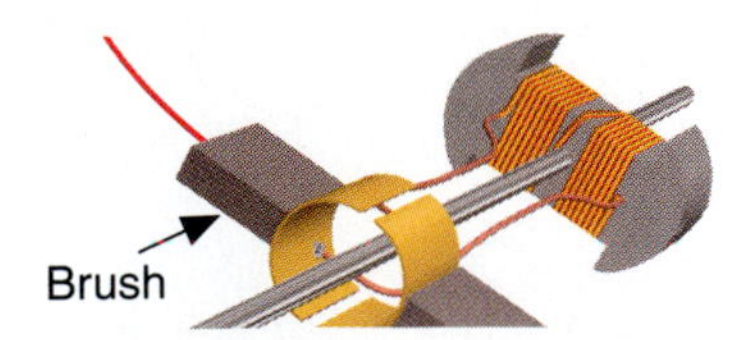

Figure 14-13 A commutator converts alternating current to direct current by switching the coils connected to the brush contacts.

DC Generator Waveforms

The voltage produced by a two-pole generator using a commutator is pulsating direct current because the voltage drops to zero twice per revolution. The pulsations represent a considerable inefficiency in the conversion of mechanical energy into electrical energy. A graph of the output voltage appears as repeated alternations of the same polarity as shown in Fig. 14-14. These waveforms maintain the same sine/amplitude relationship they would have if slip rings were used.

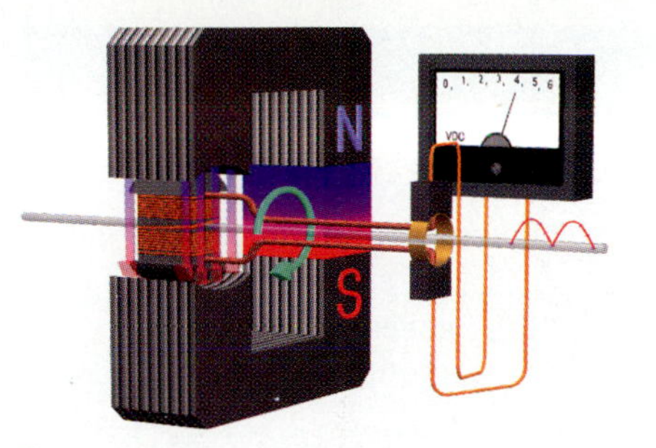

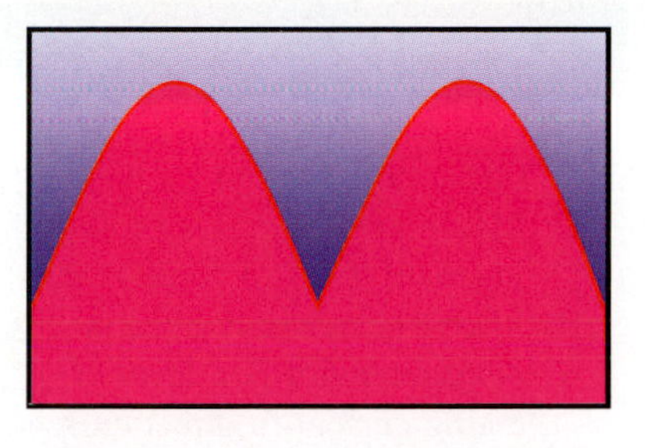

Figure 14-14 A simple two-pole generator produces pulsating direct current with each pulse resembling one sine-wave alternation.

Additional coils and commutator segments provide an acceptable, if not perfect, solution to the conversion problem by creating varying direct current. The addition of a second coil at a right angle to the first creates a four-pole armature. That allows one coil to produce the maximum voltage at the instant the other passes through zero as shown in Fig. 14-15. In this way the voltage only falls to 70.7 percent of its maximum value. The output voltage is varying direct current because its amplitude changes but does not drop to zero. Additional coils smooth the output voltage even further.

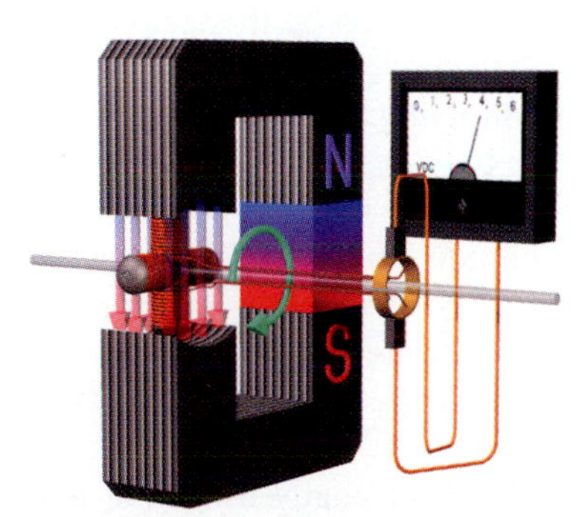

Figure 14-15 Multiple-pole generators smooth the output of the generator, reducing the output from pulsating DC to varying DC.

Brushes

Stationary contacts called brushes conduct the electric current produced by the armature to the outside world. A form of carbon called graphite is the usual brush material. Graphite has natural lubricating qualities and the ability to endure the high temperatures produced by arcing at the brush–commutator junction. Graphite brushes, as with all forms of carbon, do not melt. They slowly wear away through abrasion and, to a much lesser extent, dissipate through vaporization.

Graphite brushes tend to be brittle, but placing them into brass tubes remedies that problem. The brushes move freely inside the tubes held against the commutator or slip rings by springs. The springs usually cannot survive even moderate current levels without deforming, so a flexible copper braid completes the circuit between the brushes and the generator terminals.

REVIEW QUIZ 14-4

1. Generators convert ______________ energy into electricity.
2. Maximum voltage is produced by a generator when its coils cross the magnetic lines of force at a(n) ______________ angle.
3. Slip rings are only used in ______________ generators.
4. A commutator is a form of ______________ in that it converts alternating current to direct current.

14-5 Sine Wave Amplitude Relationships

Sine waves have four major amplitude points. They are **average**, **rms** (root-mean-squared), **peak**, and **peak-to-peak**. A graph of pure DC's amplitude over time qualifies as a waveform. Although unremarkable, a pure DC waveform is the standard by which the work potential of other more complex waveforms are gauged.

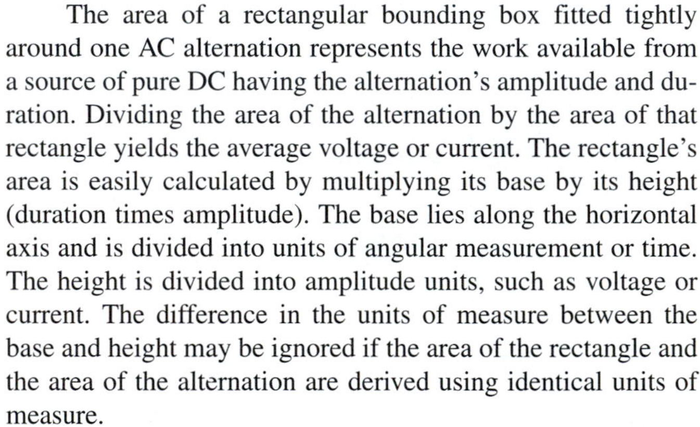

The area of a rectangular bounding box fitted tightly around one AC alternation represents the work available from a source of pure DC having the alternation's amplitude and duration. Dividing the area of the alternation by the area of that rectangle yields the average voltage or current. The rectangle's area is easily calculated by multiplying its base by its height (duration times amplitude). The base lies along the horizontal axis and is divided into units of angular measurement or time. The height is divided into amplitude units, such as voltage or current. The difference in the units of measure between the base and height may be ignored if the area of the rectangle and the area of the alternation are derived using identical units of measure.

In mathematics and electrical engineering, the unit most frequently used along the horizontal scale of a sinusoidal plot is the radian. In general, electronics technicians are not as familiar with radians as they are with degrees. The frequent graduations of the degree scale are generally more practical. The standard amplitude divisions of sinusoidal AC waveforms (average and rms) are based on calculations using the radian unit of measure.

When radians are used as the unit of angular measurement, the width of one alternation is π. The amplitude of the positive alternation has a maximum value of 1. The negative alternation has a maximum value of -1. The rectangular bounding box for one alternation would therefore have an area of π units, as determined by multiplying π times 1. Because shapes cannot contain negative areas, the absolute value of 1 is used for the negative alternation. Sinusoidal AC waveforms are a form of pure AC, so the areas under both alternations must be identical.

The area contained under a sine wave alternation is not as easily determined as the area of a rectangle, triangle, or circle. The areas of those more common geometric shapes may be calculated using only multiplication. In the case of a triangle, division may also be employed as a discretionary procedure. Calculating the area under a sine curve is more complex and best suited to calculus through the process of integration. Applying calculus procedures to determine the area under a sine curve with a base of π and a height of 1 yields an answer of exactly two square units. The illustration in Fig. 14-16 demonstrates the relative areas of the bounding box and one sine wave alternation.

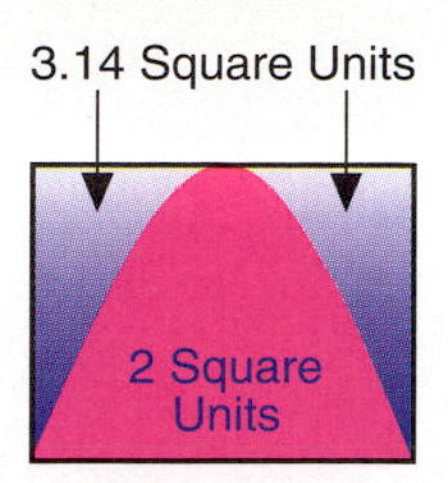

Figure 14-16 The area of the bounding box surrounding one sine-wave alternation is 3.14159, or π. The area enclosed by the alternation is exactly 2.

Average Values

The average value of any voltage, current, or power depends on the area contained by its waveform. The area of the waveform divided by the area of its bounding box is the average value. The bounding box of an alternation with a base width of π and a height of 1 contains an area of π square units. Dividing the area of a sine wave alternation by the area of its bounding box always produces the ratio $2/\pi$, or 0.637. The average value of a sine wave alternation is therefore 0.637 times the maximum amplitude.

It may seem redundant to consider the average value of an AC quantity useful, when over the entire cycle it equates to zero, but conversion to direct current makes it an important measurement. Rectification converts AC waveforms to pulsating DC, or, in the case of a multipole generator, varying DC. Meters cannot follow the rapid changes in voltage or current normally associated with those DC waveforms. Consequently, they display the changing DC values as an average. Also, the ratings of some electronic components are given as average values. Average measurements can be misleading because they do not reveal the abuses a component may experience when voltages or currents exceed recommended levels for short periods of time.

Peak Values

The peak value of a voltage or current is the maximum value attained by the waveform. Alternating current waveforms have both positive and negative peaks, which may have to be identified separately. If the waveform is sinusoidal, the alternate peaks are of equal amplitude. Peak voltages are easily measured with an oscilloscope. Peak currents are usually calculated. If the current's amplitude is large enough it may be sampled using its magnetic field. A current probe produces a voltage that is proportional to the current of a conductor. The probe senses the magnetic field around the conductor, which is proportional to the current. An alternative method is to measure the voltage developed across a resistance through which the current is flowing. Peak currents can be very destructive and difficult to measure without special equipment.

RMS (Root-Mean-Squared) Values

An rms voltage or current produces the same heating effect in resistance as pure DC. For that reason, rms is also called **effective** and the two terms are used interchangeably. The abbreviation *rms* means root-mean-squared, which is a mathematical procedure for finding the average power under a sine wave. The shape of a sine wave is the result of plotting the sines of all the angles from 0 to 2π radians or 0° to 360° against the angles from which they were derived. The rms conversion squares each of the sine values, adds them together, finds their average (mean), and then takes the square root of that average. Thus the name root-mean-squared, which is in the reverse order of the operations performed. The result is that the rms value is 0.707 of the peak voltage or current. That also equates to one-half the power produced at the peak.

The mathematical procedure used to calculate the rms value, although technically correct, is cumbersome and conceals what is actually going on. The reason for squaring the sine values is to convert the voltage or current waveform to a power waveform. Remember that power increases as the square of voltage or current. Since the sine values represent either voltage or current, squaring the sines accomplishes the conversion to a power graph. The peaks of the power graph are no greater in amplitude than the peaks of the voltage or current waveforms from which it was created. The sine values at the peaks of a sinusoidal waveform are 1 and -1. Squaring those peak values does not change their absolute value, but the waveform changes drastically.

The power waveform has two positive peaks, one for each alternation as shown in Fig. 14-17. The peaks result from the squared sine values, which regardless of their mathematical sign produce a positive result. The two positive peaks verify that work is performed during each alternation. The area under the graph represents an average value, just as it did with voltage and current.

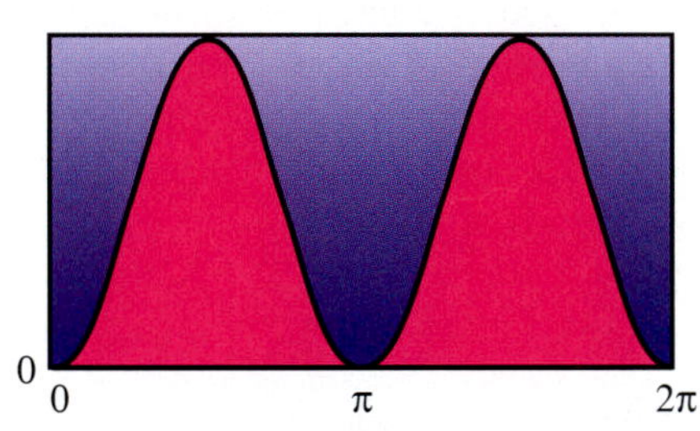

Figure 14-17 The power waveform for one cycle of sine-wave alternating current has two positive peaks. The magenta area under the graph is one-half that of this bounding box.

In the rms procedure, adding the squares of the sines and dividing the sum by the number of sines sampled produces the average area under a graph of power. That area is always 0.5 of the peak value, or half-power. Taking the square root of the average power results in a voltage or current that is 0.707 of the peak value. Unless otherwise indicated, all AC waveforms are considered to be sinusoidal, and all amplitude values are assumed to be rms. The rms values on an alternation mark the half-power points.

It is often necessary to determine the peak voltage or current when only the rms value is known. This can be done in one of two ways: The rms value can be multiplied by 1.414, as shown in Eq. 14-4, or divided by 0.707, as shown in Eq. 14-5. The method to use is a matter of preference. As an example, the typical AC outlet voltage in the United States is 120 V rms. The peak voltage (120 V $\times$ 1.414, or 120 V $\div$ 0.707) is actually 169.7 V. The multiplier of 1.414

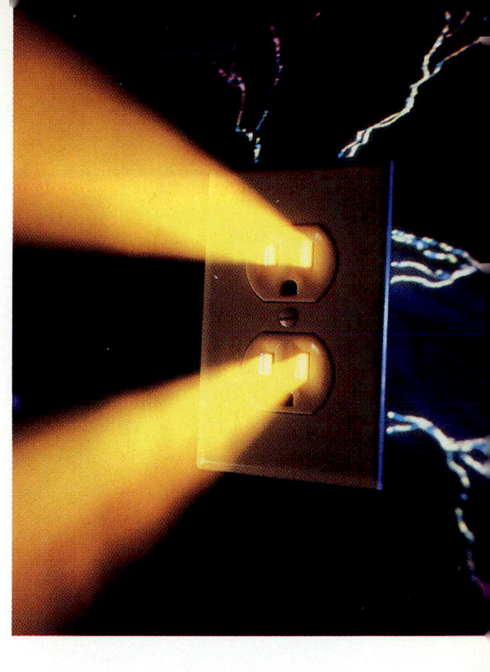

is the ratio of the peak-to-rms value. The divisor of 0.707 is the ratio of rms to peak.

Equation 14-4

$$peak = rms \times 1.414$$

Equation 14-5

$$peak = \frac{rms}{0.707}$$

Occasionally the rms value must be calculated when only the peak value is known. Again, there are two approaches to the problem. The peak value can be multiplied by 0.707, as shown in Eq. 14-6, or divided by 1.414, as shown in Eq. 14-7. Both procedures will yield similar results; the one to use is a matter of preference.

Equation 14-6

$$rms = peak \times \frac{1}{\sqrt{2}}$$

$$rms = peak \times 0.707$$

Equation 14-7

$$rms = \frac{peak}{\sqrt{2}}$$

$$rms = \frac{peak}{1.414}$$

Meters designed to measure sinusoidal AC voltages or currents typically read rms values. If alternating current is converted to direct current, a DC meter will read the average value, which is about 0.9 of rms. Although seldom used, the conversion from average to rms requires that the average value be multiplied by 1.11. The two factors are found by dividing average by rms and rms by average, respectively.

Technicians are expected to have the rms-to-peak and peak-to-rms conversions committed to memory. They should also know that the average value of a sine wave is 0.637 of peak and is equal to the area of the waveform. Figure 14-18 shows the relative relationship of peak, effective or rms, and average to each other on a sine wave.

rms = 0.707 x peak
Average = 0.636 x peak

Peak = 1.414 x rms
Average = 0.9 x rms

Peak = 1.572 x average
rms = 1.1 x average

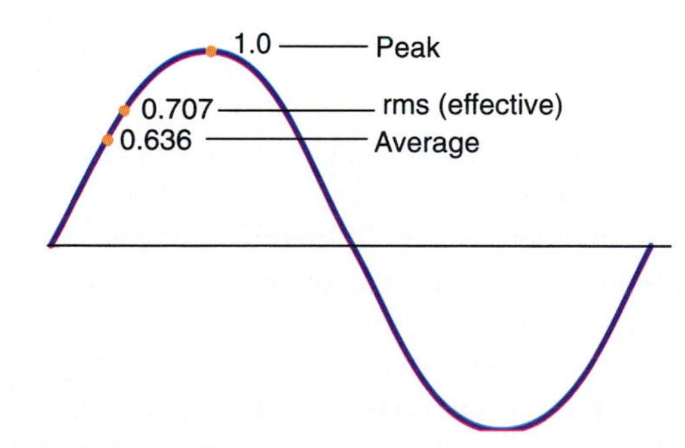

Figure 14-18 Peak, effective or rms, and average voltage or current values have constant relationships to each other that should be committed to memory.

Peak-to-Peak Values

The total range of voltage between the positive and negative peaks of an AC cycle is called **peak-to-peak**. The term applies only to voltage; the waveform does not have to be sinusoidal for this measurement to be valid. It is an important method because oscilloscopes can only display peak-to-peak voltages.

The peak-to-peak value of a sinusoidal voltage is twice that of the peak. In relation to peak-to-peak, the rms and average values are one-half of their peak relationships, but those conversions are seldom required. The conversion of peak-to-peak to peak is fairly common and is as simple as dividing the peak-to-peak voltage by 2. Figure 14-19 illustrates the peak-to-peak voltage measurement.

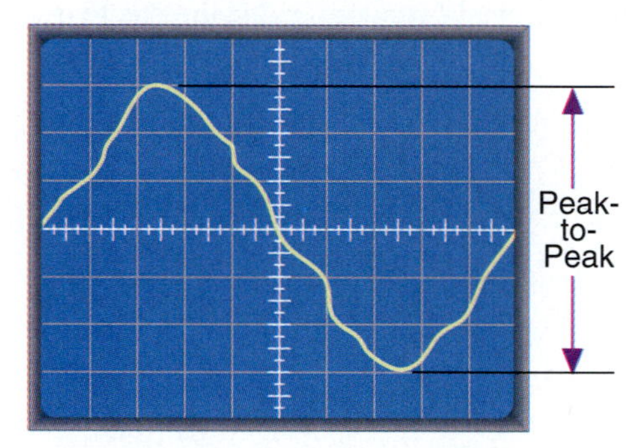

Figure 14-19 Peak-to-peak voltages are measured by oscilloscopes. These voltages do not have to be sinusoidal or even symmetrical.

REVIEW QUIZ 14-5

1. The most common unit of sinusoidal AC amplitude is ______________ .
2. The maximum value of a sinusoidal AC voltage is called ______________ .
3. The half-power points of a sinusoidal AC waveform are called ______________ .

14-6 AC Voltage and Current Measurement

Amplitude measurements of AC voltages and currents are highly dependent on the circumstances under which they are measured. The most common unit of AC measurement is rms. If there is any possibility that an AC voltage waveform is not pure and sinusoidal, an oscilloscope should be used to observe the waveform and determine its peak-to-peak value. Oscilloscopes may not be used to directly measure currents because they are strictly voltage-operated devices.

If a current flows through a pure resistance, the voltage waveform across that resistance will be a replica of the current waveform. Ohm's law could be applied to establish the current value. For nonsinusoidal waveforms, the peak current may be of the most interest. Although the voltage waveform across a resistance is the same as the current waveform, adding resistance to a circuit may cause interactions with other components that distort the current waveform beyond acceptable limits. Such a distortion can be kept to a minimum by using the smallest possible resistance that will allow reliable voltage readings.

Under limited conditions, current probes may be used to translate the intensity of the magnetic field around a current-carrying wire into a calibrated voltage. The probes depend on the current waveforms being sinusoidal and within a limited frequency range. Unless especially designed for a specific application, current probes are only useful for measuring moderate to high levels of sinusoidal current at power line frequencies.

RMS Meters

Meters, such as DMMs (digital multimeters), DVMs (Digital voltmeters), VOMs (volt-ohm-milliameters), and dedicated AC voltmeters or ammeters operate on the assumption that the AC voltages or currents they are measuring are pure, symmetrical, sinusoidal waveforms. If they are not, the readings may be incorrect or even meaningless. It is assumed and understood that general purpose AC meters are calibrated to read the rms values of presumably clean sinusoidal waveforms. Common meters may not be able to detect short duration spikes that may represent significant amounts of power.

REVIEW QUIZ 14-6

1. Oscilloscopes only measure _______________ voltages.
2. Alternating voltage or current meters assume that the waveform of the voltage being measured is _______________ .
3. Can oscilloscopes make direct measurements of current?

SUMMARY

The major difference between direct current and alternating current is that direct current always flows in the same direction. Alternating current periodically reverses direction. When alternating current reverses its direction of flow it must completely stop. A graph of the amplitude of alternating current over time produces a waveform. This waveform may be observed on an oscilloscope. The most common AC waveform is the sine wave. The sine wave relationship is derived from the trigonometric sine ratio of a right triangle. The triangle is formed by a point on the circumference of a circle, the origin of the circle, and a point on a horizontal line drawn through the center of the circle directly above or below the one on the circumference. The circle is divided into angular units of degrees, radians, or grads.

Circular motion is important in relation to alternating current because most alternating current is generated by coils of wire moving in circular paths between magnetic poles. The output voltage is greatest when the wires in the coils cut across the magnetic lines of force at a right angle; it is zero when the wires and magnetic lines of force are parallel. All generators produce alternating current, but some convert the alternating current to direct current using a commutator. The DC output of a generator is at best varying DC because of the overlapping coil waveforms.

Alternating current waveforms are equated to direct current by the power they produce. Because AC voltages and currents periodically drop to zero, they must work harder part of the time. Sinusoidal AC waveforms are normally given as effective or rms voltages and currents. The rms value is 0.707 of the peak value. Conversely, the peak value is 1.414 times the rms value.

NEW VOCABULARY

alternating current
alternation
alternators
amplitude
angular velocity
average
common ground
commutator
cosecant
cosine
cotangent
cycle
cycles per second
degrees
direct current
effective
field
frequency
grads
hertz (Hz)
hypotenuse
impure AC
irrational number
magnitude
oscilloscope
peak
peak-to-peak
period
phasor
phi
pulsating DC
pure AC
pure DC
quadrants
radians
rectified
right triangle
root-mean-squared (rms)
secant
sine
sine wave
sinusoidal
slip rings
tangent
theta
varying DC
waveform

Questions

1. What type of current flows only in one direction?
2. Does alternating current periodically drop to zero?
3. What type of current, AC or DC, is more effective for long-distance power transmission?
4. What are the four major parameters of a sinusoidal AC waveform?
5. What is the standard unit of frequency?

Problems

1. What is the period of a sine wave having a frequency of 60 Hz?
2. What is the frequency of a sine wave having a period of 1 microsecond?
3. What is the peak value of a sine wave having an rms voltage of 120 volts?
4. What is the rms value of a sine wave having a peak voltage of 10 volts?
5. What is the average voltage of one sine wave alternation having an rms value of 120 volts?

Critical Thinking

1. What is the average value of one full cycle of a sinusoidal voltage with a peak value of 100 volts?
2. What factors are affected by the rotational speed of an AC generator?
3. How can the voltage of a DC generator be varied?
4. What is the function of a commutator in a generator?
5. What is the relationship between the motion of the armature coils and the lines of force when the peak voltage is produced by a generator?

Answers to Review Quizzes

14-1
1. yes
2. no
3. yes
4. yes

14-2
1. waveform
2. hertz
3. cycle
4. period
5. height

14-3
1. degrees, radians, grads
2. Radians

14-4
1. mechanical
2. right
3. AC
4. rectifier

14-5
1. rms
2. peak
3. rms

14-6
1. peak-to-peak
2. sinusoidal
3. no

Chapter 15

Waveform Analysis

ABOUT THIS CHAPTER

This chapter provides an overview of the individual elements that compose electrical waveforms. Two pieces of test equipment—the oscilloscope and the spectrum analyzer—are used to measure and display the measurable quantities associated with waveform analysis. Nonsinusoidal waveforms are often very complex, and the math-

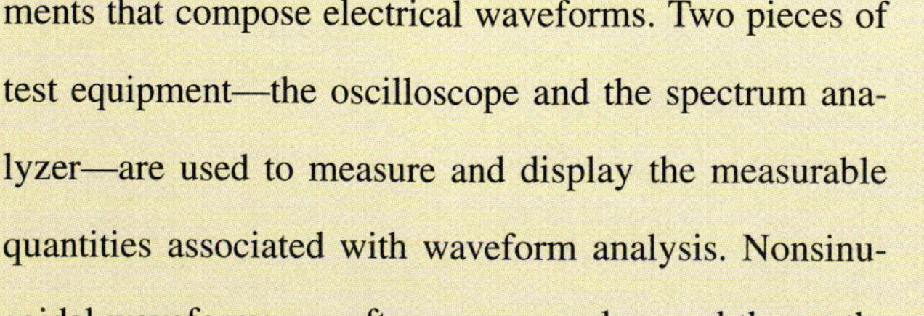

After completing this chapter you will be able to:

- Define time domain and frequency domain.
- Determine which quantities are measurable using either an oscilloscope or a spectrum analyzer.

ematics involved is beyond the scope or intent of this text. To dispense with those mathematical complexities, only three basic waveforms—the sine wave, the sawtooth wave, and the square wave—are considered. Calculus needed to work with these waveforms has been pared down to the basics to determine the amplitudes of the harmonic components of the waveforms presented.

- ▽ Explain how sine waves of different frequencies are added together to form nonsinusoidal waveforms.
- ▽ Recognize waveforms that have a DC component even though they may appear to be AC.

A good understanding of waveforms will be helpful when you are working on the latest products in electronics.

15-1 Waveform Composition

The peak-to-peak voltage and period of electrical waveforms are typically measured using an oscilloscope. Those waveforms with patterns of amplitude variations that repeat at regular intervals are said to be **periodic.** Common periodic waveforms are sinusoidal, square, and sawtooth. These three basic waveforms are shown in Fig. 15-1.

Oscilloscopes also allow the general shape of a waveform to be compared to a given standard. It is possible that a waveform may have the correct amplitude and period, but have an incorrect waveshape. A discrepancy in waveshape may be owing to a component failure or to improper circuit adjustment.

Harmonics

An oscilloscope cannot reveal the true nature of nonsinusoidal waveforms because these waveforms are formed by the summation of two or more sine waves of different frequencies. Periodic nonsinusoidal waveforms are composed of a **fundamental frequency** and multiples of the fundamental called **harmonics.** The fundamental frequency is sometimes called the **first harmonic.** In addition to the fundamental, some waveforms contain only odd harmonics, some only even harmonics, and others contain both odd and even harmonics. The shape of a periodic waveform is determined by the amplitude of its harmonic components, and the **phase** of each in relation to the others. *Phase* refers to the difference in time, if any, between the starting points of each sine wave forming the composite waveform and may be measured in degrees or radians.

Phase Shift

The normal starting point of a sine wave is at zero amplitude as it is about to move in a positive direction, as shown in Fig. 15-2(*a*). The sine wave in Fig. 15-2(*b*) leads by 90° because it begins on its positive peak, and the sine wave in Fig. 15-2(*c*) lags by 90° because it begins on its negative peak. A composite of the three sine waves is shown in Fig. 15-2(*d*). A sine wave that is 90° leading is also called a **cosine wave** because the amplitude at any point on the curve

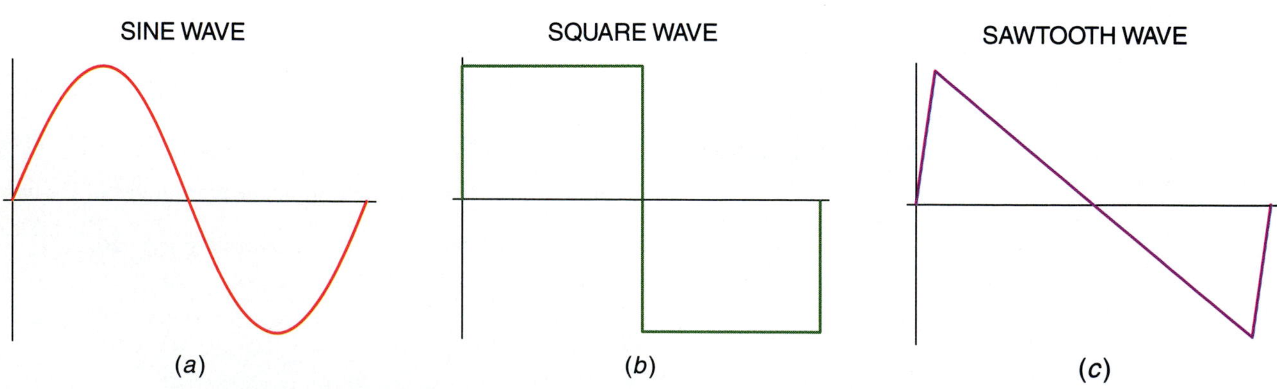

Figure 15-1 Common periodic waveforms are (*a*) sinusoidal, (*b*) square, and (*c*) sawtooth.

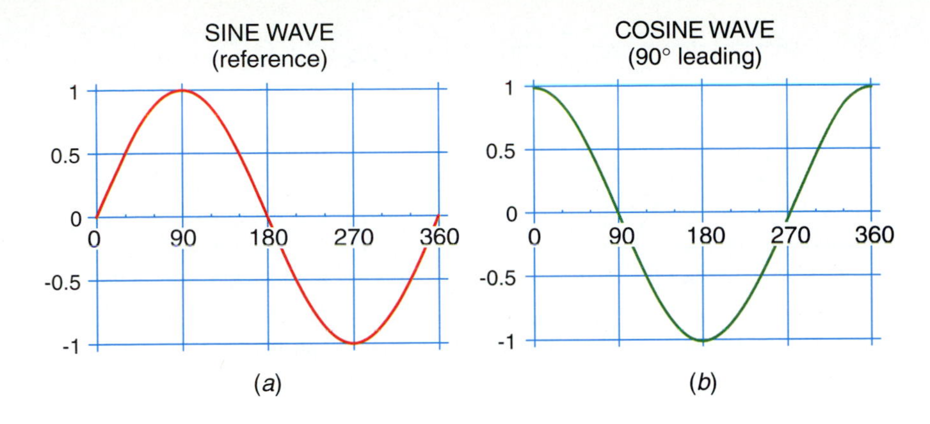

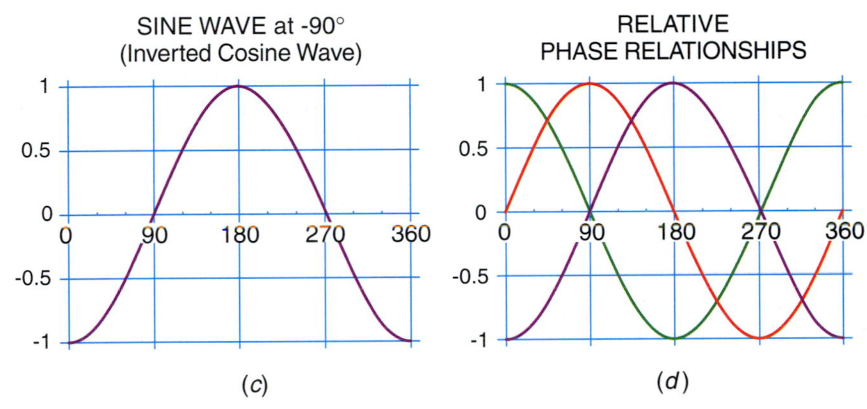

Figure 15-2 (*a*) The normal starting point of a sine wave is at zero amplitude as it is about to move in a positive direction. Other sine waves are said to be (*b*) leading or (*c*) lagging by comparing them to this established standard.

may be calculated by multiplying the peak amplitude by the cosine of the angle. Waveforms that lead or lag the reference waveform are often said to be **phase shifted.** When two waveforms are 90° out of phase, they are said to be in **quadrature.**

Time Domain

The voltage waveforms displayed by an oscilloscope reside in what is called the **time domain.** The time domain is a two dimensional plot of how voltage varies with time; it does not identify the individual frequency components that form the composite waveform. The appearance of a sine wave in the time domain does not ensure that it is not composed of other sine waves. The sine wave might have been created by the summation of multiple sine waves with identical frequencies, but not necessarily the same phase.

Figure 15-3(*a*) on the next page shows the sine wave that results from the addition of two in-phase sine waves where one has twice the amplitude of the other. When the sine waves are in phase, the amplitude of the resultant is the numeric sum of the amplitudes of the waves that created it. The summation of the two sine waves produces a sine wave with twice the amplitude of either of the

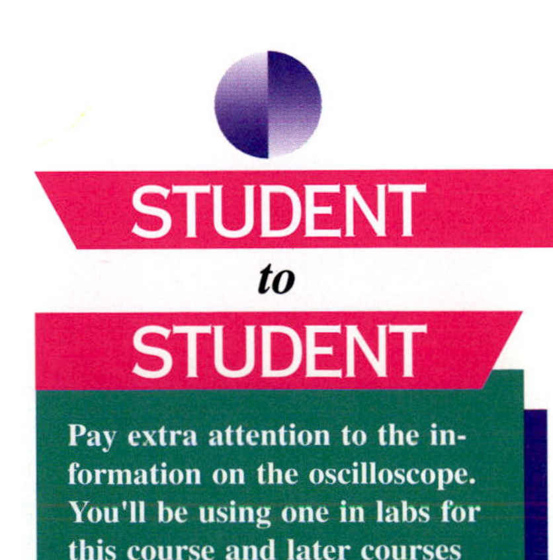

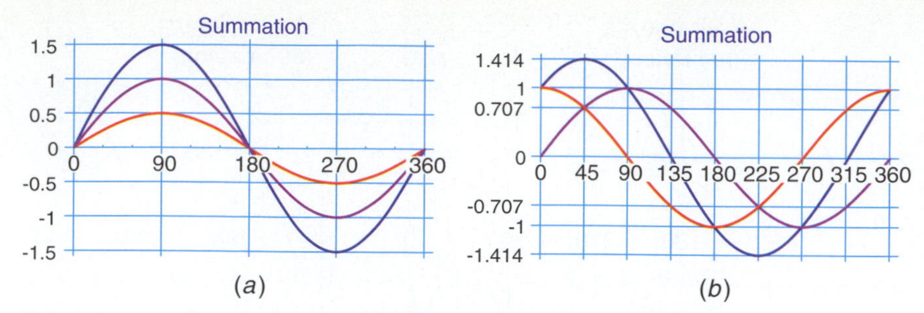

Figure 15-3 A sine wave may be composed of other sine waves of the same frequency, even if those sine waves may be different in phase.

original sine waves. Figure 15-3(*b*) shows how a sine wave is created from the summation of another sine wave and a cosine wave. These two waveforms also have equal amplitudes. In this case the resultant sine wave has a peak amplitude that is 1.414 times greater than the peak amplitude of either individual waveform. The resultant waveform leads the original sine wave by 45° and lags the cosine wave by the same amount. Once sine waves of the same frequency have been combined, there is no way to separate the resultant waveform into its component parts. As a practical matter, therefore, sine waves are generally considered to be a pure fundamental waveform regardless of how they were created.

REVIEW QUIZ 15-1

1. Can a sine wave be a composite of other sine waves?
2. A(n) ______________ is an exact multiple of a fundamental frequency.
3. All nonsinusoidal waveforms contain ______________ .
4. An oscilloscope displays a waveform in the ______________ domain.

15-2 Spectrum Analyzer Waveforms

Frequency Domain

It is sometimes useful to view only the frequency components of a waveform. This is done with a **spectrum analyzer.** In many ways a spectrum analyzer looks like an oscilloscope, but it displays waveforms in the **frequency domain.** Like an oscilloscope, the spectrum analyzer produces a horizontal line when there is no input signal. Unlike the oscilloscope, the horizontal line is calibrated in hertz, kilohertz, or megahertz. When an undistorted pure sine wave is applied to the input of a spectrum analyzer, a single vertical line is displayed, as shown in Fig. 15-4. The height of the line represents the voltage amplitude of the sine wave, just as it would on an oscilloscope. The line's position along the horizontal axis identifies its frequency. There is only one line because a pure sine wave has only one frequency component.

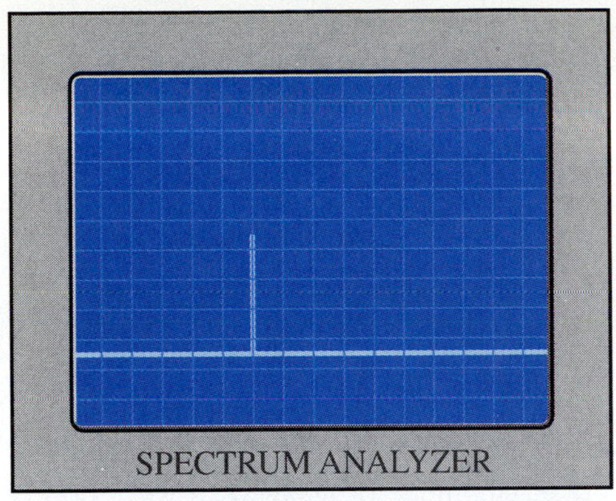

Figure 15-4 A spectrum analyzer displays only one vertical line when a sine wave is applied because a sine wave represents just one frequency element.

Spectrum

A spectrum analyzer operates on a waveform similar to the way a prism separates light into a rainbow of colors. The colors emitted by the prism are arranged according to their frequency, or wavelength. The separated and ordered frequency components form the frequency **spectrum.** A spectrum analyzer derives its name from its ability to provide frequency and amplitude information.

Sawtooth Waves

Consider the sawtooth waveform shown in Fig. 15-5(*a*). It is composed of a fundamental frequency and an infinite number of harmonics that bear a precise amplitude relationship to each other. Both the even and odd harmonics are components of the waveform, as shown in Fig. 15-5(*b*).

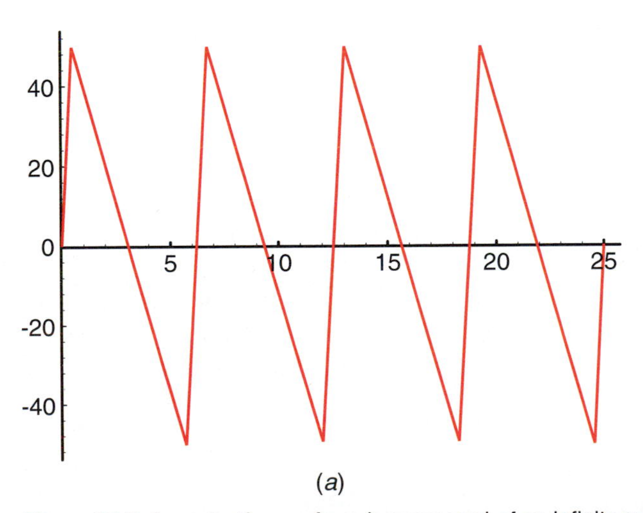

(*a*)

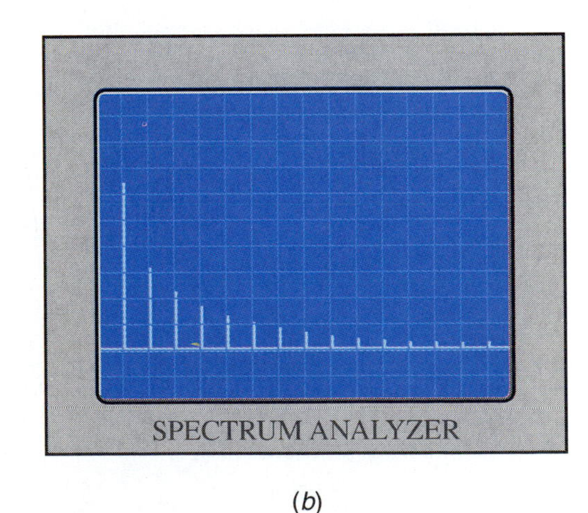

(*b*)

Figure 15-5 A sawtooth waveform is composed of an infinite number of both even and odd harmonics that bear a precise mathematical relationship to each other.

The amplitude of each harmonic component of a sawtooth waveform can be calculated by dividing the peak-to-peak amplitude of the waveform by the product of the harmonics number and π, as shown in Eq.15-1.

Equation 15-1

$$V_n = \frac{V_{\text{pp}}}{n\pi}$$

where V_n = The voltage of the harmonic identified by n.
V_{pp} = The peak-to-peak voltage of the sawtooth waveform.
π = 3.14159
n = The number of the harmonic (1, 2, 3, 4, 5, 6, . . .)

If the peak-to-peak amplitude of a sawtooth waveform equals 100%, then, using Eq. 15-1, the amplitude of the first harmonic (fundamental) is 31.83%, the second is 15.92%, the third is 10.61%, the fourth is 7.96%, and the fifth is 6.37%. In theory, Eq. 15-1 would have to be calculated an infinite number of times in order to identify the amplitudes of all the harmonics contained in a sawtooth waveform. Figure 15-6(*a*) shows how five sine wave generators could be connected in series to produce a sawtooth waveform. All of the sine waves are in phase (start at the same time), and each generator produces a sequential harmonic

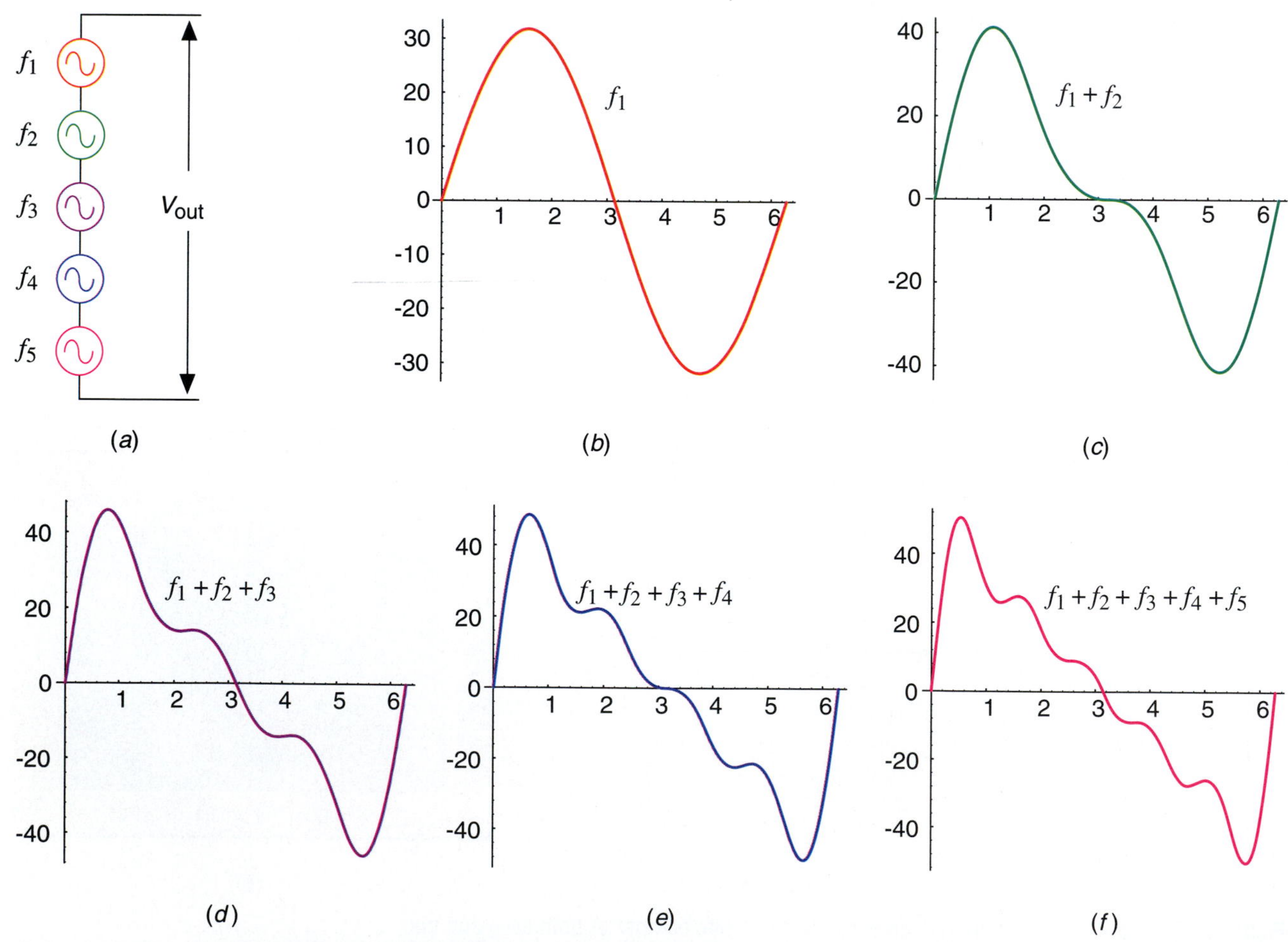

Figure 15-6 The sawtooth waveform is developed from an infinite number of both even and odd harmonics that have specific amplitude relationships to each other.

of the correct amplitude. Figure 15-6(*b*) shows only the first harmonic. Figure 15-6(*c*) shows the result of adding the first and second harmonics. Figure 15-6(*d*) adds the third harmonic, Fig. 15-6(*e*) adds the fourth harmonic, and Fig. 15-6(*f*) adds the fifth harmonic. Although several inconsistencies can be seen in the waveform, its shape is definitely sawtooth.

Square Waves

The square waveform shown in Fig. 15-7 is composed of a fundamental frequency and an infinite number of odd harmonics. Square waveforms are very common frequency in digital electronics. A knowledge of their makeup is useful when analyzing certain types of problems in digital logic circuits.

As with the harmonics that form the sawtooth waveform, the harmonics that compose the square waveform must have a precise amplitude relationship to each other. This relationship is summarized in Eq. 15-2, which requires the harmonics to be in phase.

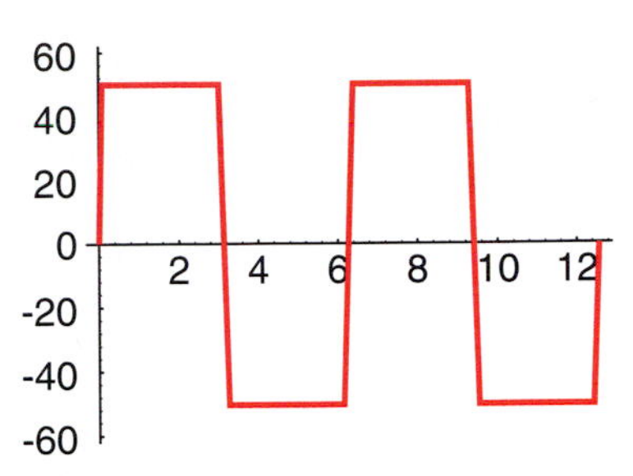

Figure 15-7 Square waveforms are produced by a fundamental frequency and an infinite number of odd harmonics that bear a precise amplitude relationship to each other.

Equation 15-2

$$V_n = \frac{2V_{pp}}{n\pi}\Big|_{n \text{ odd}}$$

where V_n = The voltage of the harmonic identified by n.
V_{pp} = The peak-to-peak voltage of the sawtooth waveform.
π = 3.14159
n = The number of the odd harmonic (1, 3, 5, 7, 9, . . .)

If the peak-to-peak amplitude of a sawtooth waveform equals 100%, then, as calculated using Eq. 15-2, the amplitudes of the first five odd harmonics (shown in Fig. 15-8) are 63.66%, 21.22%, 12.73%, 9.09%, and 7.07%. As with the sawtooth wave, the calculation would have to be made an infinite number of times in order to identify the amplitudes of all the harmonics theoretically contained in the waveform. Ten harmonic elements are usually considered necessary to yield an acceptable square waveform. Thus, for a circuit to pass a square waveform without distorting it, the circuit must be able to pass frequencies almost nineteen times higher than the frequency of the square wave.

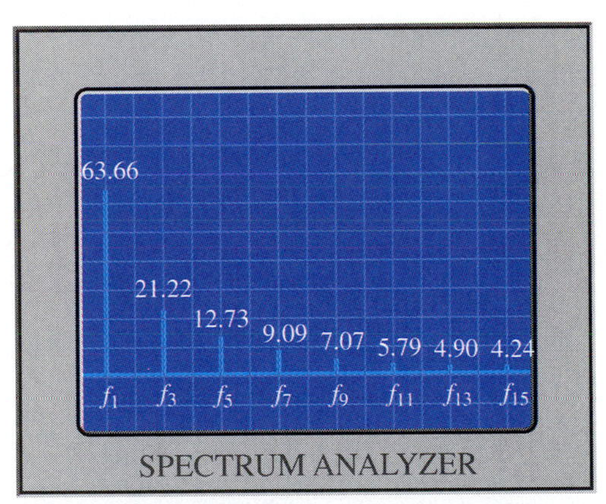

Figure 15-8 A spectrum analyzer reveals that only odd harmonics are present in a square waveform.

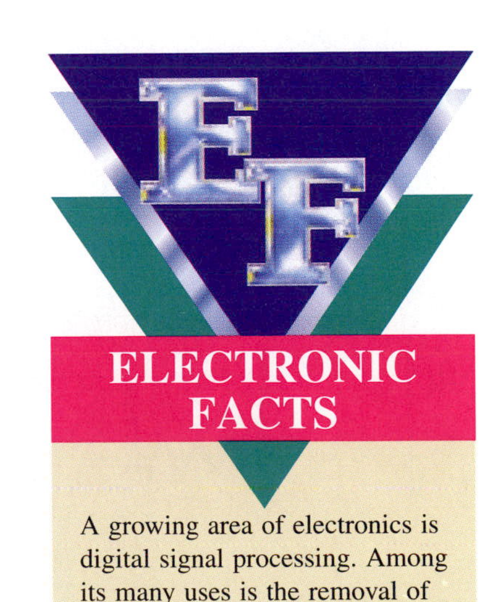

A growing area of electronics is digital signal processing. Among its many uses is the removal of unwanted noise from old recordings.

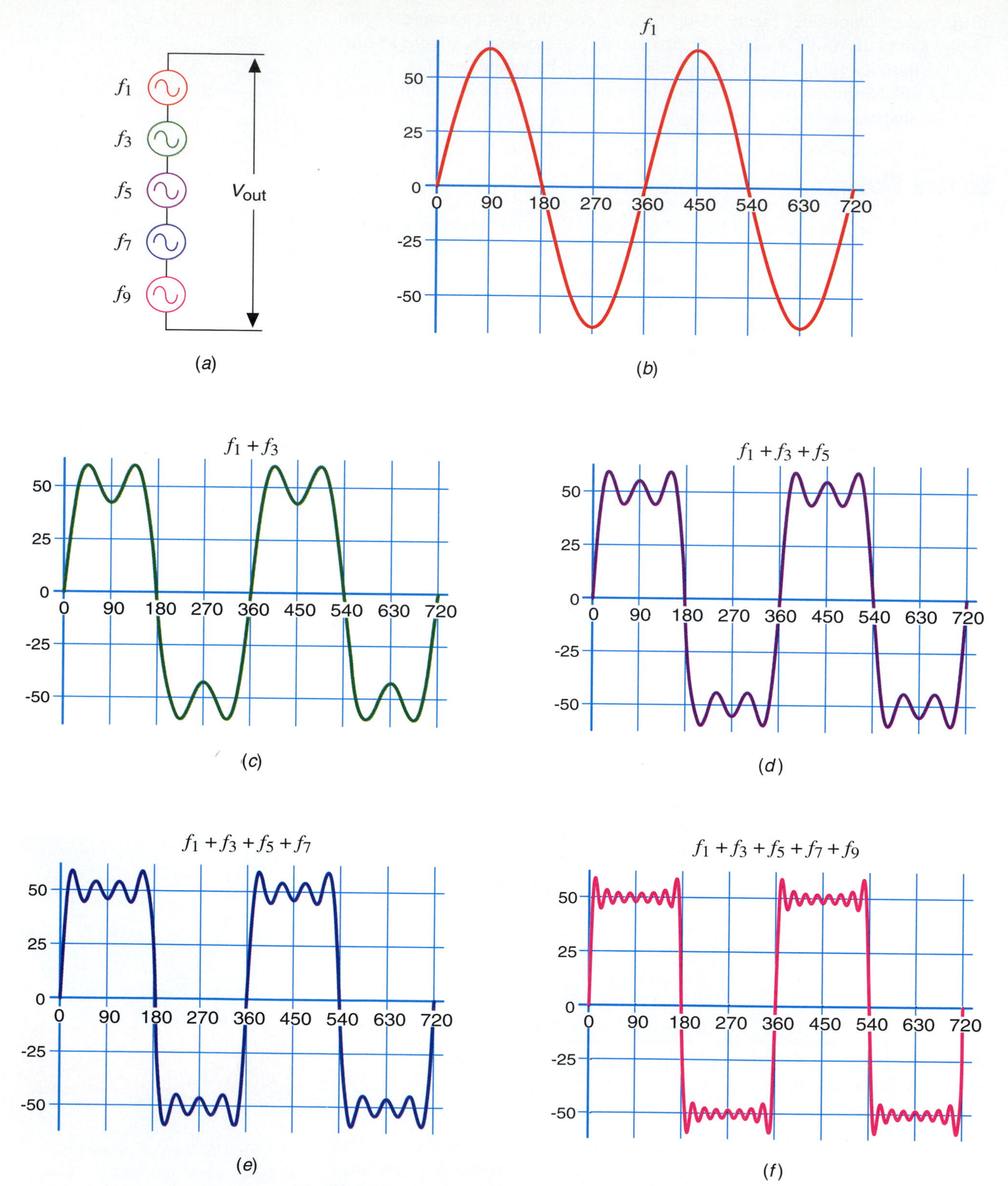

Figure 15-9 Five odd harmonic elements may not produce a square wave that is adequately defined for many applications.

Figure 15-9(a) shows how five sine wave generators are connected in series to produce a square waveform that is crude but acceptable in many situations. All five sine waves are initially in phase, meaning, they all start at the same time. The amplitude of each harmonic is defined by Eq. 15-2. Figures 15-9(b)–(f) show the progressive development of a square waveform as each of the first five odd harmonics are added. Although the composite waveform is decidedly square containing only five harmonic elements, some circuit applications may require the addition of more harmonic elements.

The sine wave–like variations at the top and bottom of the waveforms are often called **ringing.** The presence of ringing is an indication that the upper harmonic frequencies are not present. It is possible for a circuit to pass the lower frequency elements without altering their amplitudes, while completely attenuating all of the harmonics above some frequency.

REVIEW QUIZ 15-2

1. A sawtooth waveform contains______________ and ______________ harmonics.
2. A square waveform is composed of an infinite number of ______________ harmonics.
3. A spectrum analyzer displays electrical waveforms in the ______________ domain.
4. A spectrum analyzer can identify the ______________ and ______________ of each harmonic in a waveform.

15-3 Wave Symmetry

Periodic electrical waveforms are often symmetrical (balanced) and may have even symmetry, odd symmetry, or half-wave symmetry. The symmetry of a waveform may depend on when the waveform is considered to begin. For instance, an oscilloscope display of a sine wave and a cosine wave may appear similar because the oscilloscope usually begins tracing the waveforms at some predetermined voltage level. This causes the waveforms to be displayed at relatively different times—in this case, one-quarter of the period. If a dual trace oscilloscope is used, the display of the waveforms is initiated by one or the other waveform, and their true phase (time) relationship is revealed. The sine wave is symmetrical about the horizontal (time) axis, and the cosine wave is symmetrical about the vertical (amplitude) axis.

Even Symmetry

A waveform with **even symmetry**, such as a cosine wave, is symmetrical about the vertical axis, as shown in Fig. 15-10 on the next page. The vertical axis is considered to be the beginning of the period.

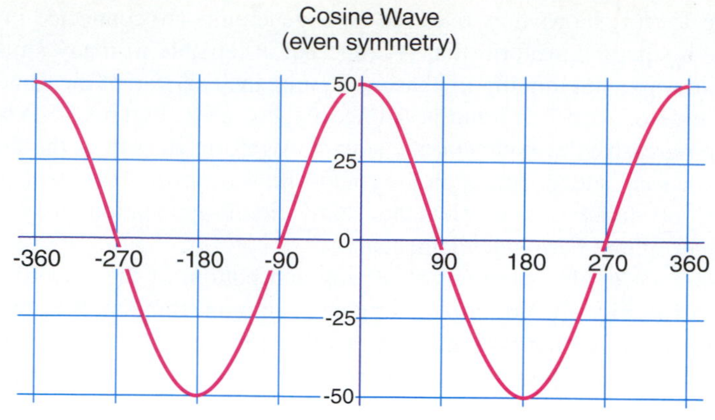

Figure 15-10 Waveforms with even symmetry are symmetrical about the amplitude axis.

Odd Symmetry

A waveform with **odd symmetry**, such as a triangular waveform, is symmetrical about a line drawn midway between the horizontal axis and vertical axis, as shown in Fig. 15-11. The line must pass through the origin formed by the intersection of the two axes.

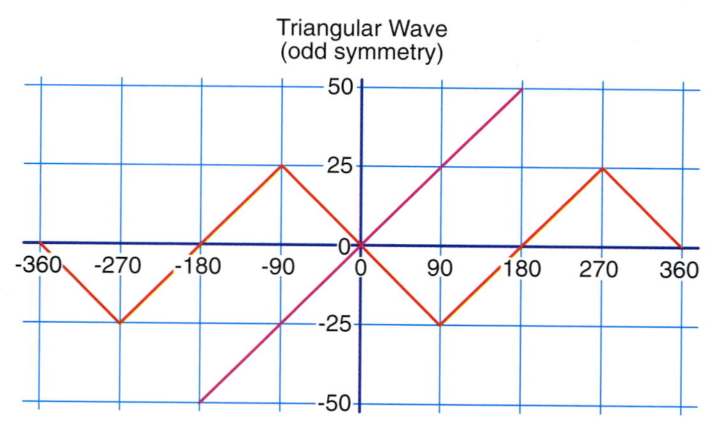

Figure 15-11 A waveform with odd symmetry is symmetrical about a line drawn midway between the horizontal axis and vertical axis. The line must pass through the origin formed by the intersection of the two axes.

Half-Wave Symmetry

If the portion of a waveform during the first half of the waveform's period is a mirror image of the portion of the waveform during the second half of the period, the waveform has **half-wave symmetry.** A square wave is an example of a waveform with half-wave symmetry, as shown in Fig. 15-12.

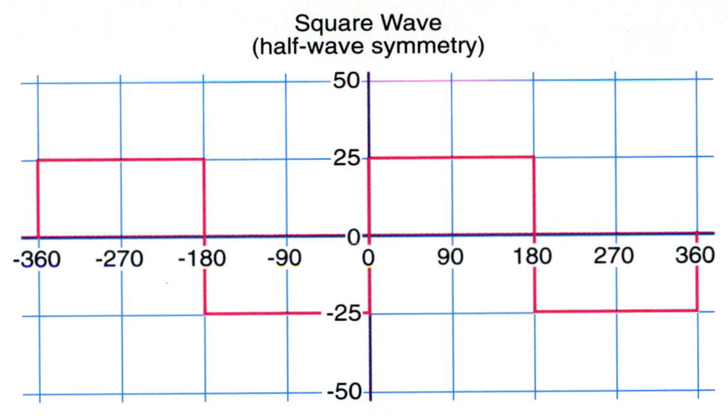

Figure 15-12 A waveform with half-wave symmetry is identical during the first and second halves of the period, except that its two half-cycles are of opposite polarity.

REVIEW QUIZ 15-3

1. Periodic waveforms may have ______________, ______________, and ______________ symmetries.

15-4 DC Component

Electrical waveforms have a **DC component** that is determined by the average amplitude value of the periodic wave. The DC component is calculated by dividing the area under one cycle by the period in seconds, as shown in Eq. 15-3.

Equation 15-3

$$V_{\mathrm{DC}} = \frac{\mathrm{A}}{p}$$

where V_{DC} = The voltage of the DC component.
A = The area under one cycle of the waveform.
p = The period of the waveform in seconds.

Figure 15-13(*a*) on the next page shows a symmetrical square wave that rises from zero to a positive peak value of 10 volts. The period of the wave is 2 seconds. The area under each half-cycle is found by multiplying its base by its height because its shape is rectangular. The area under the first half-cycle is found by multiplying 10 volts by 1 second. Thus the area is 10 volt-seconds. The voltage during the second half-cycle is zero and its period is also 1 second. The area of the second half-cycle is 0 volts times 1 second for 0 volt-seconds. The total area for the two half-cycles is found by adding them together: 10 volt-seconds plus 0 volt-seconds equals 10 volt-seconds. When 10 volt-seconds is divided by the period of 2 seconds, the units of seconds cancel, and 10 divided by 2 is 5. The average DC voltage, therefore is 5 volts, as shown in Fig. 15-13(*b*).

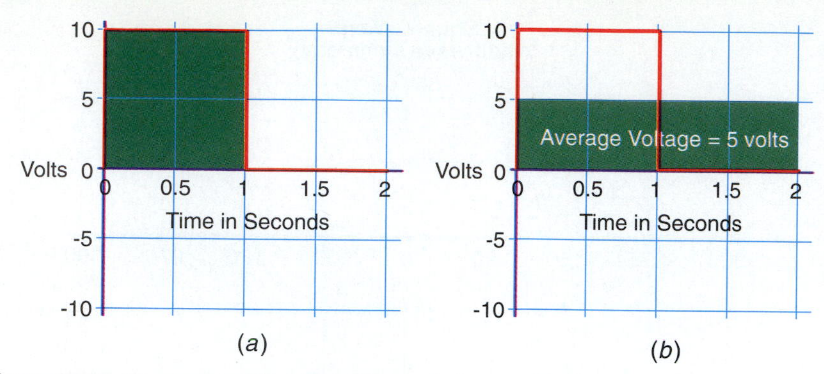

Figure 15-13 The average DC voltage of a periodic waveform is determined by adding the areas of each half-cycle and dividing the result by the period of the waveform.

The procedures for calculating the areas under each half-cycle depends on the shape of the half-cycle. When the waveshape is square, sawtooth, or triangular, the procedure is relatively easy, but it may be far more difficult for any other waveshape. If the waveform has half-wave symmetry, the DC component will be zero because the areas under each half-cycle are equal and opposite.

REVIEW QUIZ 15-4

1. If the two half-cycles of a waveform are equal and opposite, the DC component of the waveform will be ______________ .
2. When dividing the average area under one cycle of a waveform by the period, the units of ______________ cancel.
3. The DC component is calculated by dividing the area under one cycle by the ______________ .

Heinrich Rudolf Hertz

SUMMARY

Periodic, nonsinusoidal waveforms are far more complex than they appear on the display of an oscilloscope. They are composed of multiple sine waves of different frequencies that have complex mathematical relationships to each other. The oscilloscope cannot show the individual frequency components because it displays the waveform only in the time domain. Waveforms are always displayed in the time domain as a composite of its individual elements. A spectrum analyzer has a display like an oscilloscope, but it operates in the frequency domain. When a waveform is displayed in the frequency domain, the amplitude and frequency of each harmonic component are revealed. Circuits that must pass nonsinusoidal waveforms must be able to pass frequencies much higher than the fundamental frequency of the waveform; otherwise, the shape of the waveform will be distorted. Waveforms will have a DC component when the areas of the two half-cycles are not equal and opposite. As digital signal processing increases in popularity, a more thorough understanding of waveforms may become necessary to stay up-to-date in the fast-paced electronics industry.

NEW VOCABULARY

cosine wave
DC component
even symmetry
first harmonic
frequency domain
fundamental frequency
half-wave symmetry
harmonics
odd symmetry
periodic
phase
phase shifted
quadrature
ringing
spectrum
spectrum analyzer
time domain

Questions

1. What piece of test equipment is typically used to measure the peak-to-peak amplitude and period of a voltage waveform?
2. What waveform looks like a sine wave but begins on its peak positive cycle?
3. What name is given to multiples of a fundamental frequency?
4. What name is applied to the difference in time between two waveforms that is measured in degrees or radians?
5. What waveform parameters are displayed in the time domain?
6. What waveform parameters are displayed in the frequency domain?

Problems

1. If the peak-to-peak voltage of a sawtooth waveform is 30 volts, what is the voltage of the first harmonic?
2. What is the amplitude of the fifth harmonic composing a square wave if its peak-to-peak amplitude is 50 volts?
3. What is the DC component of a 100-volt square wave with half-wave symmetry and a period of 4 seconds?

Critical Thinking

1. If a square wave were applied to a circuit that rejected all frequencies above the fundamental, what would be the output waveform?
2. Why are a sine wave and a cosine wave in quadrature?
3. When a sine wave and a cosine wave of the same frequency and amplitude are added together, why does the resultant sine wave have a peak amplitude that is 1.414 of either waveform?

Answers to Review Quizzes

15-1

1. yes
2. harmonic
3. harmonics
4. time

15-2

1. even; odd
2. odd
3. frequency
4. amplitude; frequency

15-3

1. even; odd; half-wave

15-4

1. zero
2. time
3. period in seconds

Chapter 16

Inductance *and* Transformers

ABOUT THIS CHAPTER

This chapter introduces two new components: inductors and transformers. Most inductors are coils of wire that are often wound over permeable cores of ferrite, powdered iron, or laminated iron. In their simplest form, they are wound on magnetically neutral forms that allow them to be called air core inductors. Like other two-terminal components, inductors may be placed in series or parallel, but they introduce the added complexity of having

After completing this chapter you will be able to:

- Explain why inductors oppose the flow of current without consuming power, as resistors do.
- Describe how the physical properties of a coil affect its inductance value.

magnetic fields that interact and affect their values. Transformers are made from two or more inductors positioned in such a way that the magnetic lines of force produced by one cut across the other. Transformers are very useful because they can increase or decrease voltages without losing power. They can also provide DC isolation between an AC source and its load.

- ▽ Identify several applications for transformers that cannot be accomplished using other components.
- ▽ Describe how current flowing in an isolated secondary can cause current to flow in the primary.

16-1 Inductance

Inductance is the property of a circuit to oppose any change in current. The opposition stems from a self-induced voltage created by electromagnetic induction. This self-induced voltage opposes any action of the source voltage that would change the circuit current.

Inductance is most noticeable in coils of wire because coils significantly increase the flux density of the magnetic field produced by the current flowing through them. Inductance, however, is present in all conductors because conductors have the potential to produce a magnetic field regardless of their shape. Coils designed to have specific values of inductance represent a category of passive components called inductors.

In mathematical equations, the capital letter L represents inductance. The schematic symbol for an inductor is a series of open or overlapping loops, as shown in Fig. 16-1. The more modern of the two symbols is the open loop design. A capital L is also the circuit symbol for inductors and introduces their schematic identification numbers; for example, L_1, L_2, L_3, and so on.

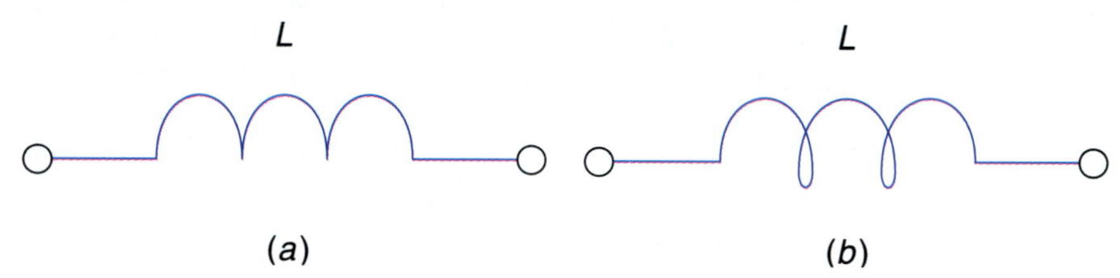

Figure 16-1 The schematic symbol for an inductor is a series of (*a*) open or (*b*) overlapping loops. The open loop version is the more modern of the two, but the overlapping loop symbol may occasionally be encountered.

Counter emf

Voltage applied across a coil initiates a flow of current. At first that current is very small, but as the current increases, a steadily intensifying magnetic field expands across the turns of wire that are responsible for its existence and induces a voltage called *counter emf,* or *back emf.* The counter emf (electro-magnetic force) pushes back against the source voltage and attempts to keep the circuit current at zero. It can never achieve this because to do so would be self-destructive. In a circuit with no resistance, the counter emf retards the increase in current, causing it to rise slowly and linearly toward infinity. The current's rate of change depends on the value of inductance in the circuit. Faster rates of change would be achieved with less inductance. Very large values of inductance could theoretically cause the increase in current to be imperceptibly small.

Induced emf

The development of a magnetic field represents the storage of energy. When a magnetic field collapses, energy either returns to the circuit that created it or is dispersed. As the magnetic field around an inductor collapses, the inductor again produces voltage, but the polarity is opposite that of the counter emf because the direction of the magnetic field has reversed. This induced emf and the source voltage are now in series-aiding. As the source voltage decreases, the inductor increases its voltage in an attempt to hold the current at its previous level. The induced voltage cannot accomplish this because it owes its existence to the changing magnetic field, which only changes when the current changes.

The Henry Unit of Inductance

The fundamental unit of inductance is the *henry*, named in honor of Joseph Henry (1797–1878). He discovered the principles on which the telegraph was founded, and did extensive research in electromagnetism. An inductor has an inductance of 1 henry when a current flowing through the inductor changing at 1 ampere per second induces a voltage of 1 volt. A capital letter *H* is commonly used as an abbreviation for the word *henry* in written inductance values, such as: $L = 1$ H.

ELECTRONIC FACTS

Since the henry is an extremely large unit of inductance, values are usually given in millihenrys, microhenrys, and nanohenrys.

Equivalent Circuits

The voltage of a source connected across an inductor and the voltage induced into the inductor by its magnetic field behave as though the voltages were connected in series across a single load resistor. For the most part that load resistor represents the resistance of the wire used to make the inductor. As the current through an inductor increases, the two voltages appear to be connected in series-opposing, as shown in Fig. 16-2(*a*). As the current through the inductor decreases, the two voltages appear to be connected in series-aiding, as shown in Fig. 16-2(*b*). The current through the coil is initially supplied by the source voltage, but as the magnetic field collapses, should the source voltage decrease, it is supplied by the coil.

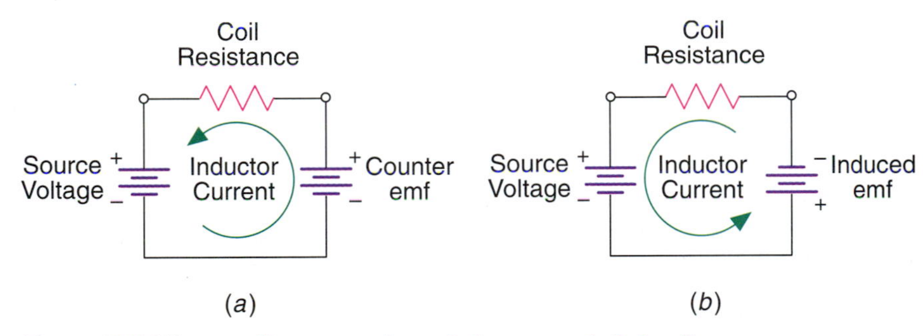

Figure 16-2 When a voltage source is applied across an inductor, it appears to be connected in series-opposing with the counter emf and connected in series-aiding with the induced emf. The source supplies the inductor current as its voltage increases. The induced emf supplies the inductor current as the source voltage decreases.

REVIEW QUIZ 16-1

1. Inductance is the property of a circuit to ______________ any change in current.
2. Counter emf is produced by a coil as the current through it ______________ .
3. The unit of inductance in the SI system is the ______________ .

16-2 Factors That Determine Inductance

The physical dimensions of an inductor and the magnetic characteristics of its core material determine its inductance value. Freestanding coils, or those wound over a support made of a nonferrous insulating material such as paper, ceramic, or plastic have an *air core*. Figure 16-3(*a*) shows the schematic symbol of an air core coil which is the symbolic coil previously introduced in Fig. 16-1(*a*). A pair of dashed lines drawn above the coil's symbol, as shown in Fig. 16-3(*b*), indicates that the coil has a *ferrite core*, or *powdered iron core*. If the lines are solid, as shown in Fig. 16-3(*c*), the coil has a *laminated iron core*.

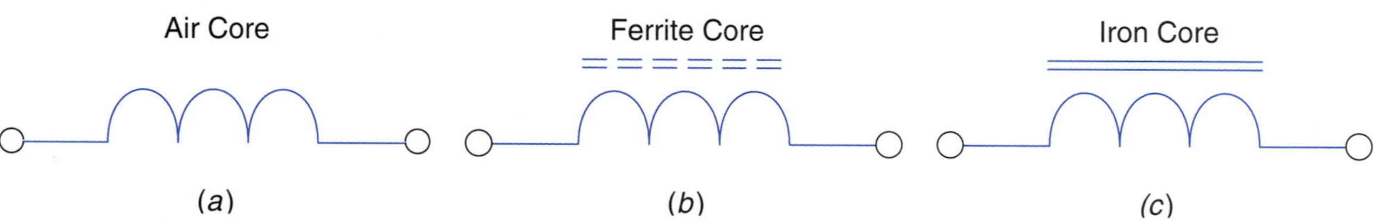

Figure 16-3 Cores may be air core, ferrite core (also called powdered iron core), or iron core. The presence of a core is symbolized by a pair of lines drawn alongside the coil's symbol. Dashed lines indicate a ferrite core, and solid lines indicate a laminated iron core.

Anything that increases an inductor's magnetic flux density increases its inductance. The five major factors that determine a coil's inductance are:

1. The number of turns in the coil.
2. The length of the coil.
3. The cross-sectional area of the coil.
4. The relative permeability of the core material compared to air or vacuum.
5. The conductivity of the core material.

Equation 16-1 allows the inductance of a coil to be determined mathematically using the first four factors. How the conductivity of the core affects the inductance of a coil is discussed later in this chapter.

Equation 16-1

$$L = \mu_r \left(\frac{N^2 \text{X} A}{l} \right) 1.26 \times 10^{-6}$$

where L = The inductance of the coil in henrys.
N = The number of turns in the coil.
A = The cross-sectional area of the coil in square meters (m^2).
μ_r = The relative permeability of the core compared to air or vacuum.
l = The length of the coil in meters.

The constant 1.26×10^{-6} is the absolute permeability of air in a vacuum in the SI system and has no unit.

The effective length of a coil is altered by the presence of a ferrous core. This is illustrated in Fig. 16-4. The overall length is not measured from the first turn to the last turn, as might be expected. Rather it is equal to the length of the core. The core increases the inductance of the coil because it is more permeable than

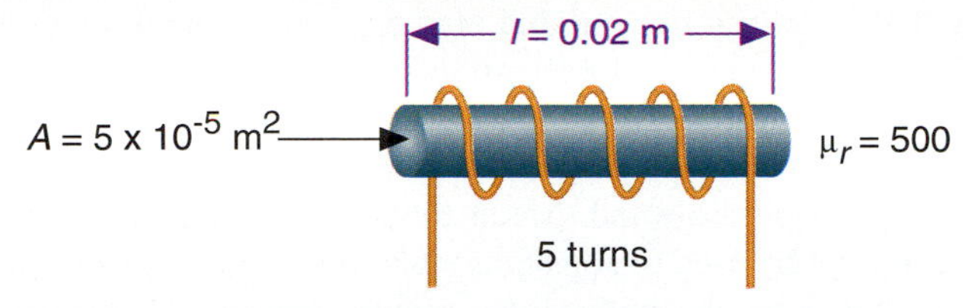

Figure 16-4 The inductance of a coil is determined by its physical dimensions and the permeability of its core material. If a core is present, the effective length of the coil is equivalent to the length of the core.

air. As a result the magnetic flux density within the coil increases. In Fig. 16-4 the core is 500 times more permeable than air. Using Eq. 16-1, the inductance of the coil in Fig. 16-4 is 39.375 μH. If the number of turns is increased from 5 turns to 10 turns without changing the coil's physical dimensions, its inductance will increase four times to 157.5 μH. All of the magnetic lines of force created by each turn cut across all of the other turns because the core extends completely through the coil.

Variable Inductors

It is often desirable to vary the inductance of a coil. This can be accomplished by changing any of its physical parameters, but the most common solution is to vary the permeability of the core. The usual method is to wind the coil on a magnetically neutral nonconductive threaded form. A threaded ferrite core, called a *slug* is then screwed into the form until a portion of it is inside the coil, as shown in Fig. 16-5(*a*). In this way the permeability can be varied over a considerable range of values. The schematic symbol of a variable inductor is often identified with a box drawn alongside the coil, as shown in Fig. 16-5(*b*), (*c*), and (*d*). The box has an arrow attached to its top, bottom, or both to indicate that the core is movable. If the arrow points up, the core is accessed from the top. If the arrow points down, the core is accessed from the bottom. If the arrow points both directions, the core is accessible from either top or bottom.There are a number of variations on this symbol, but they are usually self-explanatory. Another method

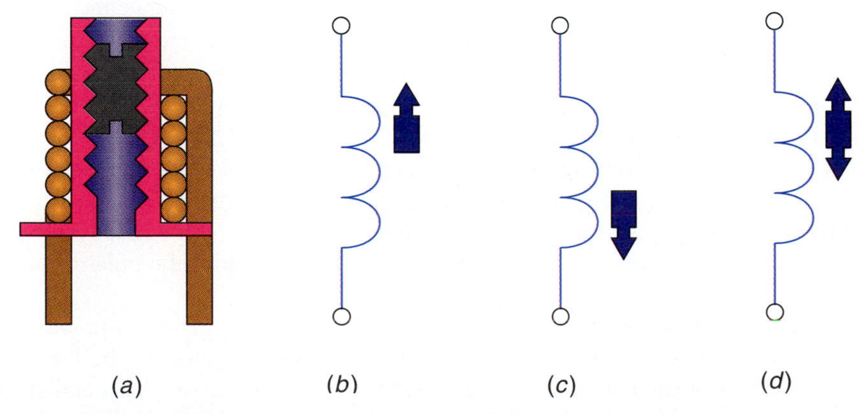

Figure 16-5 Variable inductors are usually provided with some means of changing the permeability of the core. The most common method is to use a threaded ferrite core that can be moved in to and out of the coil.

of indicating that an inductor is variable is to draw an arrow through its symbol, as is done with potentiometers and rheostats.

The cores usually have a slot to accommodate a plastic or fiberglass screwdriverlike tool, or they have a hexagonal (six-sided) hole that accommodates an appropriately shaped plastic wand. Metal tools, such as screwdrivers and allen wrenches, should not be used to adjust variable inductors because their presence inside the coil will alter the inductance, giving undesirable results. A certain amount of caution must also be exercised when adjusting such cores because they are very brittle and break easily.

Ferrite cores come in a wide range of permeabilities. It is customary for cores of vastly different permeabilities to be very similar in appearance. The permeability of a core is seldom identified with anything other than a dot of paint, and the color used is not standardized from one manufacturer to another. The potential for choosing an unsuitable replacement for a broken core dictates that extra caution be exercised when they must be replaced.

REVIEW QUIZ 16-2

1. Anything that increases the ______________ of a coil increases its inductance.
2. The permeability or air of vacuum is ______________ .
3. The effective length of a coil is altered by the presence of a(n) ______________ .
4. A threaded ferrite core used to adjust the permeability of an inductor is called a(n) ______________ .

16-3 Inductive Kickback Voltage

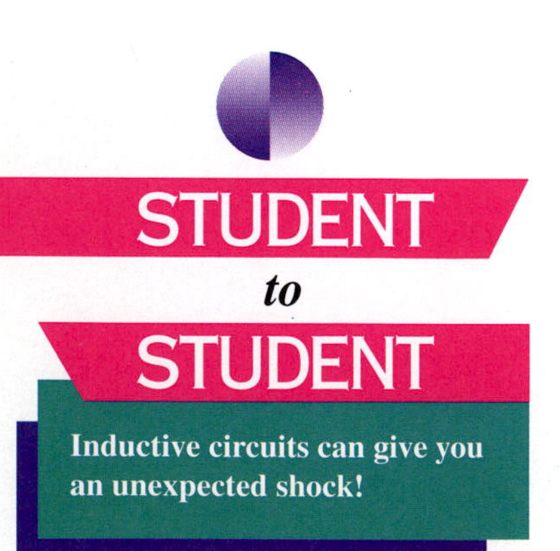

The voltage generated by an inductor as its magnetic field is collapsing may well exceed the source voltage if the source is suddenly removed. In its attempt to maintain the circuit current when turned off, the induced voltage of the inductor, also called *kickback voltage,* may be sufficient to arc across the switch contacts. Electric arcs produce very high temperatures and may melt areas of the contacts, causing them to become pitted. The sharp contours of the pits accelerate the erosion process. To avoid premature damage, switches designed for inductive loads are often plated with tungsten because tungsten has a very high melting point.

The effects of the kickback voltage and the general principles of inductance reveal themselves through an experiment using a low-voltage battery, a momentary contact switch, an iron core inductor, and a neon lamp placed across the inductor, as shown in Fig. 16-6. The neon lamp requires a minimum of 70 volts to produce light because of neon's ionization potential. The light will only be emitted from the negative terminal, thus, providing a way to identify the polarity of the applied voltage.

Connecting the neon lamp in parallel with the inductor ensures that the same voltage is applied across each of them. The battery supplies power to the lamp and coil when a momentary contact (push button) switch is pressed. This action proves uneventful because the battery's voltage is far below the ionization potential of the neon lamp. When the switch is released, however, the lamp emits

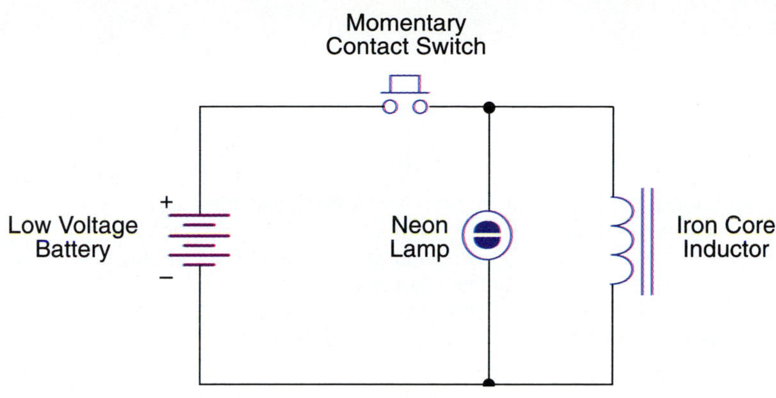

Figure 16-6 A neon lamp placed in parallel with an iron core inductor visually demonstrates that the kickback voltage can greatly exceed the applied voltage of the source. The lamp also verifies that the polarity of the kickback voltage is opposite that of the applied voltage.

a bright flash of light from the electrode that was connected to the positive side of the battery. The fact that the lamp emitted light indicates that the voltage across it and the coil must have been at least 70 volts. Because the upper electrode of the lamp produced the light, the polarity of the voltage induced into the coil by the collapsing magnetic field was opposite that of the battery. These findings are consistent with the theory of inductance.

REVIEW QUIZ 16-3

1. The kickback voltage is only present when the magnetic field is _______________ .
2. The kickback voltage can be greater than the applied voltage because the magnetic field collapses _______________ than it builds.
3. The kickback voltage and the source voltage have _______________ polarities.
4. Switch contacts designed for inductive loads are often plated with _______________, which can withstand high temperatures.

16-4 Mutual Inductance

The inductance of a coil is generally considered to increase by the square of its turns (as reflected earlier by N^2 in Eq. 16-1), but that is not always true. Each turn of wire composing a coil develops magnetic lines of force that surround it, as shown in Fig. 16-7(*a*) on the next page. If the turns are spaced far apart, very few of the lines of force created by one turn cut across the other, as indicated in Fig. 16-7(*b*). When the turns are wound very close to each other, as in Fig. 16-7(*c*), the lines of force produced by each turn cut across the other turns of the inductor. This induces a voltage into all the turns in addition to the voltage that is self-induced. The property of inducing a voltage into neighboring coils pro-

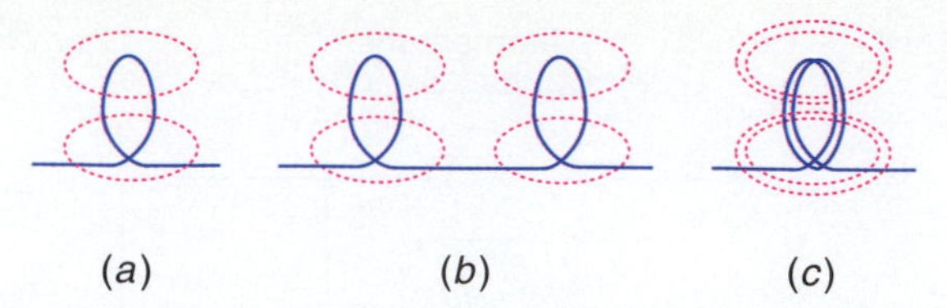

Figure 16-7 When the turns of an inductor are spaced apart from each other (*b*), their magnetic lines of force do not interact as they do when the turns are in close proximity. (*c*). The inductance value of coils whose magnetic fields do not cross each other's is owing to the sum of the self-inductance values. The inductance value of coils whose turns completely share each other's magnetic lines of force is affected by mutual inductance and increases as the square of the turns.

duces an effect called *mutual inductance.* The property of a coil to induce a voltage into itself is called *self-inductance*.

When a coil's turns are separated from each other, but share some lines of force, they are said to be *loosely coupled*. When they are wound close to each other, sharing most of their lines of force, they are said to be *tightly coupled*. When they share all of their lines of force, they are said to have *unity coupling*. In practice, unity coupling is only possible when the coil length is very short or when a coil has a permeable core.

The terms *loosely coupled* and *tightly coupled* generally refer to the degree of magnetic coupling between two individual coils. Magnetic coupling is expressed as coefficient of coupling and is stated as a decimal value from 0 to 1. The symbol for *coefficient of coupling* is a lowercase k. When all the lines of force created by one coil cut across all the turns of another coil, the coefficient of coupling is 1. If only half the lines of force are involved, the coefficient of coupling is 0.5. Two coils with mutual coupling are drawn parallel to and facing each other, as shown in Fig. 16-8.

The mutual inductance between two coils, L_M, may be calculated using Eq. 16-2 if the values of the inductors and the coefficient of coupling are known.

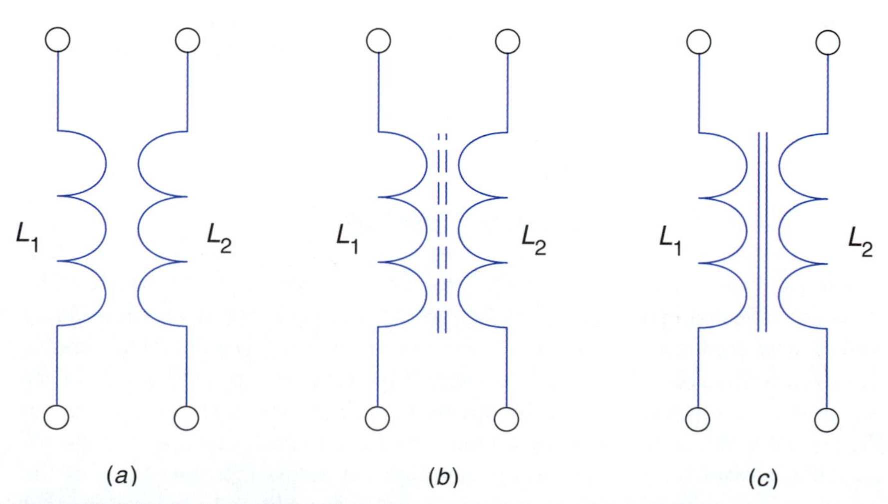

Figure 16-8 Mutually coupled coils are drawn parallel to and facing one another. Only one core symbol is shown to indicate that it is shared by both coils.

Equation 16-2

$$L_M = k\sqrt{L_1 \times L_2}$$

where L_M = The mutual inductance in henrys.
k = The coefficient of coupling as a decimal value from 0 to 1.
L_1 = The inductance of L_1 in henrys.
L_2 = The inductance of L_2 in henrys.

REVIEW QUIZ 16-4

1. Mutual coupling between coils indicates that their ______________ are shared.
2. The coefficient of coupling is between ______________ and ______________ .
3. Unity coupling is assured by the use of a(n) ______________ .
4. Mutual inductance is measured in ______________ .

16-5 Inductors in Series and Parallel

The values of inductors connected in series are added to obtain the total inductance as shown in Eq. 16-3. For instance, a 12-microhenry inductor and a 20-microhenry inductor connected in series are equivalent to a single 32-microhenry inductor. This simple numeric addition assumes a complete lack of magnetic coupling between inductors. Magnetic coupling is normally not a problem when the coils have ferrite cores because most of the magnetic lines of force developed by the coils are confined to their cores. When ferrite cores are not used, the coils are often placed at right angles to each other to minimize coupling. In that way any mutually induced voltages cancel.

Equation 16-3

$$L_T = L_1 + L_2 + L_3 \ldots$$

When two inductors connected in series have some degree of magnetic coupling, they may work together to increase the total inductance or oppose each other and decrease the total inductance. Equation 16-4 expands on Eq. 16-3 to include the effects of any mutual coupling that may exist.

Equation 16-4

$$L_T = (L_1 + L_2) \pm 2L_M$$

The $\pm$ before L_M indicates that two possibilities exist for solving Eq. 16-4. The first possibility is that the magnetic fields are additive. In that case, L_M is multiplied by 2 and then added to the sum of L_1 and L_2. The second possibility is that the magnetic fields are subtractive. Thus L_M is multiplied by 2 and then subtracted from the sum of L_1 and L_2.

For example, suppose two 500-microhenry inductors are connected in series

with a 0.5 coefficient of coupling. Using Eq.16-2, we find that their mutual inductance is 250 microhenrys. The sum of the two inductance values is 1000 microhenrys. If their magnetic fields are additive, then 2 times the 250-microhenry mutual inductance, or 500 microhenrys, is added to the 1000 microhenrys for a total of 1500 microhenrys. If their magnetic fields are subtractive, then 500 microhenry is subtracted from 1000 microhenry for a total of 500 microhenrys. Had the mutual coupling been 1.0, the inductance would have increased four times to 2000 microhenrys, or the inductances would have canceled and become zero.

Inductance is not always desirable, and the canceling of inductive effects by controlling the mutual coupling between two coils is frequently used to one's advantage. Wire-wound resistors often have undesirable values of inductance. The problem is eliminated by winding one-half of the resistor in one direction and the other half in the opposite direction, as shown in Fig. 16-9. The equal and opposite inductance values cancel because their magnetic fields oppose each other, regardless of the direction of current. Wire-wound resistors are not normally noninductive because applications that require noninductive resistors are somewhat limited and wire-wound resistors are more costly.

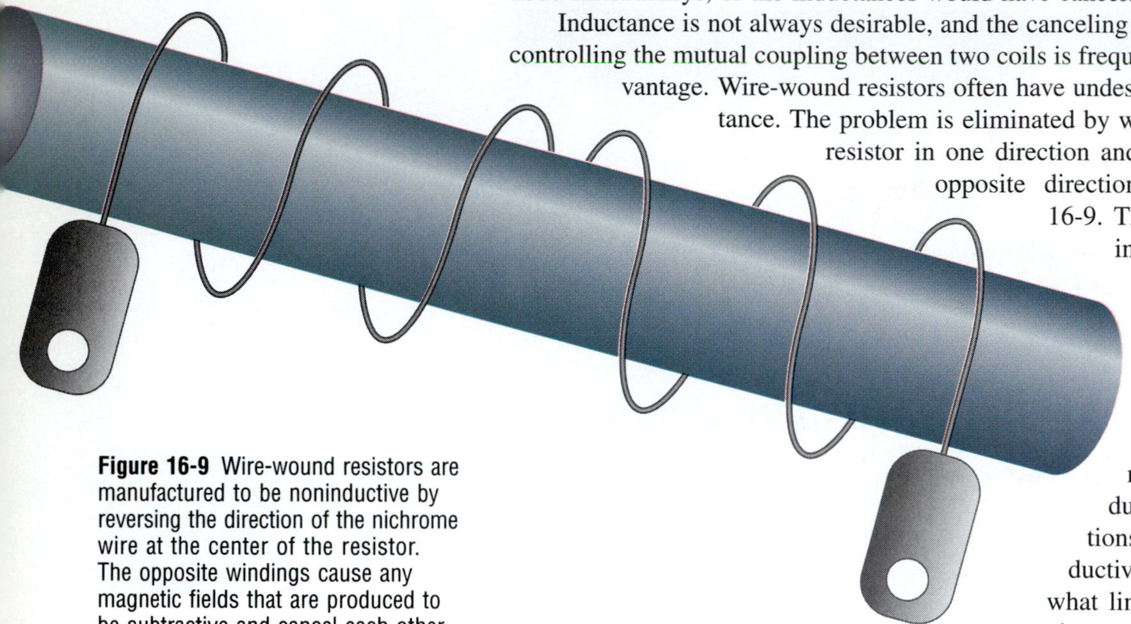

Figure 16-9 Wire-wound resistors are manufactured to be noninductive by reversing the direction of the nichrome wire at the center of the resistor. The opposite windings cause any magnetic fields that are produced to be subtractive and cancel each other.

Inductors in Parallel

The values of inductors connected in parallel are calculated in the same way as resistors connected in parallel, assuming that their magnetic fields do not interact. The equations are the same, except L is substituted for R, as shown in Eq. 16-5.

Equation 16-5

$$(a)\quad L_T = \frac{L_1 \times L_2}{L_1 + L_2}$$

$$(b)\quad L_T = \frac{1}{\frac{1}{L_1} + \frac{1}{L_2} + \frac{1}{L_3}} \cdots$$

REVIEW QUIZ 16-5

1. The total value of series-connected inductors is calculated the same way as series-connected ______________ .
2. A wire-wound resistor can be made noninductive by ______________ the nichrome resistance element at its center.
3. The magnetic fields shared between mutually coupled inductors may be ______________ or ______________ .

16-6 Inductance and Alternating Current

Inductance is normally associated with sinusoidal alternating currents because the effects of inductance on such currents are measured easily. The polarity of the voltage induced into an inductor is determined by whether the magnetic field generated by the current flowing through it is increasing or decreasing. The counter emf may never exceed the voltage of the source because to do so would be self-destructive. After all, the magnetic field that is producing the counter emf is the result of current flowing through the inductor. If that current is reduced to zero, so is the strength of the magnetic field.

The polarity reversals of the voltage induced into an inductor by its magnetic field can be visualized more clearly by comparing the individual graphs of the applied voltage, the current through the coil, and the magnetic field that surrounds it. These three quantities are compared on the composite graph in Fig. 16-10. The graph also reveals that the voltage and current of the source are

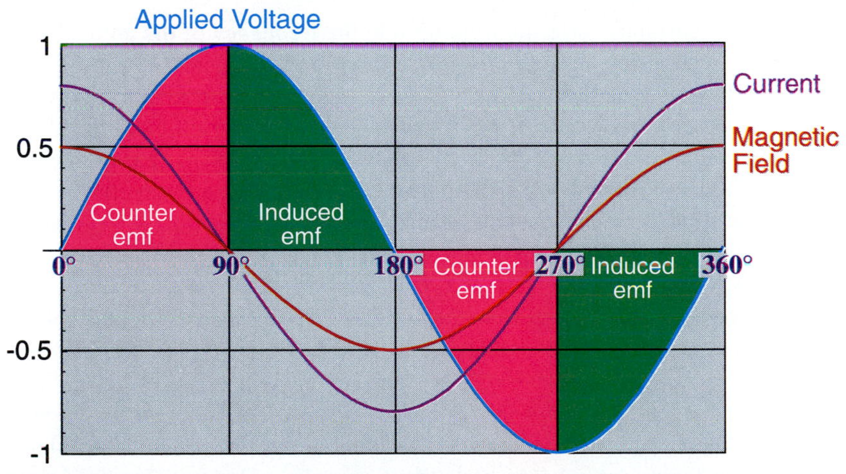

Figure 16-10 As the voltage applied to an inductor increases, the coil produces a counter emf. As the source voltage decreases, the coil produces induced emf. The voltage applied to the coil and the current through the coil are 90° out of phase with the current leading the voltage.

out of phase. The source voltage leads the source current by 90°. The phase shift between the source voltage and the source current occurs because no magnetic field is developed when the current is at its peak positive or negative values. The voltage induced by the magnetic field determines the amplitude of the source current.

Several important points should be highlighted at this time:

- The magnetic field developed around an inductor is created by the current flowing through it. The strength of that magnetic field is directly proportional to the current.
- Both the counter emf and the induced emf are dependent upon the rate of change of the magnetic field and, therefore, the rate of change of the current.

- The change in current is greatest as it passes through zero. There is no change at the instant it reaches its peak positive or peak negative amplitude.
- When a sinusoidal AC voltage is applied across an inductor, the voltage will always lead the current by 90°or $\pi/2$ radians.

Inductive Reactance

Inductors impose a limit on the current a sinusoidal AC voltage source can move through them. They do this in a way that is only remotely similar to resistance. The counter emf developed by an inductor limits the flow of current through it and thus limits the strength of the magnetic field developed around it. When the source voltage starts to decline, the induced emf of the inductor follows it down and returns the stored energy. The opposition that an inductor presents to an AC source is called *inductive reactance*. The major difference between inductive reactance and resistance is that inductive reactance does not consume power from the source. Like resistance, however, inductive reactance is measured in ohms and obeys all the forms of Ohm's law. The symbol for reactance is the capital letter X. Inductive reactance is indicated by placing a subscript capitol L after the X to form X_L.

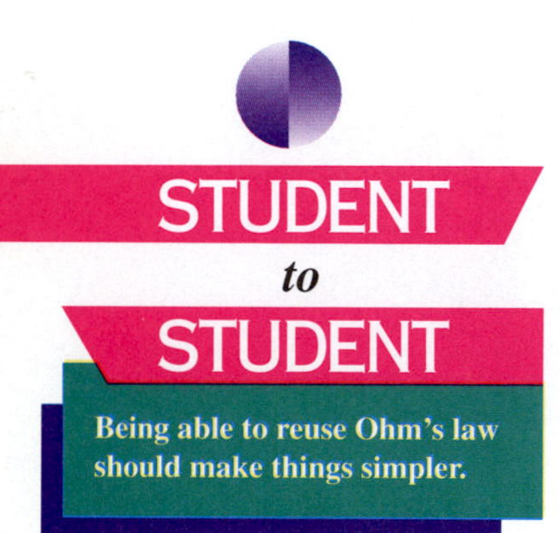

Inductive reactance varies in proportion to the frequency of the applied voltage. Higher frequencies force the magnetic field to change more rapidly. Faster changes in the magnetic field produce a larger counter emf, more reactance, and less current. The reduced current produces a weaker magnetic field, which produces less energy as the magnetic field collapses. The result is that higher frequencies produce larger values of inductive reactance, that are directly proportional to the applied frequency. The reactance of any conductor at a particular frequency will double if the frequency is doubled; it will reduce to half if the frequency is halved.

Inductive reactance depends on the angular velocity of the applied voltage in radians per second and the inductance in henrys. Each AC cycle sweeps out 2π, or 6.28, radians. Multiplying 6.28 by the frequency in hertz (cycles per second) yields the angular velocity in radians per second. This relationship appears in numerous equations as $2\pi f$. It is also written as the lower case Greek letter omega, which looks like a lowercase w with rounded sides (ω). Equation 16-6 shows the two forms of the inductive reactance equation; The form commonly used by technicians is (*a*).

Equation 16-6

$$(a)\ X_L = 2\pi f L$$

$$(b)\ X_L = \omega L$$

Reactance in Series and Parallel

The procedures for determining the reactance values of inductors in series or parallel are identical to the procedures used with resistors and inductors. Inductive reactances in series add numerically, as shown in Eq. 16-7.

Equation 16-7

$$X_{L_T} = X_{L_1} + X_{L_2} + X_{L_3} \cdots$$

Inductive reactances in parallel may be calculated using the (*a*) product over sum equation or (*b*) the reciprocals equation, as shown in Eq. 16-8.

Equation 16-8

$$(a)\ X_{LEQ} = \frac{X_{L_1} \times X_{L_2}}{X_{L_1} + X_{L_2}}$$

$$(b)\ X_{LEQ} = \frac{1}{\frac{1}{X_{L_1}} + \frac{1}{X_{L_2}} + \frac{1}{X_{L_3}}} \cdots$$

REVIEW QUIZ 16-6

1. As the applied voltage to an inductor increases, the coil produces ______________ emf.
2. The opposition to current created by an inductor is called ______________ .
3. In an inductor, the voltage will lead the current by ______________ .
4. Reactances in series or parallel behave in the same way as ______________ in series or parallel.

16-7 Transformers

A transformer is a device made from two or more mutually coupled inductors. Transformers perform the following functions:

1. Increase or decrease voltage.
2. Increase or decrease current.
3. Provide DC isolation between an AC source and a load.
4. Raise or lower the effective load resistance as "seen" from the source.
5. Provide multiple voltages from a single source.

Transformer Terminology

The basic symbol for a transformer consists of two inductors facing one another, as shown in Fig. 16-11. The inductors of a transformer are called *windings*. Transformers are designed to operate from alternating current, but they may also be operated from varying direct current or pulsating direct current. Power applied to one of the transformer's windings creates a changing magnetic field that "cuts" across the other winding, inducing a voltage into it. The winding to which power is applied is the *primary*, and the winding from which power is taken is the *secondary*.

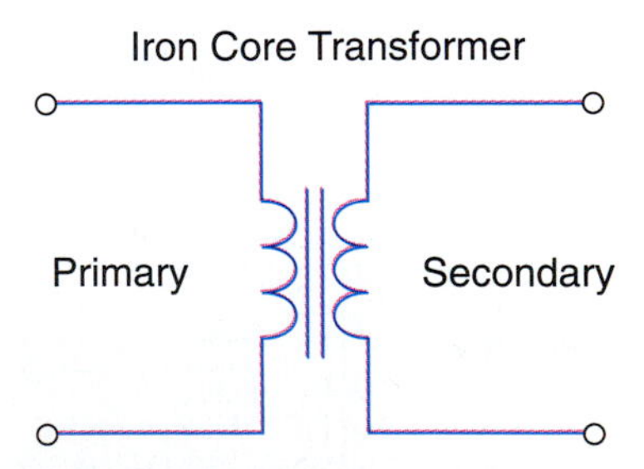

Figure 16-11 Two inductors, called windings, shown facing one another form the symbol of a basic transformer. The two windings are called the primary and the secondary.

The fact that the symbols are facing implies that the two coils share a common magnetic field. The majority of transformers have an iron core and are used at power-line frequencies (50 to 400 hertz). Transformers with ferrite cores typically operate at frequencies from 15 kilohertz to 100 kilohertz. Transformers that operate above 100 kilohertz are usually air core, although some ferrites can be operated at hundreds of megahertz.

The output voltage of a transformer with equal turns on its primary and secondary is the same as the input voltage. Such transformers are often called *isolation transformers* because they are usually employed to provide DC isolation

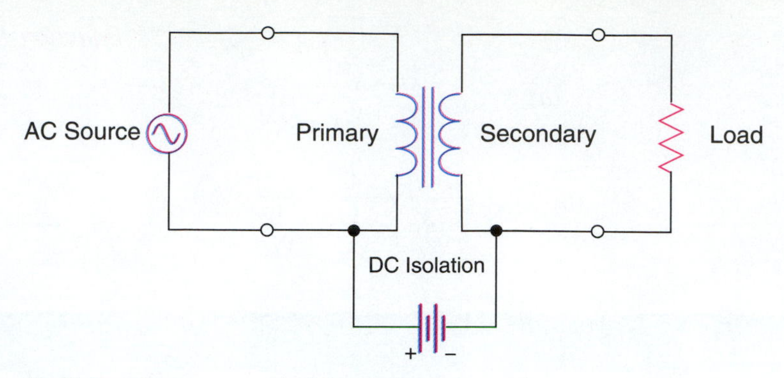

Figure 16-12 Transformers provide DC isolation between their primary and secondary windings because the windings are only connected magnetically.

of an AC source from its load, as shown in Fig. 16-12. Such isolation is particularly important when equipment is operated from the AC line because one side of the line is connected to earth ground. Other common items, such as water pipes, are also connected to earth ground. Getting between the "hot" side of the AC line and any grounded object can be lethal for people or animals and potentially harmful to electronic equipment. An isolation transformer provides safety for all.

The turns ratio and the voltage ratio of a transformer operated within its design specifications should be the same, as indicated in Eq. 16-9.

Equation 16-9

$$\frac{N_{\text{pri}}}{N_{\text{sec}}} = \frac{V_{\text{pri}}}{V_{\text{sec}}}$$

Assume that a transformer has 400 turns on its primary and 100 turns on its secondary. If 100 volts AC is applied to the primary, 25 volts appear at the secondary. The turns ratio and voltage are each four-to-one. Conversely, a transformer with 400 turns on its primary and 1600 turns on its secondary will step up a 100-volt input voltage to 400 volts. Transformers that decrease voltage are called *step-down transformers*. Transformers that increase voltage are called *step-up transformers*. The turns ratio of the transformer is sometimes written above its symbol as two numbers separated by a colon (1:1, 2:1, 4.5:1, 1:10, etc.).

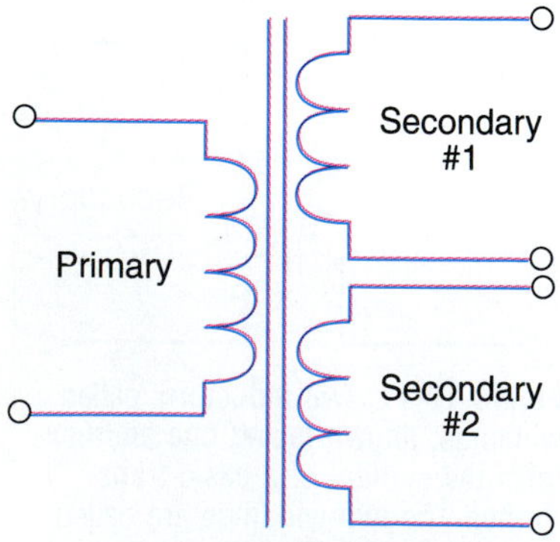

Multiple Secondaries

Many electronic circuits require several different voltages in order to operate. These voltages may be obtained from a transformer with multiple secondaries as shown in Fig. 16-13. This method has several major advantages over using resistive voltage dividers or using separate transformers. Transformers operate with great efficiency, often 95 percent or better. Resistive voltage dividers may waste more power in their resistors than they deliver to a load. Transformers provide complete DC isolation between all windings. The voltages produced by a voltage divider are interdependent and are not isolated from each other. Multiple secondary transformers are less expensive and require less space than several basic transformers used for the same purpose.

Figure 16-13 Transformers with multiple secondaries provide a cost-effective, efficient means of obtaining multiple voltages from a single source.

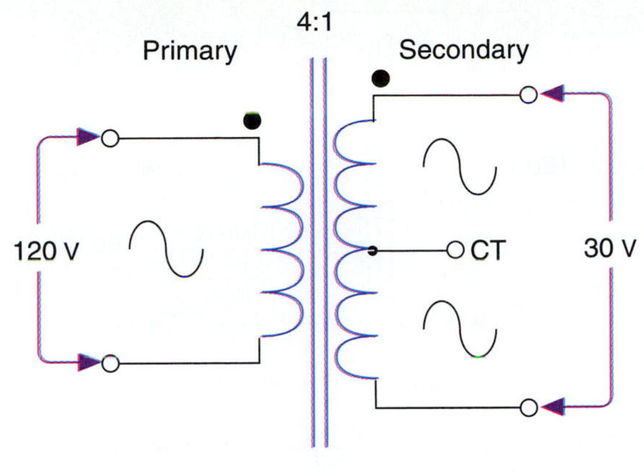

Figure 16-14 The specified voltage of any transformer winding is for the full winding. Dots are often used to indicate the phasing relationships between windings.

Sometimes transformer secondaries are connected internally. When this condition exists, the secondary is said to be *tapped*. If two identical secondary windings are internally connected in series-aiding and the connection is brought outside the transformer, the overall secondary is said to be *center tapped*. When this occurs, the tapped connection is usually labeled "CT" on schematic diagrams. The secondary voltage is supplied by the full winding.

The transformer shown in Fig. 16-14 is a 30-volt step-down transformer with a center tapped secondary. The voltage on each side of the center tap is 15 volts. It is not uncommon to have different resistance readings on each side of the center tap. One coil is often wound over the other. This causes the outer coil to have a larger diameter, which uses more wire to achieve the same number of turns. The additional wire accounts for the difference in resistance readings, but the number of turns in the two coils are equal. Sometimes the two windings forming a center-tapped secondary are wound at the same time by laying the two wires alongside each other. This allows the two windings to be as similar as possible in every respect, including resistance. The technique is called *bifilar winding*.

Other conventions are also shown in Fig. 16-14. The dots drawn at the top of the primary and secondary windings indicate that they are similarly phased. That means that if the positive alternation is being applied to the top of the primary, the top of the secondary is also producing its positive alternation. The sine waves are placed on the schematic symbol for clarification of the phasing between the primary and secondary.

Autoformers

When the main reason for using a transformer is to change the applied voltage, and isolation is not a factor, a special type of transformer is used. This type of transformer is called an *autoformer*. Autoformers are inexpensive transformers because the primary also serves as part of the secondary, saving on the cost of wire. In its simplest form, an autoformer is a tapped coil of wire. The autoformer

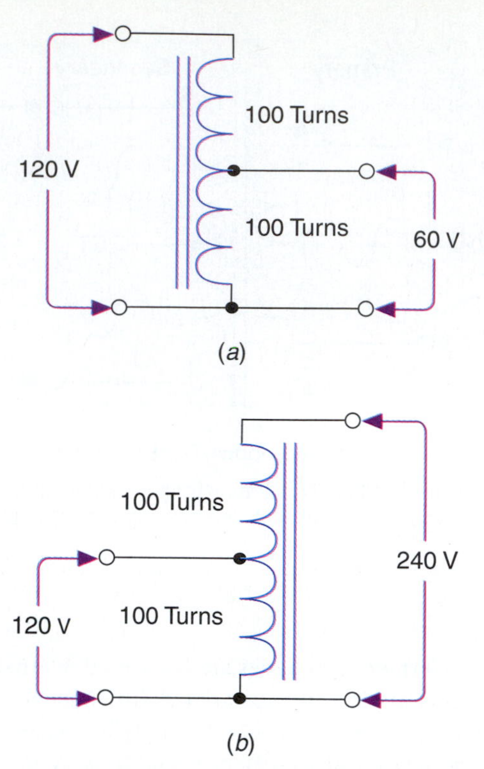

Figure 16-15 Autoformers provide an economical means of raising or lowering the input voltage: however, they do not provide isolation.

of Fig. 16-15(*a*) steps down the input voltage, whereas the autoformer of Fig. 16-15(*b*) steps up the input voltage. A special transformer called a *variac* has a contact that slides along the windings of a coil to select any voltage between zero and the source. The action is similar to that of a potentiometer.

Power Ratio

The power ratio of a transformer is very close to one-to-one (1:1). A transformer transfers power through itself and consumes very little of it in the process. If a transformer has multiple secondaries, the sum of the power supplied by the secondaries is equal to the power taken from the source by the primary. To maintain a unity power ratio, the voltage ratio and current ratio must always be opposite each other, as shown in Eq. 16-10.

Equation 16-10

$$\frac{V_{\text{pri}}}{V_{\text{sec}}} = \frac{I_{\text{sec}}}{I_{\text{pri}}}$$

The power that can pass through a transformer is largely determined by the size of its core. Larger core volumes equate to a higher power-handling capacity. A good rule of thumb for a laminated iron core transformer is that about 25 watts can be transferred for every cubic inch of core.

The Volt-Ampere

In the process of delivering power to a load, a transformer draws current from a source at some applied voltage. The product of voltage and current is normally power, but the transformer is not consuming the power—the load is. This is addressed by applying a unit called the *volt-ampere* to transformers. The volt-ampere is abbreviated *VA* and is a unit of *apparent power.* Like power measured in watts, the volt-ampere rating of a transformer is determined by multiplying voltage by current, but the VA rating should be thought of as how much power can pass through the transformer safely. The volt-ampere unit is also appropriate for inductors. An inductor draws current from an AC source in accordance with Ohm's law ($I = V/X$). Inductance, however, does not consume power. The energy stored in the magnetic field of an inductor is returned to the circuit.

Transformer Losses

Transformer losses fall into two general categories: *copper losses* and *core losses*. Copper losses refers to power that is wasted as heat because of the resistance of the copper wire used to make the transformer's windings. Core losses are of two general types: *eddy currents* and *hysteresis*. Eddy currents occur in iron core transformers because iron is a conductor. Of necessity, the iron core is at the center of the coils, where it also behaves as a shorted turn of wire. If a solid piece of iron or steel were used as a core, the eddy currents would consume most of the power available from the primary. To keep eddy currents at a minimum, the core is divided into many thin sheets of iron called laminations. The laminations are often coated with a thin layer of varnish or other insulating material to further reduce eddy currents. The concept of eddy currents in solid and laminated cores is shown in Fig. 16-16.

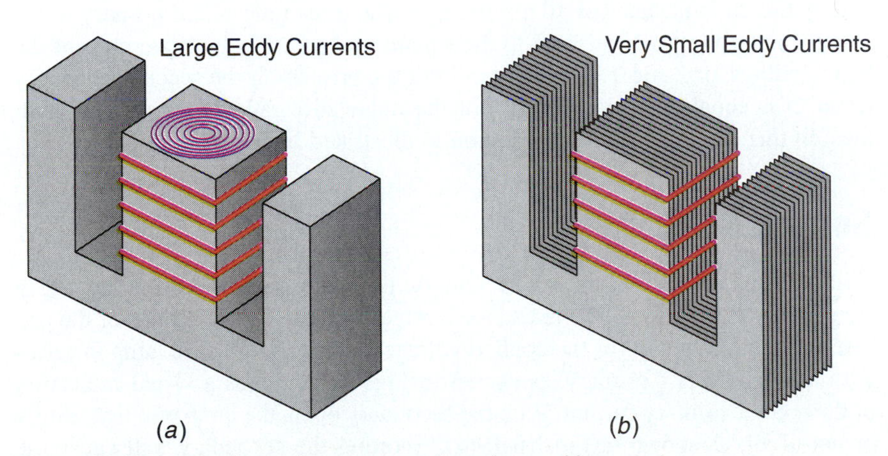

Figure 16-16 Laminating an iron core greatly reduces eddy currents.

Hysteresis, the other major source of core loss, was introduced in Chap. 13. As the magnetic polarities reverse as a consequence of the current reversals in the primary, the molecules in the iron core must follow along. They can only do so if the reversals are relatively slow. A significant reduction in the effective inductance of the transformer's primary occurs if the frequency of the applied voltage is too high for the molecules of the core material to follow. Hysteresis losses are also reduced by laminating the core because small groups of molecules are easier to coerce than large groups.

Another potential problem in transformers is saturation of the core. Once the core is saturated, the magnetism cannot be increased regardless of the primary's current. In an effort to increase the magnetism, the current in the primary can become so high that the copper wire overheats and melts. Saturation is normally prevented by placing a paper spacer between pieces of the transformer's core, as shown in Fig. 16-17. This paper must not be inadvertently removed if a transformer repair is attempted.

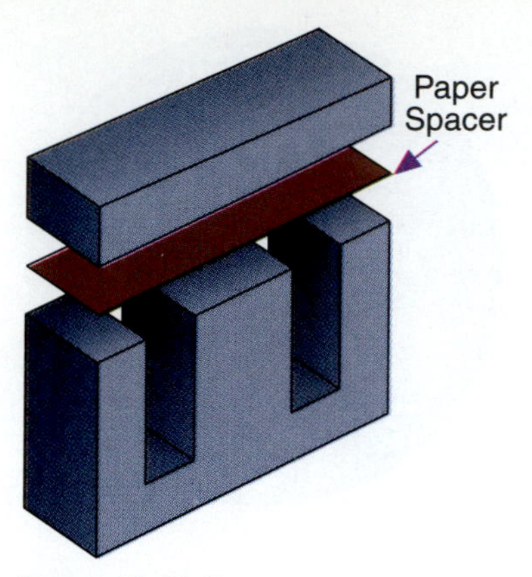

Figure 16-17 Paper spacers are often used to prevent magnetic saturation in transformer cores.

Inductance and Turns Ratios

Resistance measurements tell nothing about the voltage ratios of transformers, but inductance ratios can be used to determine them with little effort. For instance, assume a step-down transformer has a turns ratio of two-to-one (2:1). Because inductance increases as the square of the turns, the inductance of the primary is four times greater than the inductance of the secondary. Knowing this, the turns ratio is determined by taking the square root of the inductance ratios, as shown in Eq.16-11.

Equation 16-11

$$\text{turns ratio} = \sqrt{\frac{\text{primary inductance}}{\text{secondary inductance}}} = \sqrt{\frac{L_{\text{pri}}}{L_{\text{sec}}}}$$

Knowing the inductance of each transformer winding is the key to calculating the voltage ratio between any two windings. For example, assume that the primary of a two secondary transformer has an inductance of 500 millihenrys. The first secondary has an inductance of 100 millihenrys, and the second secondary has an inductance of 10 millihenrys. The turns ratio of the primary to the first secondary (ratio 1) is equal to the square root of the inductance ratio of the two windings, or 2.236:1. The turns ratio of the primary to the second secondary (ratio 2) is equal to the square root of the inductance ratio of those two windings, in this case 7.071:1. The situation is illustrated in Fig. 16-18.

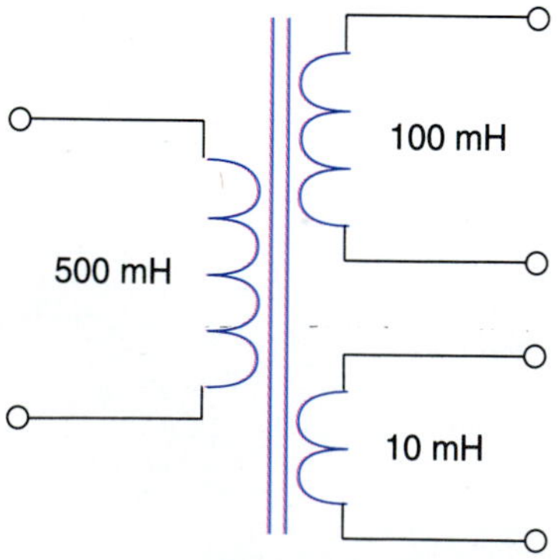

Figure 16-18 The turns ratio of a transformer can be found by taking the square root of the inductance ratio.

Reflected Impedance

With no load on a transformer's secondary, the major factor limiting current in the primary is reactance. The reactance is dependent on the inductance of the primary and the frequency of the applied voltage. Placing an electric lamp in series with a transformer's primary, as shown in Fig. 16-19, gives a visual indication of the relative primary current. With the secondary open, the lamp will light dimly or not at all, as shown in Fig. 16-19(*a*). Shorting the secondary will cause the lamp to light brightly, as shown in Fig. 16-19(*b*), almost to full brilliance. The current drawn by the shorted secondary causes the current in the primary to soar. The mechanism behind this is not immediately obvious, but it is clear that the inductance of the primary decreased because the frequency of the applied voltage has not changed.

When the secondary of the transformer delivers current to a load, the same current flows through the secondary winding, creating a magnetic field. This mag-

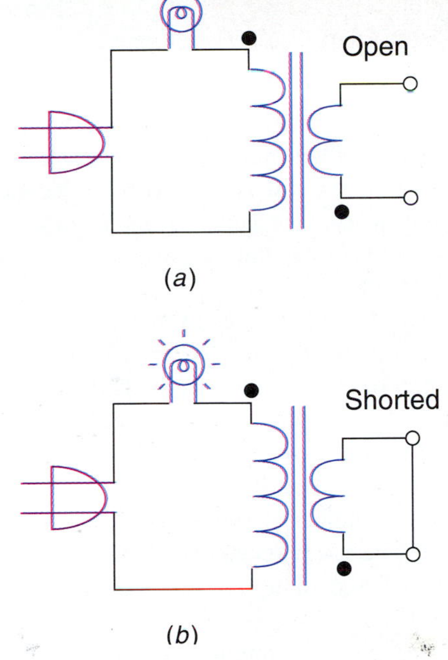

Figure 16-19 Current drawn from the secondary of a transformer causes the current in the primary to increase.

netic field will be opposite in polarity to the magnetic field created by the primary. At this point there are two ways to view the situation: (1) The magnetic field of the secondary cancels some or all of the magnetic field created by the primary, thus lowering its inductance and consequently its reactance, and (2) the magnetic field induces a counter-counter emf into the primary, which cancels some or all of its counter emf, thereby lowering its inductance. Either way, a transformer's primary draws more current from the source when the secondary is loaded because of the magnetic field created by the load current flowing through the secondary. A conductive transformer core behaves like a shorted turn. It is in this shorted turn that eddy currents flow. The overall effect is called the *shorted turn effect*.

REVIEW QUIZ 16-7

1. A transformer is a device made from two or more mutually coupled ______________ .
2. The transformer winding to which power is applied is called the ______________ .
3. The turns ratio and the ______________ ratio are identical in a properly operated transformer.
4. A transformer with a single tapped winding is called a(n) ______________ .
5. The power handling capability of a transformer is rated in ______________ .

SUMMARY

Inductance is a property mainly found in coils of wire, that opposes any change in current. The opposition is made possible by the development of two voltages within the coil. One is called counter emf, and the other is called induced emf. Counter emf is created by an increasing magnetic field, and induced emf is created by a decreasing magnetic field. The unit of inductance is the henry. One henry of inductance exists when a current changing at the rate of 1 ampere per second produces a voltage of 1 volt. It does not matter whether or not the current is increasing or decreasing.

Anything that affects the magnetic flux density of a coil changes its inductance. This includes increasing the number of turns, changing the length of the coil, changing the cross-sectional area of the coil, and changing the relative permeability of the coil's core. The conductivity of the core affects the inductance of a coil because of the shorted turn effect.

Switching inductive loads can create very high voltages as the magnetic field rapidly collapses. When the switch opens there is nothing to retard the rate at which the strength of the magnetic field changes. Because one of the major determining factors in the production of voltage from the relative motion between a magnetic field and a coil is the rate of change, surprisingly high voltages can be created in this way. In fact, this is the principle behind an automobile's ignition system.

Coils that share a common magnetic field are said to be mutually coupled. This coupling can produce additive or subtractive effects, depending on the relative polarities of the magnetic fields. Mutual coupling is used to advantage in making an electrical component called a transformer. A transformer is used to transform a voltage up or down, provide isolation between an AC source and its load, or to provide more than one voltage from a single source.

NEW VOCABULARY

air core
apparent power
autoformer
back emf
bifilar winding
center tapped
coefficient of coupling
copper losses
core losses
counter emf
eddy currents
ferrite core
henry
hysteresis
inductance
inductive reactance
inductors
isolation transformer
kickback voltage
laminated iron core
loosely coupled
mutual inductance
powdered iron core
primary
secondary
self-inductance
shorted turn effect
slug
step-down transformers
step-up transformers
tapped
tightly coupled
unity coupling
VA
variac
volt-ampere
windings

Questions

1. What is the definition of inductance?
2. Does a piece of straight wire have inductance?
3. How would the inductance of a coil be affected if more turns of wire were added to it without changing anything else?
4. List three types of cores over which inductors are wound.
5. Can mutual inductance be both additive and subtractive?
6. What name is commonly given to a threaded ferrite core used to vary the permeability of an inductor?
7. What name is given to the high-voltage spike created when the current through an inductive load is suddenly interrupted?
8. Do inductive values behave as resistor values do when inductors are placed in series and parallel with no mutual coupling?
9. How are wire-wound resistors made noninductive?
10. What is the phase relationship between voltage and current in an inductor?
11. What name is given to the opposition to AC current in an inductor?
12. List three uses of a transformer.

Problems

1. If an iron core inductor has 100 millihenrys of inductance with 100 turns, what would be the inductance if the number of turns were increased to 200?
2. What is the total inductance of a 4-millihenry and an 8-millihenry inductor connected in parallel with no mutual inductance?
3. What is the mutual inductance between a 6-millihenry coil and a 10-millihenry coil with a 0.5 coefficient of coupling?
4. What is the inductance of a 10-turn coil with a length of 0.01 meters, a cross-sectional area of 0.015 m^2, and a core with a relative permeability of 5?
5. A transformer has a turns ratio of 12:1. What is the output voltage if 120 volts is applied to the primary?
6. A transformer's primary has an inductance of 750 millihenrys and the secondary has an inductance of 10 millihenrys. What is the turns ratio?
7. What is the total inductance of two 5-microhenry coils connected in series having a mutual inductance of 2 microhenrys?
8. If a step-down transformer has an input voltage of 120 volts and is delivering 12 volts at 1 amp to a load, what is the primary current?
9. What is the inductive reactance of a 1000-millihenry inductor at 60 hertz?
10. What would be the reactance of the coil in question 9 at 120 hertz?

Critical Thinking

1. Why does the primary of a transformer draw more current when the secondary is loaded?
2. If an autoformer is tapped at 50 percent, what is the inductance ratio of the full winding compared to the inductance from the tap to either end?

3. How could a highly conductive metal slug decrease the inductance of a coil while consuming very little power?
4. Why is the voltage polarity of a transformer's secondary always opposite to the polarity of the primary when they are wound in the same direction?
5. How should two inductors be positioned in relation to each other to minimize mutual coupling?

Answers to Review Quizzes

16-1
1. oppose
2. increases
3. henry

16-2
1. flux density
2. one
3. permeable core
4. slug

16-3
1. collapsing
2. faster
3. opposite
4. tungsten

16-4
1. magnetic fields
2. zero; one
3. iron core
4. henrys

16-5
1. resistors
2. reversing
3. additive; subtractive

16-6
1. counter
2. reactance
3. 90°
4. resistors

16-7
1. inductors
2. primary
3. voltage
4. autoformer
5. volt-amperes

OPEN
CLOSED
PUSH TO OPEN
CHARGED

Chapter 17

L-R Circuits

ABOUT THIS CHAPTER

Resistance is not the only form of opposition to current. Inductors oppose current by creating counter emf. The opposition they create is called *reactance.* Circuits containing resistance and reactance require special attention because the source voltage and source current are phase-shifted so that they do not reach maximum at the same

After completing this chapter you will be able to:

- Explain how reactance is different from resistance.
- Describe how the current and voltage in an *L-R* circuit become phase-shifted.

time. In a series *L-R* circuit the voltage leads the current, and in a parallel *L-R* circuit the current lags the voltage. The voltages in a series *L-R* circuit must be added vectorially using current as a reference. The currents in a parallel circuit must be added vectorially using the voltage as a reference. The total opposition to current in a circuit owing to all effects is called *impedance.* Conductive counterparts to resistance, reactance, and impedance are introduced in this chapter, along with an introduction to vector addition and trigonometric solutions to right triangles.

- Perform vector addition and apply trigonometric solutions to right triangles.
- Give the meaning of power factor and show its effects on apparent energy consumption.

17-1 Comparisons of Resistance and Reactance

Resistors and inductors oppose alternating current using very different mechanisms. Resistors resist the flow of current by limiting the number of charge carriers. Inductors create a voltage that pushes back against a source through a process called *reactance.* Regardless of the mechanism, the ohm is the unit of measurement for both resistance and reactance. Any similarity between resistance and reactance, however, ends with their interchangeability in the Ohm's law equations.

Power and Apparent Power

Resistance radiates power as heat. The heat represents a loss of energy from the circuit that will not return. Unless the resistance is that of a heating element, the heat produced normally represents wasted power. Resistive power, which is often described as *true power,* is the product of the applied voltage and the current through the resistor. True power is measured in watts.

Reactance does not consume power from a source. It simply stores it for later use. Current is supplied to an inductor by an AC source as its voltage increases; it is supplied by the inductor as the source voltage decreases. The energy delivered by the source is stored in the inductor's magnetic field and then returned back to the source. The movement of current into and out of the inductor makes it appear that it is consuming power, but it is not. The product of the voltage across an inductor and the current through it is called apparent power, which is measured in volt-amperes (VA).

The VAR

The acronym for volt-ampere-reactive is *VAR.* The unit is the volt-ampere measurement of a pure inductance, which always has a voltage-to-current phase relationship of 90°. The VAR rating of an inductor may be determined at angles other than 90° by multiplying the peak VAR value by the sine of the angle, as shown in Eq. 17-1.

Equation 17-1

$$\text{VAR} = \text{VA} \times \sin \theta$$

Phase Angle

There is no difference in phase between the voltage across a resistor and the current through it. The two are said to be *in phase* or 0°. That 0° phase relationship remains true from DC to very high AC frequencies. It would remain true for all frequencies if resistors did not have small reactive components. Those reactive components are generally ignored for most applications and definitely ignored for the purpose of our discussion here.

When alternating current is applied across an inductor, the voltage will lead the current by 90°. This relationship remains constant regardless of the source frequency or the value of the inductor. Anytime two voltages, two currents, or a voltage and a current have a phase relationship of 90°, they are said to be in *quadrature.*

Phase relationships are often shown using a *vector* diagram. A vector is a quantity that has magnitude and direction. On vector diagrams, a vector is represented by a line with an arrowhead on one end. The length of the line indicates the magnitude, and the direction is indicated by the arrowhead. Figure 17-1(*a*)

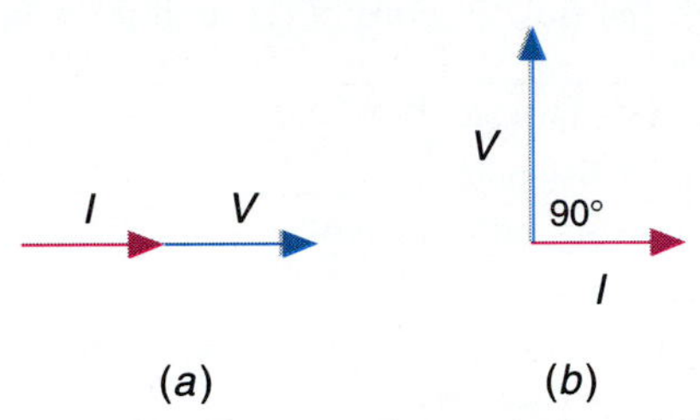

Figure 17-1 A vector is a quantity that has both magnitude and direction. It is represented by a line that has an arrowhead on one end. The length of the line indicates the magnitude, and the arrowhead indicates the direction. The voltage and current vectors are *in phase* for a resistor and in *quadrature* for an inductor.

shows a vector diagram of the voltage across a resistor and the resultant current. Figure 17-1(*b*) shows the vector diagram of the voltage and current associated with an inductor. The normal reference is along the horizontal and to the right, as was the case with a basic generator. A vector that points above the horizontal (positive) is said to be leading, and one that points below the horizontal (negative) is said to be lagging.

Frequency Effects

The value of a resistor is determined solely by its physical properties and is unaffected by frequency. Reactance, on the other hand, is extremely dependent on frequency because frequency determines the rate of change of the inductor's magnetic field. The more rapidly the magnetic field changes, the greater the counter emf will be. The counter emf ultimately determines the reactance of an inductor.

The inductive reactance equation, $X_L = 2\pi fL$, reveals that frequency has a linear effect on an inductor's reactance. Linear means that if the frequency is doubled, the reactance is doubled, or if the frequency is halved, the reactance is halved. The relationship is shown graphically in Fig. 17-2. Linear relationships

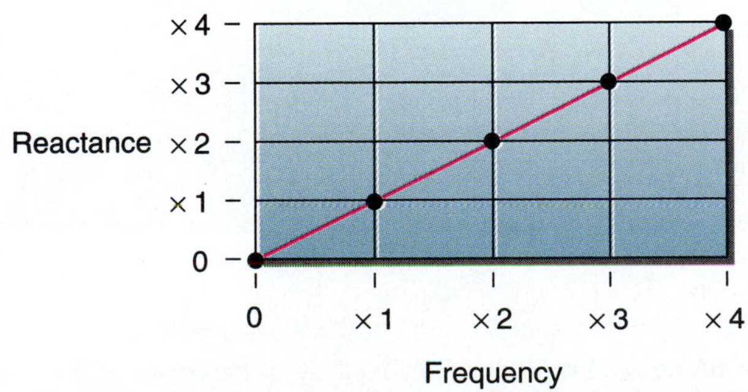

Figure 17-2 The relationship between frequency and reactance in an inductor is linear and proportional.

always graph to a straight line. The ×1, ×2, ×3, and ×4 mean times one, times two, times three, and times four, respectively.

REVIEW QUIZ 17-1

1. Resistors oppose the flow of current through them because they have a limited number of ______________ .
2. The volt-ampere is a unit of ______________ .
3. True power is only dissipated in ______________ .
4. The phase relationship between voltage and current in an inductor is ______________ .
5. Doubling the frequency of a voltage applied to an inductor will ______________ its reactance .

17-2 Series L-R Circuits

Series circuits containing both resistance and inductance clearly obey some of the series rules presented earlier, but they appear to violate others. The most notable disagreement is with Kirchhoff's voltage law. In a series circuit containing both resistance and inductance, the sum of the voltages across the resistor(s) and inductor(s) exceeds the source voltage. The reason for this is that the peak voltages across the resistor(s) and inductor(s) do not occur at the same time.

A voltage source can be thought of as being blind to what is connected across it. A source knows its voltage, the current it is delivering, and the relative phase of that voltage and current. It knows nothing about the complexity of the circuit to which it is connected. If the voltage and current are in phase, the circuit appears resistive, although it may not be a resistor. Nonsynchronous electric motors and transmitting antennas ideally appear as resistive loads. For the sake of simplicity, however, we will limit our discussions to resistors and inductors.

A source connected to an inductor is able to identify it as an inductive load because the voltage across the inductor leads the current by 90° or $\pi/2$ radians. The resultant waveform is called a *cosine wave* because the amplitude of the voltage at any point along the cycle may be found by multiplying the cosine of the angle by the amplitude of the positive peak. A cosine waveform is shown in Fig. 17-3.

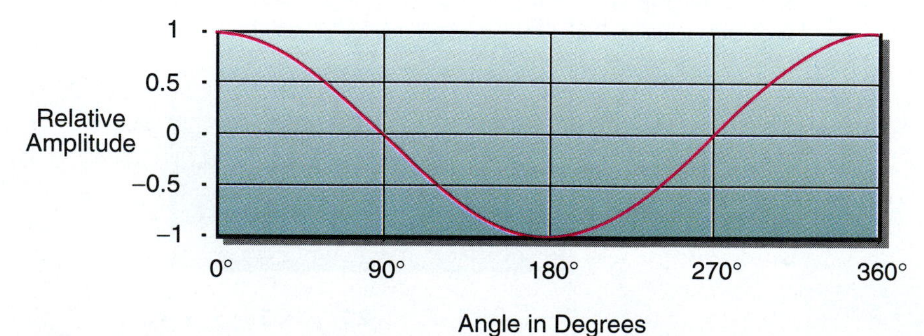

Figure 17-3 A cosine wave is generated by multiplying the cosine of the angle by the peak amplitude.

When an inductor and a resistor are connected in series, the same current flows through each of them, as in any series circuit. The rule stating that the current is the same at any point in a series circuit is not just referring to the amount of current. All aspects of the current are the same. This includes the current's amplitude, frequency, and phase. The fact that the current is the same phase through a series-connected resistor and inductor implies that the voltages across the resistor and inductor must be 90° out of phase. A basic series *L-R* circuit and the relative voltage/current relationships of an AC source are shown in Fig. 17-4.

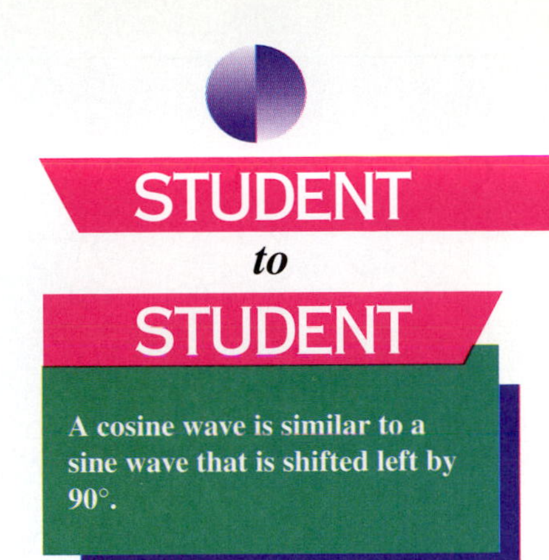

A cosine wave is similar to a sine wave that is shifted left by 90°.

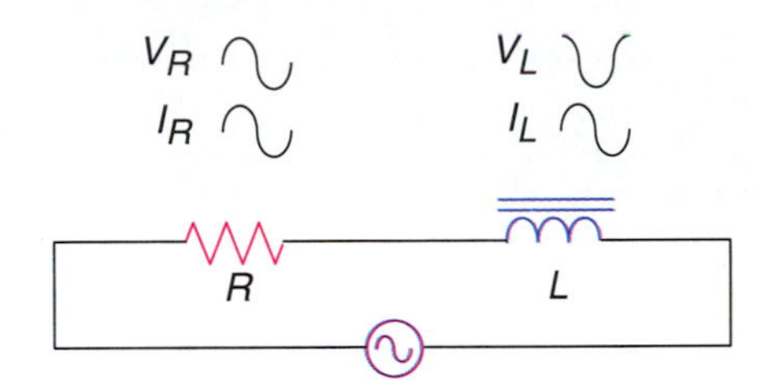

Figure 17-4 The current in a series *L-R* circuit is the same at any point, but the voltages are out of phase by 90°.

When connected across a series *L-R* circuit, an AC source sees both the inductive and resistive properties of the circuit. This is possible because the voltage of the source leads the current by some angle greater than 0° but less than 90°. Remember that the phase angle is only 0° if the circuit is purely resistive and only 90° if the circuit is purely reactive. The exact phase angle seen by the source can be determined by graphing the voltage vectors and then using a protractor, or by using trigonometry.

Vector Addition

The voltage across the resistor in a series *L-R* circuit is represented by a vector drawn horizontally and pointing to the right. The fact that the vector is horizontal indicates that its relative phase angle is 0°. The length of the vector corresponds to the voltage. The voltage across an inductor is represented by a vector drawn vertically, pointing straight up. This vector is at a right angle to the resistor's voltage vector. Again, the length of the vector is proportional to the voltage. This relationship is shown in Fig. 17-5(*a*) on the next page.

Vectors cannot be added numerically because they often have an angular relationship to each other. One method of representing the angular relationship is to draw the vectors on graph paper. It was mentioned earlier that *L-R* circuits appear to violate Kirchhoff's voltage law when viewed in the conventional sense. Kirchhoff's voltage law is obeyed if the voltages around an *L-R* circuit are added vectorially. Vector addition can be accomplished by drawing a second set of vectors parallel to the first set. These new vectors are drawn from the points of the existing vectors, as shown in Fig. 17-5(*b*). A vector representing the source voltage is formed by drawing a new vector from the point where the tails of the original vectors meet to the point where the heads of the new vectors meet. This is shown in Fig. 17-5(*c*). The length of this new vector represents the source voltage.

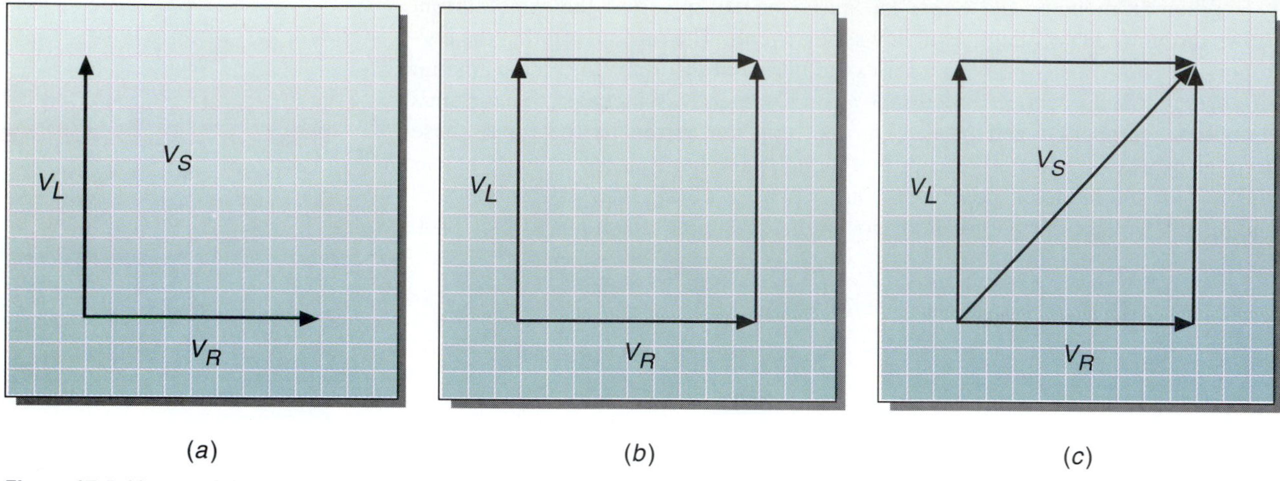

Figure 17-5 Vector addition can be done graphically, but it requires careful measurement.

Impedance

According to Ohm's law, the voltage across a resistor is the product of the current through the resistor multiplied by the resistor's value, $V_R = I_R \times R$. Reactance can be substituted for resistance in any of the Ohm's law equations. Consequently, the voltage across an inductor is equivalent to the inductor's current multiplied by the inductor's reactance, $V_L = I_L \times X_L$.

A voltage vector can also be converted to its Ohm's law equivalent if that voltage is supplying current to a circuit. If, however, the circuit contains both resistance and reactance, the circuit is not entirely reactive, but neither is it entirely resistive. The current in the circuit is limited by a combination of both called *impedance.* The capital letter Z represents impedance in mathematical equations and is used as an abbreviation for the word *impedance.* Impedance is the total opposition to current in a circuit regardless of the cause. It would be accurate to say that a 10-ohm resistor also has an impedance of 10 ohms, but the term *impedance* is usually reserved for more complex circuits. In a vector diagram, the voltage of the source, V_S, can be replaced with its Ohm's law equivalent, $V_S = I_S \times Z$.

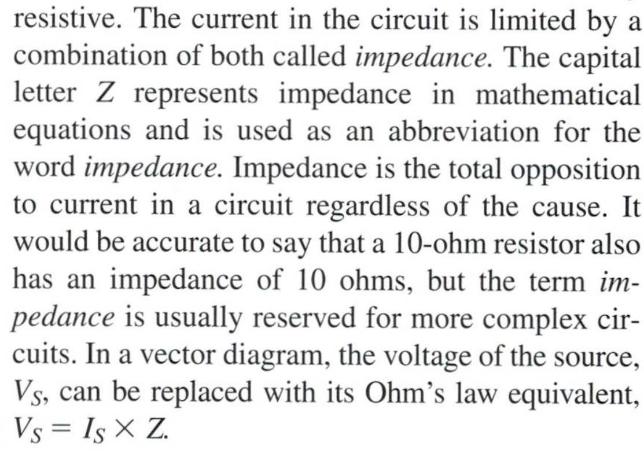

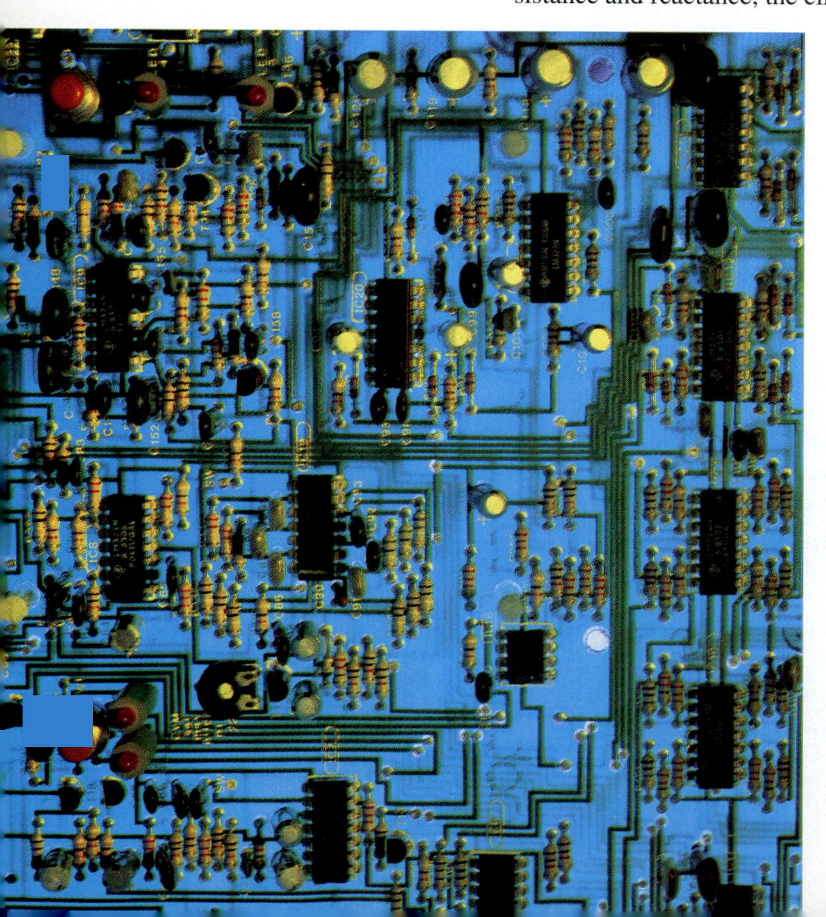

Resistance, reactance, and impedance are often more convenient to work with than the individual voltages across a circuit and its components. That is because the values of the resistors and inductors in a circuit are usually known. Of the three voltages in a simple series *L-R* circuit, two of the voltages are usually not known. Inductive reactance is calculated using the formula $X_L = 2\pi fL$, which may also be written as: $X_L = \omega L$. The voltage vectors back in Fig. 17-5(*c*) may be replaced with their Ohm's law equivalents, as shown in Fig. 17-6(*a*).

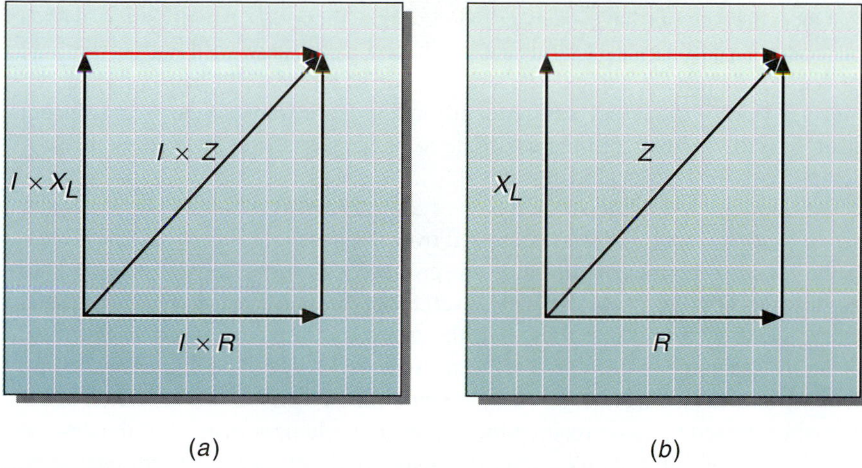

Figure 17-6 The common factor of I may be dropped from the Ohm's law equivalent of the individual voltages around a series L-R circuit.

The current is the same at any point in a series circuit. This means that I_R,I_L, and I_S are identical. If the common factor of I is dropped from the three Ohm's law equations, the vector diagram of Fig. 17-6(*a*) would be modified to contain only R, X_L, and Z. This transformation is shown in Fig. 17-6(*b*).

Angular Measurement

The vector V_R, or R, is the reference to which the other vectors are compared in a series L-R circuit. The angle between V_S and V_R or Z and R is the phase angle theta, which is represented by the Greek letter of the same name, θ. The angle symbol, $\angle$, may or may not be shown ahead of theta. Figure 17-7 shows the common usage of theta in a vector diagram. Theta represents the relative phase between the voltage of a source and the current the source is delivering to a circuit. In a series L-R circuit, the current is the reference; therefore, theta indicates by how much the source voltage leads the current.

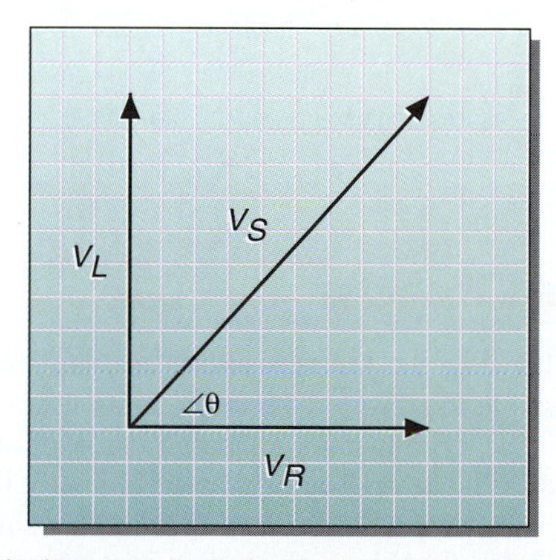

Figure 17-7 The angle theta, $\angle\theta$, lies between the voltage vector of the source and the voltage vector of the resistor. If X_L, R, and Z had been used, theta would lie between Z and R.

Once the vectors have been graphed, angle theta can be measured with a *protractor.* A protractor is a semicircular-shaped instrument with a graduated arc that is used to measure angles. The straight side of a protractor is usually marked to form a ruler. Protractors are frequently made from transparent or translucent plastic so that the lines forming the angle to be measured can be seen. Some protractors, however, are made from metal.

At midpoint along the straight side of a protractor is a pair of perpendicular lines to mark the spot that is to be placed over the junction of the two lines forming the angle to be measured. Once the protractor is in position it is rotated so that its straight edge is parallel to the reference line, or vector. This process is usually aided by a fine line drawn on the protractor from one side to the other. The angle graduations on a protractor are repeated twice. The angles read from left to right and also from right to left. Some minor caution and common sense must be exercised to avoid reading the angle's complement instead of the desired angle. It is helpful to remember that the phase angle of an *L-R* circuit cannot be less than 0° or greater than 90°. An example of angular measurement using a protractor is shown in Fig. 17-8. Although measuring angles with a protractor is relatively simple, the accuracy is limited to how accurately the lines forming the angle are drawn and how accurately the protractor's scale is interpolated.

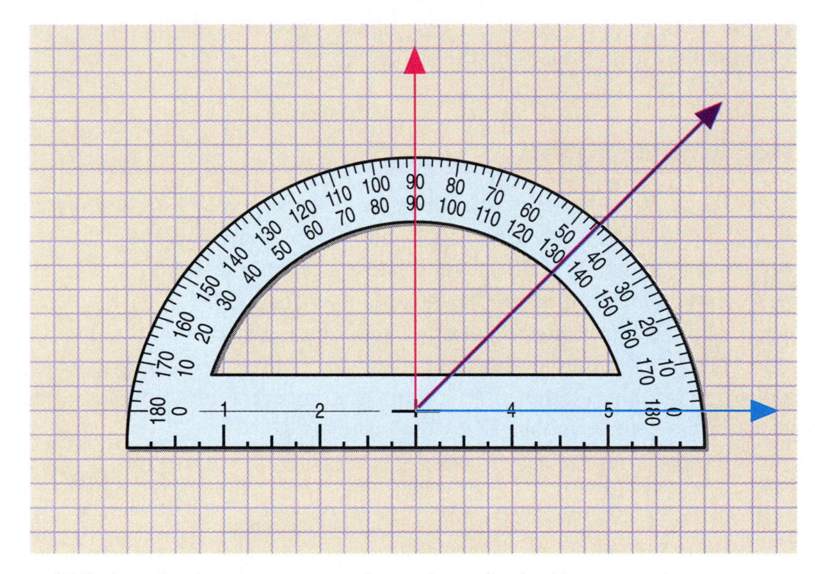

Figure 17-8 A protractor measures angles and usually doubles as a ruler.

Vectors and Right Triangles

The 90° relationship between the voltage across an inductor and the current through it makes it possible to convert a vector diagram to a right triangle. The conversion is accomplished by drawing a line from the head of one of the original vectors to the head of the other. The rules used to justify this procedure are relatively unimportant, but they revolve around the fact that similar triangles are formed when the line is drawn. A panoramic view of the procedure is shown in Fig. 17-9.

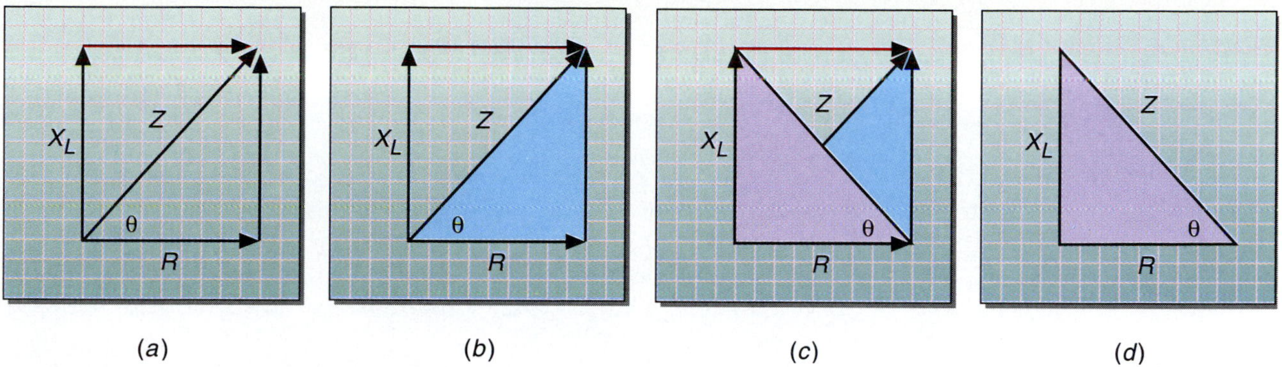

Figure 17-9 A right triangle can be developed from any of the vector diagrams derived from a series *L-R* circuit.

The Pythagorean Theorem

Once a right triangle is created, the *Pythagorean theorem* can be used to solve for an unknown impedance. The Pythagorean theorem is named after its discoverer, Pythagoras (c. 582–c. 500 B.C.), a Greek philosopher and mathematician. Pythagoras discovered that the sum of the areas formed by squaring the two sides of a right triangle are equal to the area formed by squaring the triangle's *hypotenuse.* The hypotenuse is the side opposite the right angle. A graphic example of the procedure is shown in Fig. 17-10. The sides are typically labeled "*a*" and "*b,*" with the hypotenuse labeled "*c.*"

From the geometrical representation used by Pythagoras, the formula $a^2 + b^2 = c^2$ was developed. The hypotenuse of the triangle, in this case *c,* can be determined by taking the square root of both sides of the equation. It is this square root solution for the hypotenuse, shown in Eq. 17-2, that is most commonly called the Pythagorean theorem.

Equation 17-2

$$c = \sqrt{a^2 + b^2}$$

The Pythagorean theorem can be applied to any right triangle. For that reason it and its derivations are very useful when analyzing circuits containing resistance and reactance. The following six equations provide solutions to the six parameters of *L-R* circuits presented thus far:

Equation 17-3

$$V_S = \sqrt{V_L{}^2 + V_R{}^2}$$

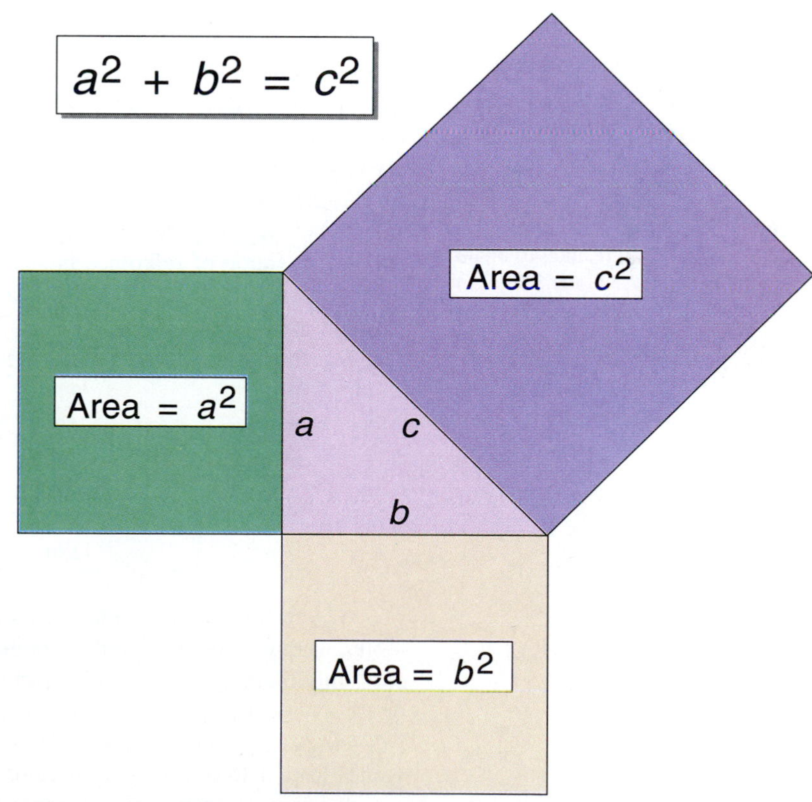

Figure 17-10 Pythagoras discovered that the sum of the squares of the two sides of a right triangle is equal to the square of the hypotenuse. This relationship is called the Pythagorean theorem.

Equation 17-4

$$V_L = \sqrt{V_S^2 - V_R^2}$$

Equation 17-5

$$V_R = \sqrt{V_S^2 - V_L^2}$$

Equation 17-6

$$Z = \sqrt{X_L^2 + R^2}$$

Equation 17-7

$$X_L = \sqrt{Z^2 - R^2}$$

Equation 17-8

$$R = \sqrt{Z^2 - X_L^2}$$

Trigonometric Right Triangle Solutions

Although the Pythagorean theorem provides a straightforward solution to finding any of the three vectors we are concerned with, it does not reveal angle theta or provide any means of finding it. This can be a serious shortcoming of the theorem when its complexities and the number of steps involved are considered.

In many cases finding angle theta is unimportant, but when it is, some method other than the Pythagorean theorem must be used. As we have already seen, the graphing method is slow, and its accuracy is dependent on the artistic and interpolative skills of the person seeking the solution. A more precise method is to use the trigonometry of right triangles. The basis of those solutions lies in the relative ratios of the sides of a right triangle. Three of the ratios are considered important enough to warrant a place on the keyboard of every scientific calculator. Those ratios are *sine, cosine,* and *tangent.* On the calculator keyboard they are respectively abbreviated to SIN, COS, and TAN.

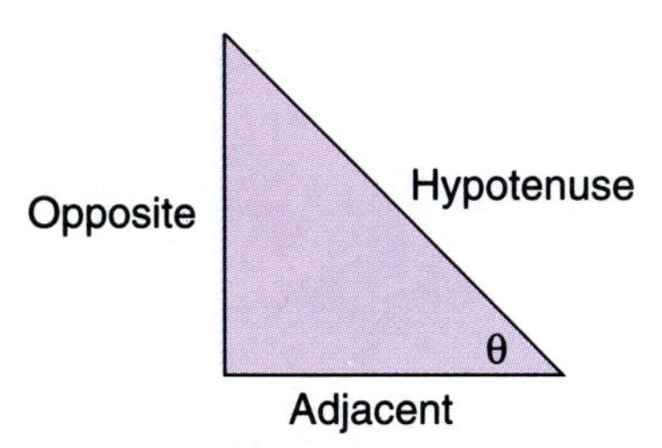

Figure 17-11 The sides of a right triangle are named in relation to angle theta. The hypotenuse is the side opposite the right angle.

Angle theta is used as a reference to establish the location of the named sides. The three sides of the triangle are called *opposite, adjacent,* and *hypotenuse.* When that frame of reference is used, the orientation of the triangle is of little or no importance. Fig. 17-11 shows a right triangle with named sides in relation to angle theta.

The three named ratios are developed as follows:

$$\sin = \frac{\text{opposite}}{\text{hypotenuse}}$$

$$\cos = \frac{\text{adjacent}}{\text{hypotenuse}}$$

$$\tan = \frac{\text{opposite}}{\text{adjacent}}$$

At one time sine, cosine, and tangent ratios were listed in printed lookup tables. Today modern scientific calculators use electronic lookup tables stored in read-only memory (ROM). When angle theta is known, pressing "SIN," "COS," or "TAN" on a scientific calculator's keyboard will look up the appropriate result and place it in the calculator's display. Sine values range from 0 to 1, cosine values from 1 to 0, and tangent values from 0 to infinity (∞).

A scientific calculator will provide angle theta if the sine, cosine, or tangent is known. Calculators have four different ways of indicating these functions: 1. Press "INVERSE" then "SIN," "COS," or "TAN." 2. Press "ARC" then "SIN,"

"COS," or "TAN." 3. A variation on the previous method has the SIN, COS, and TAN keys labeled with the alternate functions ASIN, ACOS, and ATAN. 4. The alternate functions assigned to the SIN, COS, and TAN keys are SIN^{-1}, COS^{-1}, and TAN^{-1}. The "-1" is another way of saying "inverse." (Some calculators have the reciprocals key labeled in this fashion. Instead of the usual "*1/x*," the key is labeled "x^{-1}.") Figure 17-12 consolidates the standard trig functions and their inverse functions.

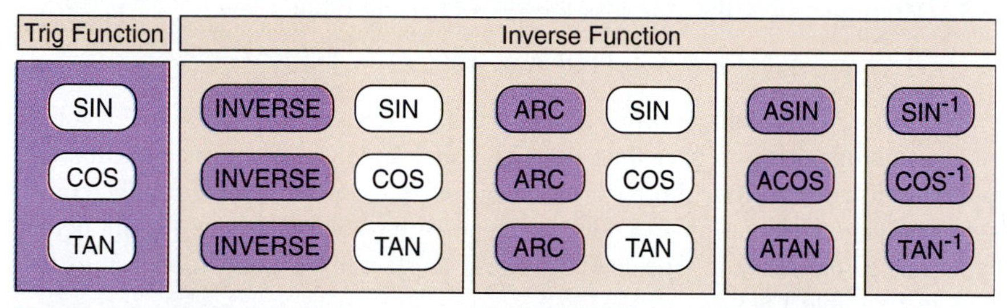

Figure 17-12 Scientific calculators generally use the same method to identify the sine, cosine, and tangent functions, but the inverse trig functions may be indicated differently from one calculator to another.

Example 17-1: Impedance and Phase Angle Assume that a 1 kilohm resistor is in series with a 500-ohm inductive reactance, as shown in Fig. 17-13. Determine the impedance of the circuit and the phase angle.

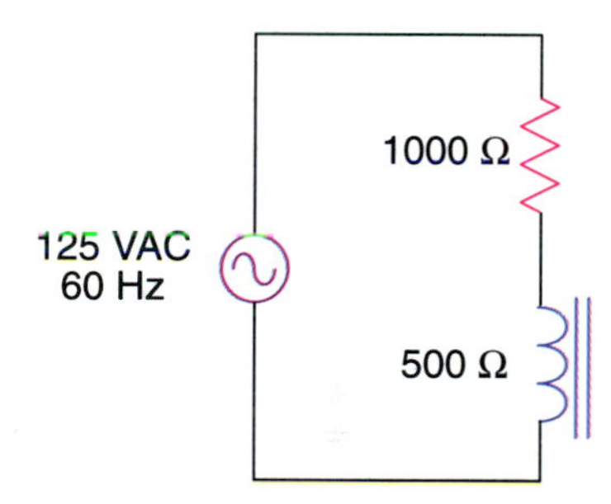

Figure 17-13

Solution

1. Determine the tangent ratio by dividing the inductive reactance by the series resistance.

$$\tan\theta = \frac{\text{opposite}}{\text{adjacent}} = \frac{X_L}{R} = \frac{500\ \Omega}{1000\ \Omega} = 0.5$$

2. Using the inverse tangent function of a scientific calculator or printed lookup tables, determine angle theta.

$$\tan^{-1} 0.5 = \theta = 26.565°$$

3. Using the sine function of a scientific calculator or printed lookup tables, determine the sine of 26.565°.

$$\sin 26.565° = 0.4472$$

4. Determine the impedance of the circuit by dividing the inductive reactance by the sine of angle theta.

$$Z = \frac{X_L}{\sin\theta} = \frac{500\ \Omega}{0.4472} = 1118\ \Omega$$

Example 17-2: Voltage Drops Now that the impedance of the circuit in Fig. 17-13 is known, determine the voltage drops across the inductor and resistor using Ohm's law.

Solution

1. Determine the circuit current using Ohm's law.

$$I_T = \frac{V_S}{Z} = \frac{125\text{ V}}{1118\ \Omega} = 0.1118\text{ A}$$

2. Determine the voltage across the inductor using Ohm's law. Remember that the current is the same at any point in a series circuit; therefore, $I_T = I_L = I_R$.

$$V_L = I_T X_L = 0.1118\text{ A} \times 500\ \Omega = 55.9\text{ V}$$

3. Determine the voltage across the resistor using Ohm's law.

$$V_R = I_T R = 0.1118\text{ A} \times 1000\ \Omega = 111.8\text{ V}$$

The sum of the voltage drops across the inductor and resistor is 167.7 volts, which is unmistakably higher than the 125-volt source. As mentioned earlier, an apparent violation of Kirchhoff's law is to be expected when measuring the component voltages around an *L-R* circuit because the voltage across the inductor leads that of the resistor. The leading phase angle can be confirmed using the steps in Example 17-3.

Example 17-3: Confirming the Leading Phase Angle

1. Divide the voltage across the inductor by the voltage across the resistor. This will yield the tangent ratio in relation to angle theta.

$$\tan\theta = \frac{V_L}{V_R} = \frac{55.9\text{ V}}{111.8\text{ V}} = 0.5$$

2. Using the inverse tangent function of a scientific calculator or printed lookup tables, determine angle theta. This angle is the same as the phase angle found using inductive reactance and resistance to establish the tangent ratio.

$$\tan^{-1} 0.5 = \theta = 26.565°$$

The phase angle between the source voltage and the source current is the same as the phase angle between the voltage across the inductor and the voltage across the resistor in a series *L-R* circuit. This is possible because there cannot be a difference in phase between the voltage across a resistor and the current through it.

L-R Time Constants

The current through a series *L-R* circuit does not instantaneously reach maximum when connected to a voltage source. The current is retarded because the counter emf of the coil most nearly matches the source voltage at the instant the source is connected to the circuit. Unlike a purely inductive circuit, the current in an *L-R* circuit will reach some maximum level. The determining factors are the source voltage and the value of the series resistor.

The time that the current takes to reach maximum depends on the values of the inductor and resistor. Increasing the value of either one will slow the buildup of current. The inductance in henrys divided by the resistance in ohms produces what is called a *time constant.* Equation 17-9 shows this mathematical relationship.

Equation 17-9

$$T = \frac{L}{R}$$

During each time constant the current will increase to about 63 percent of the difference between where it is and its eventual maximum. Theoretically the current will never reach maximum. In practice the current is considered to be maximum after five time constants. The percentage of overall current for the first five time constants are: 63 percent, 86 percent, 95 percent, 98 percent, and 99 percent. If after five time constants the source were removed and replaced with a short circuit, the current would decrease by 63 percent for each time constant. The current-time constant relationship is shown in Fig. 17-14.

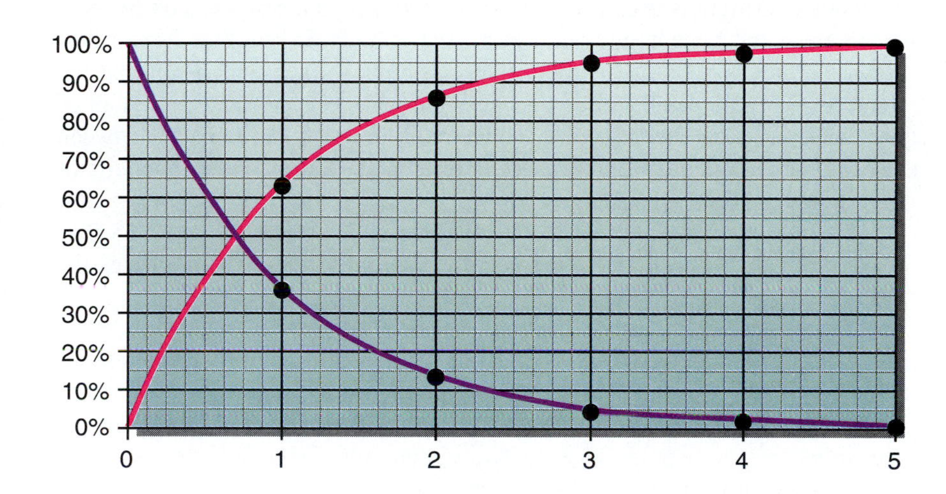

Figure 17-14 The current in an *L-R* circuit increases or decreases by 63 percent for each time constant. When decreasing it is understood that the source is replaced with a short circuit. The current produced by the collapsing magnetic field flows through the inductor, creating another magnetic field that opposes the collapse.

REVIEW QUIZ 17-2

1. The sum of the voltage drops around a series *L-R* circuit ______________ exceed the source voltage.
2. A cosine wave is the same as a sine wave shifted left by ______________ degrees.
3. The phase angle between the voltage across a resistor and the current through it is ______________ degrees.
4. The total combined opposition to current in a circuit is called ______________ .
5. The phase angle is found using ______________ .
6. In one time constant the current will increase or decrease by ______________ percent of difference between its present value and its next value.

17-3 Parallel L-R Circuits

Inductors and resistors connected in parallel obey some of the parallel rules and appear to violate others, just as they did with the series rules. The key parallel rule that can be taken at face value is that the voltage is the same across all of the elements of a parallel circuit. Once again, "same" means the same voltage, the same frequency, and the same phase.

An inductor retains the 90° phase relationship between the voltage across it and the current through it regardless of how it is used in a circuit. In a parallel *L-R* circuit the current through the inductor will lag the current through the resistor by 90°. Saying that voltage leads current implies that current is the point of reference. Current is usually the point of reference in a series circuit because the current is the same at any point in a series circuit. Saying that current lags voltage implies that voltage is the reference point. Voltage will usually be the reference point in a parallel circuit because the voltage is the same across all of the elements of a parallel circuit. It is important to note the differences between the two types of circuits because there is often a tendency to view both circuit types from a single perspective.

A vector diagram of current in a parallel circuit is shown in Fig. 17-15. The I_L vector points downward (into the negative region) to indicate that it lags the reference I_R vector.

The impedance of a parallel *L-R* circuit can be determined using Ohm's law if the currents are known or can be calculated. The current through the resistor is found by dividing the source voltage by the resistor's value ($I_R = V_S/R$). The current through the inductor is found by dividing the source voltage by the inductor's reactance ($I_L = V_S/X_L$). The circuit impedance is determined by dividing the source voltage by the vector sum of the two currents. In this case vector sum means that the two currents are added in quadrature, as shown in Eq. 17-10.

Equation 17-10

$$Z = \frac{V_S}{\sqrt{I_R^2 + I_L^2}}$$

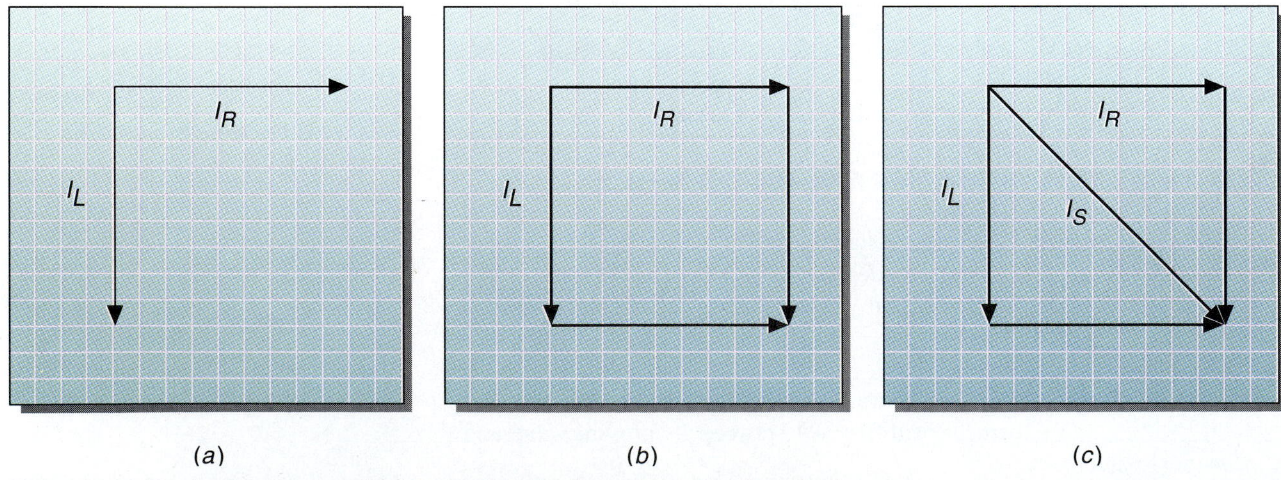

Figure 17-15 The I_L vector in a parallel *L-R* circuit is negative and therefore lags the I_R vector.

Conductance, Susceptance, and Admittance

In a series *L-R* circuit resistance and inductive reactance were added in quadrature to obtain the circuit impedance. These units cannot be added numerically or vectorially when connected in parallel. The situation is very similar to a resistive parallel circuit where the conductance values of the resistors had to be added and then converted back to resistance using the prodivism or reciprocals equation (discussed in Chap. 16).

All three forms of electrical opposition—resistance, reactance, and impedance—have their conductive counterparts. The conductive values are found by taking the reciprocal of the opposition values. The reciprocal of resistance is already known to be conductance and is symbolized by a capital G. The reciprocal of reactance is susceptance, which is symbolized by the capital letter B. The reciprocal of impedance is *admittance*, which is symbolized by the capital letter Y. In all three cases the unit is siemens, abbreviated as S. Equation 17-11(*a*) repeats the conductance equation, Eq. 17-11(*b*) is the susceptance equation, and Eq. 17-11(*c*) is the admittance equation. As with conductance, susceptance and admittance values may be converted back to their original units by taking their reciprocals.

Equation 17-11

$$(a)\ G = \frac{1}{R} \quad \text{and} \quad R = \frac{1}{G}$$

$$(b)\ B = \frac{1}{X} \quad \text{and} \quad X = \frac{1}{B}$$

$$(c)\ Y = \frac{1}{Z} \quad \text{and} \quad Z = \frac{1}{Y}$$

Conductance, susceptance, and admittance values are very useful when describing parallel *L-R* circuits because they do not require that the source voltage or individual currents be known. The values can be inserted into the Pythagorean theorem equation or substituted into the trigonometry equations. For example, assume that the 500-ohm reactance and the 1000-ohm resistance used in the series example are now connected in parallel. If we want to know the impedance of the circuit we now have several ways to do so, as shown in the following examples.

Example 17-4: Solving for Parallel Impedance A variation of the Pythagorean theorem may be used to solve for a parallel *L-R* impedance, as shown in Eq. 17-12.

Equation 17-12

$$Z = \frac{1}{\sqrt{\left(\frac{1}{R}\right)^2 + \left(\frac{1}{X_L}\right)^2}} \quad \text{or } Z = \frac{1}{\sqrt{(G)^2 + (B)^2}}$$

1. Convert the 1000-ohm resistance to conductance using the reciprocals equation.

$$G = \frac{1}{R} = \frac{1}{1000\ \Omega} = 0.001\ \text{S}$$

2. Convert the 500-ohm inductive reactance to susceptance using the reciprocals equation.

$$B = \frac{1}{X_L} = \frac{1}{500\ \Omega} = 0.002\ \text{S}$$

3. Square the conductance value.

$$G^2 = (0.001)^2 = 0.000001$$

4. Square the susceptance value.

$$B^2 = (0.002)^2 = 0.000004$$

5. Add the squared values.

$$G^2 + B^2 = 0.000001 + 0.000004 = 0.000005$$

6. Find the square root of the sum of the squared values.

$$\sqrt{G^2 + B^2} = \sqrt{0.000005} = 0.002236$$

7. Take the reciprocal of the square root of the squared values.

$$\frac{1}{\sqrt{G^2 + B^2}} = \frac{1}{0.002236} = 447.2\ \Omega$$

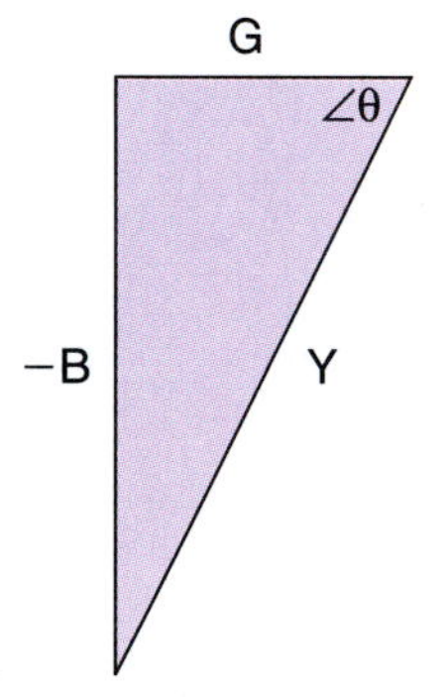

Figure 17-16 A minus sign is added to susceptance to identify the lagging phase.

Example 17-5: The Trigonometry Solution The use of Pythagorean theorem to determine the impedance in a parallel *L-R* circuit is considerably more tedious than it was for the series circuit. And as with the series circuit, only the circuit impedance was obtained. Conductance values and susceptance values may be used in the trigonometry solution to obtain the admittance. Once the admittance is known, it can be placed into the reciprocals formula to obtain the impedance. The phase angle is not affected by the units selected because it is the result of ratios that do not change from one unit system to another. A minus sign is assigned to susceptance to identify the lagging phase.

The triangle we will be working with is shown in Fig. 17-16. Note that the range of the sine, cosine, and tangent values is now 0 to −1, 1 to 0, and 0 to minus infinity (−∞). Using the same 1000-ohm resistor and 500-ohm inductive reactance, the trigonometry solution goes as follows:

1. Determine the tangent of theta by dividing the opposite side by the adjacent side.

$$\tan\theta = \frac{-B}{G} = \frac{-0.002}{0.001} = -2$$

2. Find theta using the inverse tangent function on a calculator or by using printed lookup tables.

$$\text{TAN}^{-1}\ -2 = -63.435°$$

3. Find the sine of angle theta using a scientific calculator or by using printed lookup tables.

$$\sin\theta = \sin(-63.435°) = -0.89443$$

4. Determine the admittance by dividing the susceptance by the sine of theta.

$$Y = \frac{-B}{\sin\theta} = \frac{-0.002\ \text{S}}{-0.89443} = 0.002236$$

5. Determine the impedance by finding the reciprocal of the admittance.

$$Z = \frac{1}{Y} = \frac{1}{0.002236} = 447.2\ \Omega$$

Once again the trigonometry solution has obtained more information in fewer steps. Notice also that the overall impedance is less than the smallest impedance in the parallel circuit.

REVIEW QUIZ 17-3

1. The conductive counterparts to resistance, reactance, and impedance are ______________, ______________, and ______________ .
2. The impedance of a parallel *L-R* circuit can be determined by dividing the source voltage by the vector sum of the ______________ .
3. In a parallel *L-R* circuit the source current ______________ the source voltage.

17-4 Power Factor

True power is dissipated only in resistance. In a circuit that contains both resistance and reactance, the volt-ampere product of the source is a combination of apparent power and true power. The ratio of true power to apparent power is called the *power factor*. In a purely resistive circuit the power factor is 1. In a purely reactive circuit the power factor is 0. A circuit that contains both resistance and reactance will have a power factor that is between 0 and 1. It is common to multiply the power factor by 100 so that it may be expressed as a percentage. A purely resistive circuit would then have a power factor of 100 percent, and a purely reactive circuit would have a power factor of 0 percent.

To find the power factor of the series *L-R* circuit shown in Fig. 17-17 the volt-ampere product of the source will have to be known. The power dissipated by the resistor is found using the equation $I^2 \times R$ because the current through the resistor can be found sooner than the voltage across the resistor. The power factor is determined using the steps listed in Example 17-6.

Figure 17-17 The power factor is the ratio of true power to apparent power and will be a number between 0 and 1.

Example 17-6

1. Determine the impedance of the circuit using the Pythagorean theorem or trigonometry.

$$Z = \sqrt{X_L{}^2 + R^2} = \sqrt{250{,}000 + 1{,}000{,}000} = \sqrt{1{,}250{,}000} = 1118\ \Omega$$

2. Find the circuit current.

$$I = \frac{V_S}{Z} = \frac{125\ \text{V}}{1118\ \Omega} = 0.1118\ \text{A}$$

3. Find the apparent power.

$$P_A = V_S \times I = 125\ \text{V} \times 0.1118\ \text{A} = 13.975\ \text{VA}$$

4. Find the true power dissipated by the resistor.

$$P_T = I^2 \times R = (0.1118\ \text{A})^2 \times 1000\ \Omega = 0.0125\ \text{A} \times 1000\ \Omega = 12.5\ \text{W}$$

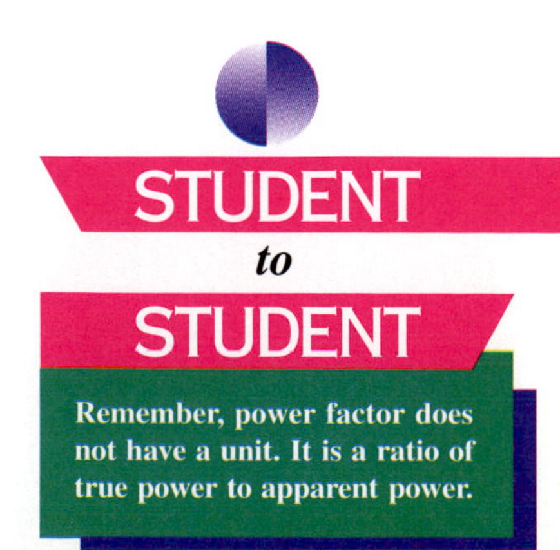

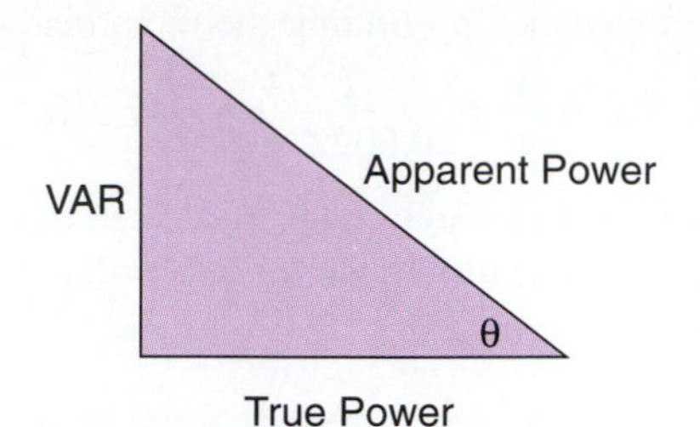

Figure 17-18 Apparent power, true power, and volt-ampere-reactance may be related to each other using a right triangle. The power factor ratio—true power over apparent power—is equivalent to the cosine of theta.

5. Calculate the power factor.

$$PF = \frac{P_T}{P_A} = \frac{12.5\text{ W}}{13.975\text{ VA}} = 0.8945 \text{ or } 89.45\%$$

Figure 17-18 shows how the relationship between apparent power and true power can be displayed graphically using a right triangle developed from the true power and VAR vectors. The power factor ratio—true power over apparent power—is seen to be the adjacent side over the hypotenuse. This ratio has been identified as the cosine ratio. The power factor ratio can then be stated as the apparent power multiplied by the cosine of theta, as shown in Eq.17-13.

Equation 17-13

$$PF = VA \times \text{Cos } \theta$$

Apparent power, VAR, and true power are all voltage multiplied by current. The significance of this is that the procedures used to calculate the power factor are identical for both series and parallel circuits. Power factor is always the ratio of true power to apparent power, which is also the cosine ratio.

REVIEW QUIZ 17-4

1. Power factor is the ratio of ______________ to ______________ .
2. The power factor is equivalent to the ______________ ratio in trigonometry.
3. The power factor of a purely resistive circuit is always ______________ .
4. The power factor of a purely reactive circuit is always ______________ .

SUMMARY

Inductors provide an opposition to current from an AC source that is called reactance. Reactance is measured in ohms and can be used in any of the Ohm's law formulas. An interesting thing about inductors is that the voltage across them and the current through them are out of phase by 90°, with the voltage leading. In a series *L-R* circuit the voltage components must be added vectorially. This is usually accomplished by converting the vector diagram to a right triangle and then using trigonometry to find the various circuit parameters. In a parallel *L-R* circuit it is the currents that must be added vectorially. These are also converted to adapt to a right triangle solution.

Circuits that contain both resistance and inductive reactance cause a shift in the phase of the source voltage in relation to the source current that is greater than 0° and less than 90°. In a series *L-R* circuit the source voltage leads the source current. The source current is the reference because current is the same at any point in a series circuit: the same amplitude, same frequency, and same phase. Voltage is the reference in a parallel circuit, causing the phase of the inductive current to lag. It is important to remember that *L-R* circuits obey two sets of rules depending on whether the circuits are series or parallel.

Power is only dissipated in resistance. No power is dissipated in reactance. When resistance and reactance are combined in a single circuit, albeit series or parallel, the volt-ampere product of the source is greater than the resistive power. The ratio of resistive power, called true power, to the apparent power of the source is called power factor. The power factor is equivalent to the cosine of the phase angle.

NEW VOCABULARY

- ACOS
- adjacent
- admittance
- ARC
- ASIN
- ATAN
- COS
- COS^{-1}
- cosine wave
- hypotenuse
- impedance
- in phase
- INVERSE
- opposite
- power factor
- protractor
- Pythagorean theorem
- quadrature
- reactance
- SIN
- SIN^{-1}
- susceptance
- TAN
- TAN^{-1}
- time constant
- true power
- vector
- volt-ampere-reactive (VAR)
- volt-amperes (VA)
- X^{-1}

Questions

1. What is the unit of apparent power?
2. What does the acronym VAR stand for?
3. When two voltages are said to be in quadrature, what is their angle of separation?
4. How does reactance vary with frequency?
5. In a series *L-R* circuit, does the voltage lead or lag the circuit current?
6. How must out-of-phase voltages of currents be added?
7. What mechanical device is used to measure angles?
8. What theorem describes how to determine an unknown side of a right triangle if the other two sides are known?
9. How is the trigonometric right triangle solution superior to the theorem developed by Pythagoras, considering both solutions use the same number of steps?
10. Is the phase of the current or the phase of the voltage shifted in a parallel *L-R* circuit?
11. Does the current lead or lag the voltage in a parallel *L-R* circuit?
12. What is the reciprocal unit of reactance?
13. What is the reciprocal unit of impedance?
14. Can the power factor be greater than 1?
15. What is the power factor of a purely inductive circuit?

Problems

1. What is the impedance, phase angle, and power factor of a series *L-R* circuit containing a 100-ohm resistor in series with a 100-ohm inductive reactance?
2. What is the source current and power factor of a parallel *L-R* circuit containing a 200-ohm resistor and a 500-ohm inductive reactance connected across a 100-volt AC source?
3. A 500-ohm resistor and a 500-ohm inductive reactance are connected in series across a 125-volt source. If the frequency of the source is doubled, what is the phase angle, and what is the voltage across the inductor?
4. What current flows when a 0.8 henry inductor is connected across a 100-volt, 60-hertz source?
5. What value resistor is required in series with a 200-ohm inductive reactance to produce an impedance of 500 ohms?

Critical Thinking

1. Why would a kilowatthour meter read incorrectly when measuring the energy supplied to an *L-R* circuit?
2. Why is the power factor independent of an *L-R* circuit's being series or parallel?
3. If an inductor had a power factor of 50 percent, what could you tell about the wire it was made from?
4. How would eddy current losses in an iron core increase the power factor of an inductor?

5. What effect would increasing the frequency of the applied voltage have on the power factor of an *L-R* circuit?
6. Explain why the numeric sum of the currents in a parallel *L-R* circuit would be greater than the source current in an apparent defiance of Kirchhoff's current law?

Answers to Review Quizzes

17-1
1. charge carriers
2. apparent power
3. resistance
4. 90°
5. double

17-2
1. will
2. 90°
3. 0
4. impedance
5. trigonometry
6. 63

17-3
1. conductance; susceptance; and admittance
2. currents
3. lags

17-4
1. true power; apparent power
2. cosine
3. one
4. zero

Chapter 18

Capacitors

ABOUT THIS CHAPTER

Capacitors are one of the most common components in electronic equipment and are second in number only to resistors. This chapter discusses the properties of capacitance and capacitors as a broad group of components. This discussion on capacitors is limited to their operating principles, physical properties, and basic electric be-

After completing this chapter you will be able to:

- Explain how capacitors store electric charge.
- Discuss how dielectric materials increase the storage capacity of capacitors.
- Account for the various differences in capacitor construction.

havior. Like inductors, capacitors store electric energy, but in an electric field rather than a magnetic field. Capacitors are more versatile than inductors because they can store their energy in a static state for long periods. Capacitors may be connected in any circuit configuration, but they often behave differently than expected. Capacitor values in series are treated as resistor values in parallel, and capacitor values in parallel are treated as resistor values in series. Capacitors are an extension of the Leyden jar so diligently studied by Benjamin Franklin.

- Explain why some capacitors are better suited for some applications than others.
- Describe how capacitors behave in series and parallel circuits.
- Describe how series capacitive circuits distribute voltage differently than series resistive circuits.

18-1 Capacitors

Capacitors are components that temporally store electric energy. They can be thought of as containers for electricity. A basic capacitor is made from two metal surfaces called *plates* that face each other but do not touch. Applying a DC voltage across the plates causes electrons to be removed from one and deposited onto the other. The plate that loses electrons becomes positively charged, and the plate that gains electrons becomes negatively charged. The separated charges result in the development of electrostatic lines of force between the plates, as shown in Fig. 18-1(*a*). The schematic symbol for a capacitor is shown in Fig. 18-1(*b*).

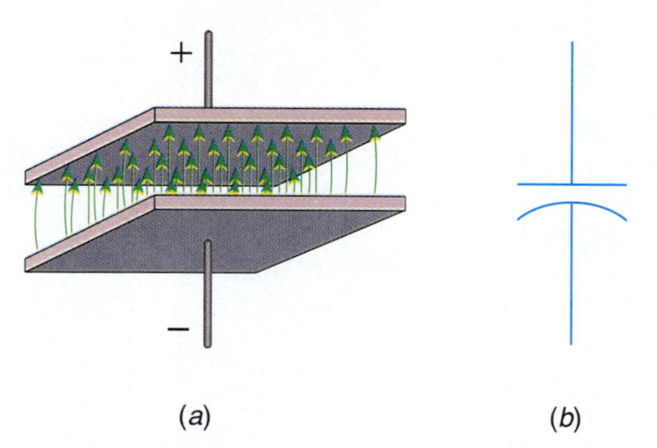

Figure 18-1 (*a*) Two electrically isolated conductors develop electrostatic lines of force between them when connected to a voltage source. (*b*) The schematic symbol for a capacitor.

Work is performed in the process of moving electrons from one plate to the other. The energy used in moving the electrons is not lost, it is present in the electrostatic field between the plates. The energy contained in the field may be calculated by multiplying the applied voltage by the charge moved, as indicated in Eq. 8-1. The work can be recovered by allowing the electrons to return to the positively charged plate through an external circuit.

Equation 18-1

$$E = V \times Q$$

where E = The energy in joules.
V = The applied voltage in volts.
Q = The charge moved from one plate to the other in coulombs.

The separation of charge on the plates produces a capacitive voltage that opposes the source voltage with like polarities. This reduces the source current by creating a pressure against the source. Even though the source voltage and capacitive voltage appear to be in parallel, the internal resistance of the source is between them. That resistance allows the voltages to be different until sufficient charge has been moved to bring the capacitive voltage equal to that of the source voltage, as indicated in Fig. 18-2. The only energy lost in this process is in heating the internal resistance of the source. The internal resistance of a capacitor is very low because of the nature of its construction and is usually ignored.

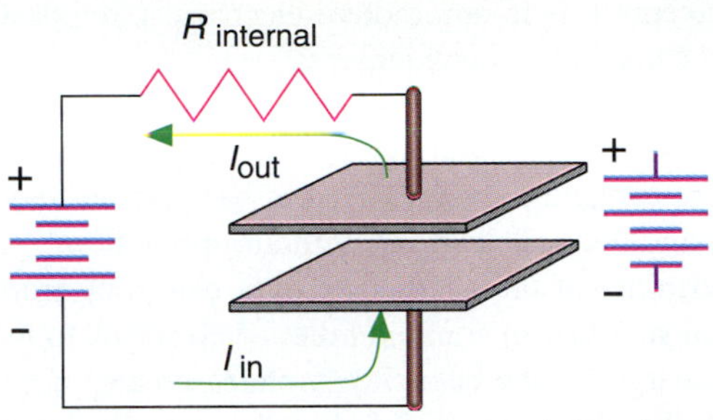

Figure 18-2 As electrons are removed from one plate and placed onto the other, a voltage is developed that will eventually equal that of the applied source. The two voltages may be different at first because of the internal resistance of the charging source.

Capacitance

Capacitance is the ability to store energy in the electrostatic field between two electrically isolated conductors with dissimilar charges. The symbol for capacitance in mathematical equations is the italic capital letter *C*.

The quantity of charge that a capacitor can hold is determined by its capacitance value and the voltage applied to it. Capacitance refers to the property of two conductors to store energy in the electric field between them when they are connected to a voltage source. Capacitance is present anytime two conductors share a common area without touching. Such conductors have the capacity to absorb, store, and deliver electric energy. Higher voltages create more pressure, thereby squeezing in more charge. Too much voltage, however, may permanently damage a capacitor.

The first capacitors evolved from experiments with Leyden jars. It was believed that Leyden jars had the ability to condense electricity. That belief was founded in the observation that electric effects were much more pronounced using a Leyden jar in conjunction with an electric machine than with the electric machine alone. Anyone unfortunate enough to discharge a Leyden jar through any part of their body became an immediate believer that the electricity was indeed concentrated. Close examinations of Leyden jars led early experimenters to the conclusion that the electric charge was concentrated in the glass between the foil plates. Consequently, early capacitors were called condensers. They later became known as capacitors because of their capacity to store charge. The two terms, *condenser* and *capacitor*, are still used interchangeably. The term *condenser*, however, is considered obsolete in almost all areas of electricity and electronics.

The one notable exception is in automotive electrical circuits where mechanics continue to refer to capacitors as condensers.

Charging a capacitor begins by connecting it to a DC voltage source. Current is delivered by the source which initially "sees" the capacitor as a short circuit regardless of the capacitor's value. That is because an uncharged capacitor can initially accept and deliver electrons without opposition.

The positive terminal of the source removes electrons from one plate, and the negative terminal supplies an equal number of electrons to the opposite plate. The separation of charge on the capacitor's plates creates a capacitive voltage that pushes back against the voltage of the source, steadily lowering the amount of current the source can deliver to the capacitor. The voltage across the capacitor will eventually equal that of the source and cause all current to stop. When that occurs the capacitor is said to be charged. If the voltage of the source is then lowered, the capacitive voltage will force current back through the source in an attempt to maintain the established voltage. The opposite actions of first opposing any increase in voltage and then opposing any decrease in voltage alternatively defines capacitance as an *opposition to a change in voltage.*

Capacitive Units of Measure

The measure of capacitance is found by taking the charge stored in the dielectric and then dividing that charge by the voltage across both plates. The unit assigned to the charge/voltage ratio, as described in Eq. 18-2(*a*), is the *farad*, abbreviated as *F*. One farad is the SI unit of capacitance that will store 1 coulomb of charge when 1 volt is applied.

The charge in the dielectric is twice that removed from the positive plate or added to the negative plate because it represents the difference in charge between the two plates. For instance, assume that 1 coulomb is removed from one plate, making it positive. It is the nature of voltage sources and capacitors to maintain an equal number of positive and negative charges at all times. Therefore, 1 coulomb will be added to the opposite plate, making it negative. The total difference in charge is 2 coulombs. The doubling of charge between the plates is reflected in Eq. 18-2(*b*), which is derived in part from Eq. 18-2(a). Note that the numerator part of Eq. 18-2(*b*) is divided by 2 because it is concerned with the charge on to and off of the individual plates.

Equation 18-2

$$(a)\ C = \frac{Q}{V}$$

$$(b)\ E = \frac{C V^2}{2}$$

where C = The capacitance in farads.
Q = The charge in coulombs.
V = The applied voltage in volts.
E = Energy in joules.

Practical Units of Capacitance

In actual practice the farad is a very large unit that is seldom encountered. Practical values of capacitance are normally given in microfarads (μF), nanofarads (nF), or picofarads (pF).

Older capacitors often used letter abbreviations to identify their capacitive values. A capacitor in the microfarad range could be marked mf, MF, mfd, MFD, uf, UF, ufd, or UFD. Those in the picofarad range were marked mmf, mmfd, uuf, or uufd. Capacitor values were seldom labeled directly in picofarads. Instead they were marked for *micro-microfarad;* a millionth of a millionth. The letters *u* or *U* were used instead of the Greek symbol μ, which is preferred for micro.

Sometimes older capacitors were marked only with a number. Decimal values were understood to be in microfarads, and whole values were in picofarads. A capacitor of one or more microfarads had the value and the unit stamped somewhere on the capacitor's body. Microfarads and picofarads were originally the only values assigned to capacitors. The farad was thought to be impossibly large, and *nano* was simply not used as a prefix.

Whole numbers without units are more common today than ever. The numbers continue to represent a capacitor's value in picofarads, but that number cannot always be taken literally. The last digit often represents the number of zeros in the value. It is used in the same way as the multiplier band on resistors. Capacitor values that do not end in zero should be suspected of being encrypted. For example, assume that the number printed on a capacitor reads "104." Its value may be one of two possibilities. The first is 104 picofarads and the second is 100,000 picofarads, which is 10 plus four zeros.

Even a number ending in zero is not a guarantee that a capacitor's value is actually the number printed on it. Suppose the number on a capacitor is "100." It would seem likely that the value is 100 picofarads, but the last number, being zero, could mean that there are no zeros after the 10. That would make the capacitor's value 10 picofarads. With very small valued capacitors, their physical sizes are usually identical over a fairly wide range of values. Size is not always a reliable indication of value.

If there is any doubt about the numbering system used on capacitors encountered while working on a piece of equipment, one of the capacitors can be measured as a sample. The value may also be confirmed by checking the schematic diagram, if one is available. It is not uncommon to have capacitors in identical pieces of equipment supplied by different vendors who use different numbering systems. The "104," or 10,000 picofarad capacitor in the previous example might also be given as ".01 μF."

Another clue to a capacitor's value is the presence of a tolerance marking. If a capacitor were actually marked to mean 104 picofarads, it would most likely be a precision value identified with a low percentage tolerance, such as ±1%, as shown in Fig. 18-3(*a*). A 100,000-picofarad capacitor is shown in Fig. 18-3(*b*). One hundred thousand picofarads is also 0.1 microfarads. Familiarity with the use of capacitors in various circuits will also help identify what value range should be present.

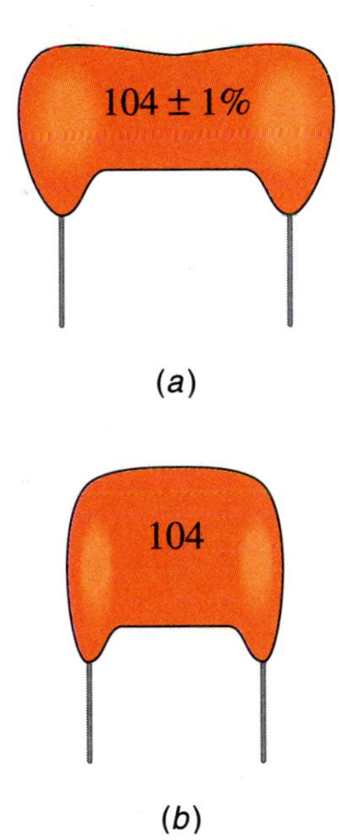

Figure 18-3 A precision tolerance rating can be an indication that a capacitor value is to be taken literally.

REVIEW QUIZ 18-1

1. Capacitance is an opposition to a change in ______________ .
2. The standard unit of capacitance is the ______________ .
3. When voltage is first applied to an uncharged capacitor, the capacitor appears to be a(n) ______________ .
4. A capacitor is a component that stores electric ______________ .
5. A capacitor marked "104" might have a value of ______________ picofarads, or ______________ picofarads.

18-2 Dielectrics

A capacitor's value can be made larger by increasing the common area between the plates, decreasing the distance between the plates, or using a *dielectric* with a higher *permittivity*. A dielectric is any nonconductor of electricity. The permittivity of a dielectric indicates the ease with which electrostatic lines of force can exist in an insulating material when compared to a vacuum. The permittivity ratio is called the *dielectric constant*. The dielectric constant is a unitless number that identifies the increase in capacitance between two conductors when a dielectric material fills the space between them. A vacuum has a dielectric constant of 1 and is used as a standard by which all other dielectric materials are compared. The dielectric constants of insulating materials range from just over 1 to several thousand. It is often more convenient to use the dielectric constant of air instead of a vacuum as the standard reference. With a dielectric constant of only 1.0006, air allows the dielectric constants of insulating materials to be evaluated easily.

The amount of energy stored in a capacitor depends on the quantity of charge moved from one conductor to the other and the applied voltage. When a dielectric material is placed between charged conductors, the molecules in the material become stretched and electrically polarized. The valence electrons of the dielectric are attracted to the positive plate, and the positive nuclei are attracted to the negative plate, as shown in Fig. 18-4. This generates polarized charges within

Figure 18-4 The molecules of a dielectric stretch under the influence of an electric field like a rubber band, requiring more work from the source to move charge from one plate to the other.

the dielectric that are opposite to the charges on the plates. The polarization in the dielectric allows the source to move more charge from one plate to the other before an equilibrium is reached. The additional energy is stored in the distortion of the dielectric in the same way that energy is stored in a stretched rubber band. The increase in the number of charges moved is the result of the capacitor's being electrically larger without increasing the area of the plates facing one another or moving them closer together.

Dielectric Strength

When a dielectric is subjected to a difference in potential, it must be able to withstand the applied voltage without breaking down and allowing unwanted current flow. All materials eventually break down and conduct if the voltage across them is raised high enough. A vacuum is not composed of atoms, so there is nothing to break down. A gas, such as air, that is used as a dielectric is not damaged when an electric arc passes through it because it only conducts temporarily. Such a dielectric is said to "heal" itself. When a gas is ionized, however, significant amounts of current can flow. Remember that ionized gases are good conductors.

An electric arc passing through a solid dielectric may bring about chemical changes in the material that cause it to permanently conduct. That is a particularly common problem with plastic dielectrics. Plastics are made from organic[1] materials that can easily be reduced to carbon when an electric arc passes through them. As might be expected, the thickness of a dielectric affects its breakdown voltage. The thicker it is the more voltage it can withstand. The dielectric strength of materials is normally given in volts per millimeter at 20° C. If the dielectric is to be operated at a higher temperature it may have to be derated because higher temperatures weaken them electrically.

The table below shows the dielectric constants and dielectric strengths of some common insulators.

[1]Organic materials are or were once living and are based on carbon.

Material	Dielectric Constant	Dielectric Strength (V/mm)
Vacuum	1.0000	∞
Air	1.00059	787
Mica	5.0	59,055
Mylar	3.1	6,000
Glass	7.5	78,740
Ceramic	25 – 5000	39,370

Table 18-1.

Dielectric Applications and Restrictions

Capacitors are usually named after their dielectric material. Common dielectrics are mylar, polyester, ceramic, mica, polystyrene, and air. Older capacitors used a waxed paper dielectric, but numerous problems arose over time and they were abandoned. Many types of plastic dielectrics are used, and care must be taken to choose a proper replacement or premature failure may occur.

The molecules of a dielectric are stretched and relaxed during the process of charging and discharging. This molecular motion creates friction that in turn creates heat. Polyester capacitors can withstand rapid changes in voltage without significant heating. Polyester and other high-performance capacitors look very similar to mylar capacitors, but mylar capacitors would rapidly fail if operated under circumstances similar to those for which polyester was designed. Mylar capacitors are far less expensive than polyester and are intended for general purpose applications. Polyester capacitors are not only more costly, they are less available. If cost were not a consideration, a polyester capacitor could be used to replace a mylar, but a mylar could not be used to replace a polyester without compromising circuit integrity.

Capacitors used at frequencies above the audio spectrum are often mica or ceramic. These capacitors have very low self-inductance values and are efficient at high frequencies. The operating range of tubular capacitors is extended into the low megahertz region because major efforts are made in their construction to cancel the natural inductance inherent in a rolled capacitor. The tolerances of tubular capacitors are often too broad for many applications.

An important consideration for capacitors used at high frequencies is that their values remain constant over a wide temperature range. A form of mica capacitor called "silvered mica" has excellent temperature stability and low losses. The plates are formed when silver vapor is deposited onto the mica dielectric. This process creates an excellent mechanical bond. The major drawbacks of silvered mica capacitors are that they are expensive and are only available in relatively small values.

An alternative to silvered mica capacitors is temperature-stabilized ceramic. These are identified *NP0,* which stands for negative–positive–zero. Sometimes temperature stabilized capacitors are labeled with the letter *N* followed by a number such as 750. The number represents the number of parts per million decrease for each degree in temperature increase. When these capacitors are encountered they must be replaced with a similar type. Ceramic capacitors offer the highest capacitance for their size than any other type of capacitor.

When alternating current is applied to a capacitor the molecules of the dielectric must follow the changes in voltage. All materials will fail to do this at some frequency. The effect is similar to hysteresis in magnetic cores, where the molecules can no longer follow the changes in current. The inability of the dielectric to follow changes in the applied voltage is called *absorption loss*. At higher frequencies absorption loss can be significant. It causes the capacitor to behave as though a small resistor were placed in series with the capacitor. That means that the current in to and out of the capacitor must flow through this resistance creating heat in the process. Capacitors should not develop noticeable amounts of heat during normal operation.

REVIEW QUIZ 18-2

1. Dielectrics is another name for a class of materials called ______________ .
2. If a material with a dielectric constant of 5 were slipped between the plates of a capacitor replacing the air, its capacitance would increase ______________ times.
3. The material with the highest dielectric constant is ______________ .
4. A common dielectric that can "heal" is ______________ .
5. The ______________ capacitor offers the maximum capacitance in the smallest physical size.

18-3 Capacitor Construction

A basic capacitor has two plates facing each other and separated by a dielectric, as shown in Fig. 18-5 (*a*). The dielectric often forms the mechanical foundation of the capacitor. Larger value capacitors use several plates, forming a dielectric sandwich, as shown in Fig. 18-5 (*b*). This flat sandwich-type construction is common

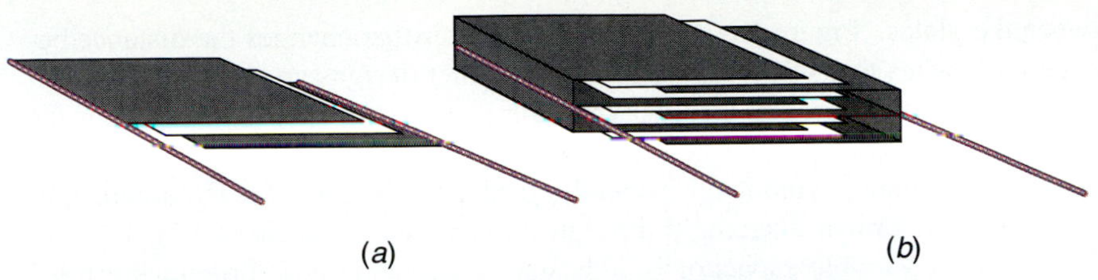

Figure 18-5 A basic capacitor is formed using two parallel metal plates separated by a dielectric. The capacitance can be increased by adding additional plates.

with ceramic capacitors. A capacitor with leads that emerge from the same side is a *radial lead* capacitor.

Tubular capacitors have been popular for decades.The basic construction of a tubular capacitor is similar to that of a jelly roll, as shown in Fig.18-6. Plastic dielectrics, such as mylar, are well suited for this type of construction. The plates are generally made from aluminum foil. When the leads extend in opposite directions from the ends of the capacitor, the capacitor is said to have *axial leads*. Tubular capacitors are also available in the radial lead configuration.

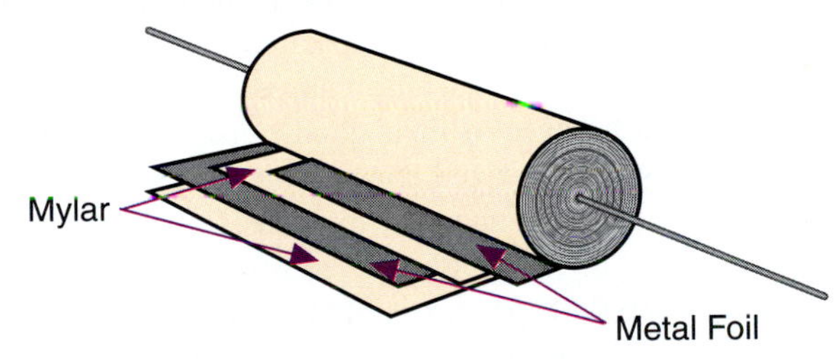

Figure 18-6 A tubular capacitor is constructed similarly to a jelly roll. Plastic dielectrics, such as mylar, are well suited for tubular construction because they are so flexible.

Variable Capacitors

Sometimes it is necessary that the value of a capacitor be adjustable. This could be accomplished by changing the area of the plates facing one another, changing the distance between the plates, or changing the dielectric constant of the material between the plates. Changing the area of the plates facing one another is the most common method. Although changing the distance between the plates is not uncommon, it is more often encountered with older variable capacitors. Fig. 18-7(*a*) shows an example of a variable capacitor that changes the area be-

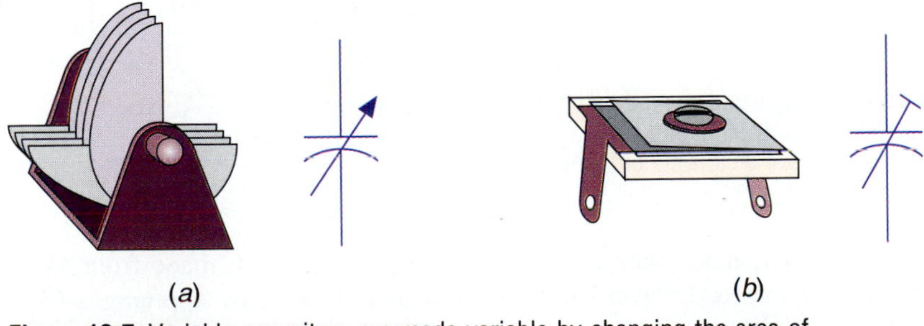

Figure 18-7 Variable capacitors are made variable by changing the area of the plates facing one another or changing the distance between the plates.

tween the plates. Figure 18-7(*b*) shows a capacitor that changes the distance between the plates through compression. Many other designs are employed, but all of them change either the area between the plates or the distance between the plates.

The schematic symbol for a variable capacitor is that of a fixed capacitor with an arrow drawn diagonally through it, as shown adjacent to Fig. 18-7(*a*). Sometimes a variable capacitor is indicated by drawing a T through the fixed schematic symbol for a fixed capacitor, as shown adjacent to Fig. 18-7(*b*). The *T* indicates that the capacitor is a *trimmer capacitor*. Trimmer capacitors are used to make small adjustments using a screwdriver or other tool. A variable capacitor using this symbol would definitely not be expected to have a knob.

Electrolytic Capacitors

Electrolytic capacitors are an exception to the "rule" that the name of a capacitor identifies its dielectric. As the name implies, an electrolytic capacitor contains a liquid conductor (electrolyte). Electrolytic capacitors are commonly made from aluminum, but they may also be made from tantalum. Several advantages are offered by tantalum, but they are offset by cost. Although tantalum electrolytic capacitors are expensive, they offer larger capacitance values in a smaller size, longer life, and less leakage current than aluminum electrolytics. Aluminum electrolytics are most common because of their good price and performance ratio.

Electrolytic capacitors are polarized components and must be operated from direct current or they will overheat and may explode. Polarized *sintered slug* capacitors *will* explode if voltage is applied in the reverse polarity direction. The positive plate of an electrolytic capacitor is made from aluminum. The negative plate is an electrolyte made from an alkali solution. In years past the solution was truly a liquid. Electrolytic capacitors had to be operated in a fixed vertical position or the electrolyte would pour out of the vent hole on top of the capacitor. The liquid has since been replaced with a paper gauze soaked in a pasty alkali solution.

Figure 18-8 (*a*) shows a cross section of an aluminum electrolytic capacitor. The negative electrode serves only as a contact to the electrolyte-soaked gauze. The special symbol that identifies an electrolytic capacitor is shown in Fig. 18-8 (*b*).

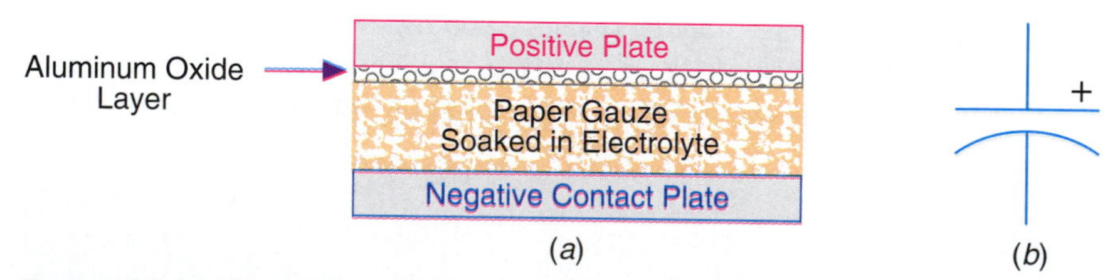

Figure 18-8 (*a*) The negative plate of an electrolytic capacitor is paper gauze soaked in an alkali paste. (*b*) The symbol for an electrolytic capacitor reinforces that it is a polarized component.

The negative contact plate and the positive plate are both made from aluminum. At first the oxide layer does not exist. It is formed by applying a DC voltage to the would-be capacitor and is the result of electrolytic action. The oxide layer is very thin, but it has relatively high breakdown voltage. Aluminum oxide is one of the best insulators known. The extremely thin aluminum oxide dielectric results in some impressive capacitance values. Modern electrolytic capacitors have values that range from 0.1 microfarads to 10 farads.

A problem with early electrolytic capacitors was that they would deform, causing the oxide layer to become thinner. The thinner oxide layer meant that the breakdown voltage of the capacitor had reduced. A piece of equipment that had been left unused for a long time was likely to fail if the full voltage were suddenly applied. It is always a good idea to increase the voltage slowly on equipment that has been out of service for an extended time, although modern electrolytics are far less prone to deforming.

Electrolytic capacitors are not precision components. Their values may range from −20 percent to +80 percent of their stated value. It is more common to find that their measured value exceeds their indicated value by a large margin. Their operating voltage, however, is a different matter. Electrolytic capacitors are usually operated very close to their rated voltage. This ensures that they will not deform. There is also the added advantage of being a guide as to what voltage can be expected if a schematic diagram is unavailable.

Electrolytic capacitors have measurable amounts of dielectric leakage. This is normal for them and does not indicate that they are defective. They may also have both series resistance and a small amount of series inductance. The series resistance results in the power factor of electrolytic capacitors being less than 100 percent. Power factors above 90 percent are considered acceptable. It is important that a capacitor tester be able to test for power factor in addition to the capacitance value. It is possible that the value of an electrolytic capacitor may measure within tolerance and still have an unacceptable power factor.

The polarity of an electrolytic capacitor is identified using a plus, a minus, or both. The positive end of an axial lead electrolytic capacitor may also be identified by a crimped seal, as shown in Fig. 18-9(*a*). The negative lead is usually

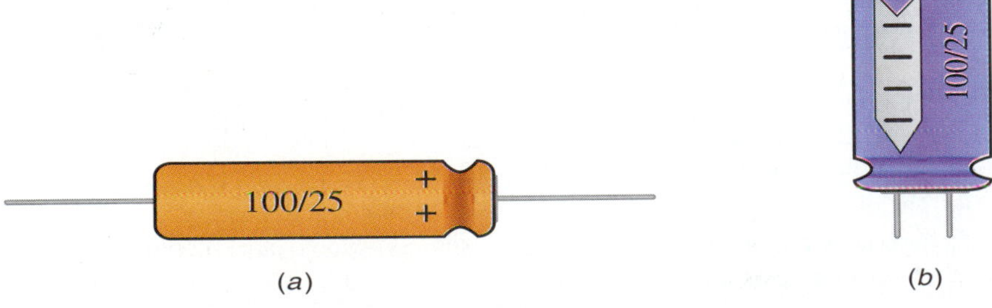

Figure 18-9 The polarity of an electrolytic capacitor is always indicated on the body of the capacitor.

identified on radial lead electrolytic capacitors with a broad stripe containing minus signs, as shown in Fig. 18-9(*b*). The capacitive value of an electrolytic is understood to be in microfarads unless specifically stated otherwise. When two numbers are separated by a slash, the first number is the capacitor's value, and the second number is its voltage rating.

Nonpolarized Electrolytics

One variety of electrolytic capacitor is said to be nonpolarized. This would at first seem to be a contradiction, but it is not. Nonpolarized electrolytics are really two separate capacitors connected in series. The negative ends of the capacitors are normally common to each other. This type of capacitor takes advantage of the fact that an electrolytic capacitor can withstand a polarity reversal for a period of one alternation. Nonpolarized electrolytic capacitors are used as inexpensive starter capacitors in electric motors and in speaker crossover systems. The value

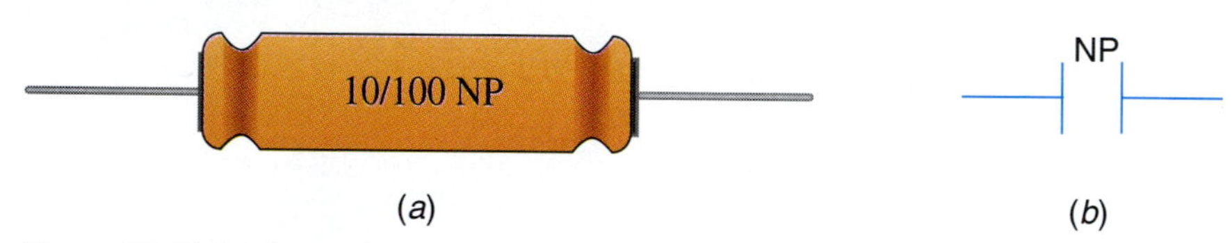

Figure 18-10 (*a*) A nonpolarized electrolytic capacitor is composed of two standard electrolytic capacitors connected in series. The negative ends are usually common. (*b*) The schematic symbol for a nonpolarized electrolytic capacitor.

and voltage of a nonpolarized electrolytic capacitor are followed by the letters *NP* (nonpolarized). Figure 18-10 (*a*) shows how a typical nonpolarized electrolytic capacitor has a crimp at each end, revealing that the positive ends of the two series-connected capacitors it is made from are exposed. The special schematic symbol for a nonpolarized electrolytic is shown in Fig. 18-10 (*b*). Nonpolarized electrolytic capacitors are sometimes found in general circuit applications. Care must be taken to replace them with another nonpolarized electrolytic or a high value mylar if space permits. New technologies have yielded large capacities in small packages for mylar capacitors.

REVIEW QUIZ 18-3

1. The plates of a capacitor are separated by a(n) ______________ .
2. A(n) ______________ capacitor is constructed like a jelly roll.
3. An arrow through a capacitor symbol indicates that it is ______________ .
4. The dielectric of an aluminum electrolytic capacitor is ______________ .
5. A power factor of ______________ is considered acceptable for an electrolytic capacitor.

18-4 Capacitors in Series and Parallel

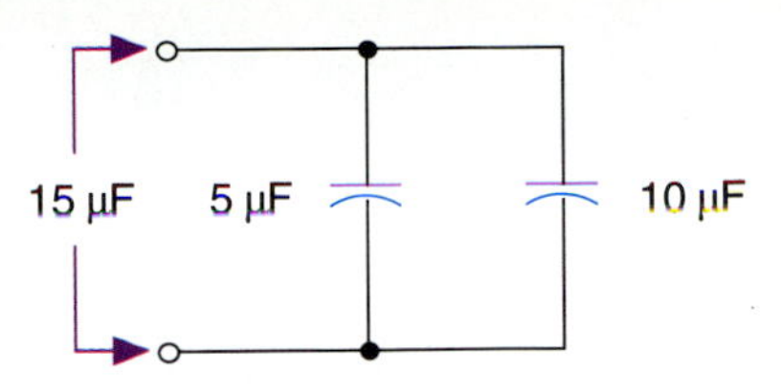

Figure 18-11 Capacitor values in series add numerically.

As with any other electrical component, capacitors may be connected in series or parallel. The values of capacitors in parallel add numerically. A 5-microfarad capacitor connected in parallel with a 10-microfarad capacitor will provide 15 microfarads of capacitance, as shown in Fig. 18-11.

The total capacitance of series-connected capacitors is calculated in the same way as resistors in parallel. If the capacitors are of equal value, the total capacitance equals the value of one of the capacitors divided by the number of capacitors, as shown by Eq. 18-3. For instance, two 10-microfarad capacitors connected in series have a total capacitance of 10 microfarads divided by 2, or 5 microfarads.

Equation 18-3

$$C_T = \frac{C}{N}$$

where C_T = The total capacitance.
C = The value of one capacitor.
N = The number of capacitors.

When series-connected capacitors are of unequal values, the reciprocals equation may be invoked. The total capacitance is equal to the reciprocal of the sum of the reciprocal of the capacitor values, as shown in Eq.18-4.

Equation 18-4

$$C_T = \frac{1}{\frac{1}{C_1} + \frac{1}{C_2} + \frac{1}{C_3} \ldots}$$

For example, assume that a 12-microfarad, a 6-microfarad, and a 28-microfarad capacitor are connected in series. The total capacitance using Eq.18-4 is 3.5 microfarads. Note that the total will always be less than the smallest value capacitor in a series circuit.

Voltage Division in Series-Connected Capacitors

When capacitors of equal value are connected in series, the source voltage is divided equally among them. When the capacitor values are not equal, the voltages across them are unequally distributed.

The voltages across the capacitors in Fig. 18-12 were calculated using ratios. The total capacitance was first calculated using Eq. 18-4 and found to be 3.5 microfarads. The total capacitance was then divided by the value of each individual capacitor, and the result was multiplied by the applied voltage, as shown in Eqs. 18-5, 18-6, and 18-7.

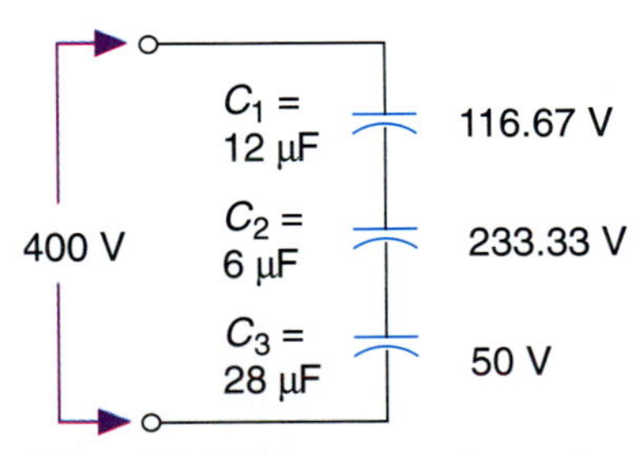

Figure 18-12 When unequal capacitors are connected in series the voltages across them are also unequal. The highest voltage will be across the smallest capacitor, and the lowest voltage will be across the largest capacitor.

Equation 18-5

$$V_{C_1} = V_S\left(\frac{C_T}{C_1}\right) = 400\left(\frac{3.5\ \mu F}{12\ \mu F}\right) = 400\ (0.292) = 116.67\ V$$

Equation 18-6

$$V_{C_2} = V_S\left(\frac{C_T}{C_2}\right) = 400\left(\frac{3.5\ \mu F}{6\ \mu F}\right) = 400\ (0.583) = 233.33\ V$$

Equation 18-7

$$V_{C_3} = V_S\left(\frac{C_T}{C_3}\right) = 400\left(\frac{3.5\ \mu F}{28\ \mu F}\right) = 400\ (0.125) = 50\ V$$

If a situation should ever arise where a capacitor value must be obtained by placing two unequal capacitors in series, remember that the voltages do not divide equally. This is easy to overlook because the stated voltage rating on the capacitor may seem adequate. For example, suppose that a 6-microfarad and a 3-microfarad capacitor are to be placed in series to obtain a value of 2 microfarads. Each of the capacitors has a voltage rating of 150 volts. The circuit voltage is 250 volts, which does not seem to be a problem. There is a tendency to consider the voltage rating of the capacitors connected in series as 300 volts and falsely assume that the voltage will be evenly divided between the two capacitors. That is only true if the capacitor values are equal. The 3-microfarad capacitor will experience twice the voltage of the 6-microfarad capacitor. Two-thirds of 250 volts is 166.7 volts. The voltage rating of the 3-microfarad capacitor would be exceeded by 11 percent.

REVIEW QUIZ 18-4

1. The values of capacitors in parallel are calculated in the same way as resistors in ______________ .
2. The total capacitance in a series is always ______________ than the smallest capacitor in the circuit.
3. When capacitors are connected in parallel, the applied voltage may not exceed the voltage rating of the ______________ rated capacitor in the circuit.
4. The highest voltage in a series capacitive circuit is across the ______________ capacitor.

SUMMARY

Capacitors store energy in the form of electric charge. That energy may be delivered quickly or slowly. Current does not flow through a capacitor; it flows into one plate and out of the other. The plates of a capacitor are separated by an insulator called a dielectric. The name of the capacitor—mica, ceramic, mylar, and so on—normally identifies the dielectric. Electrolytic capacitors are an exception. One of their plates is an electrolyte (liquid conductor). The dielectric is either aluminum oxide or tantalum oxide.

The value of a capacitor is determined by the area of the plates facing one another, the distance between the plates, and the dielectric. The material from which the plates are made or their thickness are unimportant because they do not affect capacitance.

Plastics, which give their name to the capacitor, are widely used as dielectrics in capacitors. Mylar, polyester, and polystyrene are only a few of the common plastic dielectrics. Ceramics are also widely used as dielectrics in capacitors. There are so many ceramics that it would be nearly impossible to remember the names of them all, if they even have names. Unless you were designing capacitors, the names and other features of the ceramics would not be important.

Variable capacitors are frequently employed in a variety of circuits. They usually have an air dielectric and therefore have values in the picofarad range. Variable capacitors that are designed to be adjusted with a screwdriver or other tool are frequently called trimmer capacitors. Virtually all variable capacitors change the surface of the plates facing one another or the distance between the plates. A variable capacitor could be designed that changes the permeability of the dielectric, but that would be a rare exception.

Electrolytic capacitors are polarized components. If the polarities are reversed for any length of time, an electrolytic capacitor may overheat and even explode. Electrolytic capacitors offer high values of capacitance in a small size. This factor often outweighs their disadvantages. Electrolytics are only useful at low frequencies, such as power and audio frequencies. They are not precise as evidenced by their normal tolerance of −20 percent to +80 percent.

The values of capacitors connected in parallel add numerically in the same way as resistors connected in series. The reverse is also true: The values of capacitors connected in series are calculated in the same way as resistors in parallel. Unlike resistors, the greatest voltage in a capacitive series circuit is across the smallest capacitor.

NEW VOCABULARY

absorption loss
axial lead
capacitance
dielectric
dielectric constant
electrolytic capacitor
farad (F)
micro-microfarad
permittivity
plates
radial lead
sintered slag
trimmer capacitor

Questions

1. What is the standard unit of capacitance?
2. What name is given to the insulator between the plates of a capacitor?
3. Where is the energy stored in a vacuum capacitor?
4. How does an uncharged capacitor look to a voltage source?
5. What two possible values could a capacitor that is labeled 104 have?
6. How is the dielectric constant of a material related to its permittivity?
7. What insulating material has the highest voltage rating?
8. What is meant by the label *NP0*?
9. What name is given to the inability of a dielectric to follow changes in the applied voltage?
10. What type of configuration has the leads protruding from the same end of a capacitor?
11. What type of capacitor is manufactured similar to a jelly roll?
12. What type of dielectric is normally used with variable capacitors?
13. What forms the negative plate of an electrolytic capacitor?
14. What will happen if the polarity applied to an electrolytic capacitor is reversed or its rated voltage is exceeded?
15. Should a nonpolarized electrolytic capacitor be used to replace a mylar capacitor?

Problems

1. How much energy is stored by a 10-microfarad capacitor charged to 100 volts?
2. What value capacitor will store 0.1 coulomb when charged to 50 volts?
3. A 10-microfarad, 20-microfarad, and 30-microfarad capacitor are connected in series. What is the total capacitance of the circuit?
4. What is the total capacitance when a 4-microfarad and a 6-microfarad capacitor are connected in parallel?
5. If the circuit in question 3 were placed across a 300-volt source, what would be the voltage across each capacitor?
6. If a 50 microfarad capacitor has a tolerance of −20 percent to +80 percent, what measured range of values would be acceptable if the capacitor was being tested?

Critical Thinking

1. If the plates of a charged capacitor were pulled apart, what would happen to the voltage across those plates and why?
2. Why does a DC source "see" an uncharged capacitor as a resistive load when it has been established that a capacitor does not consume energy?
3. Why may a capacitor with a plastic dielectric be unusable if the dielectric has broken down, allowing an electric arc?
4. How can the inductive tendencies of a tubular capacitor be overcome in the manufacturing process? *Hint*: Look at Fig. 18-6.
5. The compression type capacitor shown in Fig. 18-7 really uses two principles to change the capacity. The obvious one is that the distance between the plates is varied. What is the other one?

Answers to Review Quizzes

18-1

1. voltage
2. farad
3. short circuit
4. charge/energy
5. 100,000; 104

18-2

1. insulators
2. five
3. ceramic
4. air vacuum
5. ceramic

18-3

1. dielectric
2. tubular
3. variable
4. aluminum oxide
5. 90 percent

18-4

1. series
2. less
3. lowest
4. smallest

Chapter 19

R-C Circuits

ABOUT THIS CHAPTER

Placing a resistor in series with a capacitor and then connecting the combination across a DC voltage source causes the charge on the capacitor to build up over time. Higher value resistors or larger capacitors increase the time it takes for the capacitor to charge and discharge.

After completing this chapter you will be able to:

- Explain *R-C* time constants and describe how voltage is shifted in a series *R-C* circuit.
- Describe how current is shifted in a parallel *R-C* circuit.

When alternating current is applied to an *R-C* circuit, be it series or parallel, the phase of the source voltage and source current are shifted relative to each other. In a series *R-C* circuit, current is the reference; the voltage across the capacitor lags the voltage across the resistor. In a parallel *R-C* circuit, voltage is the reference; the capacitor current leads the resistor current. The source voltage or current is phase-shifted by some value greater than 0° but less than 90°.

- ▽ Understand the various methods of determining the phase angle and impedance of an *R-C* circuit.
- ▽ Describe how *R-C* circuits can alter waveshape.

19-1 R-C Time Constants

Voltage applied to an uncharged capacitor moves electrons from one plate to the other, creating a capacitive voltage in the process. That voltage will increase until it equals the source voltage. The time to do this is dependent on the resistance in series with the capacitor and the capacitor's value. As with an *L-R* circuit, the horizontal axis of the amplitude graph is segmented by time constants. In an *R-C* circuit one time constant equals the value of the capacitor multiplied by the series resistance, as shown in Eq. 19-1.

Equation 19-1

$$t = RC$$

where t = time in seconds
R = resistance in ohms
C = capacitance in farads

The Universal Charge Curve

A plot of the increase in voltage across a capacitor as it charges is shown in Fig. 19-1. The plot is similar to one of current through an inductor in a series *L-R* circuit after being connected to a voltage source. In theory the capacitor is never completely charged. In practice the difference in voltage between the charging source and the voltage across the capacitor becomes so small that it cannot be measured. A capacitor is considered fully charged after five time constants.

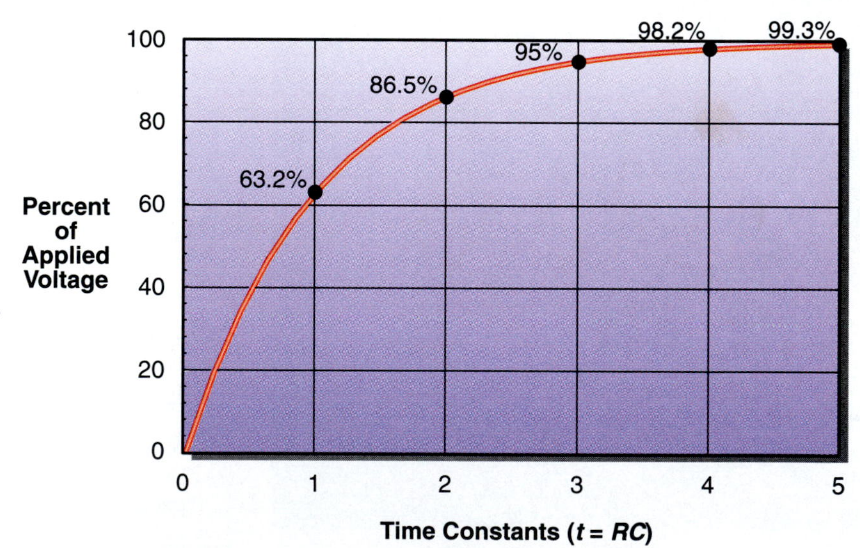

Figure 19-1 A capacitor is considered fully charged after five time constants.

The Universal Discharge Curve

A charged capacitor is often thought of as being similar to a battery in that it can supply power to a circuit. Batteries, however, maintain a fairly constant output voltage; a discharging capacitor does not. When a capacitor is discharged through a resistor, its output voltage falls rapidly at first, but the change in voltage slows with time. A graph of the output voltage plotted against the first five time constants is shown in Fig. 19-2. The discharge curve is an inverted image of the charge curve presented in Fig. 19-1. A capacitor is considered to be fully discharged after five time constants.

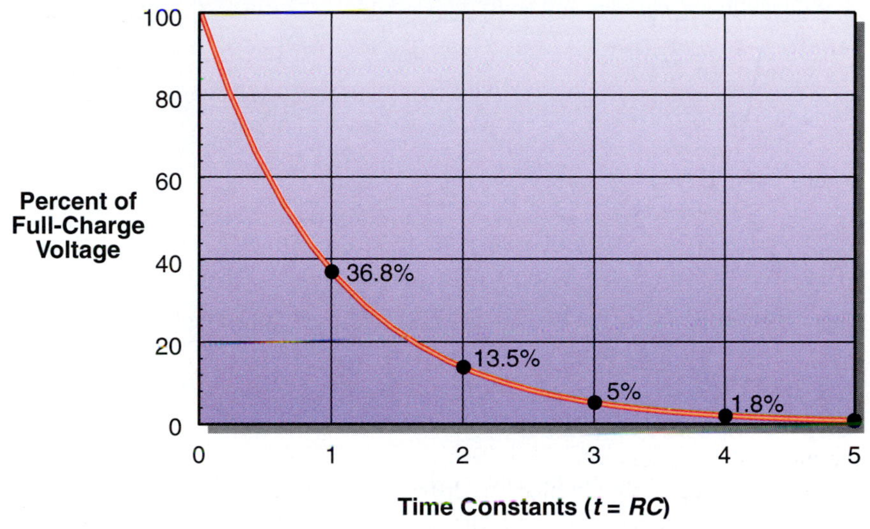

Figure 19-2 A capacitor is considered to be fully discharged after five time constants.

REVIEW QUIZ 19-1

1. A capacitor is considered to be fully charged after __________ time constants.
2. A capacitor is considered to be fully discharged after __________ time constants.
3. The greatest change in voltage occurs during the __________ time constant on both the charge and discharge curves.

19-2 Capacitors as Reactive Circuit Elements

A sinusoidal AC source "sees" a capacitor as a reactance, as it does with an inductor, but with opposite voltage-current phase relationships. Current leads voltage by 90° in a capacitive AC circuit, as shown in Fig. 19-3 on the next page. That makes sense because initially a capacitor appears as a short, and voltage cannot be developed across a short, but the current can be quite high. The opposite is true for an inductor because at the instant voltage is applied, an inductor appears to be an open circuit.

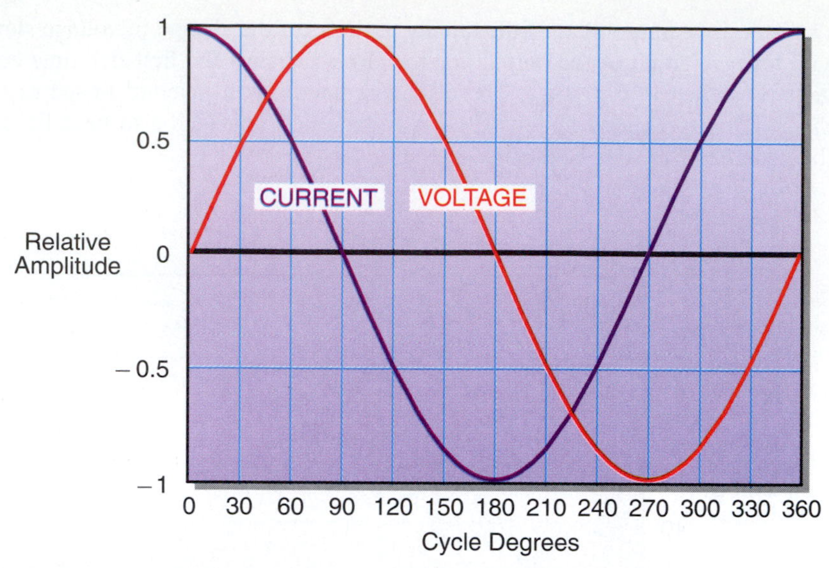

Figure 19-3 In an *R-C* circuit the current leads the applied voltage by 90°.

Capacitive Reactance

A capacitor does not dissipate energy; it stores it for future use. Capacitors oppose current from an AC source with a reactive voltage. A capacitor's opposition to current is called **capacitive reactance**. The unit of capacitive reactance is the ohm. The general symbol for reactance is a capital X. A subscript identifies reactance as inductive (X_L) or capacitive (X_C). Capacitive reactance depends only on the value of the capacitor and the frequency of the applied voltage. As with inductive reactance, the amplitude of the applied voltage is not a factor. Capacitive reactance may be calculated using either version of Eq.19-2. The difference in the two versions is how angular velocity is represented (2πf or ω).

Equation 19-2

$$X_C = \frac{1}{2\pi f C} = \frac{1}{\omega C}$$

where X_C = capacitive reactance in ohms
f = frequency of the applied voltage in hertz
C = capacitance in farads
2π = 6.283
ω = angular velocity in radians per second

Capacitive reactance is inversely proportional to frequency. That means that if a capacitor has 1000 ohms of reactance at 60 hertz, it will have 500 ohms of reactance at 120 hertz. Doubling the frequency of the applied voltage cuts the reactance of a capacitor in half. Conversely, reducing the frequency of the applied voltage to half doubles the reactance.

Capacitive Reactances in Series

When capacitors are connected in series, their reactances add numerically like resistor values. This relationship is shown in Eq. 19-3.

Equation 19-3

$$X_{C_T}=X_{C_1}+X_{C_2}+X_{C_3}\ldots$$

The voltages across capacitors connected in series can be calculated using Ohm's law when the individual reactances and circuit current are known. Chap. 18 showed how the voltages across series-connected capacitors were determined using ratios. That procedure worked well because the math involved was fairly simple and the only electrical parameter required was the applied voltage.

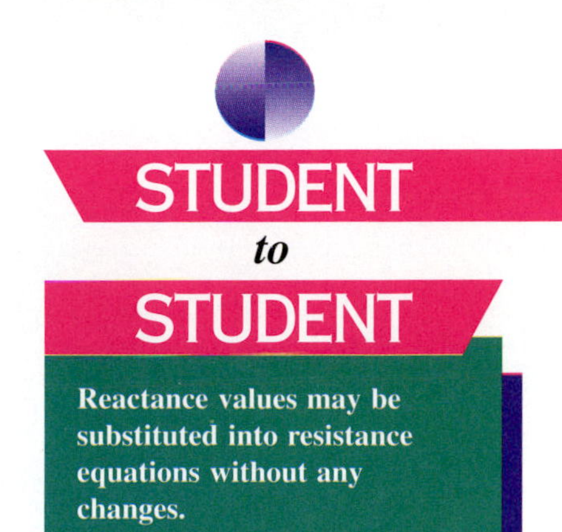

Figure 19-4 employs a circuit from Chap. 18 that contains only series-connected capacitors. This time, however, the applied voltage is alternating current and the reactance of each capacitor is provided. Using Eq. 19-3, the total reactance in the circuit is determined to be 757.89 ohms. Note that the largest reactance belongs to the capacitor with the smallest value and that this capacitor has the highest voltage across it. The voltage across each capacitor is found by multiplying its reactance by the circuit current in accordance with Ohm's law.

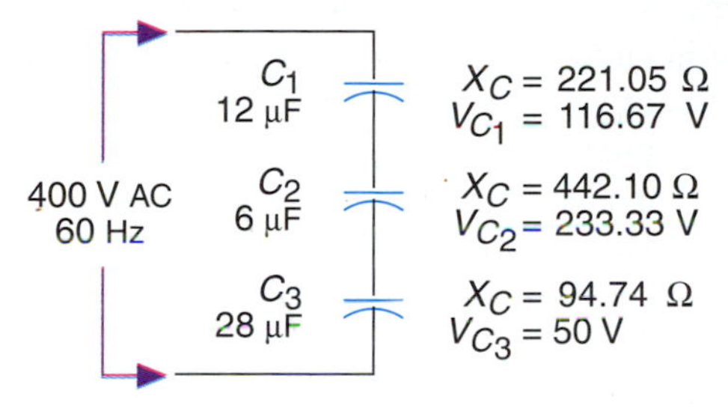

Figure 19-4 The reactances of capacitors connected in series behave in the same way as resistors connected in series.

The circuit current is obtained when the applied voltage is divided by the total reactance. The procedure is the same as for a circuit containing only series-connected resistors. Multiplying each reactance by the circuit current produces the voltages across each capacitor. The sum of these voltages equals the source voltage as predicted by Kirchhoff's voltage law.

Connecting capacitors in series does not alter the relative phase relationship between the applied voltage and circuit current. It remains at 90° and simultaneously obeys the rules of series circuits, namely, that the current is the same at any point in the circuit. The amplitude remains the same, the phase remains the same, and the waveshape remains the same. From the viewpoint of the source, capacitors connected in series behave as a single capacitor with a value that is less than the smallest capacitor in the circuit. The circuit current is the phase reference because it is the electrical quantity common to all of the capacitors. Voltage lags current by 90° in a circuit containing only series-connected capacitors. The 90° lagging phase angle is often written as −90°.

Capacitive Reactances in Parallel

The reactances of capacitors connected in parallel behave in the same way as the values of parallel resistors. The equivalent reactance may be calculated using the prodivism equation, reciprocals equation, ratio equation, or the averaging equation. In the averaging equation, the reactance of one of the parallel capacitors is divided by the number of capacitors when all of the capacitors have the same value. The four procedures are shown in Eq. 19-4.

Equation 19-4

(*a*) Prodivism:

$$X_T = \frac{X_1 \times X_2}{X_1 + X_2}$$

(*b*) Reciprocal:

$$X_T = \frac{1}{\frac{1}{X_1} + \frac{1}{X_2} + \frac{1}{X_{3} + \ldots}}$$

(*c*) Ratio:

$$X_T = \frac{X_1}{1 + \left(\frac{X_1}{X_2}\right)}$$

(*d*) Average:

$$X_T = \frac{X}{N}$$

Figure 19-5 is a schematic diagram of three capacitors connected in parallel across a 100-volt, 400-hertz, AC source. Notice that the capacitor currents are directly proportional to the capacitor values. Larger capacitors have smaller reactances, resulting in larger currents for any given voltage. The sum of the currents in the branches of a circuit containing only parallel capacitors is equal to the source current as predicted by Kirchhoff's current law. The 90° phase relationship between the applied voltage and the total capacitor current is not altered when capacitors are placed in parallel.

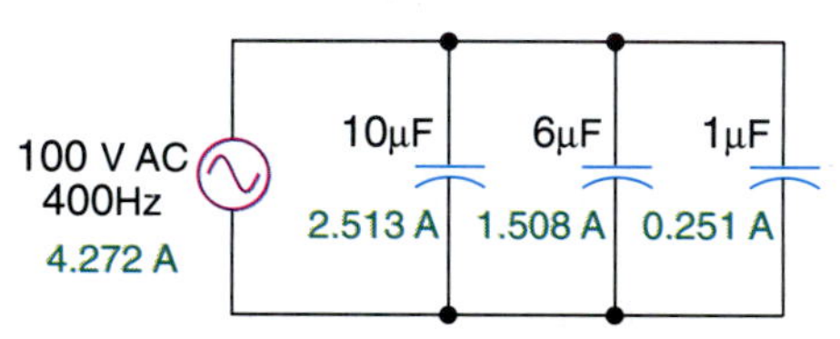

Figure 19-5 The currents in a parallel capacitor circuit add numerically in the same way that the currents through parallel resistors add.

The frame of reference for capacitors connected in parallel is voltage because the voltage is always the same across all of the elements of a parallel circuit. Therefore, in a circuit containing only parallel capacitors, the current leads the voltage by 90°. The frame of reference is normally the electrical quantity that is common to all the elements of the circuit being considered.

REVIEW QUIZ 19-2

1. The relative phase relationship between capacitive voltage and capacitive current is _______________.
2. Capacitive reactance is _______________ proportional to the frequency of the applied voltage.
3. The total capacitance of a circuit containing only series-connected capacitors is _______________ than the smallest capacitor in the circuit.
4. The total reactance of capacitors connected in series is calculated in the same way as _______________-connected resistors.

19-3 Series R-C AC Circuits

The circuit of Fig. 19-6 contains a resistor and a capacitor connected in series across an AC source. Such a circuit impedes the flow of current from the source through a combination of resistance and capacitive reactance. The source is able to identify that the circuit contains both resistance and capacitive reactance because its voltage and current are out of phase with the voltage lagging. If the circuit were purely resistive, the phase angle between the applied voltage and circuit current would be 0°, which is commonly referred to as being in phase. As already described, if the circuit were purely capacitive, the phase angle between the applied voltage and the circuit current would be −90°, or 90° lagging. When both resistance and capacitance are present, the phase angle will be somewhere between 0° and −90°.

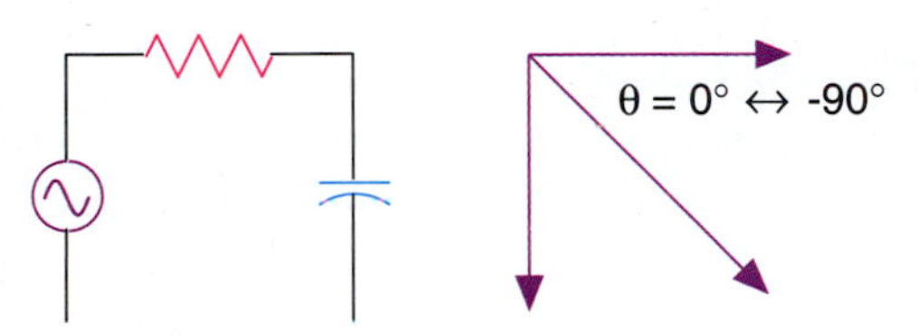

Figure 19-6 In a series *R-C* circuit, the phase angle between the applied voltage and the circuit current is between 0° and 90°.

The source voltage is the vector sum of the voltage across the capacitor and the voltage across the resistor, as shown in Fig. 19-7(*a*). Conversion of the vector (phasor) diagram to a right triangle, as shown in Fig. 19-7(*b*), allows the various circuit parameters to be determined using trigonometry. The voltages in a series *R-C* circuit are out of phase because the current is the same at any point in a series circuit. The voltage across a resistor and the current through that resistor are always in phase, so V_R is drawn along the horizontal axis. The voltage across the capacitor must therefore lag the voltage across the resistor by 90°.

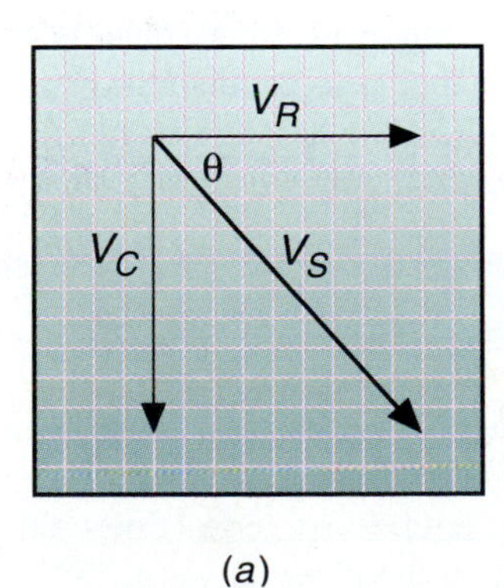

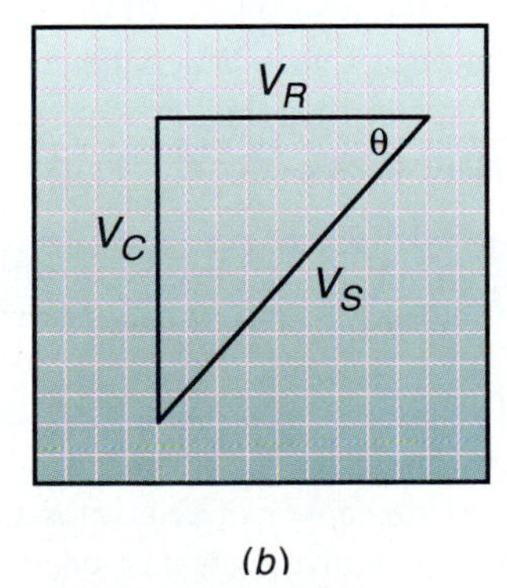

(*a*) (*b*)

Figure 19-7 The vector (phasor) diagram and resultant triangle for an *R-C* circuit are similar to those used with inductance, but they are drawn upside down because the source voltage now lags the source current.

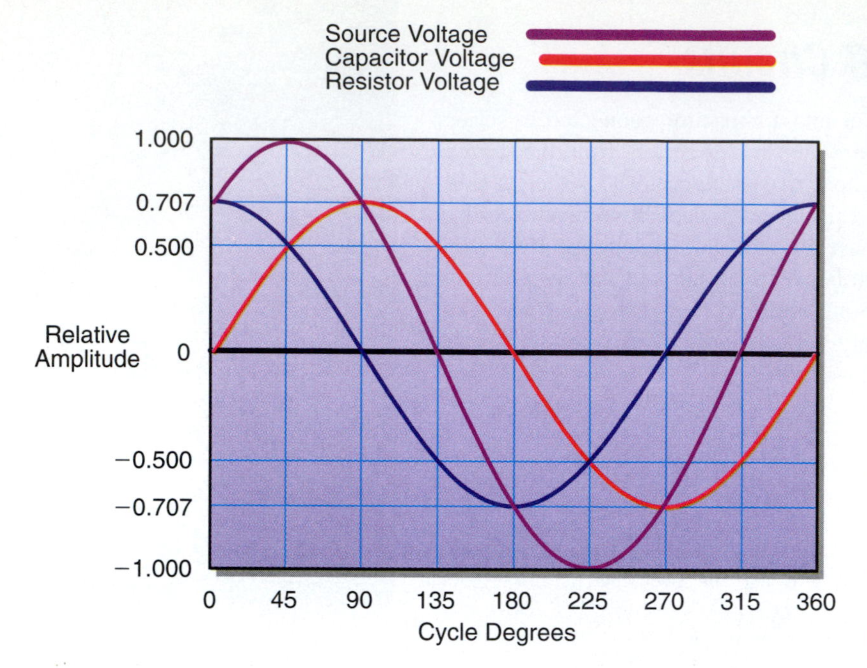

Figure 19-8 When viewed over time it is clear that Kirchhoff's voltage law is not broken by a series *R-C* circuit placed across an AC source.

Series *R-C* Voltage Sums

The voltage across the capacitor and the voltage across the resistor form the sides of a right triangle adjacent to the right angle. The length of either side is less than the hypotenuse, but the sum of the two sides is always greater than the hypotenuse. It is helpful to remember that relationship because it establishes some guidelines for the voltages they represent. For example, the voltage across either the resistor or capacitor in a series *R-C* circuit will never exceed that of the source. Moreover, the voltage of the source is always less than the sum of the voltages across the resistor and capacitor. The latter statement appears to break Kirchhoff's voltage law, but that is only because a meter cannot separate the voltage and time components. Figure 19-8 shows the true relationship between the voltage across the capacitor and the voltage across the resistor. At no time is Kirchhoff's voltage law violated. The sum of the resistor voltage and capacitor voltage is equal to the source voltage at all times.

Series *R-C* Impedance

The impedance of a series *R-C* circuit can be found using the Pythagorean theorem or trigonometry. The basic equations are the same as for an *R-L* circuit, except that a minus sign may be assigned to X_C to indicate that it is opposite of X_L. The impedance of an *R-C* circuit in relation to capacitive reactance or resistance is greatest when the capacitive reactance and resistance are equal. When this condition exists, the impedance is 1.414 times either one. The impedance of an *R-C* circuit cannot be more than 1.414 times either the larger of capacitive reactance or resistance, whichever is larger.

When the resistance and the capacitive reactance are equal, theta is 45°, as shown in Fig.19-9(*a*). When the capacitive reactance is larger than the resistance, theta is greater than 45°, as shown in Fig. 19-9(*b*). Theta is less than 45° when the capacitive reactance is less than the resistance, as shown in Fig. 19-9(*c*).

REVIEW QUIZ 19-3

1. The phase relationship between the capacitive current and resistive current in a series *R-C* circuit is ______________ degrees.
2. The impedance of a series *R-C* circuit is ______________ times either the capacitive reactance or resistance when the phase angle is 45°.
3. The hypotenuse of a right triangle is always ______________ than the longest side.

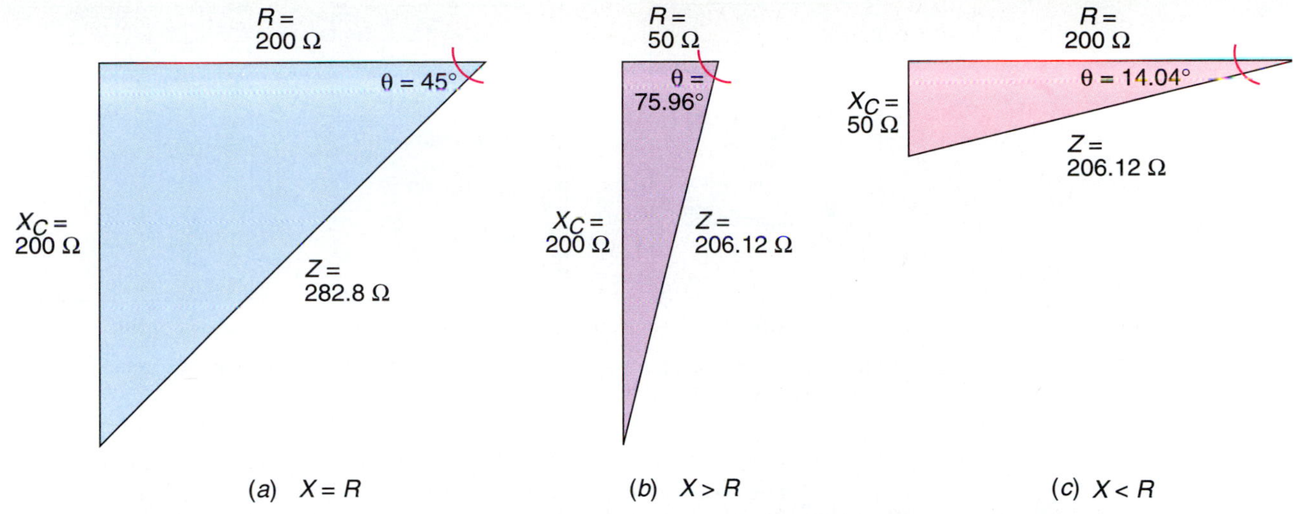

Figure 19-9 (*a*)The hypotenuse of a right triangle is never more than 1.414 times the longest side and never less than the longest side. (*b*)Theta is greater than 45° when the side opposite theta is longer than the side adjacent to theta. (*c*)Theta is less than 45° when the side opposite theta is shorter than the side adjacent to theta. These relationships are useful when approximating reactance, resistance, impedance, or the voltages across these quantities.

19-4 Parallel R-C Circuits

The voltage is the same across a resistor and capacitor connected in parallel, just as it is for the elements of any other parallel circuit. With a parallel *R-C* circuit, however, being the same means more than amplitude. The parallel connection forces the phase of the applied voltage to be the same for both the capacitor and the resistor. This situation causes the currents in the two components to be out of phase. If the circuit is mainly resistive, the phase angle will be close to zero. If the circuit is mainly capacitive, the phase angle will be close to 90°. For most *R-C* circuits the phase angle will be somewhere in between.

In a parallel *R-C* circuit, such as the one shown in Fig. 19-10, the resistance and capacitive reactance cannot be added directly or vectorially, but this situation is no different from any other parallel circuit. When components are connected in parallel, their conductive components are added, not their oppositions to current.

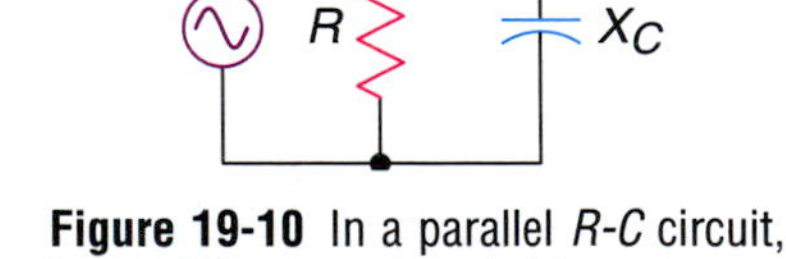

Figure 19-10 In a parallel *R-C* circuit, the resistance and capacitive reactance cannot be directly added numerically or vectorially.

Capacitive reactance must be converted to susceptance and resistance must be converted to conductance. These conversions are done using reciprocals as shown in Eq. 19-5.

Equation 19-5

$$(a)\ B_C = \frac{1}{X_C}$$

$$(b)\ G = \frac{1}{R}$$

Once calculated, conductance and susceptance can be added vectorially, as shown in Fig. 19-11(*a*) on the next page; employed in a trigonometric solution, as shown

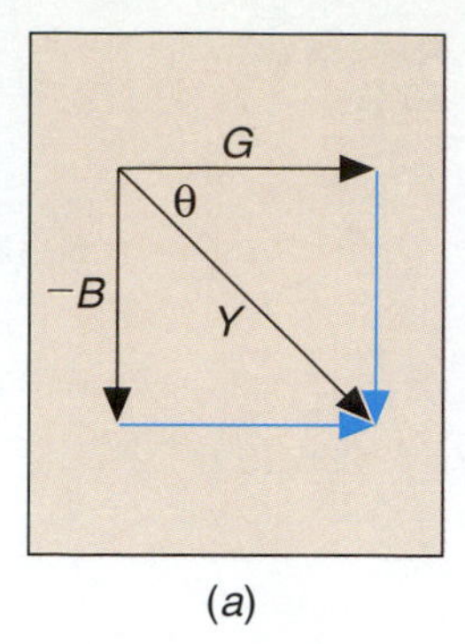

(a)

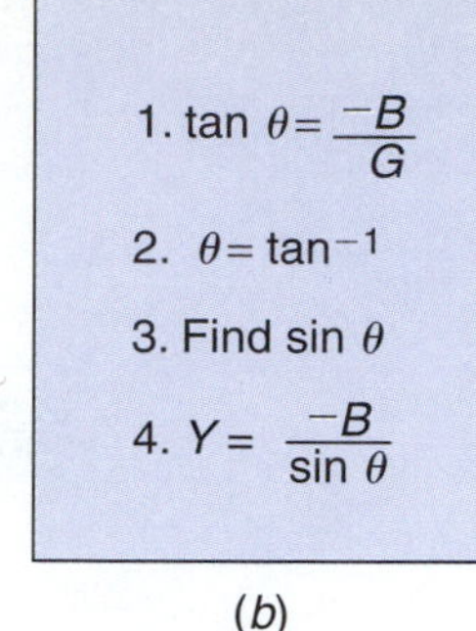

(b)

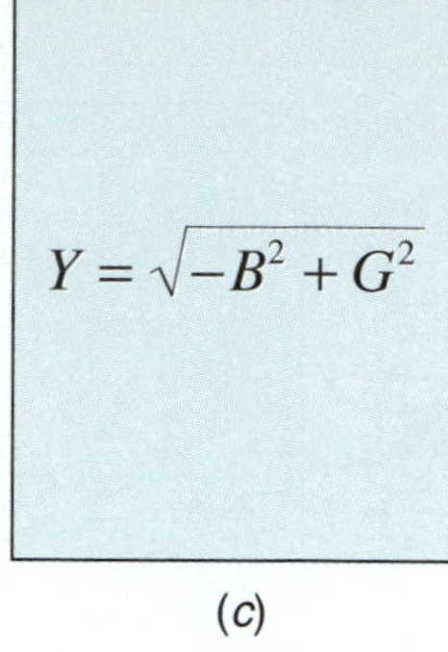

(c)

Figure 19-11 Admittance may be determined using (*a*) vector addition, (*b*) trigonometry, or (*c*) Pythagorean theorem.

in Fig. 19-11(*b*); or inserted into the Pythagorean solution, as shown in Fig. 19-11(*c*). A negative sign is placed before B to indicate that it is capacitive.

The reciprocal of admittance is the circuit impedance, as shown in Eq. 19-6.

Equation 19-6

$$Z = \frac{1}{Y}$$

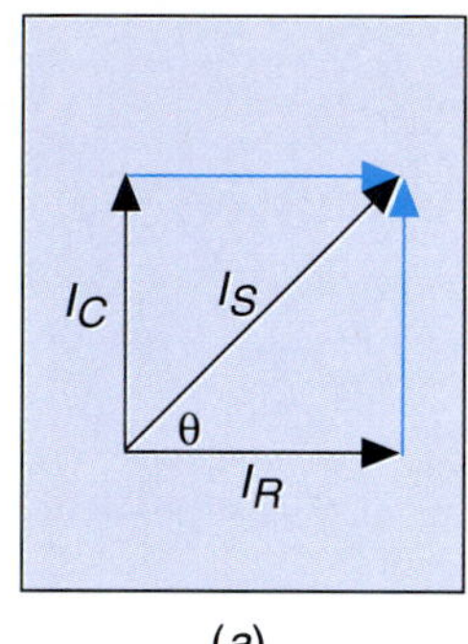

(a)

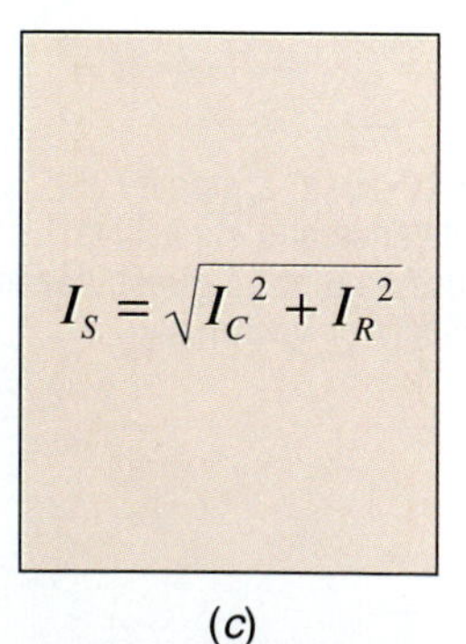

(b)

$$I_S = \sqrt{I_C^{\,2} + I_R^{\,2}}$$

(c)

Figure 19-13 The source current may be found using(*a*) vector addition, (*b*) trigonometry, or (*c*) the Pythagorean theorem.

Current in Parallel *R-C* Circuits

The source current in a parallel *R-C* circuit is the vector sum of the resistor current and capacitor current. The respective current/voltage phase relationships of each component remain unaltered. The resistor current and voltage are in phase with each other, and the capacitor current leads the capacitor voltage by 90°. Because the resistor voltage and capacitor voltage are the same, the capacitor current leads the resistor current by 90°, as shown in Fig. 19-12.

The source current may be found using vector graphing, Pythagorean theorem, or trigonometry, as shown in Fig. 19-13.

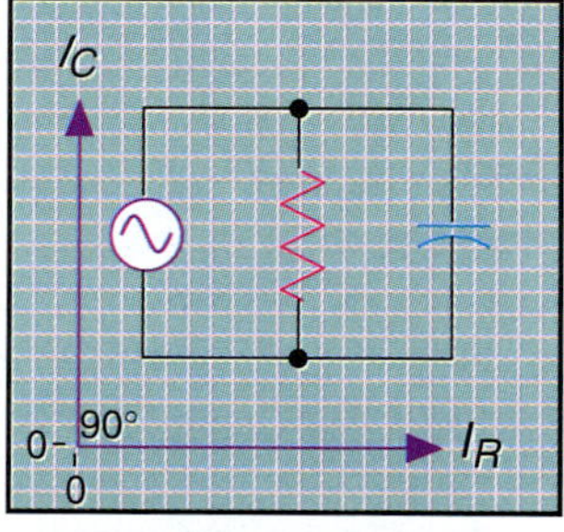

Figure 19-12 The capacitor current leads the resistor current by 90° in a parallel *R-C* circuit.

REVIEW QUIZ 19-4

1. The voltage is the same across a resistor and capacitor connected in ________________.
2. When calculating the impedance of a parallel *R-C* circuit, reactance must be converted to ________________ and resistance must be converted to ________________.
3. The source current in a parallel *R-C* circuit is the ________________ sum of I_C and I_R.

SUMMARY

When a capacitor is connected to a voltage source through a resistance, the voltage across the capacitor does not reach its maximum value immediately. The time the capacitor takes to fully charge is related to the product of the resistance in ohms and the capacitor value in farads, which is referred to as a one time constant. For practical purposes, a capacitor is fully charged in five time constants. It also takes five time constants for a capacitor to fully discharge.

The current through a series-connected resistor and capacitor is the same in both components, with the word *same* meaning both amplitude and phase. The voltage across the resistor is in phase with the current, but the voltage across the capacitor lags the current by 90°. The relative phase of the source current and source voltage is between 0° and 90°, depending on the values of resistance and capacitance. A phase angle near 0° indicates that the circuit is mostly resistive. A phase angle near 90° indicates that the circuit is mostly capacitive.

Capacitors oppose an alternating current by producing a reactive voltage. That opposition is called reactance and is measured in ohms. When capacitors are connected in series, their reactances add numerically. When capacitors are connected in parallel, the equivalent reactance is calculated using any of the procedures to calculate parallel resistances. In fact, reactance may be substituted into any of the resistance equations.

The voltages across capacitors connected in series are added together. The source "sees" a series of capacitors as a single capacitor with a value that is less than the smallest capacitor in the series. The voltage/current phase relationship is the same as for a single capacitor. The capacitive currents of capacitors connected in parallel are also added together. A voltage source "sees" a group of parallel capacitors as one capacitor with a capacity value that is higher than the largest value in the circuit. The voltage/current phase relationship is the same for parallel-connected capacitors as it is for a single capacitor.

The voltage is the same across a resistor and capacitor connected in parallel, just as it is for any other parallel circuit. The important consideration in a parallel *R-C* circuit is the phase of the resistive and capacitive currents. Because each component retains its respective voltage/current phase relationship, the resistive and capacitive currents must be 90° out of phase. In a parallel *R-C* circuit, the current will lead the voltage by some amount between 0° and 90°. When the two currents are graphed vectorially, the capacitor current is positive. When the voltages of a series *R-C* circuit are graphed vectorially the capacitive voltage is negative.

NEW VOCABULARY

- capacitive reactance
- capacitive reactances in parallel
- capacitive reactances in series
- parallel *R-C* circuits
- series *R-C* circuits
- universal charge curve
- universal discharge curve

Questions

1. An *R-C* time constant is the product of which two electrical quantities?
2. How many time constants must elapse before a capacitor is considered to be fully charged?
3. Does capacitive current lead or lag the voltage across a capacitor?
4. Is capacitive reactance directly proportional or inversely proportional to the applied frequency?
5. When capacitors are connected in series or parallel, is their equivalent reactance calculated in the same way as resistors?
6. Is the voltage vector positive or negative in a series *R-C* circuit?
7. Is the current vector positive or negative in a parallel *R-C* circuit?

Problems

1. What is the total capacitive reactance of a 4-microfarad and a 6-microfarad capacitor connected in series across a 60-hertz AC source?
2. What is the impedance of a series *R-C* circuit containing a 100-ohm capacitive reactance and a 100-ohm resistor?
3. What phase shift is created when a 400-ohm resistance is placed in series with a 300-ohm capacitive reactance?
4. What is the sum of the voltage drops across a 50-ohm resistor and 100-microfarad capacitor connected in series across a 120-volt, 60-hertz source?
5. What is the source current when a 500-ohm resistance and a 500-ohm capacitive reactance are connected in parallel across a 100-volt AC source?
6. Assuming that a 1-microfarad capacitor is fully discharged, what voltage would appear across it two seconds after being connected to a 50-volt source through a 1-megohm resistor?
7. What voltage would appear across a 1-microfarad capacitor that had been charged to 100-volts two seconds after a 1-megohm resistor was placed across it?

Critical Thinking

1. Explain why the sum of the measured voltages across a resistor and capacitor connected in series is greater than the source voltage when the graph in Fig. 19-8 indicates that Kirchhoff's voltage law is valid for series *R-C* circuits.
2. Explain why the capacitive current is zero when the voltage across the capacitor is at its peak value?
3. Why can't the relative phase angle in an *R-C* circuit ever be exactly 0° or exactly 90°?
4. If you were to measure the voltages across a resistor and capacitor in series across an AC source, how could you tell when the phase angle was at or near 45°?
5. Sketch the waveform you would expect to see across the capacitor in Fig.19-14 if a positive square wave voltage of 100 volts with a 50 percent duty cycle and a period of 10 seconds were applied to the circuit. The value of the resistor is 1 megohm, and the capacitor is 1 microfarad.

Answers to Review Quizzes

9-1
1. five
2. five
3. first

19-2
1. 90°
2. inversely
3. less
4. series

19-3
1. 0
2. 1.414
3. longer

19-4
1. parallel
2. susceptance; conductance
3. vector

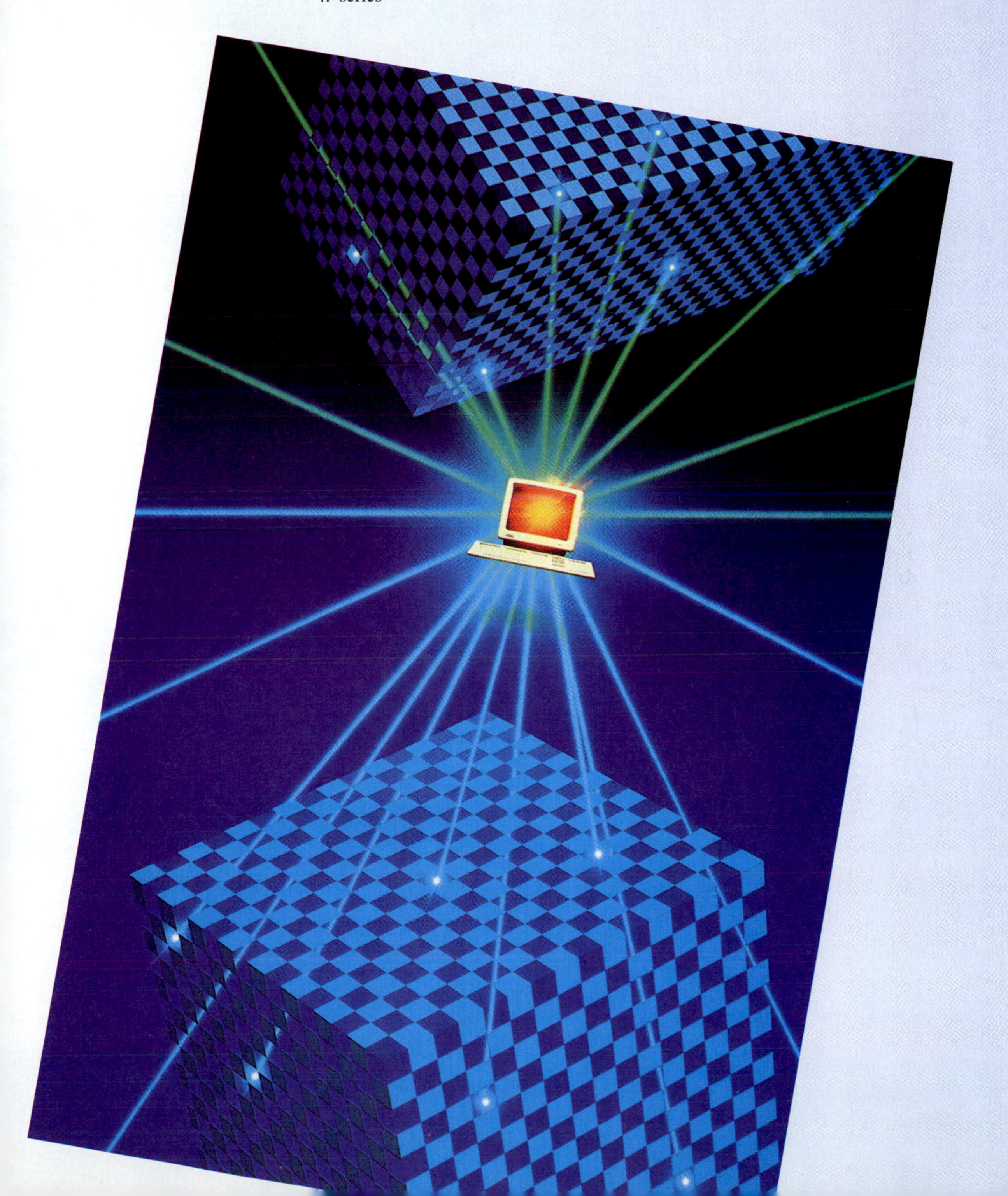

Chapter 20

R-C-L Circuits

ABOUT THIS CHAPTER

R-C-L circuits are frequency selective and will pass or reject a frequency that is at their point of resonance. They may be series resonant or parallel resonant. Series-resonant circuits pass the maximum current at resonance, and parallel-resonant circuits develop the maximum voltage

After completing this chapter you will be able to:

- Explain the basic operating principles of series and parallel *L-C* circuits in and out of resonance.
- Calculate the frequency of resonance.
- Calculate the capacitor value to use with an inductor to make a resonant circuit at a desired frequency.

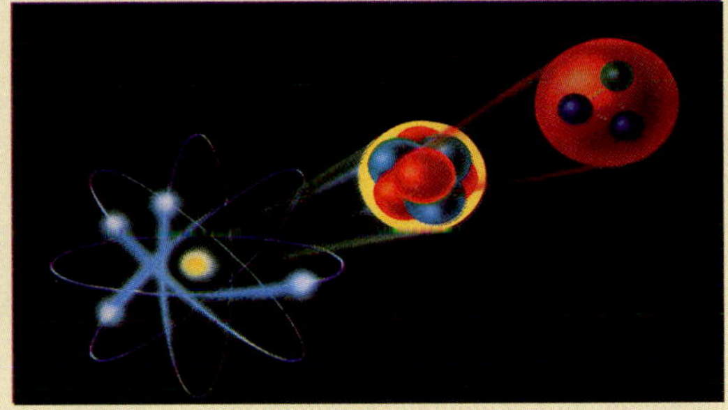

at resonance. Some interesting effects are observed with series and parallel *L-C* circuits. In series *R-C-L* circuits the reactive components cancel, allowing large currents to flow. The currents flowing through the reactances develop large voltages across them. These voltages can become a serious consideration when replacing components; they are also important from a safety perspective. Without resonant circuits, radio and electronics as we know them would not have existed.

- ▼ Calculate the inductor value necessary to complete a resonant circuit.
- ▼ Determine the Q and bandwidth of a resonant *L-C* circuit.
- ▼ Describe how a parallel *L-C* circuit is a storage tank for electric energy and how it behaves like a flywheel.

20-1 Series L-C Circuits

The reactive voltages of inductors and capacitors connected in series are 180° out of phase; therefore, they cancel each other either partially or totally. In cases where the inductive reactance and capacitive reactance are equal, the voltages across them are also equal with opposite polarities. The current in the circuit is 90° out of phase with both the inductive voltage and the capacitive voltage, as shown in Fig. 20-1.

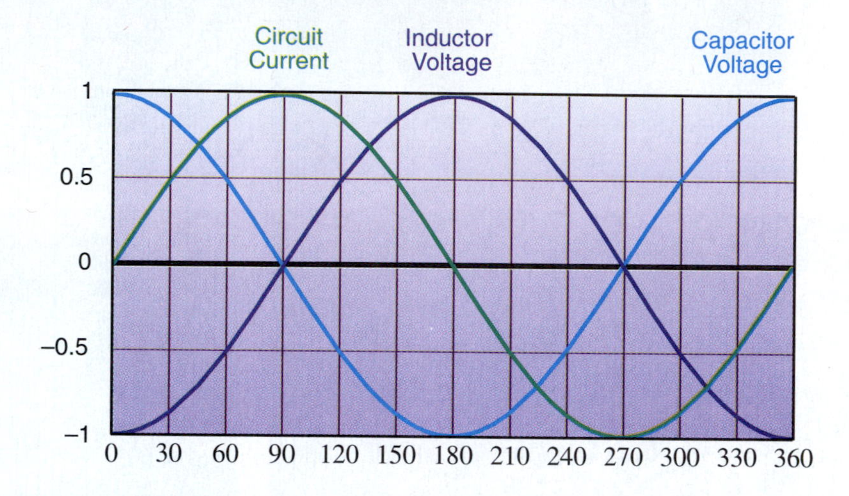

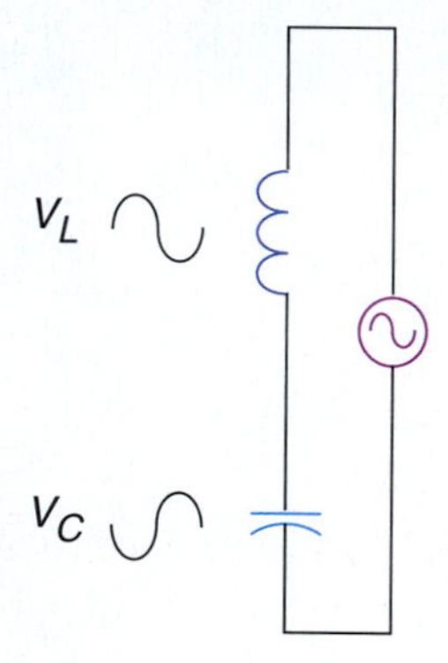

Figure 20-1 In a series *L-C* circuit, the reactive voltages are 180° out of phase with each other and 90° out of phase with the source current. The inductor and capacitor always retain their relative voltage/current phase relationships; inductive voltage leads inductive current by 90°, and capacitive voltage lags capacitive current by 90°.

The opposition to current associated with reactive components results from their reactive voltages. When those voltages are equal and opposite, there can be no opposition to current owing to reactance. Theoretically, a series *L-C* circuit becomes a perfect conductor under those circumstances. All *L-C* circuits contain resistance, however, mostly because of the wire in the inductor. Any circuit resistance is electrically in series with the inductor and capacitor, forming a series *L-C-R* circuit.

When the inductive reactance and capacitive reactance in a series *L-C-R* circuit are equal, the source "sees" the circuit as a single resistor. The current is therefore limited by the applied voltage and the circuit resistance as predicted by

Ohm's law. The voltage across the resistance is in phase with the source current and 90° out of phase with the two reactive voltages. The individual current/voltage phase relationships of the inductor and capacitor remain unaltered as always. Fig. 20-2 shows a vector diagram of the three voltages.

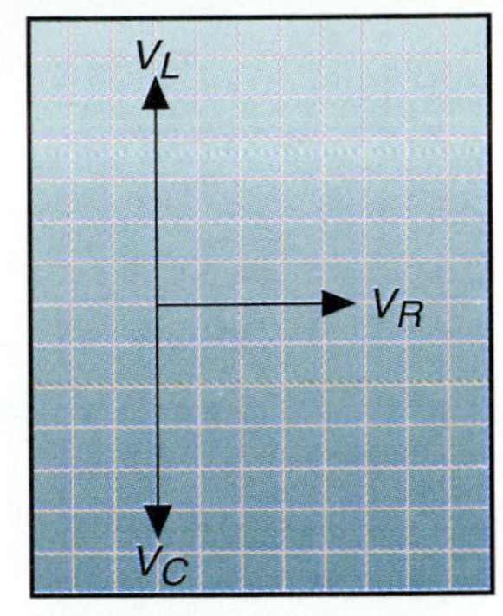

Figure 20-2 The inductive voltage vector and capacitive voltage vector are 180° out of phase with each other and 90° out of phase with the resistive voltage vector.

Series *L-C-R* Circuit Impedance

A mathematical distinction is made between inductive reactance and capacitive reactance by assigning polarities to them when they appear in a series circuit. Because inductive voltage leads inductive current, the inductive reactance becomes positive. Capacitive reactance is negative because capacitive voltage lags capacitive current. The resistance vector lies along the zero reference line at a 90° angle to both reactances as shown in Fig. 20-3(*a*). A vector diagram of the equivalent circuit results when the two reactances are added vectorially, as shown in Fig. 20-3(*b*). The addition of vectors 180° apart amounts to subtracting the length of the shorter vector from the longer one.

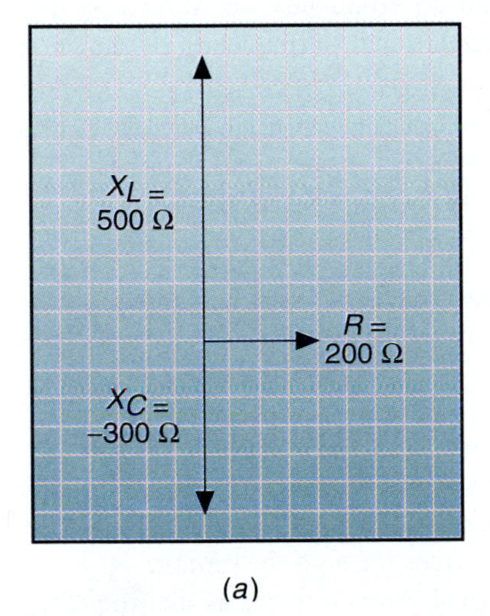

(*a*)

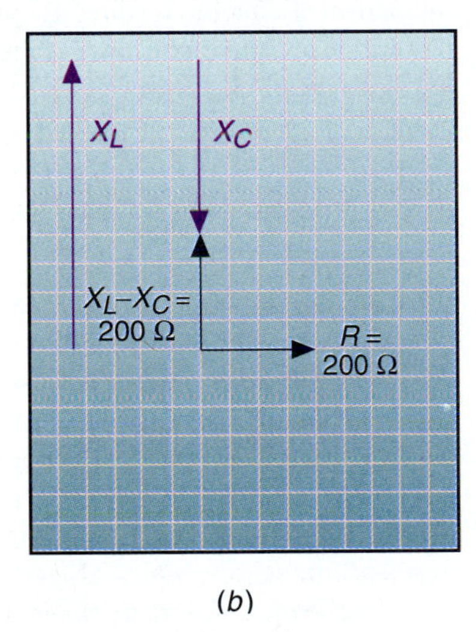

(*b*)

Figure 20-3 Inductive reactance and capacitive reactance are added vectorially by subtracting the smaller reactance from the larger.

The impedance of a series *L-C-R* circuit is the same as the circuit resistance when the inductive and capacitive reactances are equal. An *L-C-R* circuit, however, appears to be an *R-L* circuit if the inductive reactance is greater than the capacitive reactance and an *R-C* circuit if the capacitive reactance is greater than the inductive reactance. For instance, if a coil having a reactance of 500 ohms and a DC resistance of 100 ohms is placed in series with a capacitor having a reactance of 300 ohms, the effective reactance is found by subtracting 300 ohms from 500 ohms for a total of 200 ohms. The remaining 200 ohms of reactance is inductive because the inductive reactance was the larger of the two.

The impedance of a series *L-C-R* circuit is the vector sum of the equivalent reactance and the circuit resistance, as shown in Fig. 20-4. The inductor's DC resistance and any other resistance connected in series are added to obtain the total circuit resistance. A similar procedure is used when several inductances are present. The values of their inductive reactances are added to obtain the total inductive reactance. Also, all capacitive reactances are added when more than one

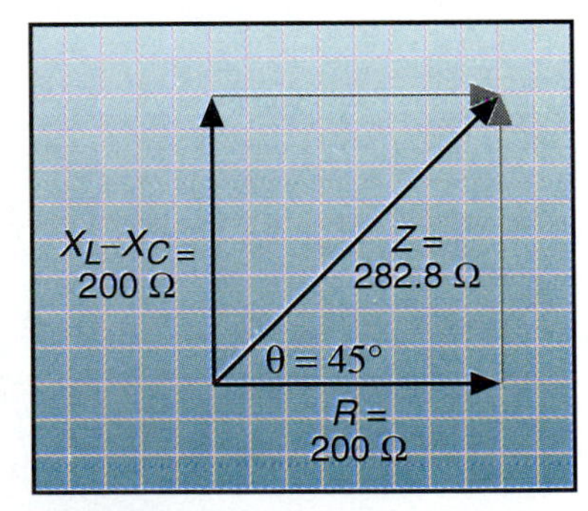

Figure 20-4 The final impedance of an *L-C-R* circuit is the vector sum of the equivalent reactance and the total resistance. Should the inductive and capacitive reactances be equal, the impedance is the resistance of the circuit.

capacitor is present. Similar types of reactance are added together first, regardless of their arrangement in the circuit.

REVIEW QUIZ 20-1

1. In a series L-C-R circuit, the voltage across the inductor and the voltage across the capacitor are ______________ degrees out of phase.
2. The overall reactance in a series L-C-R circuit is only inductive or only capacitive, whichever is ______________.
3. The current in a series L-C-R circuit is ______________ degrees out of phase with both the inductive and capacitive reactance.

20-2 Series Resonance

Inductive reactance is directly proportional to frequency, as shown in Eq. 20-1(a). Capacitive reactance is inversely proportional to frequency, as shown in Eq. 20-1(b).

Equation 20-1

$$(a)\ X_L = 2\pi fL$$

$$(b)\ X_C = \frac{1}{2\pi fC}$$

In a series L-C-R circuit there is only one frequency where the inductive reactance and capacitive reactance are equal. When that condition exists, the circuit is said to be **resonant**. At **resonance** the current through a series L-C-R circuit is maximum, being limited only by the resistance of the circuit and the applied voltage. As the frequency of the applied voltage is lowered below resonance, the capacitive reactance increases, which reduces the current through the circuit. As the frequency of the applied voltage is increased above resonance, the inductive reactance increases, which also lowers the current through the circuit.

A series L-C-R circuit changes its electrical characteristics as the frequency of the applied voltage increases from below resonance to above resonance. Below resonance a series L-C-R circuit is capacitive, at resonance it is resistive, and above resonance it is inductive. Fig. 20-5 is a graph of the relative currents through an L-C-R circuit over a wide range of frequencies.

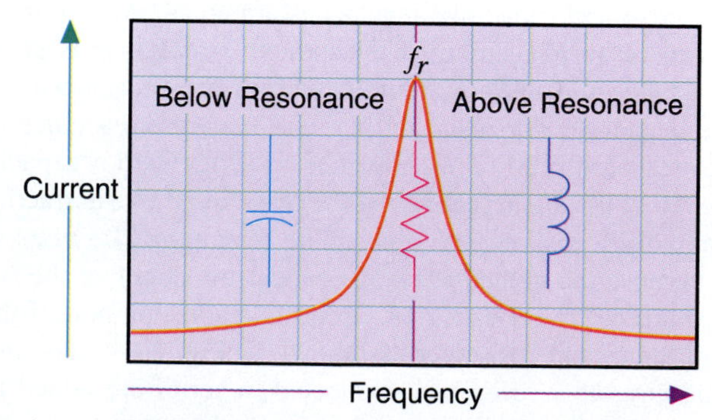

Figure 20-5 An L-C-R circuit appears capacitive at frequencies below resonance, resistive at the resonance frequency, and inductive at frequencies above resonance.

Series *L-C-R* Calculations

The condition of resonance is defined as $X_L = X_C$. This definition forms the basis for calculating the resonant frequency, as shown in Eq. 20-2(*a*). The frequency equation is sometimes simplified further by dividing the numerator, 1, by the 2π portion of the denominator. The numerator then becomes 0.159, and the 2π disappears from the denominator, as shown in Eq. 20-2(*b*).

Equation 20-2

$$(a)\ f_r = \frac{1}{2\pi\sqrt{LC}}$$

$$(b)\ f_r = \frac{0.159}{\sqrt{LC}}$$

where f_r = resonant frequency in hertz
L = inductance in henrys
C = capacitance in farads

The units of hertz, farad, and henry are too large for many practical situations. Resonant circuits are more commonly found at radio frequencies than power line frequencies. Equation 20-3(*a*) is scaled to accommodate electrical values commonly encountered with radio frequencies by changing the numerator from 1 to 1×10^3. Equation 20-3(*b*) simplifies the equation further by dividing the enlarged numerator by the 2π portion of the denominator.

Equation 20-3

$$(a)\ f_r = \frac{1 \times 10^3}{2\pi\sqrt{LC}}$$

$$(b)\ f_r = \frac{159.15}{\sqrt{LC}}$$

where f_r = resonant frequency in megahertz
L = inductance in microhenrys
C = capacitance in picofarads

Equation 20-4(*a*) also originates from the $X_L = X_C$ definition of resonance and is used to calculate the capacitance value needed to form a resonant circuit when the inductance and resonant frequency are known. Equation 20-4(*b*) is a simplification of Eq. 20-4(*a*). The simplification is realized by dividing the numerator by the $4\pi^2$ portion of the denominator.

Equation 20-4

$$(a)\ C = \frac{1}{4\pi^2 f^2 L} = \frac{0.02533}{f^2 L}$$

$$(b)\ C = \frac{0.02533}{f^2 L}$$

A simple transposition of C and L, as shown in Eq. 20-5(*a*), allows the inductance in a resonant circuit to be calcu-

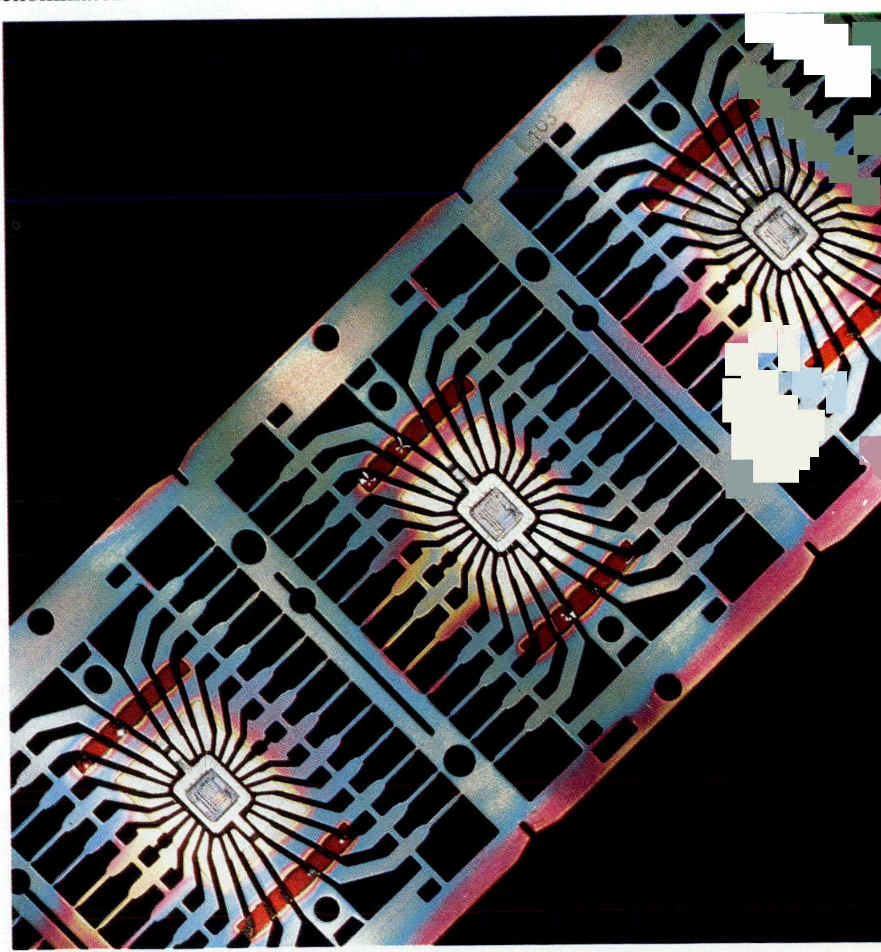

lated when the capacitance and resonant frequency are known. This equation may also be simplified by dividing the numerator by the $4\pi^2$ portion of the denominator, as shown in Eq. 20-5(*b*).

Equation 20-5

$$(a)\ L = \frac{1}{4\pi^2 f^2 C}$$

$$(b)\ L = \frac{0.02533}{f^2 C}$$

The units for Eq. 20-4 and Eq. 20-5 are as follows:

L = inductance in henrys

C = capacitance in farads

f = frequency in hertz

Once again, the units of hertz, henrys, and farads are too large for most applications. The equations can be made to accommodate the more practical units of megahertz, microhenrys, and picofarads by changing the numerators from 1 to 1,000,000 (1×10^6) as shown in Eq. 20-6(*a*) and Eq. 20-7(*a*). The numerator value 25,330 in Eq. 20-6(*b*) and Eq. 20-7(*b*) results from simplifying the equations by dividing the numerator (1×10^6) by $4\pi 2$.

Equation 20-6

$$(a)\ C = 1 \times \frac{10^6}{4\pi^2 f^2 L}$$

$$(b)\ C = \frac{25{,}330}{f^2 L}$$

Equation 20-7

$$(a)\ L = 1 \times \frac{10^6}{4\pi^2 f^2 C}$$

$$(b)\ L = \frac{25{,}330}{f^2 C}$$

L-C-R Voltages

A series *L-C-R* circuit can develop some surprisingly high voltages across the inductor and capacitor at resonance while adhering to Kirchhoff's voltage law. The voltage across the inductor, capacitor, and resistor are found using Ohm's law. The voltage across the inductor is $I \times X_L$, the voltage across the capacitor is $I \times X_C$, and the voltage across the resistor is $I \times R$. The voltages across the inductor and capacitor are excessively high because the reactive voltages that normally limit current cancel each other at resonance. As an example, the series *L-C-R* circuit in Fig. 20-6 contains a 1000-ohm inductive reactance, a 1000-ohm capacitive reactance, and a 10-ohm resistance across a 120-volt source. The two 1000-ohm reactances cancel because their reactive voltages are 180° out of phase. The only limitation on current is the 10-ohm resistance. The current in the circuit is therefore 12 amperes (120 V/10 Ω). That same 12 amperes flows in both reactive elements. The voltage across the inductor is 12,000 volts, and the voltage across the capacitor is 12,000 volts ($V = I \times X =$ 12 A; 12 A $\times$ 1000 Ω =

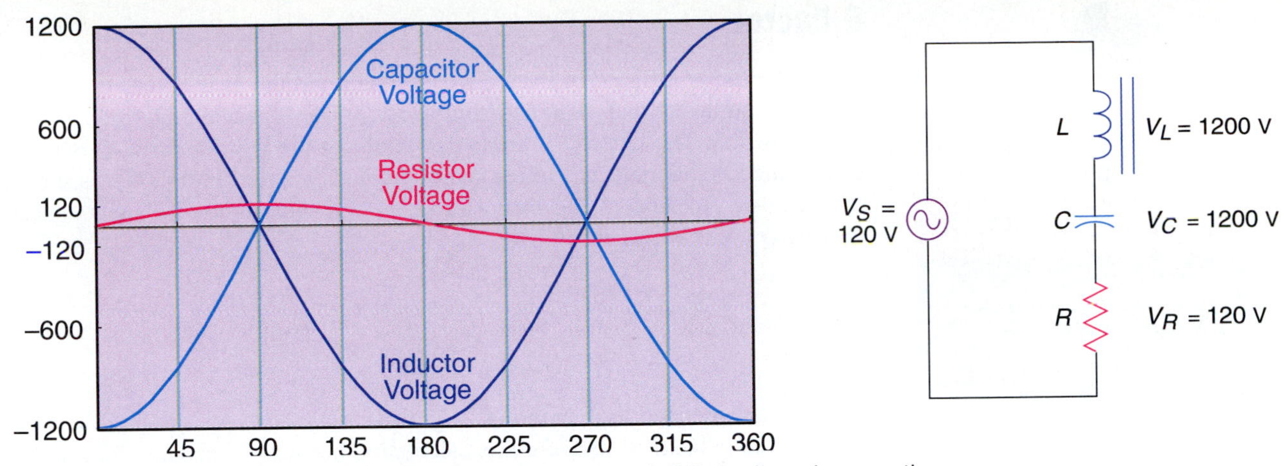

Figure 20-6 The voltages across the inductor and capacitor in an *L-C-R* circuit can be many times higher than the source voltage when the circuit is at resonance while conforming to Ohm's law.

12,000 V). Kirchhoff's voltage law is valid because the two reactive voltages cancel leaving only the resistor voltage.

Bandwidth

Theoretically, a resonant circuit should pass only one frequency, but in an actual *L-C-R* circuit frequencies near resonance also pass unimpaired. The more of these adjacent frequencies that pass through an *L-C-R* circuit, the greater the **bandwidth**. Measured between the half-power points on a graph of amplitude versus frequency, bandwidth is measured in hertz, kilohertz, or megahertz and will be the same unit as the resonant frequency. The half-power points are the same as for a sinusoidal AC waveform—0.707, or 70.7 percent, of the peak amplitude. Figure 20-7 graphically depicts the bandwidth in relation to the frequency of resonance.

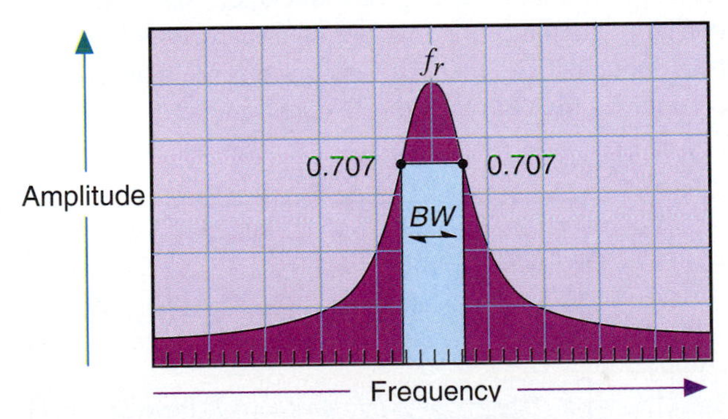

Figure 20-7 Bandwidth is measured between the half-power points on the amplitude-frequency curve in hertz, kilohertz, or megahertz.

The primary reason for the bandwidth's being wider than a single resonant frequency is the circuit resistance: The greater the resistance, the greater the bandwidth.

Q-Factor

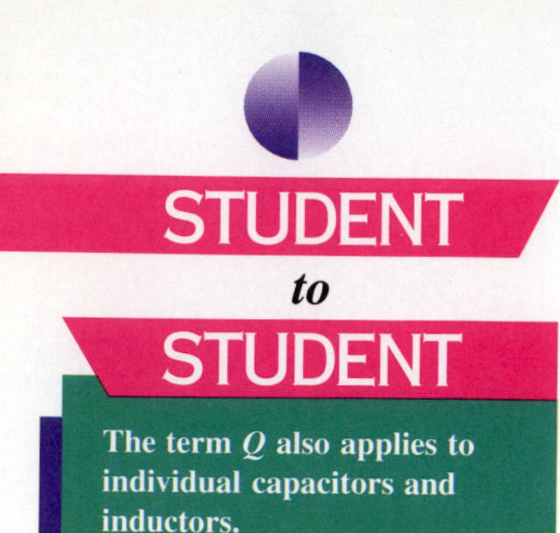

The **Q-factor** is a figure of merit for a resonant circuit. It is inversely proportional to bandwidth and identified simply as Q. The higher the Q, the narrower the bandwidth. The letter Q is an abbreviation for the word *quality*. Q is properly defined as the frequency of resonance divided by bandwidth (BW), as shown in Eq. 20-8. Because Q and bandwidth have the same units, the units cancel, making Q a unitless quantity.

Equation 20-8

$$Q = \frac{f_r}{BW}$$

Q is also defined as the ratio of inductive reactance to resistance. The resistance of the inductor is used for this calculation because it accounts for most of the circuit resistance. Any resistance present in the capacitor and wiring is usually insignificant. The alternate equation for Q is given in Eq. 20-9.

Equation 20-9

$$Q = \frac{X_L}{R}$$

The measured Q of a circuit is usually less than the calculated Q. The discrepancy results from a phenomenon called *skin effect*. High-frequency alternations of the current through the inductor force the electrons to the surface of the wire from which it is wound. The effect is the same as reducing the size of the wire, which would also increase the resistance. The effect can become so severe that some manufacturers of radio and television equipment silver-plate the coils of wire used to make the inductors.

Sometimes inductors are wound from a form of stranded wire called litz wire, where each strand is insulated from the others. The surface area of the multiple strands is greater than the surface area of a solid wire of the same gauge. Another way to improve the Q of an inductor is to use a permeable core such as ferrite. Modern ferrites have made high-Q inductors commonplace. Fewer turns of wire are required to obtain the desired inductance when they are wound over a ferrite core. The reduction in the number of turns results in a reduction in the resistance of the inductor. The reduced resistance more than offsets any adverse effects that core losses may have on the Q of the inductor.

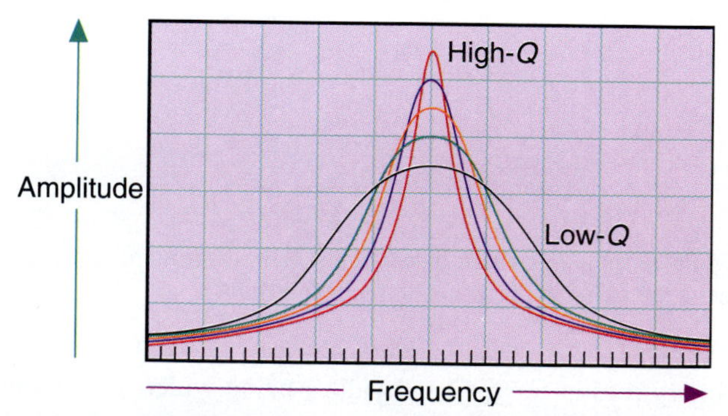

Figure 20-8 A greater change in current occurs with high-Q circuits than in low-Q circuits, allowing high-Q circuits to be more frequency selective than low-Q circuits.

The Q of an L-C-R circuit affects its bandwidth and the peak amplitude at resonance. Amplitude and bandwidth are reciprocal quantities. The narrow bandwidth associated with high-Q circuits produces a greater change in current than a broadband low-Q circuit. The effects of Q are shown in Fig. 20-8 (page 352).

REVIEW QUIZ 20-2

1. In a series resonance circuit, the capacitive reactance and inductance are ______________.
2. A series R-C-L circuit appears to the source as a(n) ______________ at resonance.
3. The general bandwidth of a series L-C-R circuit at resonance is indicated by its ______________ factor.
4. A series L-C-R circuit that is not resonant appears to the source as a(n) ______________ circuit or a(n) ______________ circuit.
5. Dividing the resonant frequency of an L-C-R circuit by the bandwidth yields the circuit ______________.

20-3 Parallel L-C-R Circuits

Inductors and capacitors connected in parallel are equally as frequency selective as series L-C circuits, but they respond to voltage rather than current. At frequencies below resonance, the reactance of the inductor is small and little voltage can be developed across it. At frequencies above resonance, the capacitive reactance is also small, retarding the development of any significant voltage. At resonance, the inductive and capacitive reactances are equal and opposite, resulting in equal and opposite currents flowing in each branch of the circuit. These currents circulate back and forth between the inductor and capacitor. The inductor and capacitor take turns charging and discharging each other. After the source has initially charged the L-C circuit, there is no further need for it to supply current. As a result, the source "sees" a parallel L-C circuit as an infinite impedance across which maximum voltage is developed, as shown in Fig. 20-9. Theoretically, a parallel L-C circuit could retain the energy placed in it by the source indefinitely by handing it back and forth between the inductor and capacitor. In a commonly used analogy, the continuous movement of energy is called the **flywheel effect**.

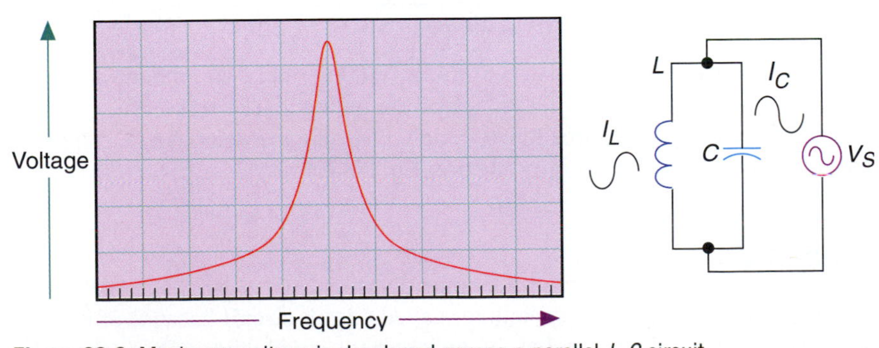

Figure 20-9 Maximum voltage is developed across a parallel L-C circuit when the inductive and capacitive reactances are equal.

Antiresonance

Parallel L-C circuits are not free from inherent resistance, which is primarily associated with the inductor. The presence of resistance in only one leg of a parallel L-C circuit compromises the previous definition of resonance. Resonance in a parallel L-C circuit occurs when the currents in the inductive and capacitive branches are equal, which is not necessarily when the reactances are equal. If a parallel L-C circuit has a Q greater than 10, the discrepancy is negligible, and the equations used to calculate the resonant frequency (Eq. 20-3), capacitance at resonance (Eq. 20-5), and inductance at resonance (Eq. 20-6) are valid. When the Q is less than 10, the equations are not accurate.

The key consideration for L-C circuits at resonance is that they appear resistive to the source. For a series resonant circuit that means maximum current, a 0° phase angle, and a power factor of 1. For a parallel resonant circuit that means minimum current, a 0° phase angle, and a power factor of 1. Parallel resonance is sometimes referred to as *antiresonance*, which distinguishes a parallel resonant circuit from a series resonant circuit.

Resistance in a Parallel *L-C* Circuit

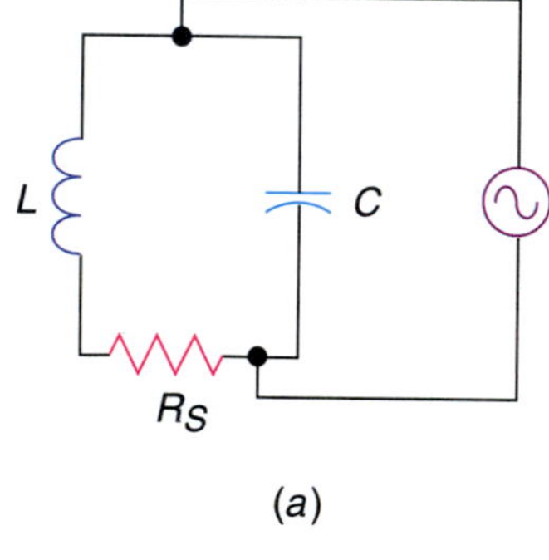

(*a*)

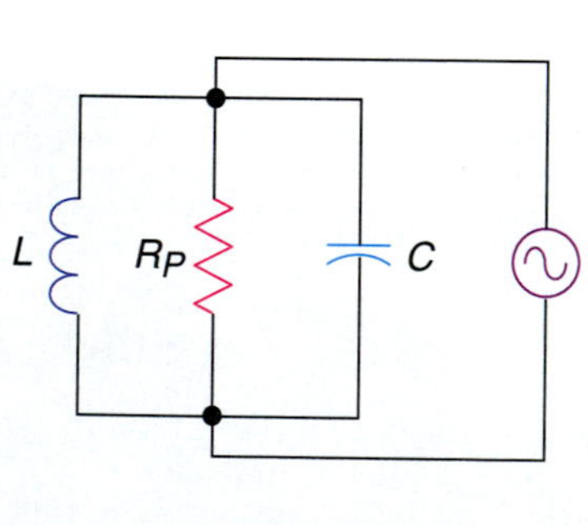

(*b*)

Figure 20-10 The series resistance of the inductor appears to the source as a higher value resistor in parallel with the *L-C* curcuit.

The effective elements of a parallel L-C circuit are shown in Fig. 20-10(*a*) as they actually occur. The resistance of the inductive branch consumes energy from the source that is replaced on each alternation. The source cannot tell that the resistance is only in the inductive branch and "sees" it as being in parallel with the inductance and capacitance, as shown in Fig. 20-10(*b*).

The value of the parallel resistance, R_P, is related to the inductor's reactance, and inherent resistance, as indicated in Eq. 20-10(*a*). The inductor's series resistance may be calculated from the inductive reactance and R_P using Eq. 20-10(*b*).

Equation 20-10

$$(a)\ R_P = \frac{X_L^2}{R_S}$$

$$(b)\ R_S = \frac{X_L^2}{R_P}$$

For example, assume that an inductor and capacitor connected in parallel each have 1000 ohms of reactance and that the inductor has 10 ohms of resistance. The procedure for finding the equivalent parallel resistance is as follows:

$$R_P = \frac{X_L^2}{R_S} = \frac{(1000\ \Omega)^2}{10\ \Omega} = \frac{1000\ \Omega \times 1000\ \Omega}{10\ \Omega} = \frac{1{,}000{,}000\ \Omega^2}{10\ \Omega} = 100{,}000\ \Omega$$

The effective parallel resistance, R_P is 100,000 ohms, an amount considerably higher than the 10-ohm resistance of the inductor. The reason for the high value is that the Q of the circuit is also high. Dividing the inductive reactance by the inductor's resistance, as prescribed in Eq. 20-10, reveals a Q of 100.

Branch Currents

A parallel resonant circuit is sometimes called a **tank circuit** because of its ability to store energy. To determine the tank current, divide the source voltage by the reactance of either the capacitor or inductor. These reactances may be low

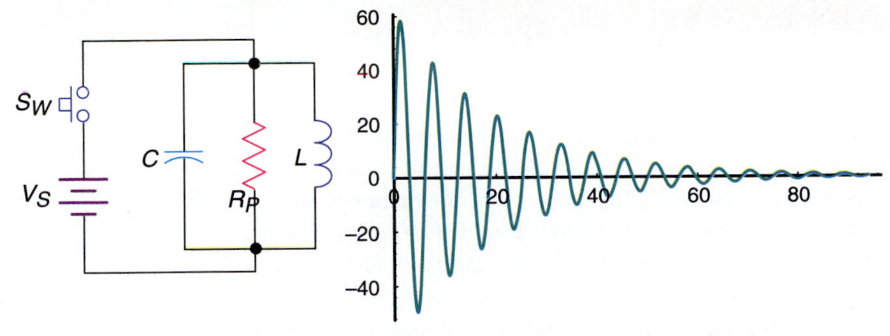

Figure 20-11 A stimulated tank circuit produces damped sinusoidal oscillations.

because it is only necessary for them to be equal to qualify as a resonant circuit. The high currents that flow in the tank circuit may seem to be a contradiction of Kirchhoff's current law, but these currents do not occur at the same time and are the result of energy being stored in the reactive components. That energy is not lost and can be delivered on demand, just like the energy in a flywheel.

Damped Oscillations

When a parallel resonant circuit is stimulated, the current in the tank circuit oscillates back and forth between the two reactive branches, producing a **damped** sinusoidal voltage across the circuit that is sometimes called **ringing**. **Damping** is a decrease in amplitude of each successive oscillation in a wave train, as illustrated in Fig. 20-11.

REVIEW QUIZ 20-3

1. The voltage across a parallel *L-C* circuit is ______________ at resonance.
2. Parallel *L-C* circuits are sometimes called ______________ to distinguish them from series *L-C* circuits because resonance occurs when the branch currents are equal, not necessarily when $X_L = X_C$.
3. An *L-C* circuit is sometimes called a(n) ______________ because of its ability to store energy.
4. The continuous oscillations of current between the inductive and capacitive branches of a parallel *L-C* circuit are known as the ______________ effect.
5. The main indicator that an *L-C* circuit is being operated at resonance is that from the source's point of view it appears to be a(n) ______________.

SUMMARY

Capacitors and inductors connected in series or parallel form frequency selective circuits. When connected in series, an *L-C* circuit passes the maximum current at its resonant frequency and rejects other frequencies by increasing its impedance very rapidly.

A parallel *L-C* circuit is just the opposite. It offers a high impedance to the resonant frequency, allowing a sharp voltage rise at resonance. The bandwidth of either circuit is determined by the amount of resistance in the circuit. In either circuit the bulk of the resistance is in the inductor. In a series *L-C* circuit, the resistance of the coil limits the maximum current the circuit may pass. Theoretically, a series *L-C* circuit with no resistance would be a perfect conductor at the resonant frequency. At frequencies below resonance, a series *L-C-R* circuit appears to the source as simply an *R-C* circuit because the capacitor offers a greater reactance than the inductor. The opposite is true at frequencies above resonance, and the circuit appears to be an *R-L* circuit. Both series and parallel *L-C* circuits are resistive at resonance. A parallel *L-C* circuit appears inductive at frequencies below resonance because more current flows in the inductive branch. At frequencies above resonance, the circuit appears capacitive.

The resistance of the inductor in a parallel *L-C* circuit appears to the source as a much larger resistor in parallel with the reactive branches. The value of the inductor's resistance is multiplied by the square of the inductive reactance to determine the parallel equivalent. Energy is stored in the reactive elements of a parallel *R-C-L* circuit, called a tank circuit. The oscillations of electric energy from one reactive component to the other are referred to as the flywheel effect. In both series and parallel resonant circuits, the bandwidth is determined by the *Q* of the circuit. *Q* is an abbreviation for the word *quality*. The higher the *Q* factor, the narrower the bandwidth. The *Q* factor is determined by dividing the frequency of resonance by the bandwidth. It is also approximated by dividing the inductive reactance by the resistance of the inductor. That resistance is usually incorrect because of the skin effect associated with alternating currents.

NEW VOCABULARY

antiresonance	damping	*Q*-factor	skin effect
bandwidth	flywheel effect	resonance	tank circuit

Questions

1. What is the relative phase difference between the voltage across the inductor and the voltage across the capacitor in a series *L-C-R* circuit?
2. Can the voltages across the inductor and capacitor in an *L-C-R* circuit at resonance be higher than the source voltage?

3. What limits the Q in a resonant circuit?
4. Can a series L-C-R circuit be resonant at more than one frequency?
5. Is the current through a series L-C-R circuit or the voltage across it maximum at resonance?
6. Does a parallel L-C circuit appear inductive or capacitive at frequencies below resonance?
7. How does lowering the Q of a series L-C-R circuit affect the change in current from resonance to nonresonance?
8. Does a parallel L-C circuit experience a voltage rise or current rise at resonance?
9. What is the effect called that reduces the amplitude of the self-oscillations generated by a parallel L-C circuit when stimulated?

Problems

1. What is the Q of an L-C-R circuit with a resonant frequency of 1 megahertz and a bandwidth of 10 kilohertz?
2. What is the resonant frequency of a 1.25-microhenry coil and a 25-picofarad capacitor connected in series?
3. What value capacitor is required in parallel with a 100-microhenry coil to resonate at 5 megahertz?
4. What value inductor is required in series with a 470-picofarad capacitor to resonate at 100 kilohertz?
5. If the inductor and capacitor forming a parallel L-C circuit each have a reactance of 500 ohms and the inductor has a DC resistance of 5 ohms, what is the effective resistance across the parallel resonant circuit as "seen" by the source?
6. What is the voltage across the inductor in a series L-C-R circuit when the applied voltage is 120 volts, the series resistance is 100 ohms, and the inductive and capacitive reactances are each 400 ohms?
7. What is the tank current in a parallel L-C circuit operated at resonance across 100 volts when the reactances are each 200 ohms?

Critical Thinking

1. Explain why the $X_L = X_C$ definition of resonance is not valid for a parallel L-C circuit with a Q less than 10.
2. Explain why there can be only one resonant frequency for a series L-C-R circuit.
3. Look at the waveform in Fig. 20-11, and explain why the dampened oscillations do not decrease literally.
4. Explain why the Q of an L-C-R circuit changes with frequency.
5. Explain what effect the capacitance between the turns of wire forming the inductor in a series L-C-R circuit could have on the circuit's overall frequency response.

Answers to Review Quizzes

20-1
1. 180
2. greater
3. 90

20-2
1. equal
2. resistor
3. Q
4. R-L; R-C
5. Q

20-3
1. maximum
2. antiresonant
3. tank circuit
4. flywheel
5. resistor

Chapter 21

Passive Filters

ABOUT THIS CHAPTER

This chapter presents passive filters as simple circuits made of inductors and capacitors. These filter circuits are used to separate different frequency components within a given circuit or to filter them between circuits. Because passive filters are made using passive components such

After completing this chapter you will be able to:

- Recognize different types of filter elements and circuits.
- Understand how filters are used to separate different parts of a signal.

as capacitors, inductors, and resistors, they are different from active filters, which are more complex and require an external DC power source. The use of passive components, in varying circuit configurations, comprises the study of passive filters discussed in this chapter.

Passive filters are especially important in communications, radio and television. Audio, video, and other AC signals are often transmitted on the same RF *carrier* frequency, so filters are required to separate them from each other for processing. Filtering, then, allows the signals to be separated for human hearing and viewing or for some other purpose.

- ▽ Apply the concepts of passive filters for testing and measuring circuits.
- ▽ Describe the role of computers in designing filters.

21-1 Passive Filter Types

The word **filter** represents a process for separating or removing one thing from another. Like a coffee filter that allows the desired liquid to pass through it, while blocking the unwanted solids, an electronic filter can separate desired signals from undesired signals. In all cases, filters perform the same basic process—they separate things.

The filters presented in this chapter are *passive* because the components used to build them—inductors and capacitors—are passive. Unlike a transistor, they do not require external DC power (bias). Active filters are usually found in more complex amplifier circuits (e.g., ICs and op amps), where they are packaged together with the amplifier for a specific type of filtering. For example, an active filter might be designed to filter and amplify a specific range of frequencies. The components of the active filter package, however, are made of the same reactive components as passive filters, especially capacitors.

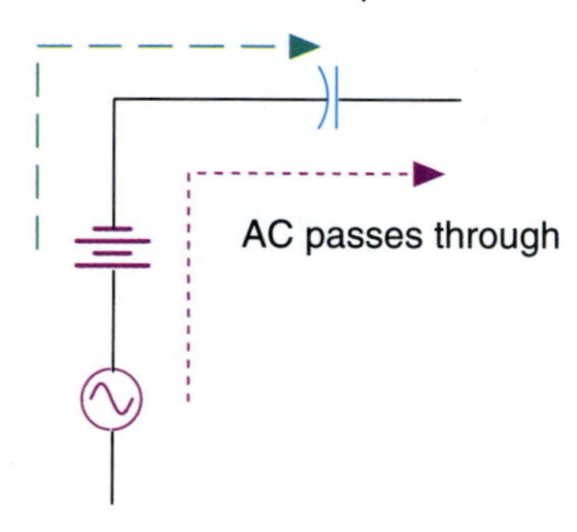

Figure 21-1 A capacitor as a filter.

The simplest type of passive filter is one that you studied earlier. Therefore, you already know something about how it works. As shown in Fig. 21-1, a capacitor is used to separate or filter direct current from alternating current.

The capacitor is an open circuit to direct current, but it looks like a short circuit to alternating current at higher frequencies. Because a capacitor can separate low frequencies from high frequencies, it can be considered a filter in this application.

A capacitor can also perform other types of filtering. For example, if you connect a capacitor between two circuits, the capacitor will allow the high frequencies to pass between the circuits while keeping any DC components separated inside the two separate circuits. Or, if you connect a capacitor across a load resistor that is connected to ground, it will short the high frequencies to ground and allow only low frequencies to pass through the load resistor. In general, the filtering action or type of filter depends upon how the capacitor is connected in the circuit.

Another filter element that you already know something about is an inductor, or **choke**. As shown in Fig. 21-2, it looks like a short circuit to low frequencies, especially direct current. But it opposes higher frequencies—it chokes them off—and looks open. If you wanted to separate low frequencies from high frequencies, you could use an inductor instead of a capacitor.

The frequency that passes from one part of the circuit to another is controlled by the type of element you use: inductor or capacitor. As Fig. 21-2 shows, if you connect an inductor in series, it prevents high frequencies from passing through. Only lower frequencies pass.

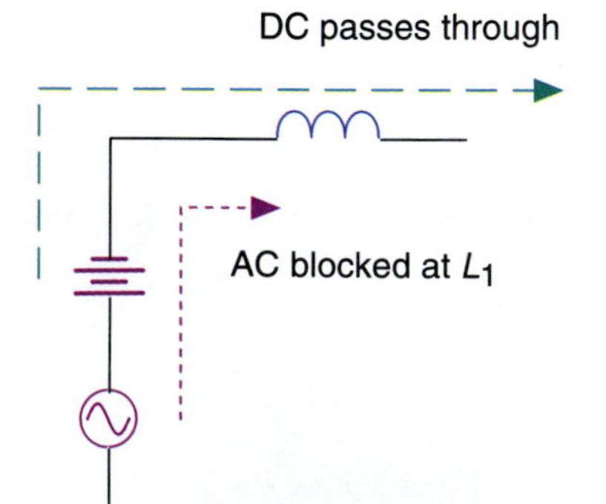

Figure 21-2 An inductor as a filter.

Another type of filter separation is performed by **coupling**. Coupling means to transfer energy from one circuit element to another. By using a capacitor to connect the output of one circuit to another, the capacitor can be used to pass or filter only the desired signals. This can also be accomplished using inductors, which, when placed in close proximity, can induce a particular voltage or frequency from one circuit to another. In this manner, coupling is also a form of filtering.

The resonant effect is also used to produce **resonant filters**. With specific values of L and C used in series or parallel combinations, the resonant effect will pass a specific band of frequencies to a load or stop the frequencies from reaching the load. For example, if you want a 20-kilohertz signal to be passed on to a load, you can use an L-C series resonant filter where the values of L and C are chosen to pass only that signal to the load. Such a filter would really pass the desired signal within some usable range, for example, 19 to 21 kilohertz. These resonant filters are used to pass or reject a frequency band.

In all filter types, the values of L and C must be carefully chosen to allow only the desired range to pass. For example, consider a capacitor of 1 microfarad. At 100 hertz, its reactance ($1/2\pi fC$) is about 1.59 kilohms. But at 1 megahertz, its reactance is less than 1 ohm. Therefore, a 1-microfarad capacitor would not be an effective short to frequencies near 100 hertz or even below 1 kilohertz. The same is true for the value of inductors.

By arranging the L and C components in various combinations, using a number of values, different filter types can be built. The components determine what type of frequencies can pass through the filter: high or low frequencies. The arrangement or configuration determines where the frequencies pass on to, and the values of L and C determine the range of frequencies that can be controlled or filtered. Also, the terms *low* and *high* are relative terms that describe the filter operation and not any particular range of frequencies.

In general, passive filters can be used to separate or pass various frequency components as one of these filter types:

- *Low-pass.* Allows only low frequencies to pass to a load.
- *High-pass.* Allows only high frequencies to pass to a load.
- *Band-pass.* Allows only a selective band of frequencies to pass to a load.
- *Band-stop.* Rejects or stops a selective band of frequencies from passing to a load.

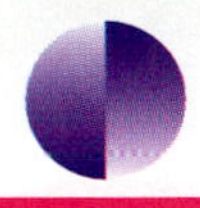

STUDENT to STUDENT

The basic building block of filters are components we already know about. This should be easy if you know how capacitors and inductors work.

REVIEW QUIZ 21-1

1. The two passive components used to build filters are ______________.
2. Passive filters do not require external ______________, like active filters, to operate.
3. Filters separate or pass signals depending upon the ______________ and ______________ of the L and C components.
4. The ______________ effect is used to build filters that pass or reject certain frequency bands.

21-2 Low-Pass Filters

The main characteristic of a low-pass filter is that it only allows lower frequencies to pass through it and on to the load. Higher frequencies simply do not get through. The elements of a low-pass filter circuit are inductors or chokes in series with the load and capacitors in parallel with the load. The circuit of Fig. 21-3 on the next page shows three basic low-pass circuits. Notice that in (*a*), the **shunt capacitor** C_1 is a short shunting the higher frequencies to ground. Only the low frequencies take the path through the load resistor. Higher frequencies take the path of least resistance, which is C_1.

The circuit (*b*) uses a choke to keep high frequencies from getting through to the load, so only low frequencies pass through the inductor. Finally, the last

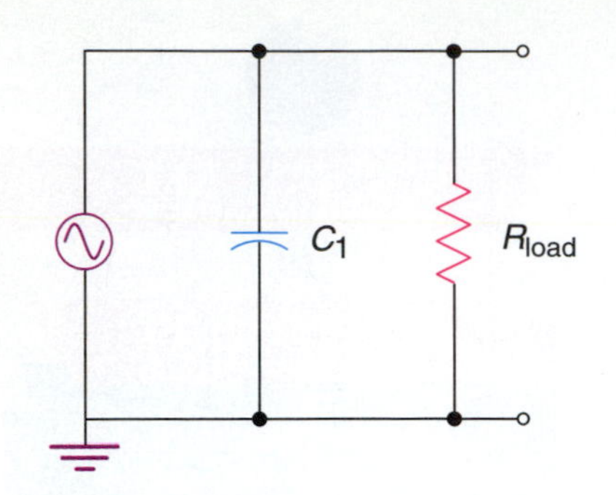

(*a*) Low-pass using shunt capacitor to ground

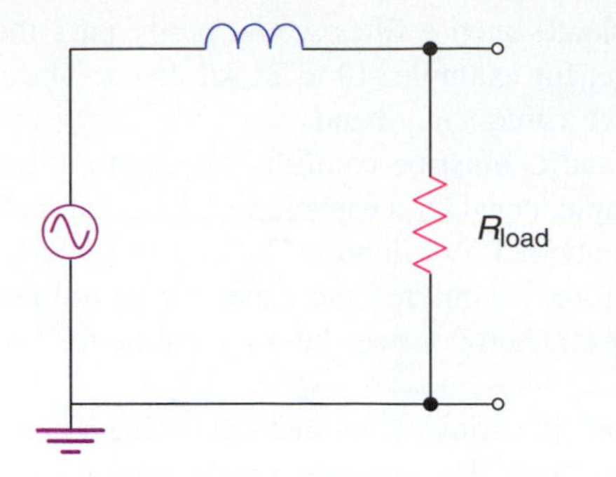

(*b*) Low-pass using series choke

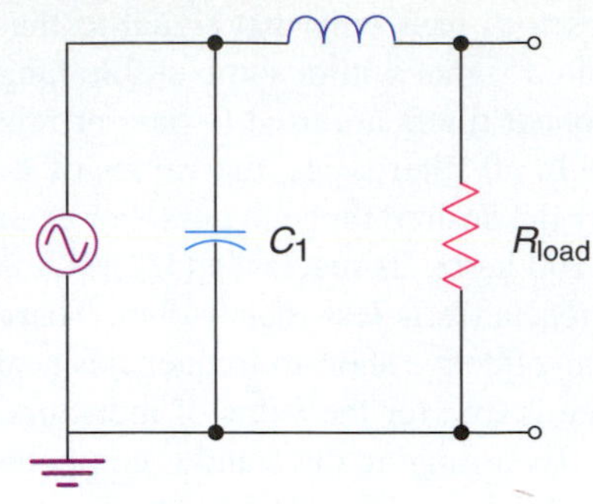

(*c*) Low-pass using shunt capacitor and series choke (*L* type filter)

Figure 21-3 Basic low-pass filters.

circuit (*c*) uses both components. The capacitor shunts higher frequencies to ground, and the inductor chokes off any high frequency components, keeping them from reaching the load.

A Series Inductor Passes Low Frequency Signals

At low frequencies, a series inductor or choke will allow the low frequencies to pass to the load because X_L is low. At higher frequencies, the inductor will have a high value of X_L, and it will choke off or prevent any high frequencies from passing to the load. This means that only the low frequencies will pass through the series inductor to the load.

A Parallel Capacitor Shorts High Frequencies to Ground

At low frequencies, a capacitor will have a high value of X_C, so it will allow low frequencies to move on to the load. At higher frequencies, X_C will be so low that it will be an effective short to ground for high frequencies, thus filtering them away from the load.

Figure 21-4 shows three other types of low-pass filters. These filters usually have sharper response curves than the basic circuit configurations shown in Fig. 21-3. They are used in power supplies, to filter rectified direct current, and in audio circuits to filter the low frequency audio signals from other higher signals.

π and *T* Type Low-Pass Filters

The π and *T type* low-pass *filters*, shown in Fig. 21-4, have a sharper response than the *L type filter*. The π type is the most efficient for producing a sharp response for low-pass filters. The RF choke in (*a*) is more effective than the resistor in (*b*), and the use of two capacitors in (*a*) is more effective than the single capacitor in (*c*). The filter in (*b*), however, is commonly used because it does not require the choke and because its series resistor can be an effective impedance matching element.

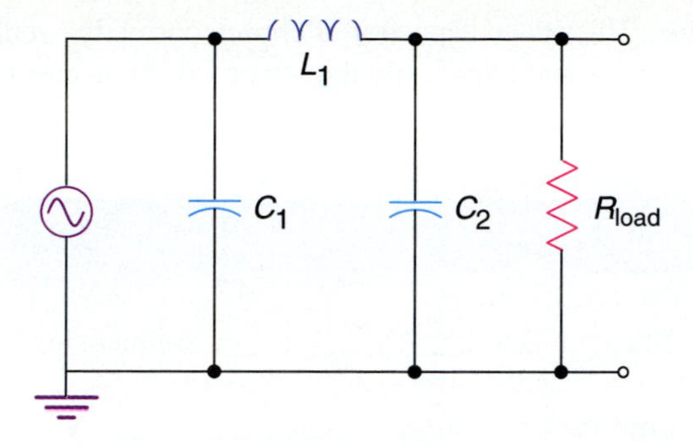

(a) π filter with choke (low pass to ground)

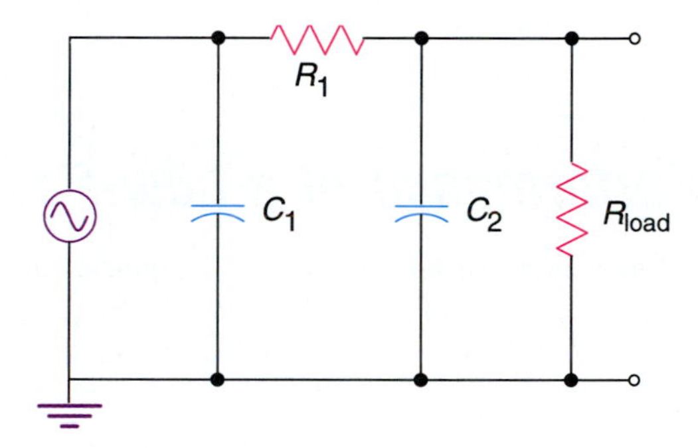

(b) π filter with resistor (low pass to ground)

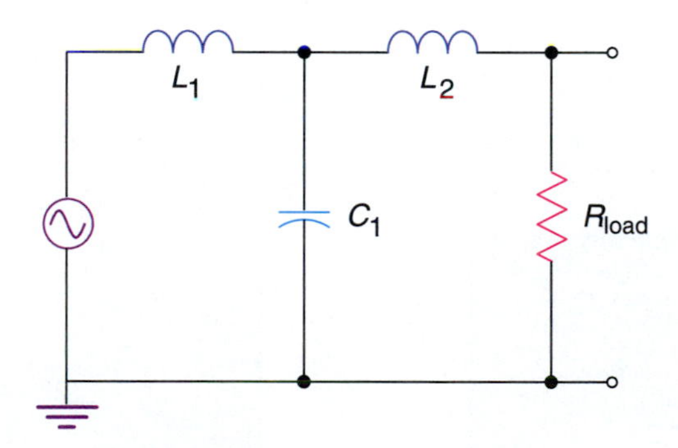

(c) T filter with 2 chokes (low-pass to ground)

Figure 21-4 T and π low-pass filters.

Example 21-1

A low-pass L type filter is operating in a circuit where only audio frequencies below 16 kilohertz are supposed to be passed on to the load. The filter, however, is currently operating such that frequencies of 100 kilohertz are also getting through. How can this problem be corrected?

The filter components should be checked to see if they are operating correctly. If they are, then their values are incorrect and the filter is improperly designed. A simple or temporary solution is to add another capacitor in parallel to create a π-type filter. This added capacitor, if chosen correctly, would cut off most of the higher frequencies and allow only the lower values, nearer to 16 kilohertz, to pass.

REVIEW QUIZ 21-2

1. A low-pass filter prevents ______________ frequencies from getting through it.
2. A low-pass filter uses inductors in ______________ with the load.
3. A low-pass filter uses capacitors in ______________ with the load.
4. The different types of low-pass configurations are ______________.

21-3 Performance of a Low-Pass Filter

The performance of any filter can be shown by its response curve. Figure 21-5 shows the typical response of a low-pass filter and the frequency range where the performance is desirable. Notice that in (*a*) the orange solid line shows the filter sharply cutting off the higher frequencies. The blue dashed line shows a filter that is performing poorly because it allows too many higher frequency components to pass. In general, the sharper the response of a low-pass filter, the more desirable or the better the performance.

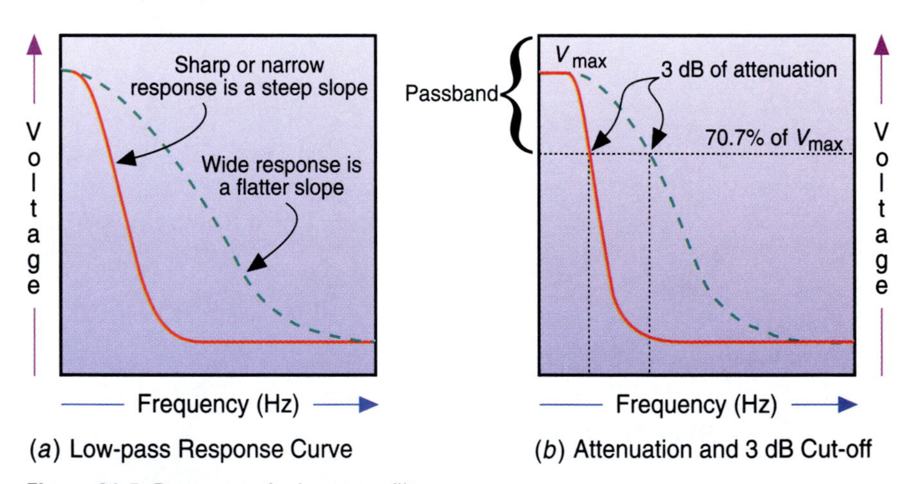

(*a*) Low-pass Response Curve

(*b*) Attenuation and 3 dB Cut-off

Figure 21-5 Response of a low-pass filter.

The solid line response passes only lower frequencies, whereas the dashed line response passes higher frequencies. That results in a wider **passband**. In both cases, the **slope** or steepness of the response curve reflects its performance. As the frequency increases, voltage decreases. This is called **attenuation**. The solid line response shows a filter with rapid attenuation, which is also called the **roll-off**. This steep slope rolls off sharply. The dashed line response shows a flatter slope, so it rolls off slowly.

The 3 dB Point or Cut-Off

In Fig. 21-5(*b*), when the response has decreased to 70.7 percent of the maximum voltage (V_{max}), the power (in watts) is one-half of the value it was at V_{max}. This is called the 3 dB cut-off or half-power point. The term *decibel* (dB), is used in all areas of electronics. A decibel is one-tenth of a *bel* and is a logarithmic unit that represents a power or voltage ratio. Here,

$$3 \text{ dB} = 20 \log \left(\frac{V_{\text{cut-off}}}{V_{max}} \right)$$

where $V_{\text{cut-off}}$ = voltage at 70.7 percent of V_{max}

Also, in terms of power (V^2/R or I^2R),

$$\text{dB} = 10 \log \left(\frac{P_{out}}{P_{in}} \right)$$

However, you need to know the power (in watts) for that equation, which requires current I or resistance R.

All frequencies above the 3 dB cut-off are said to be in the *passband,* and all frequencies below the 3 dB cut-off are outside of the passband. As a rule, most frequencies below the 3 dB cut-off have attenuated too much to be useful. Therefore, the performance is mainly measured by its passband.

Roll-Off per Octave

Another term to describe the attenuation of a filter is its roll-off per **octave**. An octave is a doubling of the frequency or an interval between two frequencies that have a 2:1 ratio, for example, 200 hertz: 100 hertz or 8 kilohertz: 4 kilohertz. Therefore, if a filter has a roll-off of 3 decibels per octave, it means that every time the frequency doubles, the power is decreased by one-half, or 3 decibels. If the power were to roll off at 6 decibels per octave, it would mean that another 3 decibels of attenuation occurred ($3 + 3 = 6$) at twice the frequency, and this would be one-fourth the power.

Example 21-2

A low-pass filter has an output which measures 6 volts maximum at about 100 hertz and rolls off to about 1 volt at 40 kilohertz. How would you determine its performance?

By calculation, 70.7 percent of V_{max} is: $6 \times 0.707 = 4.24$ volts.

By measuring the filter output voltage at increasing frequency points, the 3 dB point is established at the frequency where the output measures 4.24 volts. If this occurred at 18 kilohertz, then the passband would be established as 18 kilohertz and more measurements could be taken to draw the complete response curve.

REVIEW QUIZ 21-3

1. If the 3 dB point of a low-pass filter is at 10 kilohertz, the passband would be ______________.
2. A low-pass filter, with a flatter slope, has a(n) ______________ passband.
3. The 3 dB point means that the voltage has dropped to ______________ percent from its maximum.
4. Decibel is a(n) ______________ ratio.
5. Roll-off per octave describes the attenuation each time the frequency is ______________.
6. The power (in watts) at the cut-off point is ______________ the maximum value.

21-4 High-Pass Filters

The main characteristic of a high-pass filter is that it only allows higher frequencies to pass through it and on to the load. Lower frequencies simply do not get through. The circuit elements of a high-pass filter circuit are inductors or chokes in parallel with the load. Capacitors are placed in series with the load.

The circuit of Fig. 21-6 shows two basic high-pass circuits. Notice that in (*a*), the capacitor C_1 is a short, for higher frequencies, to the load. Only the higher frequencies take the path through C_1. Lower frequencies are blocked proportionally to the value of reactance in ohms. In general, high-pass filters are not used as often as low-pass filters because most filtering is done to remove lower frequencies from higher ones.

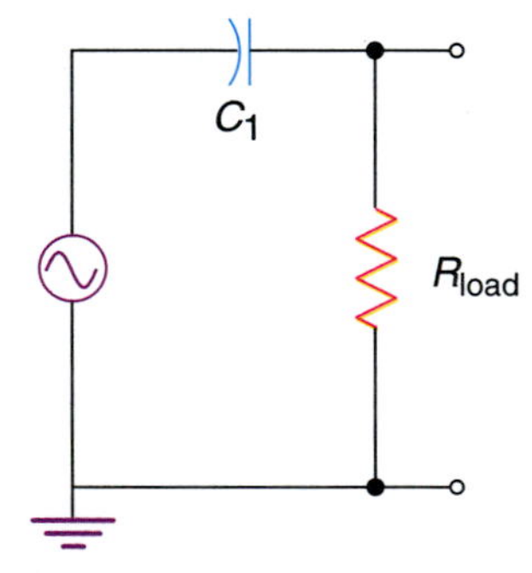

(*a*) High-pass filter blocks low frequencies

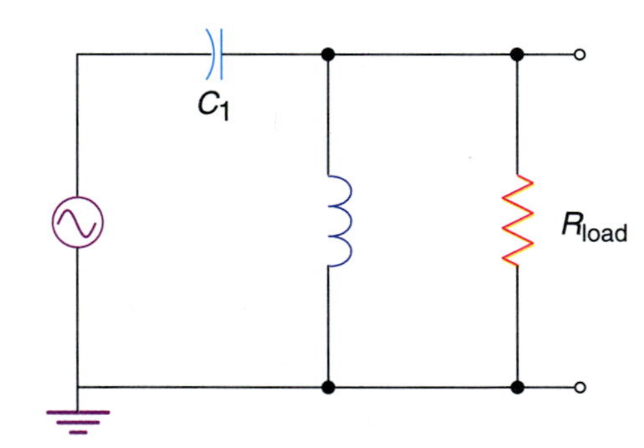

(*b*) High-pass filter shunts low frequencies to ground

Figure 21-6 Basic high-pass filters.

The circuit of Fig. 21-6 (*b*) uses the same capacitor to block low frequencies and a choke to ensure that no high frequencies take the path to ground. By using both components, low frequencies are either blocked from the load or shunted to ground. Only high frequencies pass on to the load.

A Series Capacitor Passes High Frequency Signals

At low frequencies, a capacitor will have a high value of X_C, so it will block low frequencies from moving on to the load. At higher frequencies, the series X_C will be so low that it will be an effective short and pass high frequencies directly on to the load.

A Parallel Inductor Shorts Low Frequencies to Ground

At low frequencies, a parallel inductor or choke will short the low frequencies to ground because X_L is low. At higher frequencies, the inductor will have a high value of X_L, and higher frequencies will take the path of least resistance and pass to the load. This means that only the low frequencies will pass through the parallel inductor to ground.

π and *T* Type High-Pass Filters

In order to improve the response of a basic high-pass filter, a series capacitor or a parallel (shunt) inductor can be added. This will produce a steeper response curve or slope. Figure 21-7 shows these other types of high-pass filters, which are named by their physical arrangement in the circuit.

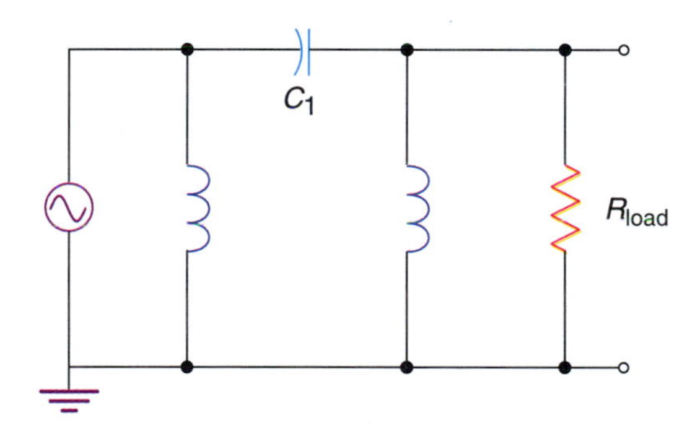

(a) High-pass π filter

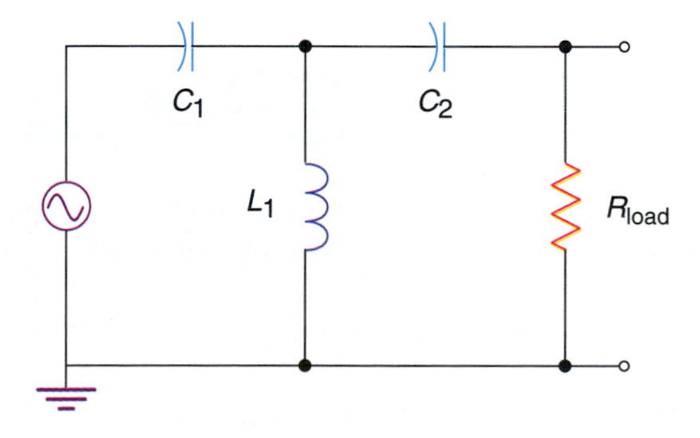

(b) High-pass T filter

Figure 21-7 T and π high-pass filters.

The shape of the curve goes from low to high. That means the voltage goes up as the frequency gets higher.

Just like their low-pass counterparts, π and T type high-pass filters are more efficient than the L type. The two RF chokes in (*a*) are effective shorts to ground for low frequencies, but the capacitors in (*b*) do an excellent job of keeping high frequencies directed to the load. Notice that a resistor cannot be used in place of a capacitor. This is because a capacitor is frequency selective; it is the most effective passive filter component used in any filter circuit.

Performance of a High-Pass Filter

Figure 21-8 shows the typical response of a high-pass filter inside and outside of its passband. Notice that in (*a*) the orange solid line shows the filter performing in a desirable manner. It has a steeper slope, rolling off sharply and cutting off the lower frequencies. This narrow response is more typical of a T or π type filter, which usually out-performs the more basic types represented by the blue dashed line or wider response in (*b*).

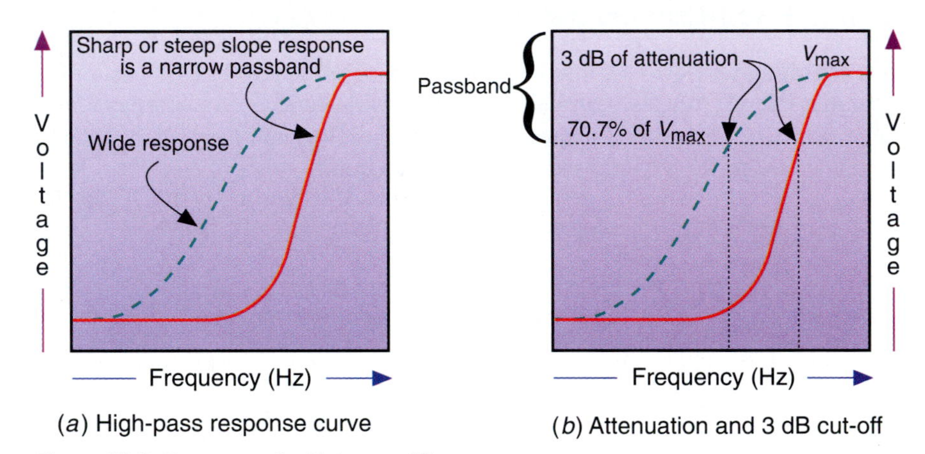

Figure 21-8 Response of a high-pass filter.

To describe the performance of a high-pass filter, refer to all of the terms used to describe the response of a low-pass filter. The roll-off or attenuation frequency, however, is in the opposite direction. Therefore, a high-pass filter will always attenuate at lower frequencies, and the passband is in the higher frequencies.

REVIEW QUIZ 21-4

1. If the 3 dB point of a high-pass filter is at 200 kilohertz, and the input frequency range is from 500 hertz to 800 kilohertz, the ideal passband would be ______________.
2. In a high-pass filter, use a(n) ______________ choke to pass high frequencies to the load.
3. In a high-pass filter, use a(n) ______________ capacitor to pass high frequencies to the load.
4. A high-pass filter with a steep slope has a(n) ______________ passband.

21-5 Band-Pass Filters

A band-pass filter is another type of passive filter (Fig. 21-9). It uses the resonant effect to pass only a specific band of frequencies to a load. The band-pass filter is made using a capacitor and an inductor in series or parallel, depending upon its orientation to the load. When a resonant circuit is tuned or designed to operate at a specific frequency, it has the effect of filtering out all signals except the resonant frequency. In practice, both series and parallel resonant circuits are used as filters. Because of the reactance of an inductor L (effective short to low frequencies) and the reactance of a capacitor C (effective short to high frequencies), both of these components are the basic building blocks of all passive filters, which separate various frequencies. By arranging specific values of series or parallel resonant circuits, different types of filters can be built.

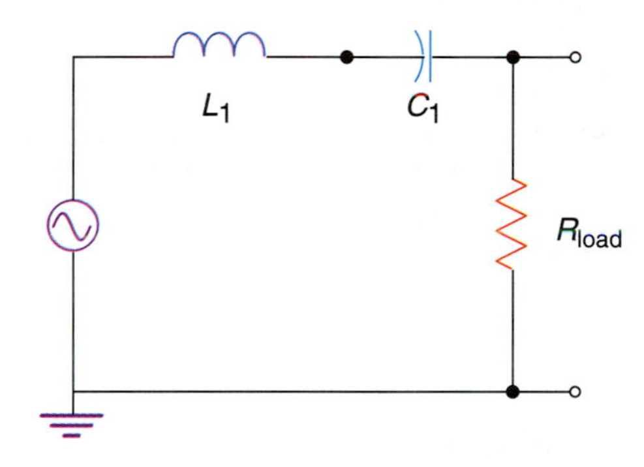

(*a*) Series-resonant band-pass filter

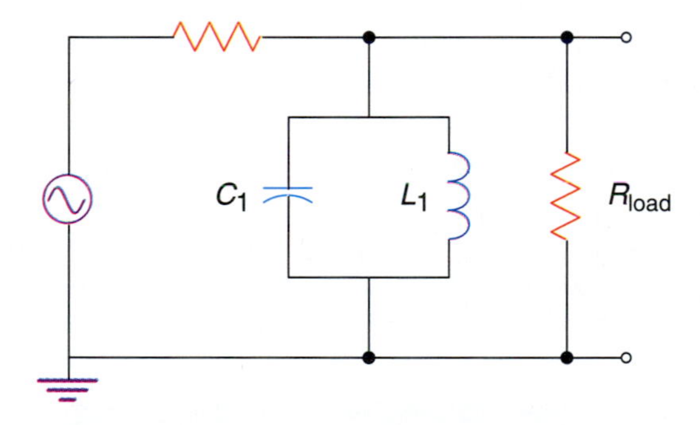

(*b*) Parallel-resonant band-pass filter

Figure 21-9 Band-pass filters.

Performance of a Band-Pass Filter

The performance of a band-pass filter is a combination of low-pass and high-pass attributes. Therefore, a band-pass filter has a lower and an upper 3 dB or cut-off point. Also, its performance is related to its center frequency, which is the resonant frequency. Thus, a band-pass filter should be well balanced on either side of the center desired resonant frequency.

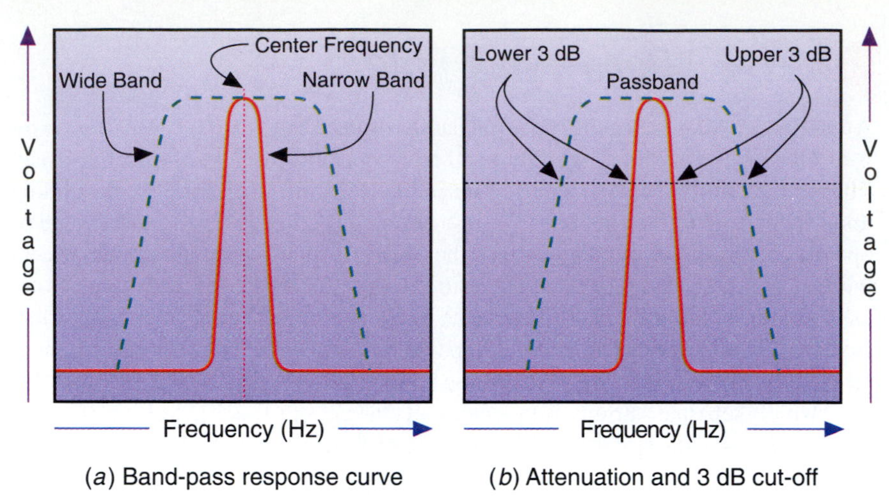

(*a*) Band-pass response curve (*b*) Attenuation and 3 dB cut-off

Figure 21-10 Response of a band-pass filter.

Figure 21-10 shows the response of a band-pass filter. The orange solid line is a narrow band, and the blue dashed line is a wider band of frequencies. In general, narrow band filters have steeper slopes, but wider bands can also have steep slopes if they are designed properly.

STUDENT *to* STUDENT

A band-pass filter looks like a high-pass filter connected to a low-pass filter at the center frequency.

Example 21-3

A band-pass filter is required to have a passband of 10 kilohertz. Is this a narrow or a wide passband?

It all depends upon the center frequency and the range of frequencies the filter operates over. For example, a 20-megahertz band-pass filter may not need to be as narrow as a 20-kilohertz filter. If a 20-kilohertz filter has a passband of 10-kilohertz, it is most likely too wide to be of any effective filtering. This is because it will have its 3 dB points at 15 kilohertz and 25 kilohertz. But a 10 kilohertz bandwidth for a filter with a 20-megahertz center frequency would be extremely narrow.

Another way to look at this is to use percentages. A 10-kilohertz passband is ± 25 percent of a 20-kilohertz center frequency band-pass filter (5 kilohertz on either side). But 10 kilohertz is only ± 0.03 percent of a 20-megahertz filter.

REVIEW QUIZ 10-2

1. A band-pass filter uses the ______________ effect to pass a desired band of frequencies.
2. Band-pass filters are either ______________ or ______________ *L-C* circuits.
3. The narrower the passband in a band-pass filter, the ______________ the 3 dB points on either side of the center frequency.
4. The center frequency of a band-pass is also the ______________ frequency.

21-6 Band-Stop Filters

A band-stop filter is another type of passive filter. It uses the resonant effect to stop or reject a specific band of frequencies from passing to a load. It is often called a **band-rejection** filter or a *notch* filter. Instead of having a passband, it has a **stop-band** or **notch-band**. The band-stop filter is constructed using either a series resonant or parallel (tank) resonant circuit.

Figure 21-11 shows the two types of band-stop filters. In (*a*), the series components L_1 and C_1 are designed to stop a band of frequencies from reaching the load by shunting them (short circuit) to ground. In (*b*), the tank circuit of L_1 and C_1 is designed to be a high impedance path for the band to be rejected; therefore, the frequencies cannot pass to the load.

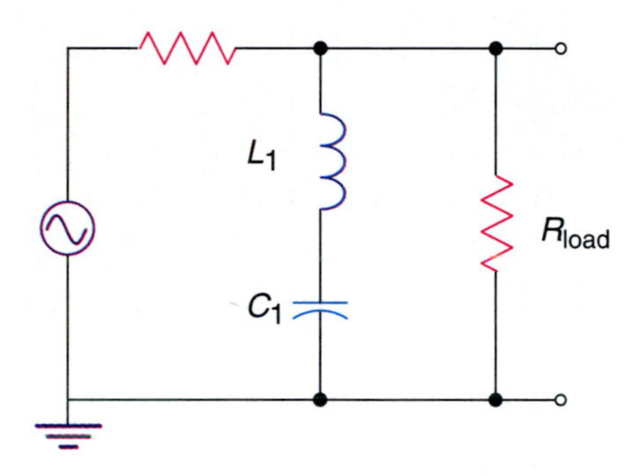

(*a*) Series-resonant band-stop filter

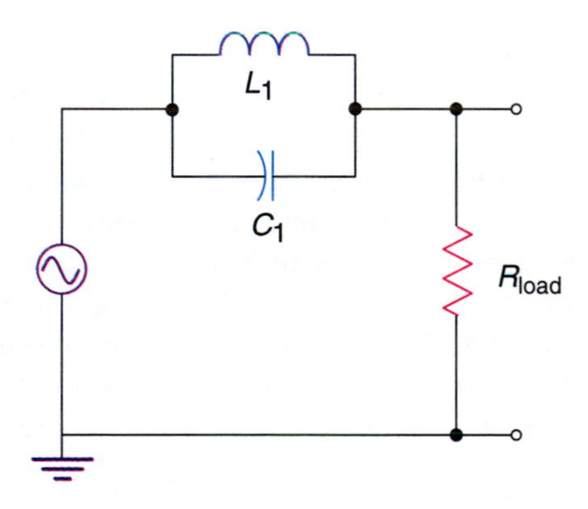

(*b*) Parallel-resonant band-stop filter

Figure 21-11 Band-stop filters.

The choice between the two configuration types usually depends upon matching the internal resistance (impedance) of the source or load to the filter. In either case, both band-stop and band-pass filters are often designed so that their impedances match across the passband to the load.

STUDENT to STUDENT

The band-stop filter has a stop-band. All the others have passbands.

Performance of a Band-Stop Filter

The performance of a band-stop filter is also a combination of low-pass and high-pass attributes. Therefore, a band-stop filter has a lower and an upper 3 dB or cut-off point. Also, its performance is related to its center frequency, which is the resonant frequency. Therefore, like a band-pass filter, a band-stop filter should be well balanced on either side of the center desired resonant frequency.

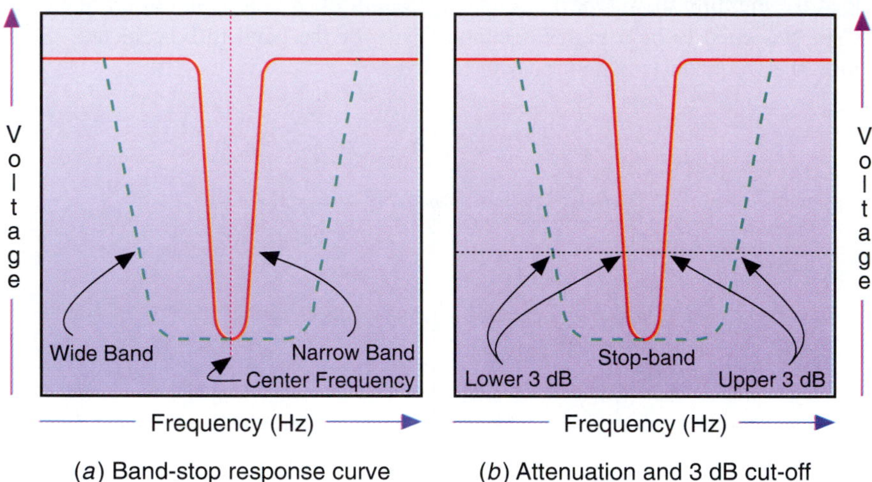

Figure 21-12 Response of a band-stop filter.

Figure 21-12 shows the response of a band-stop (also called a notch or band-rejection) filter. The orange solid line is a narrow band or notch, and the blue dashed line is a wider band or notch. The response of a band-stop filter is the opposite of a band-pass filter. Also, the higher the Q (ratio of X_L to R coil at resonance) of the resonant filter, the narrower the response. In part, this is because a choke with less resistance will allow more current to flow at resonance. For a series resonant circuit, this will mean less shunt resistance to ground. For a parallel resonant circuit, this will mean more voltage developed across the tank at resonance. Therefore, the greater the reactance and the lower the coil resistance, the narrower the response.

Example 21-4

A band-stop filter is designed to reject a particular band of frequencies. Suppose that a range of frequencies from 100 hertz to 100 megahertz was present in a circuit. How could you remove the bands (*a*) from 40 kilohertz to 60 kilohertz and (*b*) from 40.4 megahertz to 40.6 megahertz?

In this case, two separate filters could be cascaded together. In fact, you could use two tank circuits, two series resonant circuits, or a combination. Such filters, however, require more design because the reactance of one filter can interact with the reactance of the other.

REVIEW QUIZ 21-6

1. A band-stop filter is also called a(n) ______________ or a(n) ______________ filter.
2. Band-stop filters are either ______________ or ______________ *L-C* circuits.
3. The band-stop filter response is called the ______________ instead of the passband.
4. The center frequency of a band-stop is also the ______________ frequency.

21-7 Other Forms of Filtering

Bypass Capacitors

Capacitors provide an effective short-circuit or *bypass* for high frequencies around a load or some other component. In this way, bypass capacitors are low-pass filters.

The general rule is that the bypass X_C should be one-tenth or less of the value of R being bypassed at the desired frequency. By using this rule, no more than one-tenth of the voltage will be developed across the bypassed component or load. The smaller the load resistance, the greater the value of capacitor you will need to act as a bypass. For example, a 1-kilohm load needs X_C to be 100 ohms or less at the desired frequency. Because X_C is calculated as $1/(2\pi fC)$, a greater value of C must be used in the equation to produce the smaller value of X_C. Also, the value of C must be based on the lowest frequency you want to bypass.

Figure 21-13 shows a bypass capacitor C_1 used as an effective short circuit to some frequencies around a load resistor. All AC frequencies, however, are developed across the other resistors.

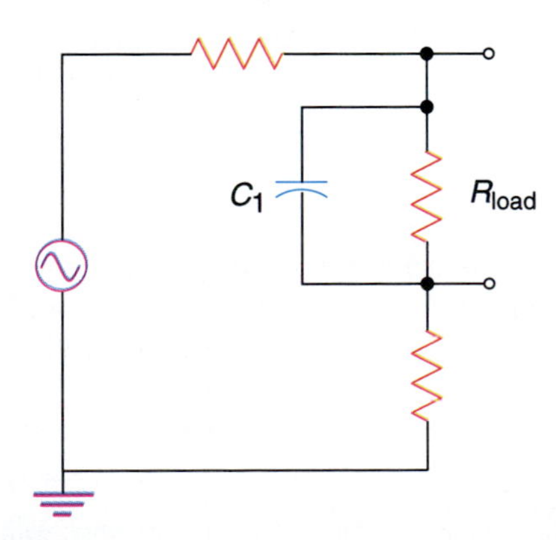

Figure 21-13 A bypass capacitor.

Example 21-5

A circuit has a frequency range of 100 hertz to100 kilohertz. There is a 330-ohm load resistor that develops the voltage for the circuit; however, only audio frequencies below 16 kilohertz can be allowed to be developed across the load. What value of bypass capacitor will be an effective bypass to keep higher frequencies from developing across the load?

Because the load resistor is 330 ohms, X_C must be 33 ohms or less above 16 kilohertz. Therefore, the equation must begin as:

$$33\ \Omega = \frac{1}{2\pi f C} = \frac{1}{6.28 \times 16{,}000 \times C}$$

This converts to solving for C which can be done as:

$$C = \frac{1}{6.28 \times 16{,}000 \times 33} = 0.3\ \mu\text{F}$$

At 17 kilohertz, C would decrease to 31 ohms and, as the frequency increased further, it would continue to decrease. This would effectively bypass all higher frequencies around the load.

Capacitive Coupling

Capacitive **coupling** is the most common method used to transfer or couple a signal from one circuit to another. When capacitors are used, their low reactance to high frequencies makes them effective high-pass filters.

Coupling means that only the desired output of one circuit is connected to the input of another circuit. For example, in a multistage amplifier circuit, the amplified output of one stage is coupled to the next stage. In this process, the DC power in each stage must remain separate, and only the desired signal must be coupled to the next stage. Therefore, a capacitor is used to couple an AC input signal into the amplifier while blocking the DC component. Also, the value of the capacitor can be chosen to cut off any other low frequencies from coupling.

Figure 21-14 shows a coupling capacitor used to couple or filter only the desired AC signal from one amplifier circuit to the next. Also, the separate DC voltages remain in their respective circuits. The triangular schematic symbols represent operational amplifiers (op amps). It is not necessary for you to know the circuitry of op amps, but you do need to know that they are used to process the AC signal in some way (make it bigger or change its shape). Notice that these amps each have their own separate DC power supplied (+15 volts and −10 volts).

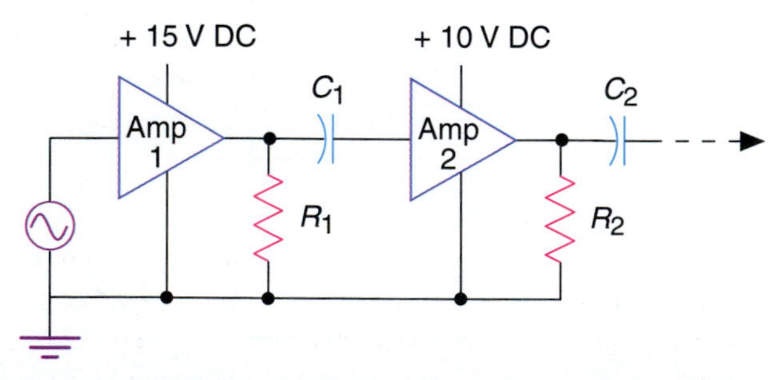

Figure 21-14 Capacitive coupling. Amps 1 and 2 are op amps with separate DC voltages. C_1 couples the AC output of Amp 1 to Amp 2, and C_2 couples the output of Amp 2 on to another stage.

Inductive Coupling

When current varies in one inductor, it can induce a secondary voltage in another inductor if the components are in close proximity. This is possible because an induced voltage is the result of a varying current and its magnetic field causing electron flow in a secondary inductor. But DC current has no inductive effects in the secondary because there is no fluctuating (expanding and collapsing) magnetic field. This also means that if the current in the primary inductor is a combination of alternating current and direct current (alternating current riding on a DC level), the induced voltage in the secondary is only induced by the AC portion. This means that inductive coupling also blocks or separates DC components between circuits as effectively as a coupling capacitor. In simple terms, only AC voltages in the primary induce AC voltage in the secondary.

In Fig. 21-15, inductive coupling is shown using two inductors. Notice that the DC voltages are kept separate because an inductor does not couple direct current. The magnitude of the AC voltage depends upon the turns-ratio of the primary-to-secondary. The secondary voltage usually has the opposite polarity of the primary voltage (180° out of phase). The direction of the windings, however, can be altered to produce different phases.

STUDENT *to* STUDENT

All these other forms of filters use the same principles. If you know how an inductor or a capacitor works, then it all makes sense.

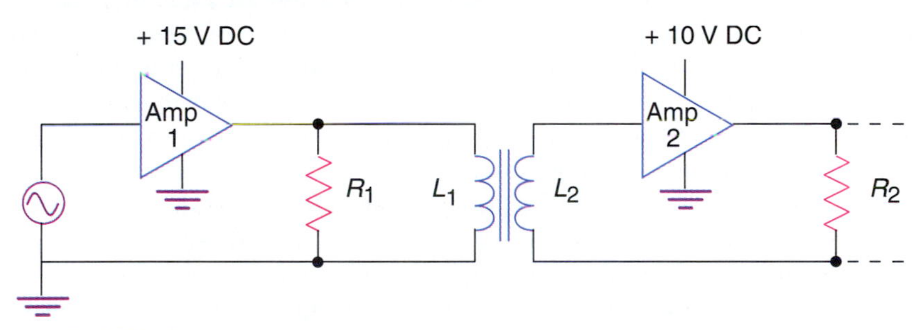

Figure 21-15 Inductive coupling. Amps 1 and 2 are op amps with separate DC voltages. L_1 couples the AC output of Amp 1 to L_2, which is in series with Amp 2. L_1 and L_2 also separate the two DC voltages that power the amps.

REVIEW QUIZ 21-7

1. A bypass capacitor is used to pass ______________ frequency signals around a component.
2. Coupling capacitors are like ______________ filters that also block ______________.
3. Inductors are used as ______________ filters to transfer energy from one circuit to another.

21-8 Filter Design

Today, most filters (with several elements) are designed by computers because the calculations take too long to perform by hand. The computers and their software are part of a business known as computer-aided engineering (CAE), which is used to design all types of circuits. Prior to CAE tools, two basic analysis techniques—**constant-*k*** and ***m*-derived** filters—were used to calculate the *L* and *C* components of a passive filter.

Constant-k and m-Derived Filters

In general, the constant-k filters have a high rejection of unwanted frequencies over a wide range (medium Q). The m-derived filters are usually sharper in their response (higher Q). The Q is the quality or sharpness of the filter's response where Q is equal to X_L divided by R.

The constant-k type gets its name from the parallel reactances in the L, T, and π configurations, where the value of series reactance multiplied by the parallel reactance is constant over a wide frequency range. The m-derived filter gets its name from the ratio of the reactance of its resonant elements to the comparable resonant elements in the constant-k filter. Therefore, a conversion factor is required to obtain the m-derived component values after calculating the constant-k values. Complete calculations for filter configurations can be found in electronic math books.

Design Calculations

In the constant-k method, the values of L and C for low-pass and high-pass filters are based on the value of the load (R in ohms) and a *cut-off frequency*. Above or below the cut-off frequency there is a gradual roll-off that gives the filter its shape. The general formulas for low- and high-pass filters apply to band-pass filters also. Band-pass filters, however, require more calculations because the L and C components are in parallel and also because two frequencies are required in the calculations—the passband frequency (series inductance and shunt capacitance) and the equivalent frequency (series capacitance and shunt inductance).

Equation 21-1 shows the calculation for the design of a typical low-pass L type (two element) filter using the constant-k technique. The value of reactance is calculated, based on the value of the load:

Equation 21-1

$$(a)\ \text{High-pass: } X_L = \frac{R_{\text{load}}}{2} \text{ and } X_C = 2R_{\text{load}}.$$

$$(b)\ \text{Low-pass: } X_L = 2R_{\text{load}} \text{ and } X_C = \frac{R_{\text{load}}}{2}$$

The values of L and C are calculated, based on the desired cut-off frequency:

$$L = \frac{X_L}{2\pi f} \text{ and } C = \frac{1}{2\pi f X_C}$$

Using the formulas in Eq. 21-1, an example filter can be calculated. For this example, the constant-k calculations for a low-pass filter (L type) with a cut-off frequency of 400 kilohertz and a load resistance of 100 ohms would be as follows:

$$X_L = 2 \times 100 = 200\ \Omega \text{ and } X_C = \frac{100}{2} = 50\ \Omega$$

$$L = \frac{200}{(6.28 \times 400 \times 10^3)} = 0.1\ \text{mH}$$

$$C = \frac{1}{(6.28 \times 400 \times 10^3 \times 50)} = 7.96 \times 10^{-9} \cong 0.008\ \mu\text{F}$$

SUMMARY

On the surface, the study of passive filters may seem basic because of the few elements used (capacitors and inductors). Passive filter design, however, is one of the most challenging fields in modern electronics. Since the introduction of cellular and hand-held communication devices, passive filters have become increasingly important as more and more audio transmissions fill the air. In fact, passive filters are being designed to operate at higher frequencies all the time. Such high-frequency filters are made of microcircuits where thin strips of metal, in odd-looking geometric forms, are deposited on top of a dielectric material. Yet, even these circuits share the same concepts as those presented in this chapter.

On the other hand, active filters—mostly in op amps—use only resistive and capacitive components because they operate at higher frequencies and large inductors, used as chokes, would be impractical. Only in more modern passive filter design does inductance play a role, especially in resonant filters built on substrates.

In general, you can recognize the basic filter types by remembering the following applications:

- Inductors are used to attenuate high frequencies and pass low frequencies.
- Capacitors are used to attenuate low frequencies and pass high frequencies.
- Series *L-C* resonant circuits are used to pass a specific band of frequencies.
- Parallel *L-C* resonant circuits are used to stop a specific band of frequencies.

Moreover, by using π and *T* type filters, you will get a steeper response curve than filters with only two components. More inductors and more capacitors will result in a sharper response curve, although filters using only capacitors are less expensive. The value of *L* and *C* for these filters can be calculated using CAE tools or the constant-*k* or *m*-derived methods. Passive filters, however, are also required to match the input and output impedance of other networks, especially in communications circuits.

NEW VOCABULARY

attenuation
band-rejection
band-stop
choke
constant-*k*
coupling
cut-off frequency
decibel (dB)
L type filter
m-derived
notch
octave
passband
π type filter
resonant filter
roll-off
shunt capacitor
slope
stop-band
T type filter

Questions

1. What are the four basic filter types?
2. What is the difference between a passive filter and an active filter?
3. Why are inductors and capacitors the building blocks of passive filters?
4. How is a series inductor similar to a parallel capacitor in a low-pass filter?
5. How is a series capacitor similar to a parallel inductor in a high-pass filter?
6. Why is the 3 dB point important in describing the performance of a filter?
7. What is meant by *dB roll-off per octave*?
8. What is a bypass capacitor used for?
9. How does capacitive coupling work?
10. How does inductive coupling work?

Problems

1. Draw the response curve for a low-pass filter with the following specifications:

   Input frequency range: 0 (DC) to 100 kilohertz
   Input voltage: 0 to 25 volts, where V_{max} = 20 volts
   Passband: 0 (DC) to 20 kilohertz

   Calculate and label the 3 dB point, label frequency every 10 kilohertz, label voltage every 5 volts, and show where the cut-off frequency intersects the 3 dB voltage.
2. A low-pass filter rolls off at 3 decibels per octave, from the 3 dB point (12 volts) at 14 kilohertz. What is the frequency and voltage when the filter has rolled-off 6 decibels?
3. A band-pass filter uses a 0.01-microfarad capacitor and a 33-millihenry choke. What is the center frequency for this filter? Use the resonant formula to determine the answer.
4. Draw the circuit and the response curve for the circuit in problem 3.
5. In order to bypass a 47-kilohm resistor at 20 kilohertz, what value of *C* is required?
6. A coupling capacitor is required to couple signals above 1 megahertz to a circuit with an input resistance of 1 kilohm. What value of *C* is required?

Critical Thinking

1. Could an 18-to 22-kilohertz notch filter be used in place of a band-pass filter if it were a series resonant type? Explain how this could work and what the effects would be.
2. What troubleshooting methods could be used to check a filter that was not operating as expected?

Answers to Review Quizzes

21-1
1. capacitors and inductors
2. DC power
3. configuration (or arrangement); value
4. resonant

21-2
1. high
2. series
3. parallel
4. L, T, and π

21-3
1. 10 kilohertz
2. wider
3. 70.7
4. voltage (or power)
5. doubled
6. one-half

21-4
1. 600 kilohertz (200 kilohertz – 800 kilohertz)
2. parallel
3. series
4. narrow

21-5
1. resonant
2. series; parallel
3. closer
4. resonant

21-6
1. notch; band-rejection
2. series; parallel
3. stop-band
4. resonant

21-7
1. high
2. high-pass; direct current
3. band-pass (or high-pass)

Chapter 22

Complex *L-C-R* Circuits

ABOUT THIS CHAPTER

In previous chapters you learned how to use phasor diagrams for analyzing series and parallel *L-C-R* circuits. This chapter discusses some of the mathematical tools that will (1) help you to understand *L-C-R* circuits bet-

After completing this chapter you will be able to:

- Describe the main components of polar and rectangular notation for AC phasor diagrams.
- Demonstrate how to multiply and divide phasors in polar form.
- Show how to add and subtract phasors in rectangular form.

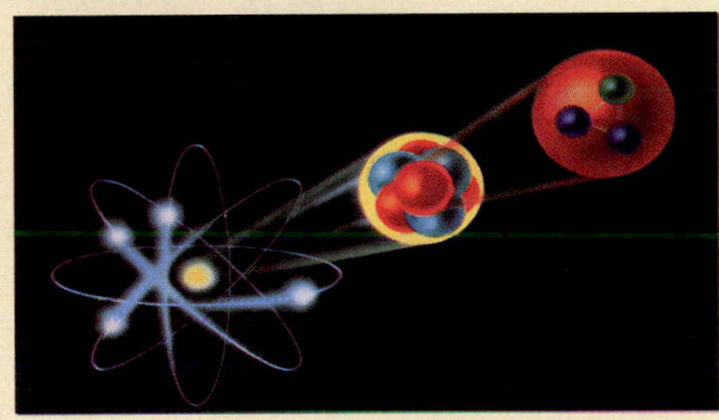

ter, (2) make it simpler for you to solve *L-C-R* circuits, and (3) make it possible for you to estimate voltages for components in complex combination *L-C-R* circuits that are impossible to analyze with purely graphical methods.

- Convert phasor notation from polar to rectangular form, and from rectangular to polar form.
- Solve series and parallel *L-C-R* circuits using phasor notation.
- Solve combination *L-C-R* circuits using phasor notation.

22-1 Introduction to Complex Numbers and Phasor Notation

Figure 22-1 shows some of the phasor diagrams with which you are already familiar. Figure 22-1(*a*) is an impedance diagram of a series *L-C-R* circuit. The resistance *R* runs along the horizontal axis. The overall reactance (which is inductive in this example) runs along the vertical axis. The combined impedance of the two elements is the hypotenuse of the right triangle formed by the resistive and reactive sides. Angle θ of this right triangle is the phase angle. Because the angle is positive (running counterclockwise from the horizontal axis), the impedance component is said to lead the resistive component.

Figure 22-1 (*b*) is the voltage phasor diagram for a series *L-C-R* circuit. The sides of the right triangle are the voltage across the resistive components, V_R, and the voltage across reactive components, V_X, which are capacitive in this example. The hypotenuse of the right triangle represents the voltage applied to the circuit. Angle θ indicates that the voltage phase of V_T lags the voltage phase of V_R.

Figure 22-1 (*c*) shows a current phasor diagram for a parallel *L-C-R* circuit. The sides of the right triangle represent the resistive and reactive currents, and the hypotenuse represents the combined current for the two components. (Because I_T is leading I_R in this example, it follows that the circuit is capacitive.)

Figure 22-1 does not show a current phasor diagram for a series *L-C-R* circuit because the current magnitude and phase are the same throughout. For the same reason, the figure does not show a voltage phasor diagram for parallel *L-C-R* circuits. Moreover there is no phasor diagram for impedance in a parallel *L-C-R* circuit because it is impossible to portray impedance in parallel *L-C-R* circuits by means of phasor diagrams. This is where a purely graphical (or phasor-diagram) analysis of *L-C-R* circuits begins to break down—and the place where other techniques described in this chapter begin to pay off.

The key to setting aside cumbersome phasor-diagram analysis is complex numbers. A **complex number** is a mathematical expression that completely defines a phasor diagram. Because of this close association of complex numbers and phasor diagrams, complex-number notation is often called **phasor notation**. Phasor diagrams can still be used to help clarify your analysis (which we will do throughout this chapter), but complex numbers (and phasor notation) are required to solve the simplest and most complicated types of *L-C-R* circuits with the necessary degree of accuracy.

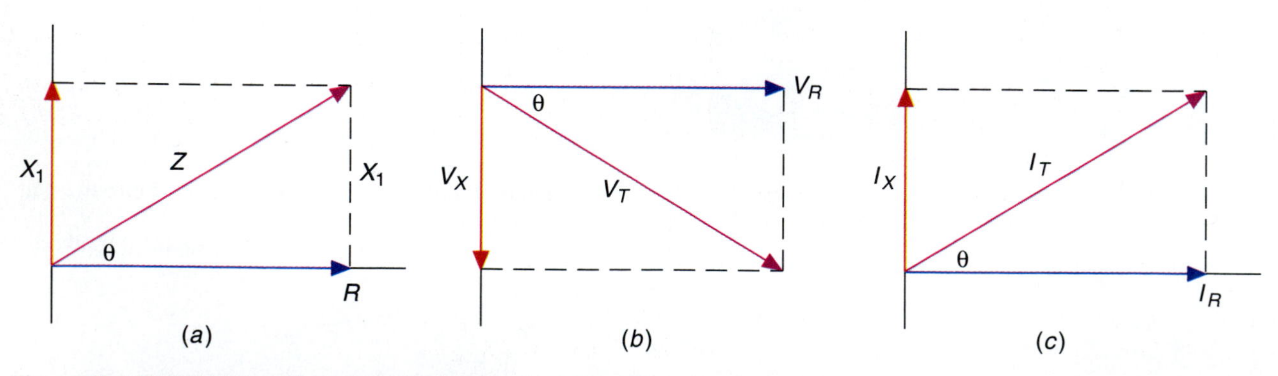

Figure 22-1 Typical phasor diagrams for *L-C-R* circuits. (*a*) Impedance in a series circuit. (*b*) Voltage in a series circuit. (*c*) Current in a parallel circuit.

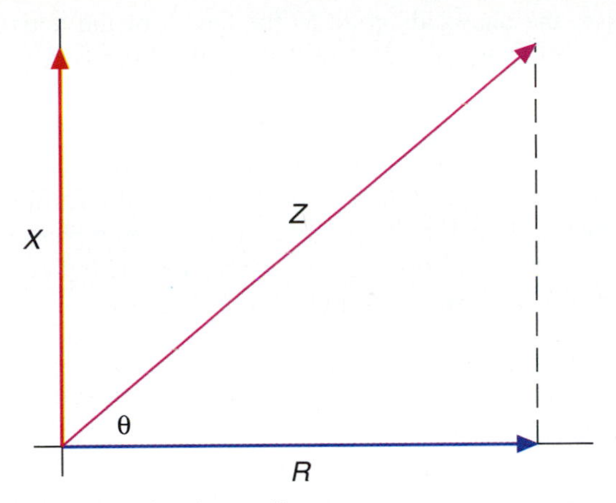

Figure 22-2 Basic elements of a phasor diagram.

There are two forms of phasor notation for analyzing *L-C-R* circuits. One is better suited for multiplication and division operations, whereas the other is preferred for addition and subtraction operations. Figure 22-2 shows how the two forms of phasor notation relate to a typical phasor diagram. Although Fig. 22-2 is an impedance diagram for a series *L-C-R* circuit, it could have been any of the three types illustrated in Fig. 22-1.

A phasor diagram is made up of three phasors and an angle. Complex-number notation takes advantage of the fact that you don't need all four of those elements in order to specify the triangle completely. In fact, you only need to know two of the values.

Suppose you are given only the length of the hypotenuse and the size of the phase angle (which are Z and θ, respectively). On the basis of that information alone, you could reconstruct the right triangle and determine the values of R and X. One of the forms of complex-number notation specifies a phasor diagram in terms of the length of its hypotenuse and its phase angle. This is known as the *polar form* of complex-number notation. The polar version of a phasor diagram is written as the length of the hypotenuse followed by the symbol $\angle$ and the angle. The polar expression in Fig. 22-2 is $Z\angle\theta$. The polar expression for the voltage phasor of a series circuit, shown in Fig. 22-1(*b*), is $V_T\angle-\theta$, and the polar expression for the current in a parallel circuit shown in Fig. 22-1 (*c*), is $I_T\angle\theta$.

Another way to specify a phasor diagram in terms of only two of its elements is by means of the **rectangular form** of complex-number notation. In the rectangular form, you are given the lengths and directions of the two sides. In Fig. 22-2, this would be the phasors for R and X. Given the length and direction of those two phasors, you could completely reconstruct the triangle. Rectangular notation expresses the phasor diagram as the length of the horizontal side, followed by a $+j$ or $-j$ operator, and then the length of the vertical side. The rectangular form of the phasor in Fig. 22-2 is $R + jX$. The j term is positive because the X phasor runs in the positive direction. Whenever the X term runs in the negative direction, the j term is negative. The phasor in Fig. 22-1(*b*), for example, would be written in rectangular form as $V_R - jV_X$. The term for the horizontal component is always plotted in the positive direction on a phasor diagram, so it follows that the first term in the rectangular notation is always positive. Therefore, it does not need a j-operator to indicate its direction.

Example 22-1: (*a*) What are the polar and rectangular forms of the phasor diagram in Fig. 22-3(*a*)?

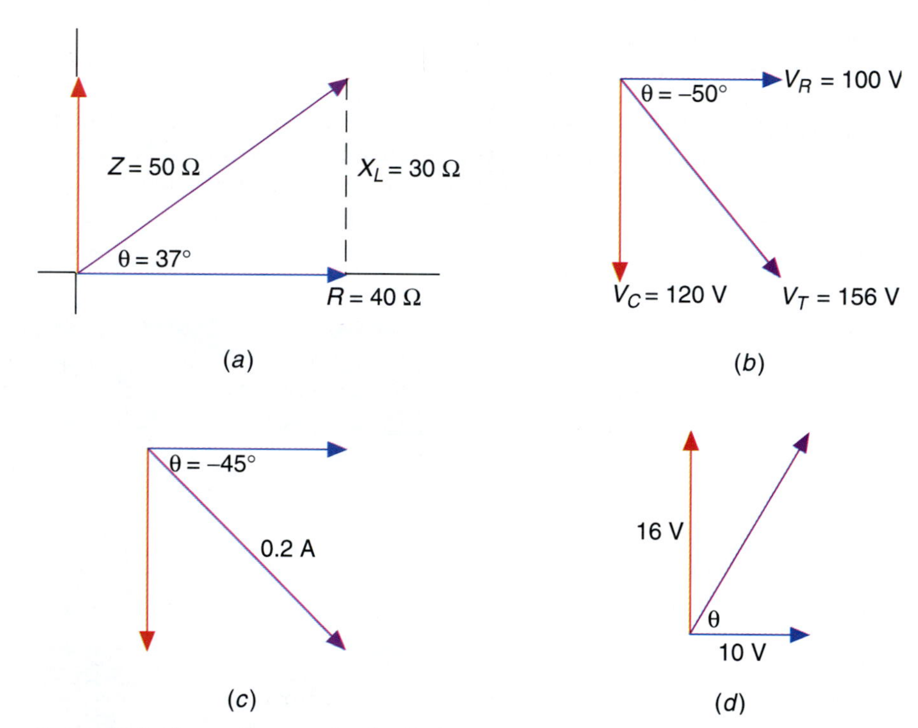

Figure 22-3 Phasor diagrams for Ex. 22-1.

Solution

The polar form is represented as $Z\angle\theta = 50\angle 37°\ \Omega$.

The rectangular form is represented as $R + jX_L = 40 + j30\ \Omega$.

(b) What are the polar and rectangular forms of the phasor diagram in Fig. 22-3(*b*)?

Solution

The polar form is $V_T\angle\theta = 156\angle -50°$ V.

The rectangular form is $V_R - j\,V_C = 100 - j120$ V.

(*c*) Show the phasor diagram for the polar expression $0.2\angle -45°$ A.

Solution

See Fig. 22-3(*c*).

(*d*) Plot the phasor diagram for the rectangular expression $10 + j16$ V.

Solution
See Fig. 22-3(*d*).

The two terms in the polar form of a complex number are called the **magnitude term** and the **angle term**. The two terms in the rectangular form of a complex number are called the **real term** and the **j-term**.

Multiplying and Dividing Polar Complex Numbers

Complex numbers in polar form are used where you must multiply or divide complex numbers. This is most often necessary in electronics when you want to solve Ohm's law for *L-C-R* circuits. For example, you can use the Ohm's law formula $V = IX$ to determine the voltage across a reactance, and $I = V/Z$ to find current in AC circuits *as long as all terms are in a phasor form.*

Suppose the current in a series circuit is given as $2\angle 20°$ A and the reactance of a certain component by $12\angle -90°\ \Omega$. Because these terms are in a phasor form (specifically as a polar complex number), you can multiply them to determine the voltage across the component:

$$V = (2\angle 20°\text{ A})(12\angle -90°\ \Omega) = 24\angle -70°\text{ V}$$

To determine the current through a parallel branch in an AC circuit with the voltage across the branch as $12\angle 0°$ V and the impedance of the branch as $20\angle 30°\ \Omega$, you can apply Ohm's law because the quantities are in a phasor form:

$$I = (12\angle 0°\text{ V})/(20\angle 30°\ \Omega) = 0.6\angle -30°\text{ V}$$

Now that you see the reason for multiplying and dividing phasors in polar form, let's look at a few simple rules.

1. To multiply phasors in polar form:
   - First multiply the magnitude terms.
   - Then sum the angle terms.

Equation 22-1 expresses this rule for multiplying two phasors in polar form where M is the magnitude and θ is the angle.

Equation 22-1

$$M_1\angle\theta_1 \times M_2\angle\theta_2 = (M_1 \times M_2)\angle(\theta_1 + \theta_2)$$

Example 22-2: (*a*) Multiply $4\angle 10°$ by $6\angle 12°$.

Solution

$$4\angle 10° \times 6\angle 12° = (4\times 6)\angle(10° + 12°) = 24\angle 22°$$

(*b*) $3\angle 18° \times 7\angle -14° = ?$

Solution

$$3\angle 18° \times 7\angle -14° = (3 \times 7)\angle[18° + (-14°)] = 21\angle 4°$$

Dividing two phasors in polar form is just about as easy as multiplying them. In fact the general procedure is almost identical.

2. To divide phasors in polar form:
   - First divide the magnitude terms.
   - Then subtract the angle terms.

Equation 22-2 expresses this rule for dividing two phasors in polar form, where M is the magnitude and θ is the angle.

Equation 22-2

$$M_1\angle\theta_1 \div M_2\angle\theta_2 = (M_1 \div M_2)\angle(\theta_1 - \theta_2)$$

Example 22-3: (*a*) Divide $12\angle 80°$ by $6\angle 45°$.

Solution

$$12\angle 80° \div 6\angle 45° = (12 \div 6)\angle(80° - 45°) = 2\angle 35°$$

(*b*) $(6\angle -18°)\ /\ (8\angle 14°) = ?$

Solution

$$\frac{(6\angle -18°)}{(8\angle 14°)} = \left(\frac{6}{8}\right)\angle(-18° - 14°) = 0.75\angle -32°$$

(*c*) Let $Z_T = V_A/I_T$, where $V_A = 12\angle 0°$ and $I_T = 0.2\angle 30°$. $Z_T = ?$

Solution

$$Z_T = \frac{V_A}{I_T} = \frac{12\angle 0°}{0.2\angle 30°} = \left(\frac{12}{0.2}\right)\angle(0° - 30°) = 60\angle -30°$$

Problem (*c*) in Example 22-3 actually solves Ohm's law for determining the total impedance of an *L-C-R* circuit, given the source voltage (12 volts of alternating current with no phase shift) and total current (0.2 amperes with a phase shift of 30°). The total impedance is 60 ohms and 30° lagging reactance. The circuit, in other words, is capacitive. Figure 22-4 illustrates the situation.

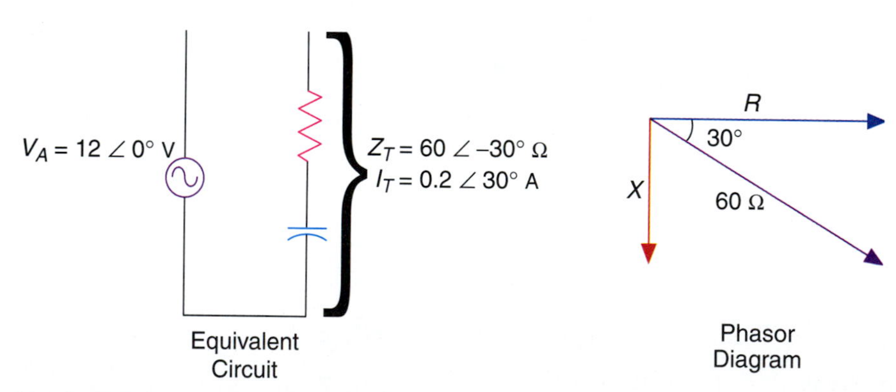

Figure 22-4 Equivalent circuit and phase diagram for Problem (*c*) of Example 22-3.

Adding and Subtracting Rectangular Complex Numbers

Just as you can apply Ohm's law to AC *L-C-R* circuits as long as the terms are in phasor form, you can apply Kirchhoff's voltage and current laws—as long as the terms are in phasor form. Recall that Kirchoff's voltage law for series circuits is:

$$V_T = V_1 + V_2 + V_3 + \dots$$

and Kirchhoff's current law for parallel circuits is:

$$I_T = I_1 + I_2 + I_3 + \dots$$

Solving Kirchhoff's laws for AC *L-C-R* circuits requires the addition and subtraction of phasors. This is usually done with phasors in rectangular form. Suppose you are given two voltage phasors for a series *L-C-R* circuit: $2 + j20$ V and $4 + j5$ V. The total voltage is the phasor sum of the two:

$$V_T = (2 + j20 \text{ V}) + (4 + j5 \text{ V}) = 6 + j25 \text{ V}$$

The total current for a parallel circuit with two branch currents of $(0.5 + j2 \text{ A})$ and $(1 - j \text{ A})$ would be

$$I_T = (0.5 + j2 \text{ A}) + (1 - j \text{ A}) = 1.5 + j \text{ A}$$

3. To add phasors in rectangular form:
   - First sum the real terms.
   - Then sum the *j*-terms.

Equation 22-3 expresses this rule, where S is the real term and jC is the *j*-term.

Equation 22-3

$$(S_1 + jC_1) + (S_2 + jC_2) = (S_1 + S_2) + j(C_1 + C_2)$$

STUDENT *to* **STUDENT**

It is possible to add and subtract phasors in polar form, but the steps are much more complicated than adding and subtracting phasors in rectangular form.

Example 22-4: (*a*) Add $3 + j5$ to $7 + j3$.

Solution

$$(3 + j5) + (7 + j3) = (3 + 7) + j(5 + 3) = 10 + j8$$

(*b*) $(2 + j4) + (4 - j3) = ?$

Solution

$$(2 + j4) + (4 - j3) = (2 + 4) + j(4 - 3) = 6 + j$$

(*c*) Combine $(6 + j2) + (2 - j5)$.

Solution

$$(6 + j2) + (2 - j5) = (6 + 2) + j(2 - 5) = 8 - j3$$

Subtraction of phasor terms becomes necessary when you are solving one of Kirchhoff's laws and the total value is known. If you have a parallel circuit with two complex *L-C-R* branches, and you know the total current and the current in one branch, you can use the form $I_2 = I_T - I_1$. The rule for subtracting rectangular forms of phasors is a simple variation of the rule for adding them. However, be careful about keeping track of the changes in signs.

4. To subtract phasors in rectangular form:
   - First subtract the real terms.
   - Then subtract the *j*-terms.

This rule is expressed in Eq. 22-4.

Equation 22-4

$$(S_1 + jC_1) - (S_2 + jC_2) = (S_1 - S_2) + j(C_1 - C_2)$$

Example 22-5: (*a*) Subtract $5 + j$ from $7 + j3$.

Solution

$$(7 + j3) - (5 + j) = (7 - 5) + j(3 - 1) = 2 + j2$$

(*b*) $(4 + j3) - (2 - j4) = ?$

Solution

$$(4 + j3) - (2 - j4) = (4 - 2) + j[3 - (-4)] = 2 + j(3 + 4) = 2 + j7$$

Converting Between Polar and Rectangular Forms

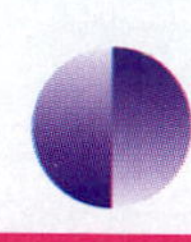

STUDENT to STUDENT

If your phasors are in polar form and you have to add or subtract them, first convert them to their rectangular form. If your phasors are in rectangular form and you have to multiply or divide them, first convert them to polar form.

You have seen how to use the polar form to multiply and divide phasors, and how to use the rectangular form to add and subtract phasors. In the process of analyzing complex *L-C-R* circuits, you have to multiply or divide at some steps along the way and add or subtract at other steps. Thus it becomes necessary to convert between polar and rectangular forms.

The polar form of a complex number, $M\angle\theta$, can be converted to the corresponding rectangular form, $S + jC$, by means of the conversion shown in Eq. 22-5.

Equation 22-5

$$S + jC = M\cos\theta + jM\sin\theta$$

This equation can be solved in two basic steps:

1. Determine the S term by multiplying M by the cosine of angle θ.
2. Determine the $\pm jC$ term by multiplying M by the sine of angle θ.

Figure 22-5 shows how this conversion works. The polar form shows the length of the hypotenuse (M) and the phase angle (θ). The rectangular form shows the lengths and directions of the sides (S and $\pm jC$). Equation 22-5 is taken from basic trigonometry where $S = M\cos\theta$ and $jC = M\sin\theta$.

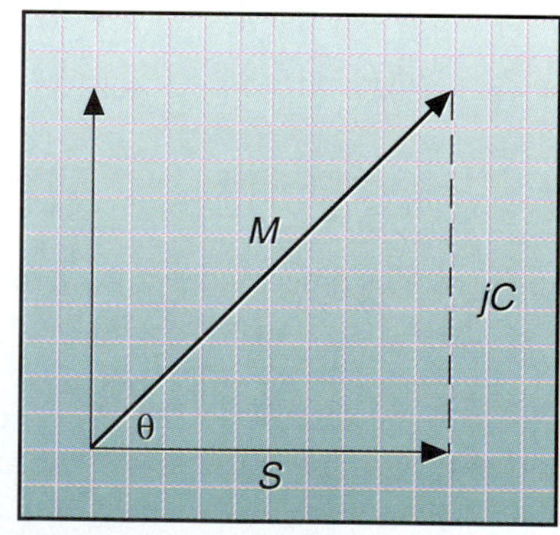

Figure 22-5 Conversion of polar and rectangular forms.

Example 22-6: (*a*) Convert $4\angle 10°$ to rectangular form.

Solution

$$S = M\cos\theta = 4 \times \cos 10° = 4 \times 0.98 = 3.9$$
$$C = M\sin\theta = 4 \times \sin 10° = 4 \times 0.17 = 0.68$$

Therefore, $4\angle 10° = 3.9 + j0.68$.

(*b*) Convert $5\angle -45°$ to rectangular form.

$$S = M\cos\theta = 5 \times \cos -45° = 5 \times 0.71 = 3.6$$
$$C = M\sin\theta = 5 \times \sin - 45° = 5 \times -0.71 = -3.6$$

Therefore, $5\angle 45° = 3.6 - j3.6$.

Equation 22-6 shows how the rectangular form of a phasor, $S + jC$, can be converted to the polar form, $H\angle\theta$.

Equation 22-6

$$M\angle\theta = \sqrt{S^2 + C^2}\angle\arctan\left(\frac{C}{S}\right)$$

This equation is executed in two steps:

1. Determine M by applying the Pythagorean theorem to rectangular values S and C.
2. Determine θ by taking the arctangent (sometimes called the inverse tangent and shown as $\tan^{-1}$) of the ratio C/S.

You can see in Fig. 22-5 that M is the hypotenuse of a triangle that is composed of sides having lengths C and S. This is why the Pythagorean theorem is used for determining the value of M in polar notation. Basic trigonometry will also show you that $\tan\theta = C/S$; the inverse of that function yields the angle: $\arctan(C/S) = \theta$.

STUDENT *to* STUDENT.

You don't have to be an expert mathematician to work these equations if you have a scientific or engineering calculator. Those calculators have keys for all the trigonometry functions mentioned in this book.

Example 22-7: (*a*) Convert $2 + j4$ to polar form.

Solution

$$M = \sqrt{S^2 + C^2} = \sqrt{2^2 + 4^2} = \sqrt{20} = 4.5$$
$$\theta = \arctan\left(\frac{C}{S}\right) = \arctan\left(\frac{4}{2}\right) = 63°$$

Therefore, $2 + j4 = 4.5\angle 63°$.

(*b*) Convert $6 - j3$ to polar form.

Solution

$$M = \sqrt{S^2 + C^2} = \sqrt{6^2 + 3^2} = \sqrt{45} = 6.7$$
$$\theta = \arctan\left(\frac{C}{S}\right) = -\arctan\left(\frac{3}{6}\right) = -27°$$

Therefore, $6 - j3 = 6.7\angle -27°$.

REVIEW QUIZ 22-1

1. True or false? A single complex number can completely represent the phasor diagram for an L-C-R circuit.
2. The magnitude term in a polar complex number represents the ______________ of a right triangle.
3. The angle term in a polar complex number represents the ______________ of a right triangle.
4. The real term in a rectangular complex number represents the ______________ of a right triangle.

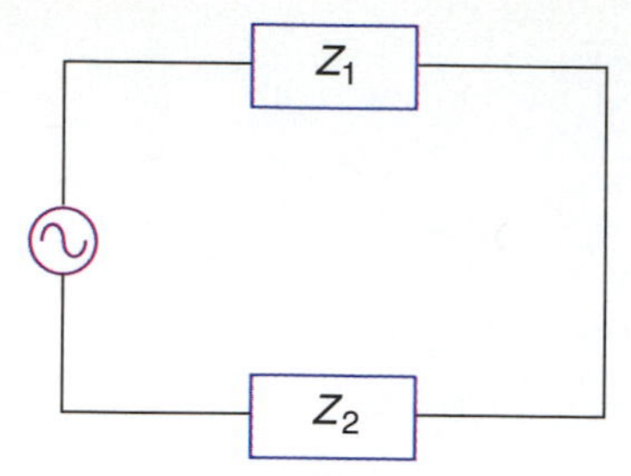

Figure 22-6 Basic series *L-C-R* circuit.

22-2 Series L-C-R Circuits

Figure 22-6 represents a generic series *L-C-R* circuit. It is composed of two impedances, Z_1 and Z_2, and an AC voltage source, V_T. A series *L-C-R* circuit could have one of these general impedances or more than two. Each of the impedance boxes could be made up of something as simple as a single resistor, capacitor, or inductor, or each could be a series/parallel combination of *L-R-C* devices. No matter what is included in the impedance boxes, three general rules of electronics apply to the circuit—as long as the values are represented in phasor form:

1. The total impedance of a series *L-C-R* circuit is equal to the sum of the individual impedances. In equation form:

$$Z_T = Z_1 + Z_2$$

2. Current flow is the same through every impedance in the circuit:

$$I_T = I_1 = I_2$$

3. The total voltage in a series circuit is equal to the sum of the voltage across the individual impedances (Kirchhoff's voltage law):

$$V_T = V_1 + V_2$$

STUDENT to STUDENT

It is important to remember that these general rules for a series *L-C-R* circuit apply only when every term is specified in phasor form, either *polar* or *rectangular* form.

When solving series *L-C-R* circuits, you usually know the two impedance phasors and the amount of applied voltage. Your objective, then, is to determine the total impedance, the total current, and the voltage drop across each impedance. Refer to the simple series circuit in Fig. 22-6. Suppose you are given the values of Z_1, Z_2, and V_T. How would you determine Z_T, I_T, V_1, and V_2? The general approach to solving the circuit is as follows:

1. Use the equation $Z_T = Z_1 + Z_2$ to determine the total impedance. Because this is a summation operation, the expressions for Z_1 and Z_2 must be in rectangular form.
2. Use Ohm's law ($I_T = V_T / Z_T$) to determine the total current. Because this is a division operation, V_T and Z_T must be in their polar form.
3. Solve for V_1 and V_2 using the equations $V_1 = I_T Z_1$ and $V_2 = I_T Z_2$. Because the I_T and Z terms are multiplied, they must be available in polar form.
4. Check your work by applying Kirchhoff's voltage, $V_T = V_1 + V_2$. Because this is a summation operation, V_1 and V_2 must be in rectangular form. You know your work is correct if the value you find for V_T very nearly matches the value supplied in the original data.

Example 22-8: Solve the circuit in Fig. 22-6, letting $V_T = 20$ V, $Z_1 = 4.5 - j10\ \Omega$, and $Z_2 = 3.2 + j4\ \Omega$.

Solution

1. Determine the total impedance.

$$\begin{aligned} Z_T &= Z_1 + Z_2 = 4.5 - j10° + 3.2 + j4° \\ &= 7.7 - j6\ \Omega \\ &= 9.76\angle -37.9°\ \Omega \end{aligned}$$

2. Determine the total current.

$$\begin{aligned} I_T &= V_T / Z_T = \frac{20\angle 0°}{9.76\angle -37.9°} \\ &= 2\angle 37.9°\ \text{A} \\ &= 1.58 + j1.23\ \text{A} \end{aligned}$$

3. Determine the voltage drops across the individual impedances.

$$V_1 = I_T Z_1 = (2\angle 37.9°)(10.9\angle -65.8°)$$
$$= 21.8\angle - 27.9° \text{ V}$$
$$= 19.3 - j10.2 \text{ V}$$
$$V_2 = I_T Z_2 = (2\angle 37.9°)(5.12\angle 51.3°)$$
$$= 10.2\angle 89.2° \text{ V}$$
$$= 0.14 + j10.2 \text{ V}$$

Check by:

$$V_T = V_1 + V_2 = 19.3 - j10.2 + 0.14 + j10.2 = 19.4 \text{ V}$$

Because 19.4 V is close to the original $V_T = 20$ V, the calculations are correct.

Component-Level Series *L-C-R* Circuits

Often when students are studying complex *L-C-R* circuits for the first time, they lose track of the practical side of the matter. This is the time to relate the theory material to actual components in a circuit. Figure 22-7 shows a basic series *L-C-R* circuit that contains both types of reactances as well as a pure resistance. Take note of the fact that this circuit uses phasor notation. The reactance of a pure inductor is $+ jX_L$ or $X_L\angle 90°$, the reactance of a pure capacitor is $-jX_C$ or $X_C\angle -90°$, and the resistance of a pure resistor is $R + j0$ or $R\angle 0°$.

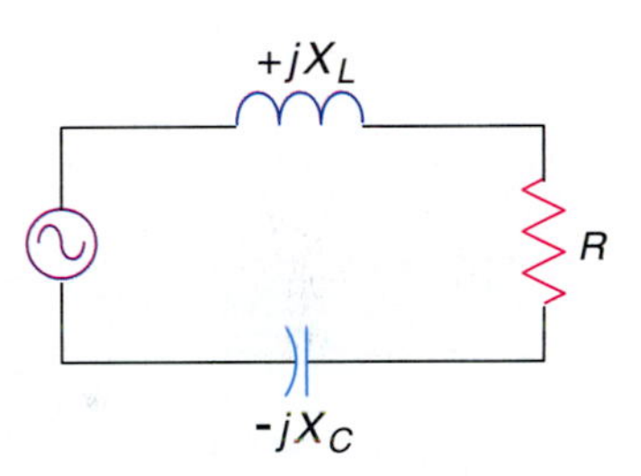

Figure 22-7 A component-level representation of an *L-C-R* circuit.

Solving a series circuit of this type usually begins with your knowing the values of V_T, X_L, R, and X_C. Solving the circuit requires finding the total impedance, the total current, and the voltage dropped across each component. As noted earlier for general series *L-C-R* circuits, you check your work by summing the voltage phasors to make sure the result is equal to the total voltage.

Example 22-9: Solve the circuit in Fig. 22-7, letting $V_T = 12$ V, $Z_L = +j20\ \Omega$, $Z_R = 5\ \Omega$, and $Z_C = -j15\ \Omega$.

Solution

1. Determine the total impedance.

$$Z_T = Z_L + Z_R + Z_C = +j20 + 5 - j15$$
$$= 5 + j5$$
$$= 7.1\angle 45°\ \Omega$$

2. Determine the total current.

$$I_T = V_T / Z_T = 12\angle 0° / 7.1\angle 45°$$
$$= 1.7\angle -45° \text{ A}$$
$$= 1.2 - j1.2 \text{ A}$$

3. Determine the voltage drops across the individual impedances.

$$V_L = I_T Z_L = (1.7\angle -45°)(20\angle 90°)$$
$$= 34\angle 45°$$
$$= 24 + j24$$
$$V_R = I_T Z_R = (1.7\angle -45°)(5\angle 0°)$$
$$= 8.5\angle -45°$$
$$= 6 - j6$$

$$V_C = I_T Z_C = (1.7\angle -45°)(15\angle -90°)$$
$$= 25.5\angle -135°$$
$$= -18 - j18$$

Check by:

$$V_T = V_L + V_R + V_C = 24 + j24 + 6 - j6 - 18 - j18$$
$$= 12 + j0$$

Because this exactly matches the given value of V_T, the calculations are correct.

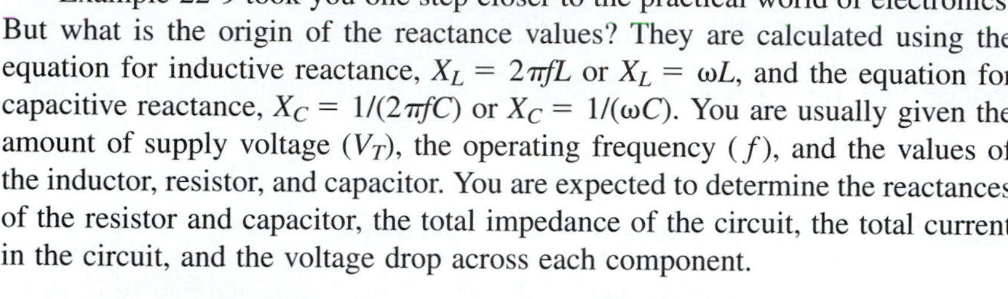

Example 22-9 took you one step closer to the practical world of electronics. But what is the origin of the reactance values? They are calculated using the equation for inductive reactance, $X_L = 2\pi fL$ or $X_L = \omega L$, and the equation for capacitive reactance, $X_C = 1/(2\pi fC)$ or $X_C = 1/(\omega C)$. You are usually given the amount of supply voltage (V_T), the operating frequency (f), and the values of the inductor, resistor, and capacitor. You are expected to determine the reactances of the resistor and capacitor, the total impedance of the circuit, the total current in the circuit, and the voltage drop across each component.

STUDENT *to* STUDENT

You can solve the circuit in Example 22-10 without resorting to polar and rectangular phasor notation. In a previous chapter, you solved circuits graphically. Now, however, you are trying to develop a skill that will make it easier for you to work with complex analyses that cannot be handled by purely graphical means.

Example 22-10: For the series *L-C-R* circuit in Fig. 22-7, let $V_T = 20$ V, $f = 100$ kHz, $C = 0.05$ μF, $R = 100$ Ω, and $L = 60$ μH. Use phasor notation to determine the total impedance, total current, and the voltage across each component.

Solution

1. Determine the inductive reactance.

$$X_L = 2\pi fL = 2\pi(100 \text{ kHz})(60 \text{ μH}) = 37.7 \ \Omega$$
$$Z_L = +j37.7 \ \Omega = 37.7\angle 90° \ \Omega$$

2. Determine the capacitive reactance.

$$X_C = \frac{1}{2\pi fC} = \frac{1}{2\pi(100 \text{ kHz})(0.05 \text{ μF})} = 31.8 \ \Omega$$
$$Z_C = -j31.8 \ \Omega = 31.8\angle -90° \ \Omega$$

3. Determine the total impedance.

$$Z_T = Z_L + Z_R + Z_C$$
$$= +j37.7 + 100 - j31.8$$
$$= 100 + j5.9$$
$$= 100.2\angle 3.38° \ \Omega$$

4. Determine the total current.

$$I_T = \frac{V_T}{Z_T} = \frac{20\angle 0°}{100\angle 3.38°}$$
$$= 0.2\angle -3.38° \text{ A}$$
$$= 0.2 - j0.01 \text{ A}$$

5. Determine the voltage drops across the individual impedances.

$$V_L = I_T Z_L = (0.2\angle -3.38°)(37.7\angle 90°)$$
$$= 7.54\angle 86.6°$$
$$= 0.45 + j7.53 \text{ V}$$
$$V_R = I_T Z_R = (0.2\angle -3.38°)(100\angle 0°)$$
$$= 20\angle -3.38°$$
$$= 20 - j1.18 \text{ V}$$

$$V_C = I_T Z_C = (0.2\angle -3.38°)(31.8\angle -90°)$$
$$= 6.36\angle -93.4°$$
$$= -\ 0.38 - j6.35 \text{ V}$$

Check by:

$$V_T = V_L + V_R + V_C = 0.45 + j7.53 + 20 - j1.18 - 0.38 - j6.35$$
$$= 20 \text{ V}$$

Because this matches the original value of V_T, the calculations must be correct.

Practical Inductors

Every inductor, or coil, includes at least a little internal resistance (r_i); no inductor is perfect. The quality (Q) of a coil is determined by the ratio of the inductor's reactance to its internal resistance, X_L/r_i.

The internal resistance of a coil in a series *L-C-R* circuit is sometimes large enough to play a significant role in determining the amount of phase shift the inductor causes. The actual impedance of an inductor can be expressed in rectangular form as $r_i + jX_L$, or $r_i + j\mu L$.

Example 22-11: What is the total impedance of the circuit in Fig. 22-8 if $X_C = -j30\ \Omega$, $X_L = +j40\ \Omega$, and $r_i = 10\ \Omega$.

Solution

The total impedance is given by:

$$Z_T = X_C + X_L + r_i = -\ j30 + j40 + 10$$
$$= 10 + j10\ \Omega$$
$$= 14.1\angle 45°$$

Electrically speaking, the circuit can be replaced with 10 ohms of pure resistance in series with a pure inductance having a reactance of 10 ohms.

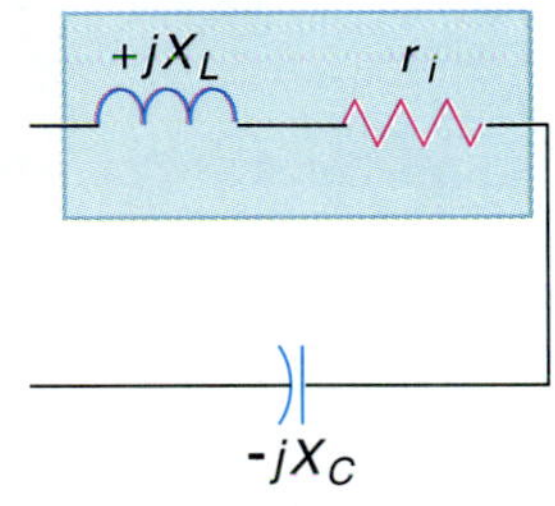

Figure 22-8 A series *L-C-R* circuit where the internal resistance of the inductor is taken into account.

REVIEW QUIZ 22-2

1. The equation $Z_T = Z_1 + Z_2 + Z_3 + \ldots$ determines the total ______________ of a series AC circuit, provided the terms are expressed in ______________ notation.
2. Kirchhoff's voltage law for series AC circuits states that $V_T =$ ______________, provided the terms are expressed in ______________ notation.
3. The expression $I_T = I_1 = I_2 = I_3 = \ldots$ means that the current in an AC circuit is ______________, provided the terms are expressed in ______________ notation.
4. Ohm's law for AC circuits has the general form $V_T = I_T Z_T$. If you are given phasors for I_T and Z_T, you would use the ______________ form of the complex number to do the multiplication.
5. To sum the voltages in a series AC circuit, use the ______________ form of the voltage phasors.

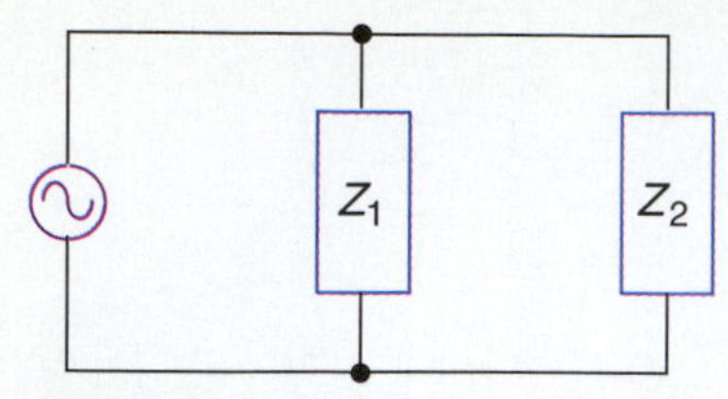

Figure 22-9 Basic parallel *L-C-R* circuit.

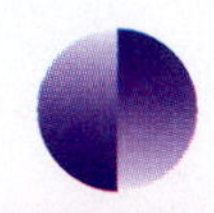

STUDENT *to* STUDENT

Don't forget that these equations apply to AC circuits, but only when every term is specified in a polar or rectangular phasor form.

22-3 Parallel L-C-R Circuits

Figure 22-9 shows the simplest possible parallel *L-C-R* circuit. There can be more branches than shown in this example, but not fewer. As with the impedance boxes in the generic series circuit, the impedance boxes in parallel circuits might contain a single component or any sort of series-parallel *L-C-R* circuit. For the time being, however, we will continue to treat them as simple impedance boxes.

The general rules for parallel circuits are:

1. The total current in a parallel *L-C-R* circuit is equal to the sum of the currents in the individual branches (Kirchhoff's current law):
$$I_T = I_1 + I_2$$
2. The voltage is the same across each branch and equal to the total voltage applied to a parallel circuit:
$$V_T = V_1 = V_2$$
3. The total impedance of a parallel circuit has the basic form:
$$\frac{1}{Z_T} = \frac{1}{Z_1} + \frac{1}{Z_2}$$

When you are expected to solve a parallel *L-C-R* circuit, you are generally given the amount of supply voltage and the phasors for all the impedances. Given that information, you can determine the current through each branch, the total current, and the total impedance. Suppose you are given the values of Z_1, Z_2, and V_T for the parallel circuit in Fig. 22-9. The task is to determine Z_T, I_T, V_1, and V_2. The general approach to solving the circuit is as follows:

1. Solve for the individual branch currents using the equations $I_1 = V_T/Z_1$ and $I_2 = V_T/Z_2$. Because the V_T and Z terms are divided, they must be available in polar form.
2. Use the equation $I_T = I_1 + I_2$ to determine the total current. Because this is a summation operation, the expressions for I_1 and I_2 must be in rectangular form.
3. Use Ohm's law ($Z_T = V_T/I_T$) to determine the total impedance. Because this is a division operation, V_T and I_T must be in polar form.
4. Check your work by applying an alternate procedure for total impedance. If there are just two impedances in parallel, you can use the product-over-sum equation $Z_T = Z_1 \| Z_2$. Or you can use $1/Z_T = 1/Z_1 + 1/Z_2 + 1/Z_3 + \ldots$. Both procedures require combinations of polar and rectangular phasors. The total impedance you find with one of these methods should match the value you find when dividing total voltage by total current.

Example 22-12: For the circuit in Fig. 22-9, determine the current through each branch, the total current, and the total impedance if $Z_1 = 4 + j2\ \Omega$, $Z_2 = 10 - j6\ \Omega$, and $V_T = 12$ V.

Solution

1. Solve for the individual branch currents.
$$I_1 = \frac{V_T}{Z_1} = \frac{12\angle 0^\circ}{4.47\angle 26.6^\circ}$$
$$= 2.68\angle -26.6^\circ \text{ A}$$
$$= 2.4 - j1.2 \text{ A}$$

$$I_2 = \frac{V_T}{Z_2} = \frac{12\angle 0°}{11.7\angle -31°}$$
$$= 1.02\angle 31° \text{ A}$$
$$= 0.87 + j0.53 \text{ A}$$

2. Determine the total current.

$$I_T = I_1 + I_2 = 2.4 - j1.2 + 0.87 + j0.53$$
$$= 3.27 - j0.7 \text{ A}$$
$$= 3.34\angle -12.1° \text{ A}$$

3. Determine the total impedance.

$$Z_T = \frac{V_T}{I_T} = \frac{12\angle 0°}{3.34\angle -12.1°}$$
$$= 3.59\angle 12.1° \ \Omega$$
$$= 3.51 + j0.75 \ \Omega$$

Check your work:

$$Z_T = Z_1 \| Z_2 = \frac{(4.47\angle 26.6°)(11.7\angle -31°)}{4 + j2 + 10 - j6}$$
$$= \frac{52.3\angle -4.4°}{14 - j4} = \frac{52.3\angle -4.4°}{14.6\angle -15.9°}$$
$$= 3.58\angle 11.5° \ \Omega$$

The check is very close to the total impedance calculated in the previous step by a different method.

Using Admittance and Susceptance

The total resistance of DC resistive circuits can be determined by the equation:

$$\frac{1}{R_T} = \frac{1}{R_1} + \frac{1}{R_2} + \frac{1}{R_3} + \ldots$$

Because this formula is often awkward to use, the notion of conductance, $G = 1/R$, becomes important. Using conductance instead of resistance in a parallel resistive circuit, the formula simplifies to:

$$G_T = G_1 + G_2 + G_3 + \ldots$$

Similar terms are available for use in AC parallel *L-C-R* circuits. As long as impedance is expressed in phasor form, the following equation is perfectly valid:

$$\frac{1}{Z_T} = \frac{1}{Z_1} + \frac{1}{Z_2} + \frac{1}{Z_3} + \ldots$$

Just as conductance (G) can replace the $1/R$ terms, **admittance** (Y) can replace the $1/Z$ terms in AC circuits. Thus, total admittance of a parallel *L-C-R* circuit can be expressed as shown in Eq. 22-7.

Equation 22-7

$$Y_T = Y_1 + Y_2 + Y_3 + \ldots$$

Impedance expresses a circuit's opposition to AC current, whereas admittance expresses a circuit's ability to pass AC current. Impedance, as you already know, is measured in ohms. Admittance, like conductance, is expressed in mhos or siemens (S).

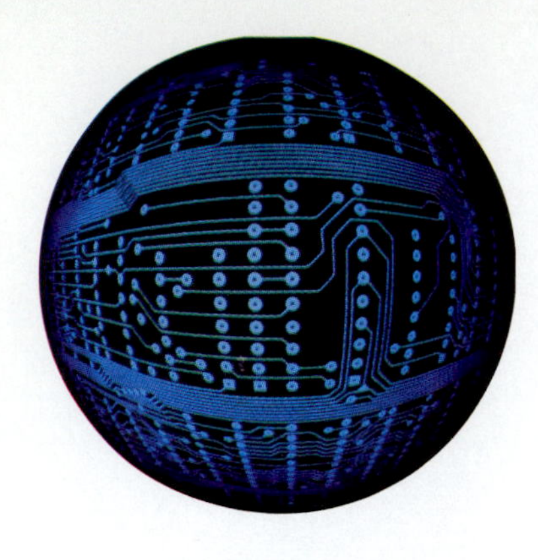

Example 22-13: What is the total admittance of a parallel circuit having two branches, where $Z_1 = 10 + j2$ and $Z_2 = 5 - j4$?

Solution

1. Convert to polar forms.

$$Z_1 = 10 + j2 = 10.2\angle 11.3°$$
$$Z_2 = 5 - j4 = 6.4\angle -38.7°$$

2. Convert to admittances.

$$Y_1 = \frac{1}{Z_1} = \frac{1}{10.2\angle 11.3°}$$
$$= 0.098\angle -11.3° \text{ mhos}$$
$$= 0.096 - j0.019 \text{ mhos}$$
$$Y_2 = \frac{1}{Z_2} = \frac{1}{6.4\angle -38.7°}$$
$$= 0.15\angle 38.7° \text{ mhos}$$
$$= 0.12 + j0.094 \text{ mhos}$$

3. Use Eq. 22-7 to solve for the total admittance:

$$Y_T = Y_1 + Y_2 = 0.096 - j0.019 + 0.12 + j0.094$$
$$= 0.22 + j0.075 \text{ mhos}$$

Conductance is the inverse of resistance for AC and DC circuits, and admittance is the inverse of impedance for AC circuits. We have to account for one more term—reactance in AC circuits. Reactance (X) is the opposition of current flow offered by an inductor or capacitor in an AC circuit. The inverse of reactance is **susceptance** (B). The relationship between reactance and susceptance is a simple one:

$$B = \frac{1}{X}$$

Susceptance is measured in mhos or siemens.

So far in this chapter, you have been working with the notion that the phasor form of an impedance is expressed as:

$$R \pm jX \quad \text{(rectangular form)}$$

where $+j$ = an inductive X
$-j$ = a capacitive X

$$Z\angle \pm \theta \quad \text{(polar form)}$$

where a positive θ denotes an inductive circuit
a negative θ indicates a capacitive circuit

The admittance form of that outline is:

$$R \pm jB \quad \text{(rectangular form)}$$

where $+j$ = a capacitive X
$-j$ = an inductive X

$$Y\angle \pm \theta \quad \text{(polar form)}$$

where a positive θ denotes a capacitive circuit
a negative θ indicates an inductive circuit

Admittance notation is often used when analyzing the action of transistors and transistor amplifier circuits.

Leaky Capacitors

A pure capacitor allows no current to flow through it, and its resistance to electron flow is purely reactive, $-jX_C$ or $-j/(\omega C)$. When a capacitor begins to show the defect of leakage, however, resistive current is allowed to flow through it. The equivalent circuit for a leaky capacitor is a capacitor in parallel with a resistor (Fig. 22-10).

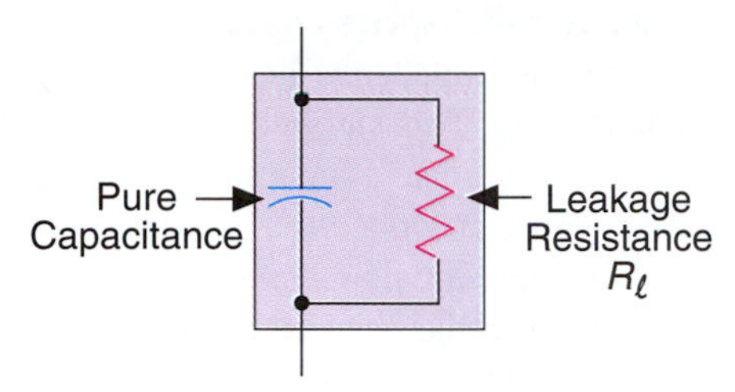

Figure 22-10 Equivalent circuit for a leaky capacitor.

For a good capacitor, the value of leakage resistance R_ℓ is so much greater than the reactance of the capacitor X_C that the leakage current is negligible. Leakage becomes a significant factor when the amount of leakage resistance is less than a hundred times the amount of capacitive reactance.

Example 22-14: (*a*) What is the impedance of the leaky capacitor circuit (Fig. 22-10) when the leakage is very small: $X_C = 100\ \Omega$ and $R_\ell = 100\ \text{M}\Omega$?

Solution
Since the value of R_ℓ is more than a hundred times greater than X_C, estimate the total impedance to be equal to X_C: $-j100$ or $100\angle-90°$.

(*b*) What is the impedance of the leaky capacitor circuit when the capacitor deteriorates to the point where $X_C = 100\ \Omega$ and $R_\ell = 1\ \text{k}\Omega$?

Solution
Using the product-over-sum equation, $Z_T = X_C \| R_\ell$, the total impedance is $99.5\angle-84.3°\ \Omega$.

REVIEW QUIZ 22-3

1. The equation $I_T = I_1 + I_2 + I_3 + \ldots$ determines the total ______________ of a series AC circuit, provided the terms are expressed in ______________ notation.
2. Admittance in an AC circuit is equal to 1 over the ______________. The equation for this relationship is ______________ .
3. The mathematical symbol for susceptance in an AC circuit is ______________ . The equation that relates susceptance to reactance is ______________ .
4. The two acceptable units of measure for conductance, admittance, and susceptance are ______________ and ______________ .
5. The product-over-sum expression for the total impedance of two impedances in parallel applies to AC circuits as long as the impedances are expressed in ______________ notation.

22-4 Combination L-C-R Circuits

The same methods and equations you have already learned for solving combination series/parallel resistor circuits can also be used for solving complex *L-C-R* AC circuits. You simply have to remember to use phasor notation for the values.

Figure 22-11 (*a*) represents a complex AC circuit which is essentially a parallel circuit that happens to have a series of two impedances in one of the branches. This is sometimes called a **parallel-series circuit**. The general approach to solving such circuits is to combine impedance boxes a step at a time until you are left with a single impedance box. This single box represents the total impedance of the circuit. From there you can use Ohm's law to determine the total current through the circuit. Then you can work your way backwards through the diagrams, finding currents in the individual branches and, finally, the voltage dropped across each of the impedances in the original circuit.

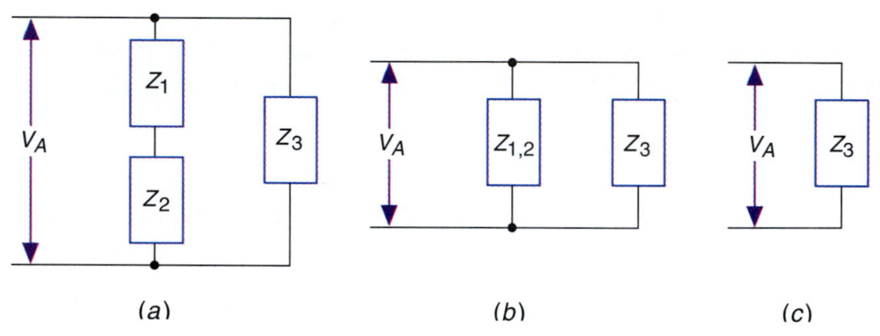

Figure 22-11 A parallel-series AC circuit.

Example 22-15: For the circuit in Fig. 22-11, suppose you are given V_A, Z_1, Z_2, and Z_3. Find the total impedance, total current, the current through each impedance, and the voltage across each impedance.

Solution

1. Use the equation $Z_{1,2} = Z_1 + Z_2$ to combine series impedances Z_1 and Z_2. See Fig. 22-11(*b*).
2. Use the equation $Z_T = Z_{1,2} \| Z_3$ to combine parallel impedances $Z_{1,2}$ and Z_3. See Fig. 22-11(*c*). This will give you the total impedance of the circuit.
3. Use the equation $I_T = Z_T / V_A$ to determine the total current for the circuit.
4. Use the equation $V_A = V_{Z_{1,2}} = V_{Z_3}$ to determine the voltage for impedance Z_3. See the parallel circuit in Fig. 22-11(*b*).
5. Use the equation $I_{Z_3} = V_{Z_3} / Z_3$ to determine the current through Z_3. See Fig. 22-11 (*c*).
6. Use the equation $I_{Z_{1,2}} = V_{Z_{1,2}} / Z_{1,2}$ to determine the current through $Z_{1,2}$. See Fig. 22-11(*c*).
7. Use the equation $I_{Z_1} = I_{Z_2} = I_{Z_{1,2}}$ to determine the current through impedances Z_1 and Z_2. See the series branch in Fig. 22-11(*a*). This will give you the current through Z_3.
8. Use the equation $V_{Z_1} = I_{Z1} Z_1$ to determine the voltage across impedance Z_1, and use $V_{Z2} = I_{Z_2} Z_2$ to determine the voltage across impedance Z_2. See Fig. 22-11(*a*).

This completes the solution of the parallel-series circuit in Fig. 22-11. Reviewing the eight steps, you can see that the procedure determines everything that can be known about the circuit.

Total voltage V_T (given)	Total impedance Z_T (step 2)
Total current I_T (step 3)	Impedance Z_1 (given)
Impedance Z_2 (given)	Impedance Z_3 (given)
Voltage across Z_1 (step 8)	Voltage across Z_2 (step 8)
Voltage across Z_3 (step 4)	Current through Z_1 (step 7)
Current through Z_3 (step 7)	Current through Z_3 (step 5)

The circuit in Fig. 22-12 represents a second class of combination *L-C-R* circuits. This one is essentially a series circuit that has a set of parallel impedances. It is called a **series-parallel circuit**.

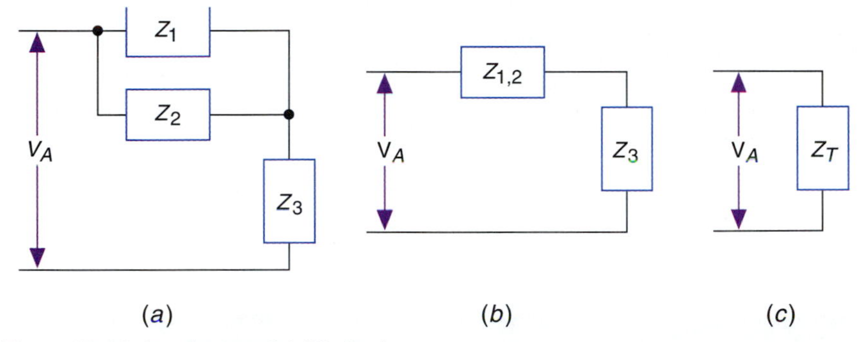

Figure 22-12 A series-parallel AC circuit.

The general approach to solving this type of circuit begins by combining impedance boxes until only one remains, as shown in Fig. 22-12(*c*). That box represents the total impedance of the circuit. From there you can easily determine the total current through the circuit if you are given the total voltage. Knowing the total current allows you to determine the distribution of voltages through the series circuit [Fig. 22-12(*b*)], as well as the currents through the parallel section of the original circuit.

Example 22-16: Given the values of V_A, Z_1, Z_2, and Z_3 in Fig. 22-12(*a*), determine the total impedance, the total current, the current through each impedance, and the voltage across each impedance.

Solution

1. Use the equation $Z_{1,2} = Z_1 \| Z_2$ to combine parallel impedances Z_1 and Z_2. See Fig. 22-12(*b*).
2. Use the equation $Z_T = Z_{1,2} + Z_3$ to combine series impedances $Z_{1,2}$ and Z_3. See Fig. 22-12(*c*).
3. Use the equation $I_T = Z_T / V_A$ to determine the total current for the circuit.
4. Use the equation $I_T = I_{Z_{1,2}} = I_{Z_3}$ to determine the current for impedance Z_3. See the series circuit in Fig. 22-12(*b*).
5. Use the equation $V_{Z_3} = I_{Z_3} Z_3$ to determine the voltage across impedance Z_3.

6. Use the equation $V_{Z_{1,2}} = I_{Z_{1,2}} Z_{1,2}$ to determine the voltage across $Z_{1,2}$. See Fig. 22-12(*b*).
7. Use the equation $Z_{1,2} = Z_1 = Z_2$ to determine the voltage across impedance Z_1 and across impedance Z_2. See the parallel circuit in Fig. 22-12 (*b*).
8. Use the equation $I_{Z_1} = V_{Z_1}/Z_1$ to determine the current through impedance Z_1, and use $I_{Z_2} = V_{Z_2}/Z_2$ to determine the current through impedance Z_2. See Fig. 22-12(*a*).

Here is what you know about this circuit:

Total voltage V_T (given)

Total impedance Z_T (step 2)

Total current I_T (step 3)

Impedance Z_1 (given)

Impedance Z_2 (given)

Impedance Z_3 (given)

Voltage across Z_1 (step 7)

Voltage across Z_2 (step 7)

Voltage across Z_3 (step 5)

Current through Z_1 (step 8)

Current through Z_2 (step 8)

Current through Z_3 (step 4)

REVIEW QUIZ 22-4

1. Which form(s) of phasor notation are most useful for determining the impedance of a series branch of a parallel-series AC circuit? ______________
2. You would use $Z_1||Z_2$ to determine the combined ______________ of two impedances connected in ______________ with one another.
3. If you are given the phasors for two or more impedances connected in parallel, which form(s) of phasor notation are most useful for determining their total impedance? ______________
4. $+jX$ represents the ______________ form of a ______________ reactance. The corresponding polar form is ______________.

SUMMARY

The main purpose of phasor, or complex-number, notation is to make it possible for you to use all the principles you have learned for DC resistor circuits and apply them directly to *L-C-R* AC circuits. The laws, rules, and methods of DC circuit analysis carry over directly to AC circuits as long as you remember to express all AC quantities in phasor form.

There are two forms of phasor notation: polar and rectangular. Use polar notation when you have to multiply or divide phasors, and use rectangular notation for addition and subtraction. This means you have to be able to convert from one form to the other in order to solve practical problems that include addition, subtraction, multiplication, and division. (It is possible to do all math operations with one form of notation or the other, but the procedures are far more cumbersome than the processes for converting from one form to the other, as required.)

Describing a circuit in terms of phasor or complex-number notation conveys a great deal of information. For instance, if the total impedance of an AC circuit is described as $20\angle 10°$ and $19.7 + j3.47$, you can see that (1) the impedance is 20 Ω, (2) the DC resistance is 19.7 Ω, (3) the reactance is 3.47 Ω, (4) the phase angle is 10°, and (5) the circuit is inductive.

NEW VOCABULARY

admittance
angle term
complex number
j-term
magnitude term
parallel-series circuit
phasor notation
polar form
real term
rectangular form
series-parallel circuit
susceptance

Questions

1. Two elements of a phasor diagram for the total impedance of a series *L-C-R* circuit are directly portrayed in a polar complex number. Which two elements are represented?
2. Two elements of a phasor diagram for the total impedance of a series *L-C-R* circuit are directly portrayed in a rectangular complex number. Which two elements are represented?

3. Under what conditions should you apply Kirchhoff's voltage law for AC circuits? Which form of phasor notation is more appropriate for dealing with this law?
4. Under what conditions should you use Kirchhoff's current law for AC circuits? Which form of phasor notation is more appropriate?
5. What is the AC version of conductance?
6. What happens to the direction of a phase angle when its phasor is divided into 1?
7. How do you indicate a lagging voltage with polar notation? Rectangular notation?
8. How do you indicate a leading current with polar notation? Rectangular notation?
9. Where are you likely to find references to admittance in your future studies of electronics?
10. True or false—Any AC circuit can be reduced to a single impedance?

Problems

1. Calculate: $12\angle 15° \times 0.25\angle 45°$
2. Calculate: $6\angle 45° \div 1.25\angle -45°$
3. Calculate: $(2 + j12) + (32 - j15)$
4. Calculate: $(4 - j6) - (15 + j2)$
5. Convert $12\angle 15°$ to rectangular form.
6. Convert $8 - j$ to polar form.
7. Write the polar and rectangular forms of a voltage phasor where the phase angle is $+45°$.
8. Solve for the total impedance of a series circuit where the impedances are described as $2 + j4\ \Omega$ and $5 + j3\ \Omega$.
9. Determine the total impedance of a circuit where the total voltage is described as $20\angle -15°$ V and the total current as $0.5\angle 5°$ A.
10. Solve for the total impedance of the following impedances connected in parallel: $15\angle 60°\ \Omega$ and $6\angle -40°\ \Omega$.

Critical Thinking

1. What happens to the impedance phase angle as a circuit becomes more inductive?
2. What happens to the voltage phase angle across a device as it becomes more inductive?
3. When an impedance is composed of a pure inductance in series with a pure resistance, what happens to the current phase angle as the resistance increases?
4. A certain impedance shows the voltage lagging the current. What should be done to the resistive component in order to reduce the amount of lag?
5. How much reactance is present in a series *L-C-R* circuit that is operating at its resonant frequency?
6. What is the admittance of a parallel *L-C-R* circuit that is operating at its resonant frequency?

Answers to Review Quizzes

22-1

1. true
2. hypotenuse
3. angle
4. horizontal side

22-2

1. impedance; phasor (or complex-number)
2. $V_1 + V_2 + V_3 + \ldots$; phasor (or complex-number)
3. the same through each impedance, phasor (or complex-number)
4. polar
5. rectangular

22-3

1. current; phasor (or complex-number)
2. impedance; $Y = 1/Z$
3. B; $B = 1/X$
4. mhos; siemens
5. phasor (or complex-number)

2-4

1. rectangular
2. impedance; parallel
3. both polar and rectangular
4. rectangular; inductive; $X\angle+\theta$ or $0 + jX$

Chapter 23

Diodes *and* Common Diode Circuits

ABOUT THIS CHAPTER

Building from the earlier chapters that discussed circuits and devices, this chapter expands on the concept of doping and the creation of a junction. Here, the diode is presented as a device that allows current to flow in only one direction. The use of diodes as half-wave and full-wave rectifiers in power supplies is discussed. Other basic diode circuits, like the clipper and the clamper, are also introduced. Zener diodes and the use of diodes in voltage multiplier circuits are described to further complement the material.

After completing this chapter you will be able to:

- Recognize diodes and common diode circuits.
- Describe how diodes operate and how they function in common circuits.

The study of diodes is ongoing in the electrical engineering world. Faster, more powerful and reliable, and less expensive diodes are always being sought because diodes are used in power supplies, mixers, receivers, and a variety of circuits that are critical to modern communication and technology.

Because solid state diodes are made from semiconductors such as silicon and germanium, they provide the ability to control the flow of electrons along a path of solid material rather than in a vacuum tube. As semiconductors, diodes are small, capable of operating at high frequencies, and based on the formation of a junction of two specially treated materials.

▼ Apply basic theory to practical schematics and circuit problems.

23-1 Diode P-N Junction Theory

Unlike resistors, capacitors, and inductors (passive components), diodes change their behavior depending on the voltage or current applied to them. They can do this because they are semiconductors. The name **semiconductor** means that devices, like diodes and transistors, conduct current somewhere between full and no conduction, that is, semiconduction. This makes them useful for conducting current only when certain conditions exist.

Semiconductor Material

As semiconductor devices, diodes consist of two materials that form a junction. Typically, the material is either germanium or silicon. The most common type of diode, used in power supplies, is made of silicon material. But in either case, the diode is formed by a junction that allows current to flow only under certain conditions.

As discussed earlier in Chapter 5, the outer orbit of an atom of copper contains valence electrons that can easily free themselves from their orbit. Because they move easily, they are good conductors. Other materials, like glass and paper, have no free (valence) electrons. Unless these materials are raised to extreme temperatures, no electrons leave their orbit. That is why such materials make good insulators. There are some materials, however, that have only a few valence electrons, in between conductors and insulators, and they form the basis of semiconductors. These materials are made by a process called **doping** (see Fig. 23-1).

STUDENT *to* STUDENT

The scientific abbreviation for Germanium is Ge and for silicon it's Si.

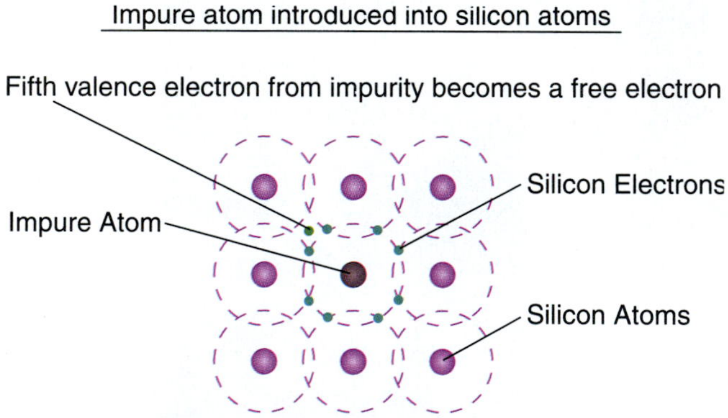

Impure atom introduced into silicon atoms

Fifth valence electron from impurity becomes a free electron

Impure Atom

Silicon Electrons

Silicon Atoms

Figure 23-1 Doping makes semiconductor material.

Semiconductor Doping

Both silicon and germanium have four valence electrons and both semiconductors can be grown as pure crystals in a laboratory. If, however, a very tiny amount of an impurity such as arsenic or phosphorus is introduced during the manufacturing process, a special crystal is formed and is called N material. The N means that the material is altered to be more negative; it has more free electrons. The process can also be used to make a material more positive—P material—using an impurity such as gallium or indium. The process of making a material with more or less charge than the original material is called doping.

Here is how doping works. Because phosphorus has five valence electrons, the result of doping silicon with it is a crystal that has an occasional atom of five valence electrons. Although this one atom may be surrounded by a million silicon atoms, it has an unusual effect upon the resulting crystal material. The rare occurrence of the phosphorus atom makes the silicon a much better conductor under certain conditions. In fact, when a voltage (static field) is applied across the doped material, it acts more like a conductor than a semiconductor because of the extra doping atom. Conversely, doping silicon with a material such as boron, which has three valence electrons, produces P material, resulting in a crystal with an occasional extra hole into which electrons can move.

In general, doped silicon is more resistive than germanium, which is doped with arsenic (As) to become the N material or gallium (Ga) to become the P material. Silicon is used in power supply diodes because its greater resistance can withstand greater temperatures. Germanium, however, is still used, and another newer combination, gallium arsenide, is a favorite semiconductor for use in modern microwave devices that operate at extremely high frequencies.

P-N Junctions

Figure 23-2 shows a P-N junction silicon diode and a *P-N junction*. Notice the area where some of the negative electrons have been attracted across the boundary to the P material side. This area is called the **junction barrier**. It is also the **depletion region**. It is like a zone which has been neutralized and which has become a barrier because it has been depleted of any charges.

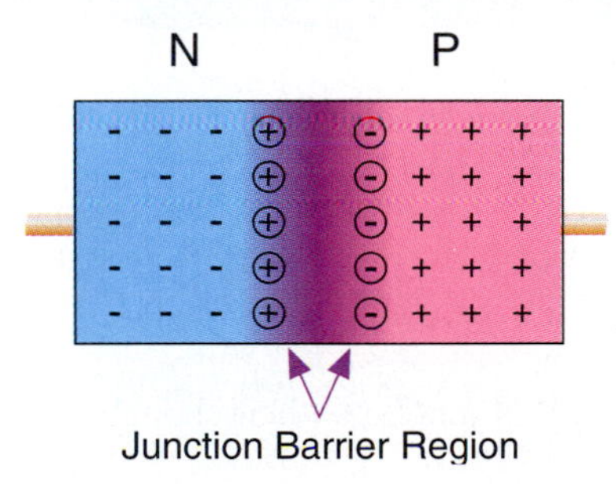

Figure 23-2 P-N junction.

The P-N junction is the basis of the diode. By growing the N and P materials together, a junction is formed where they meet. The result is that some of the free electrons in the N material migrate over to the holes in the P material, and a **barrier potential** is formed. The barrier is somewhat stable; no further flow of electrons results unless the barrier is overcome or broken down by an applied voltage or potential difference. In the case of germanium, about 0.3 volt will break down the barrier, and about 0.6 or 0.7 volt will break down a silicon-based P-N junction. The *exact* value of voltage is not important for most applications; the exact value varies anyway, depending upon the manufacturer. For example, a diode that can be overcome by 0.68 volt can be replaced by one that has a forward voltage drop of 0.71 volt if you have to replace one on a circuit board.

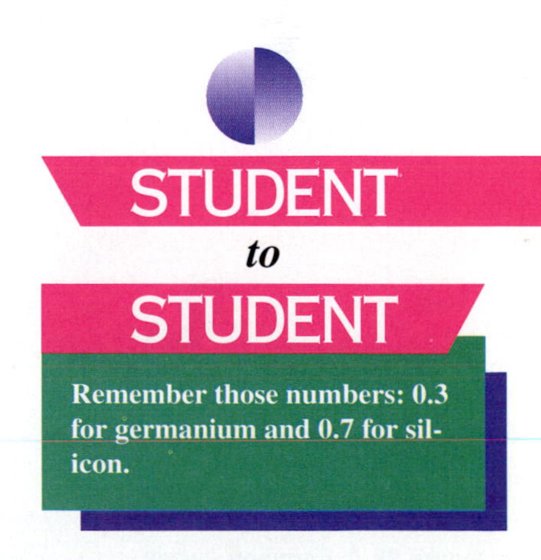

When enough voltage, called a bias voltage, is applied across a P-N junction, the barrier breaks down and current flows through the semi-conductor material as if it were a conductor. The only voltage that remains dropped across the

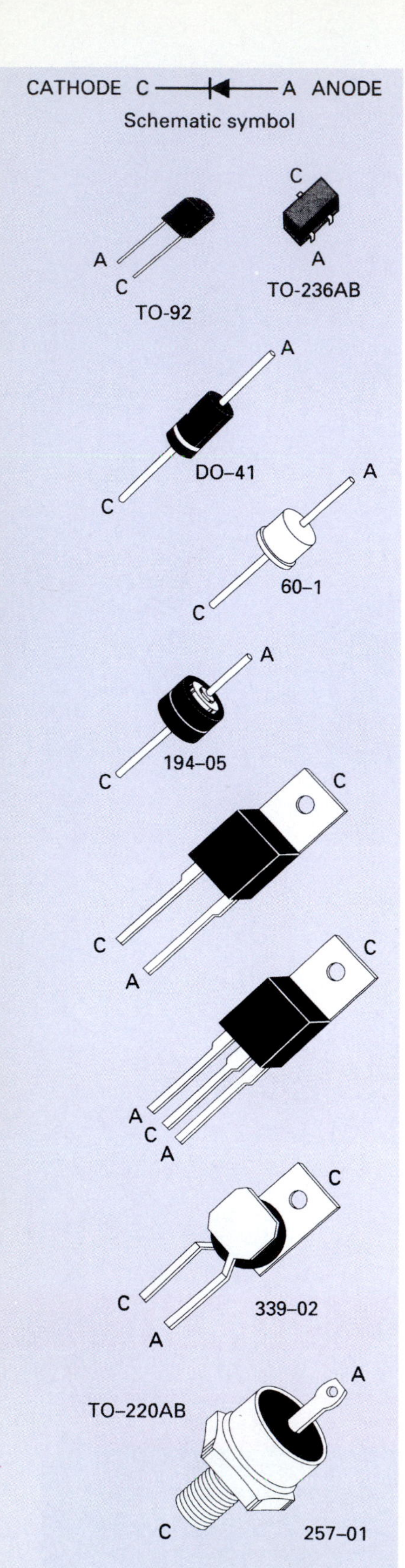

Figure 23-4 Common diodes.

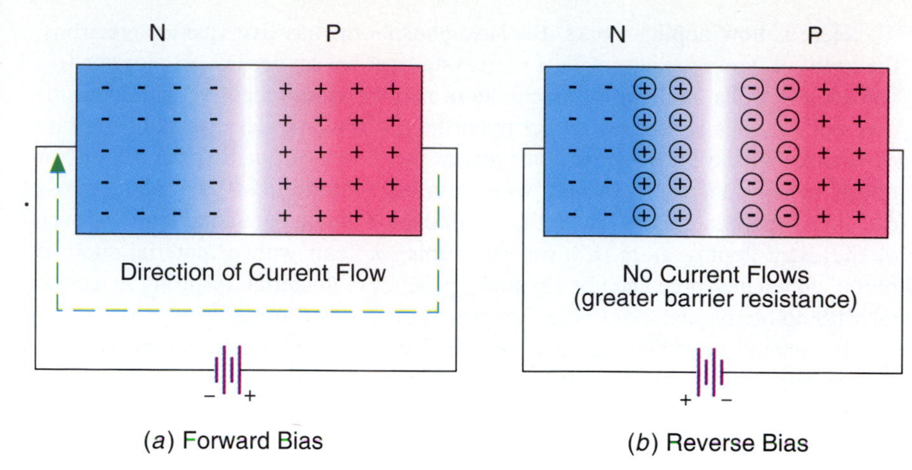

Figure 23-3 Bias voltage on a P-N junction.

diode is the voltage required to break down the barrier. Thus, the barrier has an I_R voltage drop similar to any resistance. If the applied voltage is not the correct polarity, however, the barrier will not break down, and no current will flow. Figure 23-3 shows how both forward bias and reverse bias affect the barrier junction of a silicon diode.

REVIEW QUIZ 23-1

1. _______________ and _______________ are common semiconductor materials.
2. Semiconductors are _______________ with impurities to allow greater conduction.
3. When semiconductor materials have _______________ charges, they form a P-N junction.
4. Current flows in a P-N junction when enough _______________ voltage is applied.

23-2 Types of Diodes

Diodes come in many types and sizes, and they have a variety of uses. Although diodes differ, they all share some common characteristics. For example, the positive side of a diode is called the **anode**, and the negative side is called the **cathode**. In order to connect the diode in the forward bias position, the cathode must be on the negative side of the voltage source and the anode must be on the positive. The result of forward-biasing most diodes is that the diode has little resistance in the forward direction; this allows the maximum amount of circuit current to flow. When most diodes are reverse biased, there is a very high resistance and no current flows, except for some small amount of leakage current in the microamp range. There are some special diodes manufactured to operate in the reverse direction. Of course, if enough voltage is applied in either direction, the diode can be destroyed. Figure 23-4 shows some typical diodes.

Rectifier Diodes

Rectifier diodes are the most common diodes used throughout the electronics industry. They are not special diodes, but they are called **rectifier** diodes because they are most often used to rectify (change) alternating current into direct current, mostly in power supplies. Therefore, most technicians refer to common silicon diodes as rectifier diodes.

STUDENT to STUDENT

Remember that the anode is positive and the cathode is negative.

Rectifier diodes have an anode and a cathode and, therefore, two leads. Because they are made of silicon P-N junctions, they operate in the forward direction when about 0.7 volt is applied across them. Because silicon diodes are used in power supplies, their voltage and power ratings are important. Technicians often check the diode ratings on the manufacturer's specification sheet to be sure that the diode is capable of handling the applied voltage and power.

The most common information that you should remember about a silicon diode is its characteristic curve. That curve shows the forward and reverse bias voltages and the resulting current you can expect when the diode is operating. Figure 23-5 shows the forward and reverse bias characteristics for a typical rectifier diode, including the schematic symbol and the leads. Notice that in the forward direction, the rectifier diode turns on at about 0.7 volt and current begins to flow. Any increase in applied voltage has little effect upon current flow. In fact, only that area around the **knee** voltage is very sensitive. Later, when you study transistors, you will see that the knee voltage area of a P-N junction is the key to controlling current in a transistor. As you might imagine, even very slight changes in voltage (between 0.5 and 0.8 volt) can result in great changes in current flow. Increasing the applied voltage (above the manufacturer's rating) can eventually destroy the diode.

Figure 23-5 Typical rectifier diode curve and symbol.

If the diode is reverse biased, no current will flow in the circuit. Figure 23-6 on the next page shows a reverse biased diode. If increasing voltage is applied to the diode in the reverse direction, the diode will eventually reach and exceed its rated breakdown voltage. This will result in maximum current flowing through the diode, causing it to burn up. For this reason, do not exceed the reverse-bias ratings of a diode.

In the reverse direction, as voltage is applied, no current flows. A few microamps of leakage current, however, may flow until so much voltage is applied that it breaks down the diode, destroying the barrier, and current **avalanches** like a short circuit. Of course, if a diode is breaking down in a real circuit, it may begin to melt, and smoke may appear.

Example 23-1: Imagine a curve, similar to the one in Fig. 23-5, that shows 10 milliamps of current flowing when the diode is forward biased at exactly 0.6 volt.

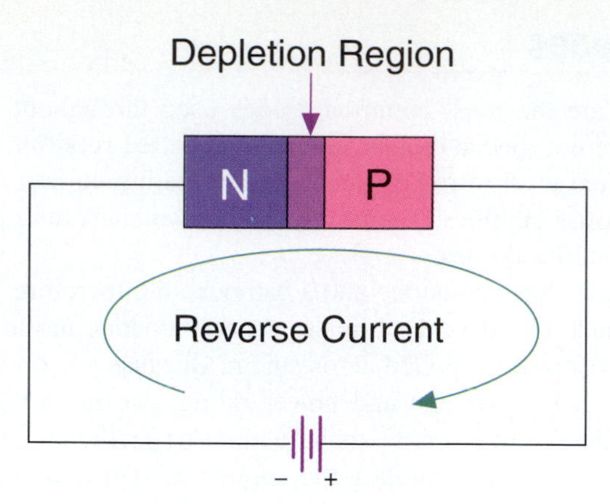

Figure 23-6 Reverse-biased diode.

If the curve shows a slope where 10 milliamps of current flows for each tenth (0.1) of a volt above 0.6 volt forward bias, how much current will flow when 0.75 volt are applied in the forward direction? Why is this important?

With just 0.15 volts more applied, 0.75 volt forward bias instead of 0.6 volt, a total of 25 milliamps of current flows through the diode instead of 10 milliamps. This is important because it means that very slight increases in forward bias voltage may result in large increases in current.

Figure 23-7 shows two silicon rectifier diodes, of opposite polarities, connected across a DC power supply. The diodes are in parallel but they are also in series with a load resistor. Because the reverse-biased diode D_1 is effectively an open circuit, no current passes through the load resistor ($R_{load,\ 1}$). Diode, D_2, however, is forward biased, and the I_R voltage drop across the $R_{load,\ 2}$ must obey the rules of Ohm's law.

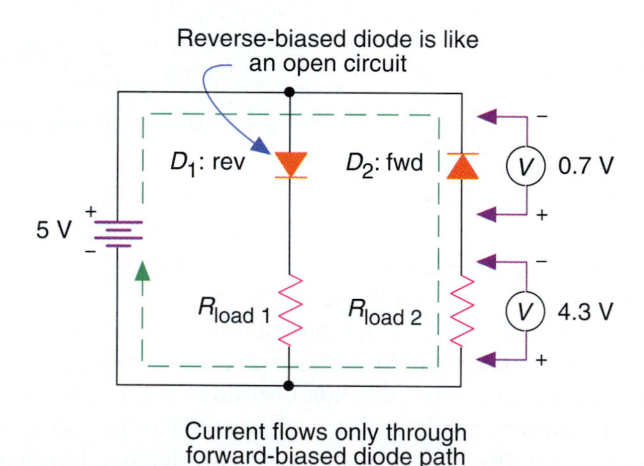

Figure 23-7 Direct current applied to rectifier diodes.

Applying alternating current to a rectifier diode results in only one-half of the alternation passing through the diode after it reaches a peak value of 0.7 volt. In other words, a 2-volt peak-to-peak sine wave would only have current passing through the diode in one direction, with 0.7 volt dropped across the diode and the rest dropped across any load.

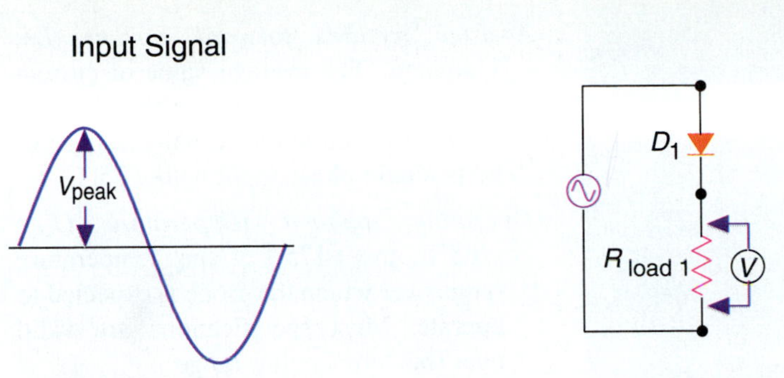

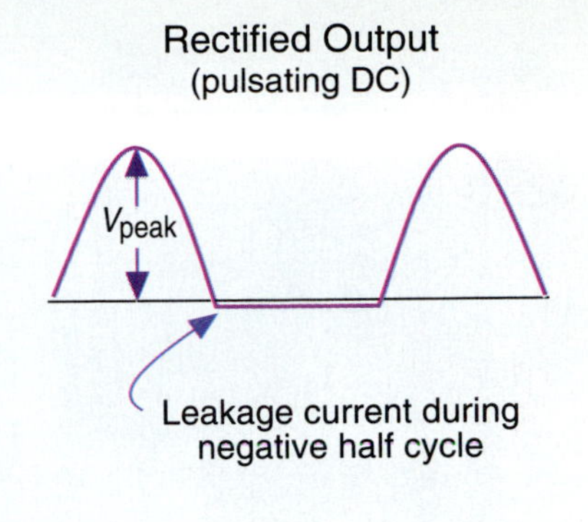

Figure 23-8 AC input and positive rectified output.

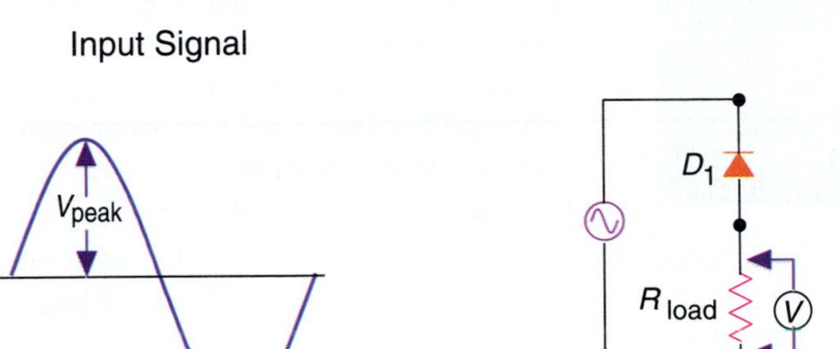

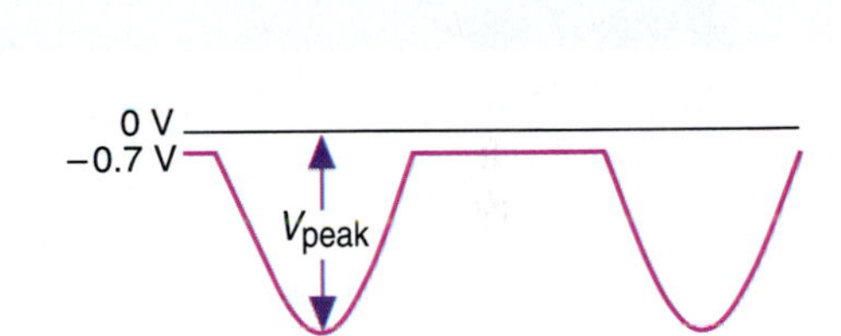

Figure 23-9 AC input and negative recitfied output.

Figures 23-8 and 23-9 show a diode connected in series with a signal generator and a resistor. Notice that the output waveform, across the load resistor, shows only one-half of the AC signal. This means that current flows only in one direction because the diode is rectifying the AC signal. The result is a DC signal that is really a peak value commonly called **pulsating DC**. Also, outputs are shown, depending on how the diode is connected in the circuit.

Rectifier Diode Specifications

Depending on the type of diode being used, the specifications will vary. You can learn a lot about components from reading and asking questions about the specifications on a data sheet. In fact, data sheets contain the results of laborious testing and are invaluable for repairing circuits where component specifications are critical. There are hundreds of books and manuals with component specifications, but the manufacturer's data book is always the best to consult.

For the type of typical rectifier diodes that technicians see in power supplies and use in lab experiments, there are some data sheet parameters with which you may want to be familiar. Some of the specifications for a typical 1N4001 rectifier diode are as follows.

Peak reverse voltage (V_{RM}): 50 volts. The maximum value of peak voltage that can be applied in the reverse direction across the diode.

Steady state reverse voltage (V_R): 50 volts (25°C to 75°C). The maximum value of reverse voltage that can be continually applied before breakdown.

Average rectified forward current (I_o): 1 ampere. The average value of current that can be passed through the diode in the forward direction, usually using 60-hertz single phase input below 75°C.

Operating ambient temperature (T_A): −65°C to +175°C. The temperature range over which the diode is expected to operate. Most specifications are valid over this temperature range.

Peak forward surge current ($I_{FM\ surge}$): 30 amperes. The maximum value of current that can be passed over a giv-en time period. This current is always much greater than I_o because it is the current expected to surge at the time when a circuit is first switched on.

Power (P): about 1 watt. Power dissipation is not always specified directly for diodes when the current and voltage ratings are given; however, 1N4001 diodes are usually used for rectification where about 1 watt is the typical value.

Light-Emitting Diodes

Today many display circuits use light-emitting diodes (LEDs). An LED is similar to a rectifier diode because it also has a P-N junction. You have probably seen many LEDs—they look like little light bulbs (red, green, yellow) with two leads. Also, the schematic symbol is like a rectifier diode, but it has two arrows pointing away from the diode to indicate light. The difference is that an LED emits light when it is properly biased because of the materials it is made of—especially phosphorus. In fact, depending on the materials, LEDs emit different colors. As the free electrons change energy levels across the junction, they dissipate energy by radiating light rather than heat. Also, an LED can switch on and off at a much faster rate than a rectifier diode. You have probably seen LEDs on hand-held calculators and as lights on computers, especially showing disk drives operating.

Zener Diodes

Zener diodes are specifically manufactured to operate and be used in the reverse direction. In the forward direction, they operate like rectifier diodes because of the forward bias barrier potential of about 0.7 volts. Zener diodes reverse breakdown characteristics are different because they are manufactured to breakdown at a specific voltage, also called the zener voltage. In other words, a zener diode is meant to be connected in the reverse direction so that electron flow goes into the anode.

For example, a 5-volt zener will conduct only when it has 5 volts across it in the reverse direction. Below 5 volts, no current will flow. When the zener reaches the breakdown voltage, any increase in applied voltage will *not* result in an increased voltage drop across the zener, but there will be an increase in current. Because of their ability to keep a constant voltage drop across themselves, regardless of current flow, zeners are used as regulators. That is, any component in parallel with a zener will be regulated by the voltage across the zener. Figure 23-10 shows the characteristic curve for a 5-volt zener diode and its schematic symbol.

Because zener diodes are more heavily doped than rectifier diodes, they are also more sensitive to extreme temperatures. In fact, a zener diode with a breakdown voltage below 5 volts has a negative **temperature coefficient.** In other words, the breakdown voltage decreases with increased temperature. Zeners above 5 volts, however, have a positive temperature coefficient, and their breakdown voltages actually increase slightly as temperature rises.

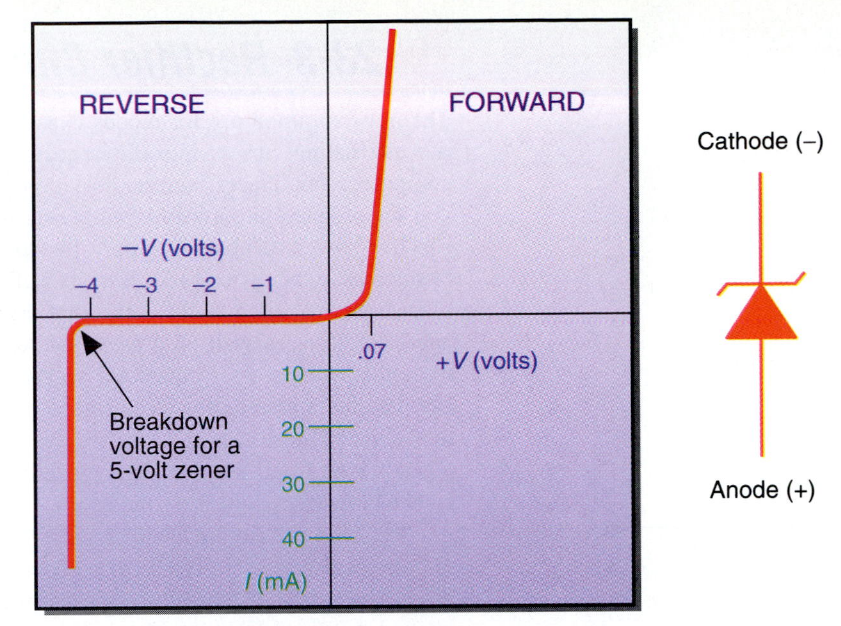

Figure 23-10 Zener diode curve and symbol.

Example 23-2: How is a zener diode used to regulate the amount of voltage across a load? As shown in Fig. 23-11, the zener diode is connected in series with the source voltage and a series resistor. The zener is also connected in parallel with the load resistor. The output across the load resistor (V_{out}) is equal to the value of the zener diode. In other words, as long as the zener is operating within its normal reverse mode, the voltage across the load will be regulated (equal to V_{zener}) even though the input voltage (V_{in}) may vary.

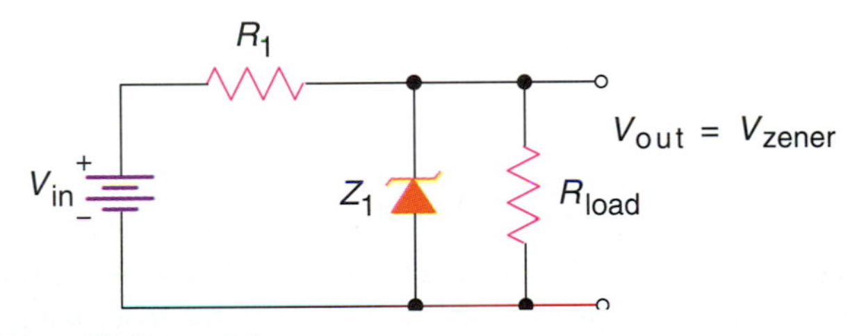

Figure 23-11 Zener diode regulator.

REVIEW QUIZ 23-2

1. The positive side of the diode is called the ______________, and the negative side is called the ______________.
2. When a rectifier diode is forward biased at about 0.7 volt, it is at its ______________ voltage.
3. When the barrier junction is destroyed, a(n) ______________ of current flows.
4. In the reverse direction, rectifier diodes are like ______________ circuits.
5. Zener diodes should be operated in the ______________-bias mode.
6. The output of a rectifier diode results in ______________ DC.

23-3 Rectifier Circuits

The most common use for diodes is as rectifier circuits. Diodes are used to rectify alternating current into direct current in all types of electronic equipment, computers, appliances, and audio-video products. In fact, every electrical cord you see plugged into a wall is supplying AC power, which is rectified by diodes. Electric power companies supply line power to customers in their homes and businesses, typically as 110-120 volts wall outlets. Of course, in Europe and other places, this voltage may be somewhat different. Regardless of the value of voltage or the type of wall plug used, the AC line power is often transformed from its higher voltage to a lower AC voltage by a step-down transformer that also isolates the appliance from the line. Then the lower voltage is converted to direct current by diodes in a rectifier circuit. The final step in the rectification process is to filter or *smooth* the pulsating DC so that a steady value of voltage, such as a perfect battery, is the result.

Rectifier circuits are made of one, two, or more diodes (usually four) arranged to convert either or both half cycles (positive and negative) of an AC signal into direct current. Then, capacitors and even inductors are used to filter out any unwanted AC fluctuations that remain.

Half-Wave Rectifier

The half-wave rectifier is the simplest type of circuit for converting alternating current to direct current. By placing a diode in series with an AC power source, only one-half of the sine wave input passes through the diode during the time it is forward biased. Figure 23-12 shows a half-wave rectifier circuit with the input and the resulting output. During the positive half cycle, the diode is forward biased, but during the negative half cycle it is not. The result (dotted line) is a rectified DC voltage, also called pulsating DC. Because a **smoothing capacitor** is connected across the output, however, the result (smooth line) is a DC voltage with a **ripple** component. The greater the value of capacitance, the less the ripple. During the positive half-cycle, the capacitor charges. During the negative half-cycle, it discharges. The DC voltage level depends on the value of the load resistance and the value of the capacitor. In general, half-wave rectifiers, even with smoothing capacitors, produce an output across the load that is equal to the average value (0.318) of the peak voltage from the transformer secondary or, as an approximation, about one-third of the peak value of the secondary output.

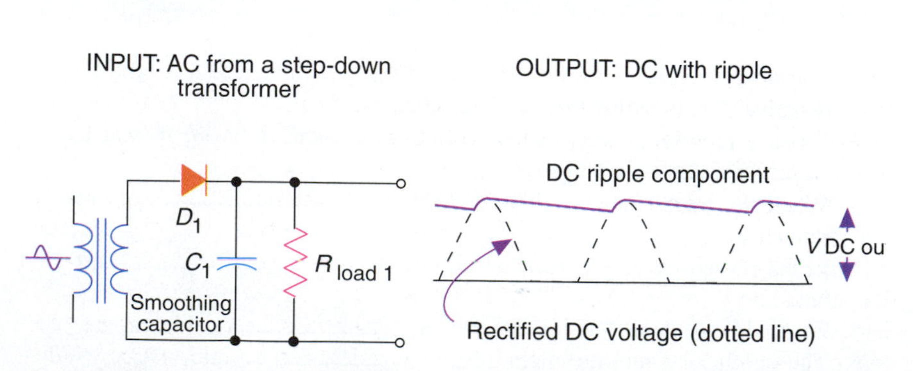

Figure 23-12 Half-wave rectifier circuit.

Full-Wave Rectifier

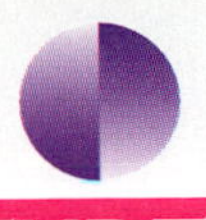

STUDENT *to* STUDENT

Bridge rectifiers are the most common full wave rectifiers. You might even have to fix one some day, especially if the circuit is not rectifying both half cycles.

A full-wave rectifier uses the full AC cycle. Because of the unique arrangement of the diodes, both half waves (positive and negative) can be converted or rectified into pulsating DC. Each diode conducts only during the time it is forward biased. Full-wave rectifiers can be built using either two or four diodes. Figure 23-13 shows the most commonly used configuration: the bridge rectifier. For practical purposes, the load can be any circuit that requires a DC voltage, including the power supply for a television, a computer, or a microwave oven. Here, it is simply represented as a resistance with current flowing in only one direction.

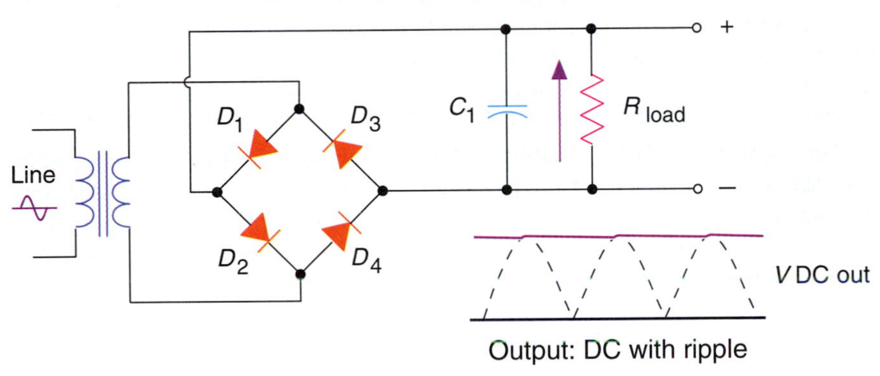

Figure 23-13 Full-wave bridge rectifier circuit.

Many modern bridge rectifiers come in packaged devices (like chips) but still require discrete filtering or smoothing capacitors. Notice that the AC input is always connected between two diodes of opposing polarity, and that the DC output is always taken from diodes of the same polarity. You can trace the path of current flow through the resistor and verify that the resistor polarity is determined by current flow. Regardless of which direction current flows into the bridge circuit, it can only pass through the load in one direction.

Thus, the load resistor has only direct current passing through it, and the total amount of available voltage is very close to the peak voltage of the step-down transformer's secondary output. This is because the diodes only drop 0.7 volt and the rest of the secondary voltage passes. For example, if the secondary of Fig. 23-13 were a 12-volt peak step down, then about 10.6 volts would pass through the rectifier: 12 V−2 (0.7 V). This would be valid approximation for any rectifier circuit.

Example 23-3: For a bridge rectifier similar to the one in Fig. 23-13, the output across the load is about 4 volts of direct current measured with a DC voltmeter. Because the secondary is meant to produce a 12-volt peak output, however, the load voltage seems too low. What is the problem?

This is a typical rectifier problem. If you think about it, the load voltage is less than half of what it should be. Actually, it's about a third of the peak value out of the secondary, but it should be closer to 12 volts than 4 volts. After all, if the step-down transformer is at 12 volts peak, then the load voltage should be close to it, but it is not. So, you would have to come to only one conclusion. That is, you would suspect that one of the diodes in the bridge is not functioning. That would account for the output's having about the same value as a half-wave rectifier.

Percent (%) Ripple

Smoothing or filtering removes the AC ripple component of a rectified output. If you connect a capacitor in parallel with the load, the capacitor keeps the DC voltage charged to the peak value during the time it would normally drop to zero. Because most loads have low resistance (large current demands), the *RC* time constant is slow and requires a larger value of capacitance to keep the charge. This is why many power supply boards have several large filter capacitors to ensure that the ripple component is minimized.

In addition to capacitive filtering, an inductor can be placed in series with the load so that its out-of-phase (with the capacitor) voltage can supply current that produces an even smoother DC output. Ideally, pure direct current (no ripple) would be a perfect supply, such as an ideal battery. Rectified alternating current, however, always has some amount of ripple. Therefore, the amount of ripple produced by a rectifier circuit becomes a measure of its quality. If 100 percent of the ripple could be removed or smoothed, the rectifier circuit would be perfect.

In practice, rectified alternating current should not have any AC ripple greater than about 1 or 2 percent of the peak value of DC output after smoothing. For that reason, you will see a variety of filter components (capacitors and inductors) used for that purpose. The percent of ripple is calculated using the measured values across the rectifier output. The formula is:

$$\text{Percent (\%) ripple} = \frac{V_{\text{rms}}\ \text{ripple}}{V_{\text{DC}}} \times 100$$

Treat the ripple component like an AC signal. Then calculate the rms value of the ripple (0.707 × peak) and divide it by the average DC value. You can use an oscilloscope to see the ripple and the average DC value (zero axis of the ripple component).

REVIEW QUIZ 23-3

1. Half-wave rectifiers produce an output load voltage equal to about ______________ of the peak input voltage or secondary transformer output.
2. Full-wave rectifiers produce an output load voltage equal to about ______________ of the peak input voltage or secondary transformer output.
3. The output from a full-wave rectifier is taken where the diodes have the ______________ polarity.
4. A filter capacitor used in a rectifier output is called a(n) ______________ capacitor.
5. A full-wave rectifier is more efficient than a half-wave rectifier because it produces more output ______________ for the same input ______________.
6. The quality of a rectifier circuit is determined by its ______________.

23-4 Clipper, Clamper, and Multiplier Circuits

Clippers and **clampers** are used in radar applications, in computer products, and in various radio frequency (RF) systems. In general, clippers and clampers are used to control or remove a particular part of a signal, or to shift the level of direct current upon which a signal is riding. Clippers do what their name implies—they remove or clip off part of a signal. Clampers are so named because they are most often used to clamp or fix a signal to a particular level.

Clipper Circuit (Limiter)

A clipper (also called a **limiter**) is a device that automatically limits the peak or instantaneous value of a signal. In other words, it *clips* or limits the output of a circuit. Diodes make excellent clippers because they can effectively clip the positive or negative half-cycle of a sine wave. Figure 23-14 shows a clipper circuit where a silicon rectifier diode is used to clip the negative half-cycle. Of course, if the diode were reversed, the output would be a clipped positive voltage. Hence, the name *positive* or *negative* clipper is sometimes used because the entire half-cycle is clipped.

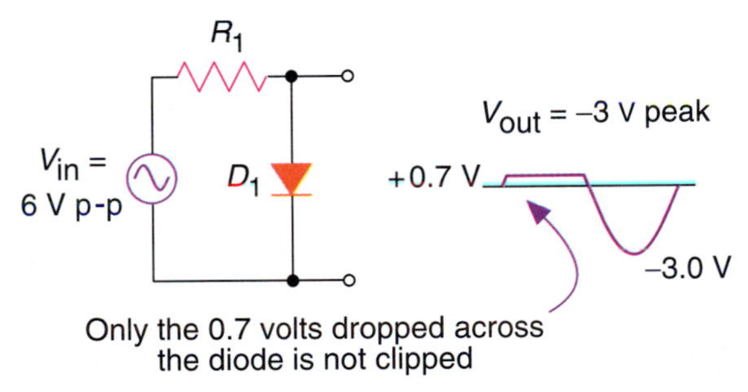

Figure 23-14 Positive clipper circuit.

The circuit in Fig. 23-14 is sometimes called a **shunt clipper** because the diode is in parallel with the output. In other words, the diode shunts the output voltage. If, however, you exchange R_1 and D_1 in the circuit, the output changes slightly because the 0.7 volt dropped across the diode is no longer in parallel with the output. Figure 23-15 shows this circuit, which is sometimes called a positive series clipper. Note that the diode is in series with the output.

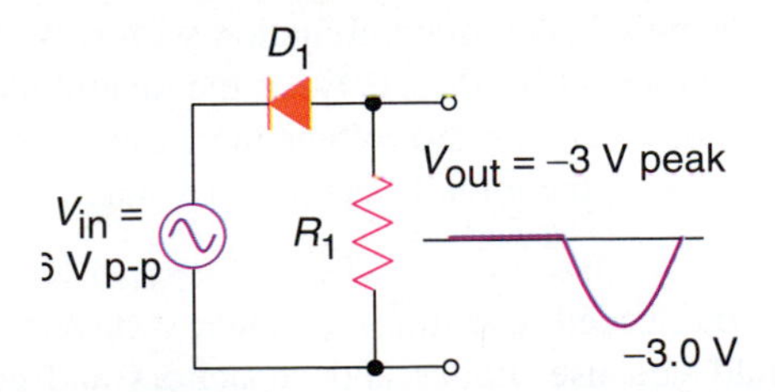

Figure 23-15 Positive series clipper.

Clamper Circuit

A clamper (also called a **DC restorer**) is a circuit that adds a fixed bias voltage to a signal. This is done so that a particular signal or some part of it is *clamped* or held at a particular value. Sometimes signals are clamped when they reach a certain value or shape. Clampers are usually positive or negative, depending on the type of bias voltage used in the circuit. Figure 23-16 shows a positive clamper circuit where a diode and a capacitor are used to clamp or shift the sine wave input above zero. Notice that on the negative half-cycle of V_{in}, the diode is on and current flows. When this happens, the capacitor (C_1) charges to the peak value of V_{in}. Therefore, the negative cycle never appears across the output resistor, except for the −0.7 volt that is across the diode during conduction. On the positive half cycle, the peak value of V_{in} appears across the load. Because the output voltage is the sum of V_{in} and V_C (capacitor voltage), however, the output is effectively shifted above zero. Of course, if the diode were reversed, the sine wave would be clamped below zero, and the circuit would be a negative clamper. The arrow of the diode symbol points in the direction that the DC level of the signal shifts.

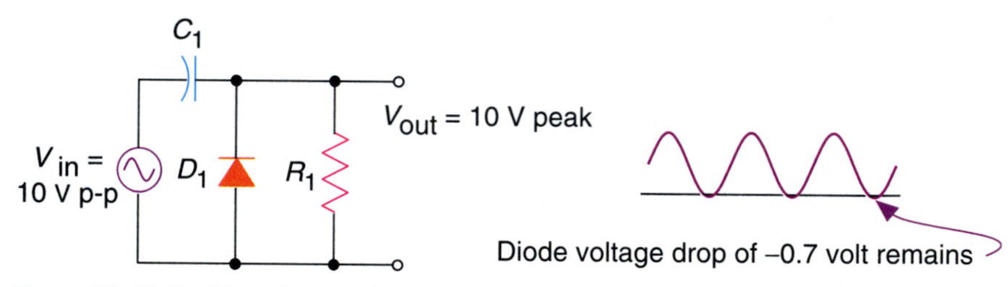

Figure 23-16 Positive clamper circuit.

Multiplier Circuits

Multiplier circuits are used to increase voltage. Diodes are also used along with capacitors to form voltage multiplier circuits. Wherever a high voltage (low current) is required in a DC circuit, doublers are commonly found. For example, the CRT of a computer or television often uses high-voltage doublers and triplers. The basic principle used is similar to the clamper. In other words, the diode conducts during one-half cycle of the sine wave input, and the capacitor charges to the peak value of V_{in}. Then, on the other alternation, the capacitor discharges to double the peak value. The difference between the clamper and the voltage multiplier (doubler) is the addition of a second series capacitor–diode combination connected in parallel.

Figure 23-17 shows two diodes in combination with two capacitors to form a voltage doubler. The output across R_{load} is twice the magnitude of V_{in}. For each combination of a diode and capacitor, the voltage increases once again. By adding another peak rectifier section, the circuit becomes a voltage tripler.

STUDENT to STUDENT

Voltage multipliers look like two half-wave rectifiers in series. The difference, however, is that the doubler has the capacitor connected first.

Example 23-4: If you needed four times as much voltage as a signal generator could supply, could you use diodes and capacitors and connect a voltage quadrupler circuit to the output?

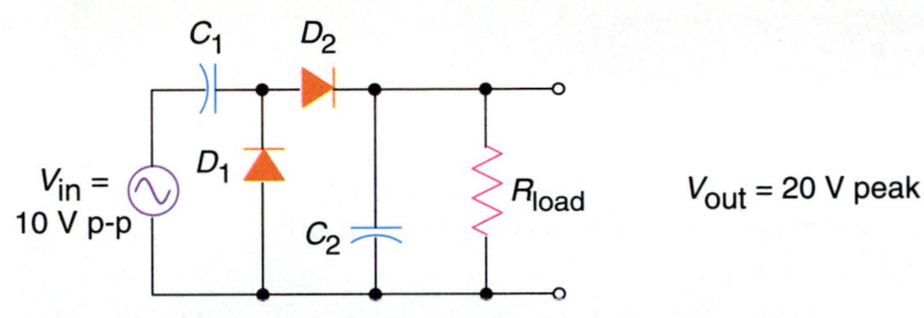

Figure 23-17 Voltage doubler circuit.

The answer is yes . . . and no. In fact, to build the quadrupler, you would double the configuration in Fig. 23-17 so that a series of four capacitors and four diodes would be connected to the signal generator output before the load resistor. The problem, however, is that the output ripple keeps getting worse as you add stages. Because of this, you usually only see doublers and triplers used. Anything beyond that will have a very large ripple, and the output current will be very low. So, the answer to the question of making a quadrupler is still yes . . . and no.

REVIEW QUIZ 23-4

1. A(n) ______________ circuit is practically the same as a(n) ______________ rectifier circuit.
2. To hold an AC voltage below zero volts, you would use a(n) ______________ circuit.
3. Voltage multipliers use only two components: ______________ and ______________.

23-5 Voltage Protection Circuits

Another example of common diode circuits is the use of zener diodes as protection devices. Two zener diodes in a back-to-back configuration create a dual clipping action so that no peak-to-peak voltage greater than the zener value is allowed to pass. Figure 23-18 shows an over-voltage protection circuit that operates like a clipper in both directions.

Notice in Fig. 23-18 that each zener diode is rated at 3 volts. This means that when the input voltage exceeds the peak value of 3 volts, one zener goes into breakdown (at 3 volts) while the other is forward biased at 0.7 volt. In this way, the zeners protect the load from any voltage above 3.7 volts peak.

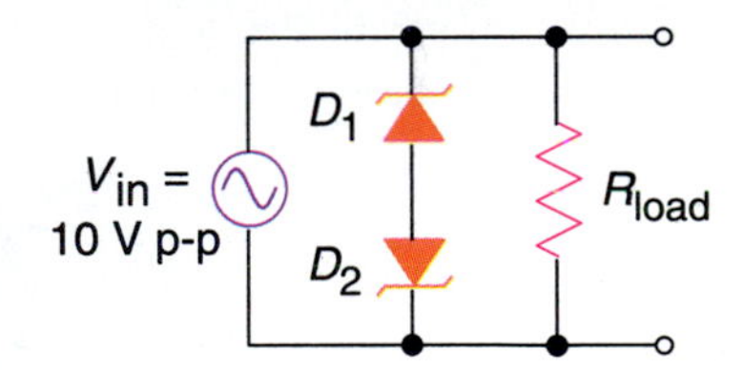

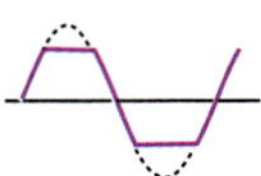

Note: D_1 and D_2 are both 3-volt zener diodes. V_{out} = about 6 V p-p

Figure 23-18 Zener voltage protection circuit.

REVIEW QUIZ 23-5

1. A voltage protection circuit uses ______________ diodes.
2. The diodes are used in a(n) ______________ configuration.

SUMMARY

Diodes are P-N junctions where two semiconductor materials meet. Because the two materials have different charges, a junction is formed where they come together. The charges result from doping, which is the mixing of a very tiny amount of some other element with the semiconductor material.

Whenever you have a P-N junction, it can be overcome when enough voltage is applied with the proper polarity. The applied voltage is called the bias voltage. In fact, a diode is like a switch that is closed when enough voltage is applied in the forward direction. For silicon, it only takes about 0.7 volts. In the other direction, a reverse biased diode is just like an open switch that allows no current to flow. Only a zener diode is meant to be used in the reverse direction because it is manufactured for a specific value of reverse breakdown voltage.

Because most silicon rectifier diodes withstand heat better than other materials and allow current to flow with less than a volt applied, they are most often used in power supply circuits. As their name suggests, they rectify or change AC signals to direct current. It only takes one diode to block half an AC signal, resulting in a half-wave rectifier. But more diodes are required to rectify the full wave of alternating current. The most common is the bridge rectifier, which has four diodes that are used to route the AC current in one direction as DC through a load.

Regardless of the AC signal and the diode arrangement, rectifier circuits do not produce pure DC voltage. For this reason, smoothing capacitors are necessary to smooth the pulsating DC output of a rectifier circuit. Smoothing capacitors are usually the biggest components on a power supply board, except for the transformer.

Other circuits made using diodes are clippers, clampers, multipliers, and voltage protection and regulation circuits. Clippers are like half-wave rectifiers designed to clip off part of a signal voltage. Clampers are designed to clamp or hold a signal at a certain DC level. Multipliers use at least two capacitors and rectifier diodes to multiply the input voltage level but not the available current. Special semiconductor diodes are also used in logic circuits where their ability to turn off and on is appropriate for use in binary logic. LEDs are diodes that dissipate light rather than heat. Finally, zener diodes are used to regulate voltage and protect other parts of a circuit.

NEW VOCABULARY

anode
avalanche
barrier potential
bias
cathode
clamper
clipper
DC restorer
depletion region
doping
junction barrier
knee
limiter
P-N junction
pulsating DC
rectifier
ripple
semiconductor
shunt (clipper)
smoothing capacitor
temperature coefficient
zener

Questions

1. What makes a material a semiconductor?
2. What is doping?
3. Name three types of diodes.
4. What is the basic difference between a rectifier diode and a zener diode?
5. Why is the knee area of the diode so important?
6. Explain what is meant by the term *pulsating DC*.
7. When is the temperature coefficient of a diode important?
8. What is the basic difference between a half-wave and a full-wave rectifier?.
9. Why is the percent of ripple a measure of the quality of a rectifier circuit? Identify the components most responsible for determining this quality.
10. Among all these circuits—clamper, clipper, multiplier, regulator—which ones use zener diodes and which ones use rectifier diodes?

Problems

1. A half-wave rectifier has a 100-volt AC input applied, and there is a smoothing capacitor large enough to keep the ripple below 2 or 3 percent. How much voltage and current should there be for a 330-ohm load resistor?
2. For the circuit in Fig. 23-19, what are the effective voltages across R_{load}, D_1 and Z_1?

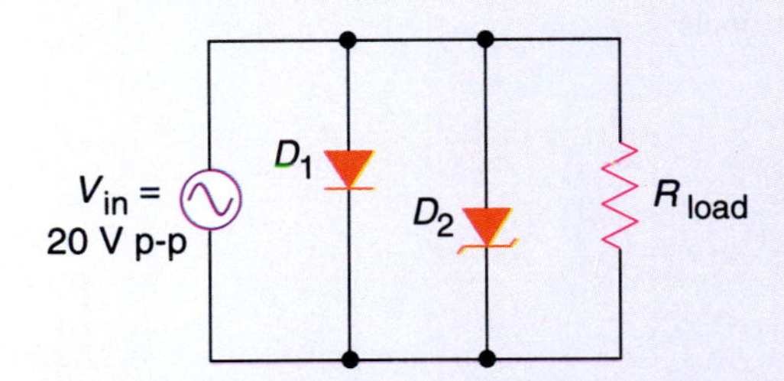

Note: D_2 is a 5-volt zener diode. Also assume the source is protected from shorting out.

Figure 23-19

3. A full-wave bridge rectifier is connected to the output of a 20-volt peak secondary transformer with no smoothing capacitor. Draw the output waveform.
4. Draw a full-wave bridge rectifier circuit and show the current through the diodes and the load resistor.
5. A 12-volt zener diode has a positive temperature coefficient of 0.1 percent per °C at its ambient temperature 25°C. If the zener is operating at 170°C, what is the effect on the zener breakdown voltage?
6. The output of a bridge rectifier has an average DC output of 15 volts. The ripple

Critical Thinking

1. If you connected ten rectifier diodes in series, would they act like a single 7-volt zener diode? Assume that each is a rectifier with exactly 0.7 volts required for forward conduction.
2. In the circuit of Fig. 23-19, can you see any practical application of a zener and a rectifier diode in that configuration? What would happen if the three components were in series with both diodes in the same direction?
3. On a circuit board, a packaged bridge rectifier (chip) is destroyed. Could it be replaced using four discrete rectifier diodes? If so, what difficulties could there be?
4. The output of a bridge rectifier has too much ripple, but a smoothing capacitor is already connected on the circuit. Could you connect another capacitor of the same value to see if it decreases the ripple? If so, how would you connect it?

Answers to Review Quizzes

23-1
1. silicon; germanium
2. doped
3. opposite
4. forward bias

23-2
1. anode; cathode
2. knee
3. avalanche
4. open
5. reverse
6. pulsating

23-3
1. one-third (or 0.318)
2. 0.7 volt
3. same
4. smoothing
5. voltage; voltage
6. percent ripple

23-4
1. clipper; half-wave
2. negative clamper
3. capacitors; rectifier diodes

23-5
1. zener
2. back-to-back

TAIWAN 8149 AG
SN54LS74AJ
TAIWAN 8249CG
SN54LS10J
HEF4011BP
HEF4528BP
LC FILTER
MIXING OSC.
979.560.05

Chapter 24

BJT *and* FET Transistors

ABOUT THIS CHAPTER

In this chapter you will learn about two different types of transistors: the bipolar junction transistor (BJT) and the field-effect transistor (FET). As you will see, each type of transistor has its own unique construction and operating characteristics. In general, both transistors are used when it is necessary to amplify voltage, current, and power. With a small AC signal applied as an input to either type of transistor, the output can be hundreds or even thousands of times larger than that input. Also, both transistor types

After completing this chapter you will be able to:

- Describe the construction and operation of BJTs and MOSFETs.
- Express the relationship between the base, collector, and emitter currents in a BJT.
- List several ratings for BJTs. Calculate DC voltages and currents in transistor circuits.

can be used as an electronic switch that can be either ON or OFF. Unless indicated otherwise, the term *transistor* is usually used only with bipolar junction transistors and not field-effect transistors. When discussing field-effect transistors, the acronym *FET* is almost always used.

In 1947 three American scientists—John Bardeen, Walter Brattain, and William Schockley, all of whom worked for Bell Labs—developed the first transistor. In 1956 they were awarded the Nobel Prize in Physics for their outstanding contribution. The invention of the transistor has had an enormous impact on the field of electronics, especially in the area of computers.

- Describe the construction and operation of a junction field-effect transistor (JFET).
- Identify the different types of charge carriers in both NPN and PNP transistors.
- Give the characteristics of both JFETs and MOSFETs. Calculate the drain current I_D in an FET.

24-1 Construction of a BJT

The construction of an NPN **bipolar junction transistor** (BJT) is shown in Fig. 24-1(*a*). The transistor is constructed with a thin layer of P-type material sandwiched between two larger N-type regions. The center region is called the base. The base region is made very thin and is lightly doped. Doping is a process where you add other materials called impurities to the transistor in order to change its electrical characteristics. The N region on the left side of the base is heavily doped and called the emitter. Its job is to emit or inject free electrons into the P-type base. The N region on the right side of the base is called the collector. The collector is only moderately doped. Its job is to collect electrons that diffuse across the base collector junction. It is the largest of the three regions within a transistor because it must dissipate the most heat.

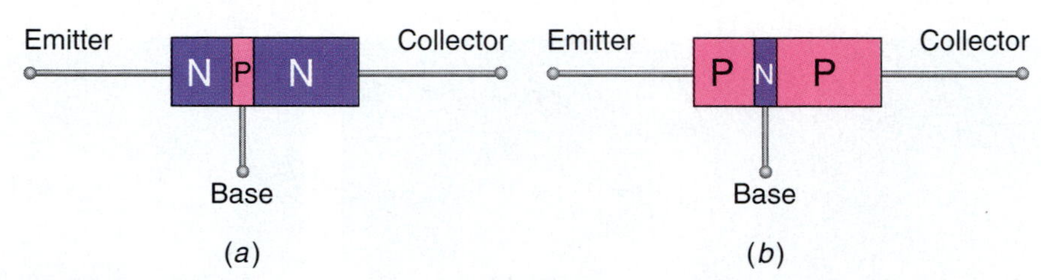

Figure 24-1 Transistor construction. (*a*) NPN transistor. (*b*) PNP transistor.

Figure 24-1(*b*) shows the construction of a PNP transistor. Notice that the P and N regions are arranged in a manner that is opposite to that of the NPN transistor in Fig. 24-1(*a*).

For the NPN transistor in Fig. 24-1(*a*), free electrons are the **majority current carriers** in both the emitter and collector regions. In the base, however, holes are the majority current carriers. The opposite is true for the PNP transistor shown in Fig. 24-1(*b*). Here, holes are the majority current carriers in both the emitter and collector region, whereas free electrons are the majority current carriers in the base.

Both NPN and PNP transistors have the same ability to amplify voltage, current, and power. Basically the operation of both types is the same, but they require opposite voltage polarities to operate.

Figure 24-2 shows the schematic symbols used for both the NPN and PNP bipolar junction transistors. A lead is connected to each region within the tran-

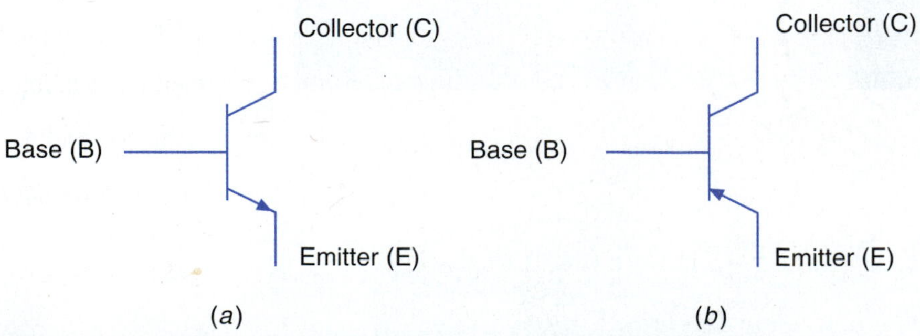

Figure 24-2 Schematic symbols for transistors. (*a*) NPN transistor. (*b*) PNP transistor.

sistor, thereby making it a three-terminal device. The schematic symbol for the NPN transistor is shown in Fig. 24-2(*a*), and Fig. 24-2(*b*) shows the schematic symbol for a PNP transistor. On both symbols, *C* stands for collector, *B* for base, and *E* for emitter. Notice that for both schematic symbols there is an arrow on the emitter lead. For the NPN transistor in Fig. 24-2(*a*) the arrow points outward, and for the PNP transistor in Fig. 24-2(*b*) the arrow points inward. The arrow points in the direction of conventional current flow for both types.

ELECTRONIC FACTS

In a transistor there are two PN junctions: the emitter-base (EB) junction and the collector-base (CB) junction.

PN Junctions

Refer back to Fig. 24-1. Notice that two PN junctions exist. The PN junction formed by the emitter and base regions is called the emitter-base (EB) junction. The PN junction formed by the collector and base regions is called the collector-base (CB) junction.

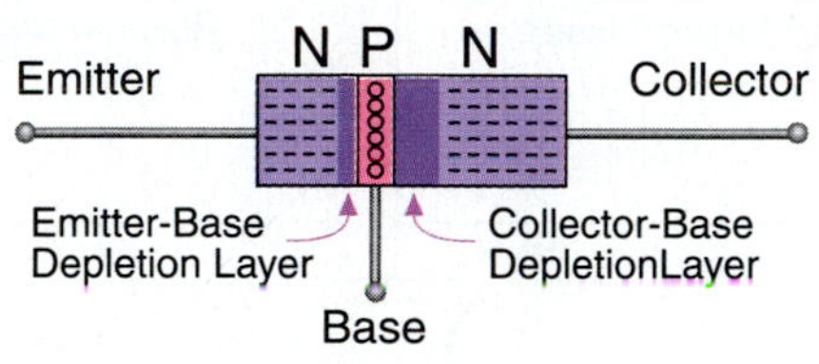

Figure 24-3 Depletion regions in an unbiased NPN transistor.

Figure 24-3 shows the depletion regions that exist in an unbiased NPN transistor. The diffusion of free electrons from both N-type regions into the P-type base causes a barrier potential, V_B, to exist for both PN junctions. For silicon (Si), the barrier potential V_B is approximately 0.7 V, whereas for germanium (Ge), the barrier potential is about 0.3 V. Because silicon transistors are much more common than germanium transistors, only silicon transistors will be discussed throughout the rest of this text. It is important to note the different widths of the EB and CB depletion regions in Fig. 24-3. The different widths can be attributed to the different doping levels in the emitter and collector regions. Because the emitter is heavily doped and many free electrons are available, the penetration into the N material is minimal. Because there are fewer free electrons available on the collector side, the penetration of the depletion region is deeper than that in the emitter region. In Fig. 24-3 dash marks are used in the N-type emitter and collector regions to denote the large number of free electrons available. Small circles are used to indicate the holes in the P-type base.

REVIEW QUIZ 24-1

1. The base region of a transistor is made very __________ and is __________ doped.
2. In an NPN transistor, __________ are the majority current carriers in the emitter and collector.
3. In a transistor, which depletion region is wider, the EB depletion region or the CB depletion region?
4. In a transistor, which region is the most heavily doped?

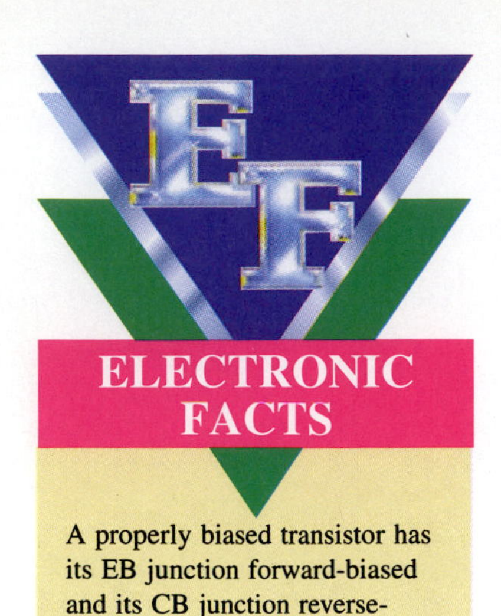

24-2 Analyzing Transistor Operation

All transistors require the use of DC voltages to operate properly. Figure 24-4 shows the proper biasing arrangement for both NPN and PNP transistors. The EB junction in both cases is forward-biased, and the CB junction is reverse-biased. For our discussion here, we will analyze what happens inside the NPN transistor. The operation of the PNP transistor is basically the same except that the current directions are reversed.

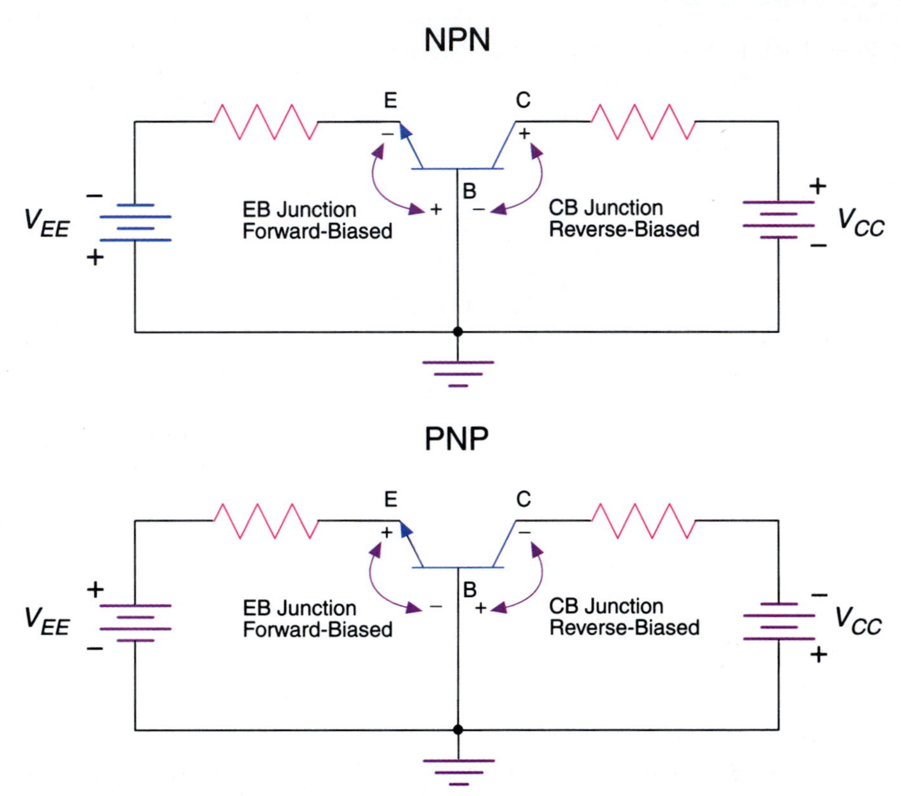

Figure 24-4 Biasing transistors. (*a*) Proper polarity of V_{EE} and V_{CC} for forward-biasing the EB junction and reverse-biasing the CB junction in an NPN transistor. (*b*) Proper polarity of V_{EE} and V_{CC} for forward-biasing the EB junction and reverse-biasing the CB junction in a PNP transistor.

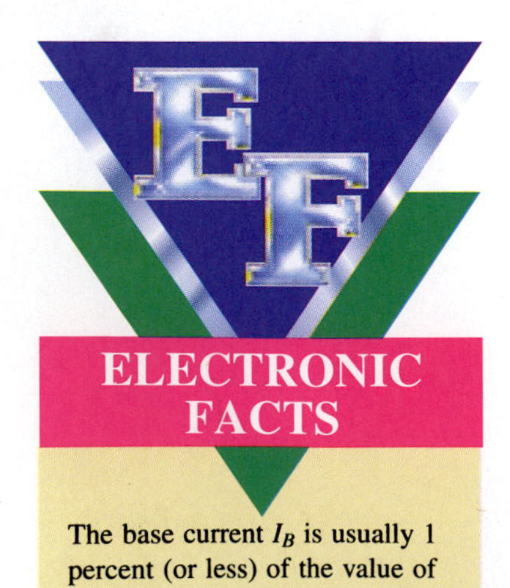

Look at Fig. 24-5. The emitter supply voltage V_{EE} forward biases the EB junction, and the collector supply voltage V_{CC} reverse biases the CB junction. Because the EB junction is forward-biased, its depletion region is narrowed. Similarly, the reverse-bias voltage across the CB junction widens its depletion region.

In Fig. 24-5, the negative terminal of V_{EE} repels free electrons into the N-type emitter of the transistor. The current flow in the emitter is called emitter current and is designated I_E. Because the EB junction is forward-biased, free electrons are injected into the thin and lightly doped P-type base. Because the P-type base is so thin and lightly doped, there are only a few holes available for the incoming free electrons to combine. Thus, only a very small percentage of all electrons entering the base can combine with an available hole. The result is that the current flowing out of the base lead, called the base current (I_B) is very small.

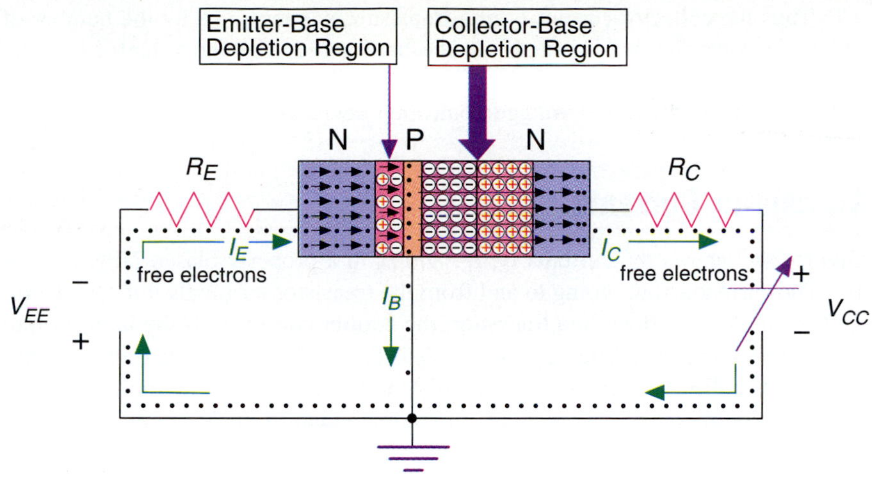

Figure 24-5 Current flow in an NPN transistor.

The base current I_B is usually 1 percent (or less) of the value of the emitter current I_E. Most of the free electrons injected into the base from the emitter diffuse across the CB junction and into the collector. The current in the collector region is called collector current and is designated I_C. In Fig. 24-5 the positive ions in the CB depletion region act as an attraction to the free electrons in the base. It is important to note that the collector current I_C is mainly dependent on the number of current carriers injected into the base and not on the actual value of reverse-bias voltage across the CB junction.

Important Transistor Facts

Only a small amount of reverse-bias voltage across the CB junction is needed to create an electric field strong enough to collect almost all of the free electrons that are injected into the P-type base region. After the CB reverse-bias voltage reaches a certain level, increasing it further will have little or no effect on the number of free electrons collected by the collector. In fact, after the voltage source V_{CC} in Fig. 24-5 is slightly above zero, full collector current I_C is achieved. This is depicted by the graph in Fig. 24-6. If, of course, the reverse bias voltage V_{CC} exceeds the breakdown voltage rating of the CB junction, breakdown will occur as shown.

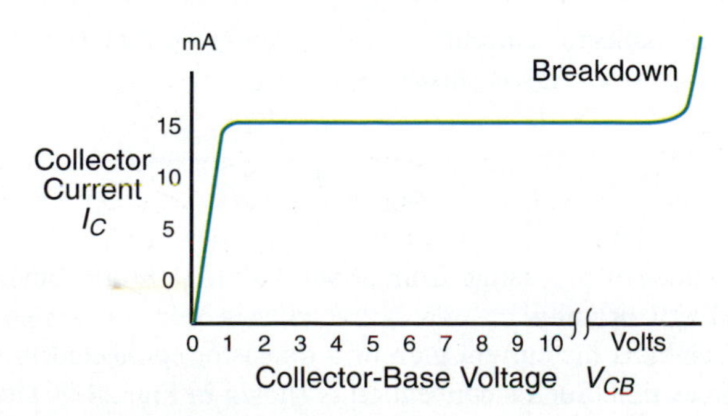

Figure 24-6 Graph of V_{CB} versus I_C.

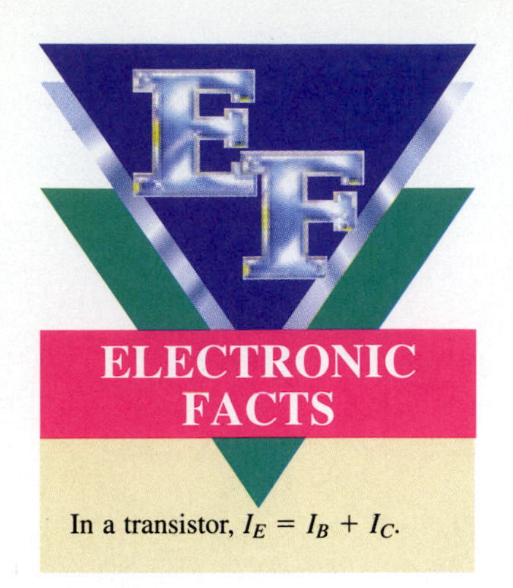

ELECTRONIC FACTS

In a transistor, $I_E = I_B + I_C$.

Thus the collector current I_C in a transistor is determined by the number of current carriers that are injected into the base and not by the value of reverse-bias voltage across the CB junction. Because of this fact, transistors are current-controlled devices and not voltage-controlled devices.

Transistor Currents

Figure 24-7 shows the currents I_E, I_C, and I_B in a properly biased NPN transistor. The current arrows going to and from the transistor terminals indicate the direction of electron flow. In a transistor, the emitter current I_E is the largest of all the transistor currents. The base current I_B is the smallest of all the transistor currents. The collector current I_C has nearly the same value as the emitter current I_E. I_C is less than the value of I_E by an amount equal to the base current I_B. In a transistor, the currents are related as follows:

Equation 24-1

$$I_E = I_B + I_C$$

Rearranging, we have the other forms of this equation.

Equation 24-2

$$I_C = I_E - I_B$$

Equation 24-3

$$I_B = I_E - I_C$$

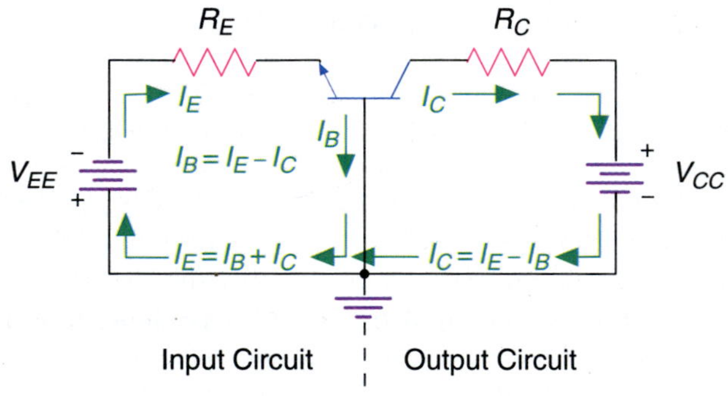

Figure 24-7 Transistor currents in a properly biased NPN transistor. The transistor shown is connected in the common-base configuration.

DC Alpha (α_{DC})

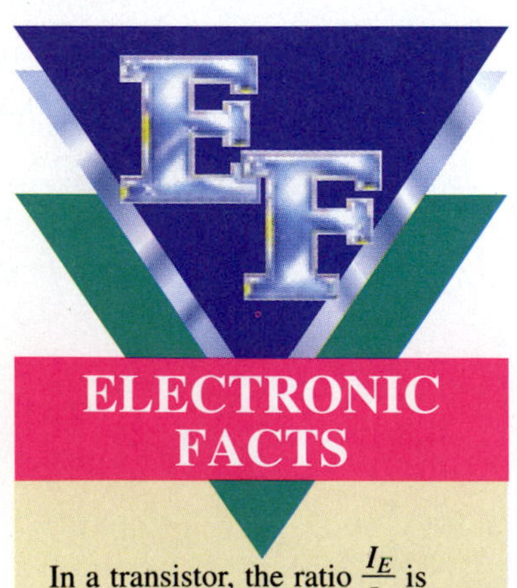

ELECTRONIC FACTS

In a transistor, the ratio $\frac{I_E}{I_B}$ is called the DC alpha, designated α_{DC}. In most cases, α_{DC} equals approximately 1 or unity.

The ratio of the collector current I_C to the emitter current I_E is the **DC alpha** (α_{DC}) of the transistor. This is shown as:

Equation 24-4

$$\alpha_{DC} = \frac{I_C}{I_E}$$

Typical values of α_{DC} range from about 0.95 to 0.99. In some cases, α_{DC} is approximated as 1 or unity.

The DC alpha is the current gain of a transistor connected in the common-bias (CB) connection. Such a connection is shown in Fig. 24-7. Here the base is common to both the power supplies, V_{EE} and V_{CC}. For this type of connection an AC input signal would be applied to the EB junction, and the output signal would be taken from the CB junction. Because the base is common to both the input and output sides of the circuit, we called it a common-base connection or configuration.

Common-Emitter (CE) Configuration

Figure 24-8 shows an NPN transistor connected in a common-emitter configuration. Here an AC input signal would be applied to the BE junction, and the output signal is taken across the collector-emitter (CE) region.

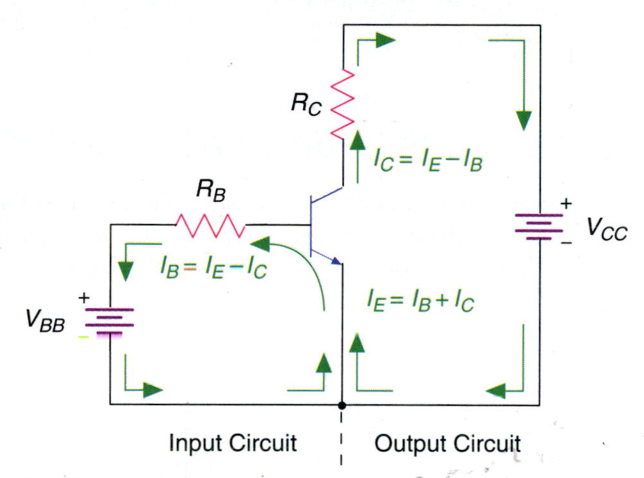

Figure 24-8 Transistor currents in the common-emitter configuration.

In Fig. 24-8, it is important to note that the relation of the DC currents is the same as that shown in Fig. 24-7, even though the transistor is arranged differently. Notice the current arrows for I_E, I_C, and I_B. These arrows indicate the direction of electron flow in the circuit. In Fig. 24-8, Eqs. 24-1, 24-2, and 24-3 still apply to the transistor currents.

The DC current gain of a transistor in a common-emitter configuration is called the **DC beta,** designated β_{DC}. The DC beta is expressed in Eq. 24-5.

Equation 24-5

$$\beta_{DC} = \frac{I_C}{I_B}$$

Typical values of β_{DC} range anywhere from 50 to 300.

In a transistor, the ratio $\frac{I_C}{I_B}$ is called the DC beta, designated β_{DC}. For most transistors, β_{DC} ranges from about 50 to 300.

Relating β_{DC} and α_{DC}

If β_{DC} is known, then α_{DC} can be calculated as shown:

Equation 24-6

$$\alpha_{DC} = \frac{\beta_{DC}}{1 + \beta_{DC}}$$

Also, if α_{DC} is known, β_{DC} can be calculated. This is shown as:

Equation 24-7

$$\beta_{DC} = \frac{\alpha_{DC}}{1 - \alpha_{DC}}$$

REVIEW QUIZ 24-2

1. In a properly biased transistor, the EB junction is _______________ biased and the CB junction is _______________-biased.
2. The BJT is a(n) _______________-controlled device.
3. In a transistor, the base current I_B is usually very _______________.
4. In a transistor, the emitter current I_E = 2 milliamps and the collector current I_C = 1.98 milliamps. What is α_{DC}?
5. In a transistor, I_C = 5 milliamps and I_B = 20 microamps. What is β_{DC}?
6. A transistor has β_{DC} = 125. What is α_{DC}?

24-3 Transistor Operating Regions

Figure 24-9(*a*) shows a common-emitter circuit where V_{BB} is providing forward bias for the base-emitter (BE) junction, and V_{CC} provides the required polarity to reverse-bias the CB junction. Because V_{BB} is adjustable, it can control the DC base current I_B. If $\beta_{DC} = I_C/I_B$, then $I_C = I_B \times \beta_{DC}$.

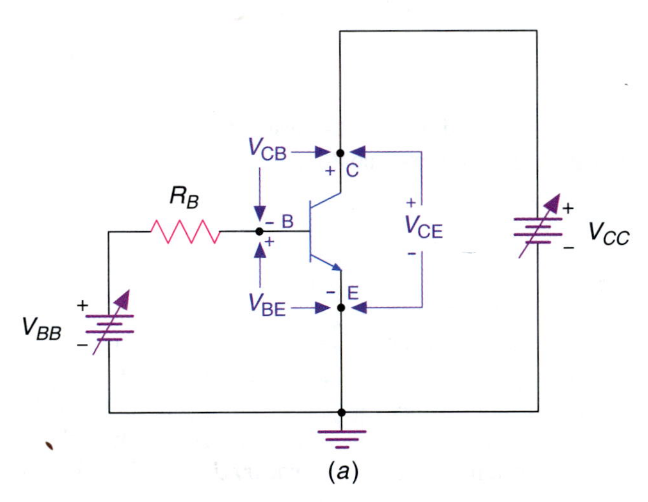

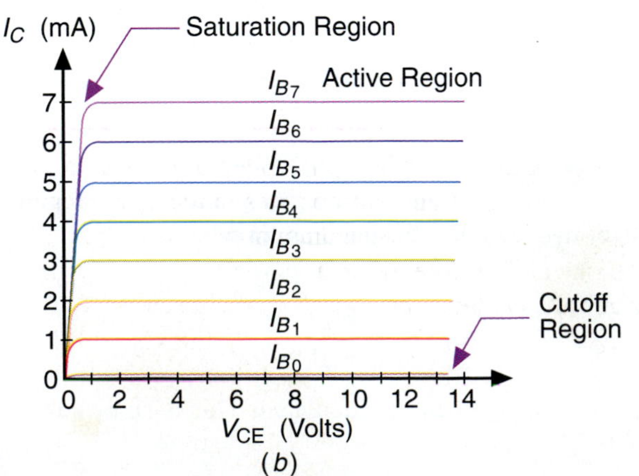

Figure 24-9 Common-emitter configuration. (*a*) Circuit. (*b*) Graph of *IC* versus V_{CE} for different base currents.

ELECTRONIC FACTS

When a transistor is saturated, I_C is not controlled solely by I_B.

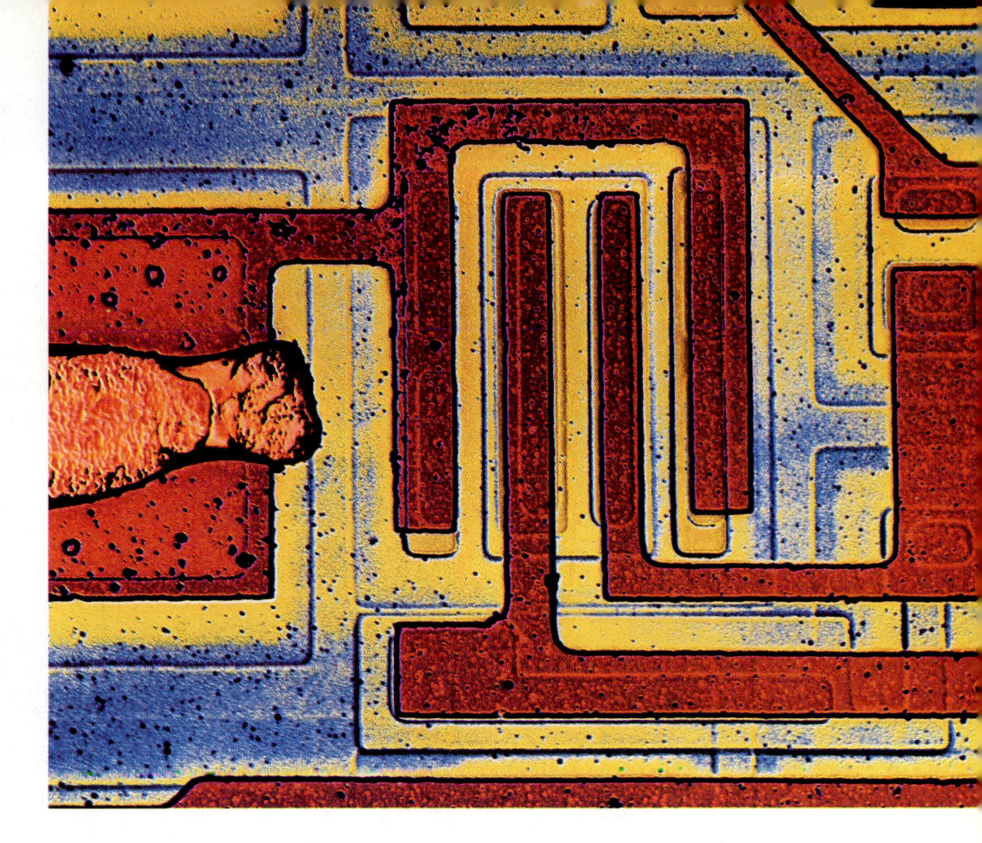

Note the polarity of the transistor terminal voltages. Also note the following voltage relationship in Fig. 24-9(a):

$$V_{CE} = V_{CB} + V_{BE}$$

When V_{CB} is just a few tenths of a volt above zero, the CB junction is reverse-biased and $I_C = I_B \times \beta_{DC}$.

Figure 24-9(b) shows a set of collector characteristic curves that can be generated from the circuit shown in Fig. 24-9(a). Each separate collector curve corresponds to a specific value of base current I_B.

In Fig. 24-9(a), assume that V_{BB} has been adjusted to provide a base current of 100 microamperes. If $\beta_{DC} = 200$, then I_C can be calculated as follows:

$$I_C = I_B \times \beta_{DC}$$
$$= 100\ \mu\text{A} \times 200$$
$$= 20\ \text{mA}$$

As long as the CB junction remains reverse-biased, I_C remains at 20 milliamps. This is true because the only factor controlling the collector current is the base current.

Saturation Region

As shown in Fig. 24-9(b) the collector current IC is zero when the collector-emitter voltage V_{CE} is zero. This is because the CB junction is not reverse-biased. As a result, it cannot draw or collect electrons across the CB junction. When V_{CE} increases from zero, however, IC increases. The vertical portion of the collector characteristic curves (near the origin) is called the **saturation region.** When a transistor is saturated, I_C is not controlled solely by the base current I_B.

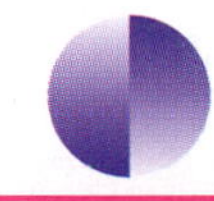

STUDENT *to* STUDENT

When a transistor is operating in the active region, I_C is controlled solely by I_B, assuming that β_{DC} is constant. The value of V_{CC} and R_C have no effect on the value of I_C!

Active Region

The **active region** of a transistor is where the collector curves are nearly horizontal; see Fig. 24-9(b). When a transistor operates in the active region, the collector current I_C is controlled solely by the amount of base current I_B, assuming that β_{DC} remains constant. Therefore, when a transistor operates in the active region, the collector circuit acts like a current source whose output current is calculated as follows: $I_C = I_B \times \beta_{DC}$.

Cut-Off Region

Notice that the $I_B = 0$ curve in Fig. 24-9(b) is only slightly above the horizontal axis of the graph. In the **cut-off region,** a transistor has only a very small collector current. For silicon transistors, this current is very small! A transistor is said to be cut off when its collector current I_C is zero.

Breakdown Region

Although not shown in the collector curves in Fig. 24-9(*b*), the CB junction will break down if V_{CE} exceeds the breakdown voltage rating of the CB junction. The result would be a rapid increase in the collector current I_C. For normal operation, the transistor is never operated in the **breakdown region.** This means that V_{CE} should always be below the breakdown voltage of the CB junction.

DC Equivalent Circuit

Figure 24-10 shows the **DC equivalent** circuit of a transistor operating in the active region. The BE junction will act like any forward-biased PN junction with a current I_B. With silicon transistors, the base-emitter voltage, V_{BE}, is 0.7 volts.

Notice in Fig. 24-10 that the collector circuit is replaced with a current source. The current source has an extremely high internal impedance with an output current, I_C equal to $I_B \times \beta_{DC}$. The blue arrow in the current source points in the direction of conventional current flow. Electron flow is indicated by the green dashed lines.

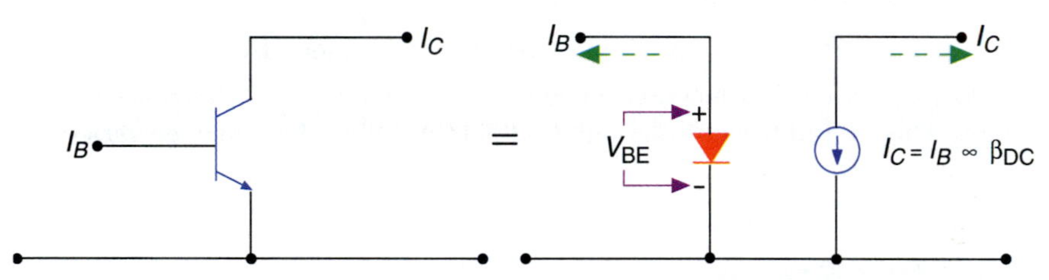

Figure 24-10 DC equivalent circuit of a transistor.

REVIEW QUIZ 24-3

1. When a transistor is operating in the active region, the collector acts like a(n) _______________.
2. A transistor has a DC beta of 150. If I_B = 25 microamps, then I_C = _______________.
3. A transistor should never be operated in the _______________ region.
4. When the collector current I_C is zero, it is said to be _______________.
5. For a silicon transistor, the BE junction has a voltage drop of about _______________.

24-4 Maximum Transistor Ratings

A transistor has maximum ratings that cannot be exceeded. If they are exceeded, the transistor will most likely fail and need to be replaced. For most transistors, the maximum ratings usually include the maximum allowable collector-emitter voltage designated V_{CEO}, the maximum allowable collector-base voltage designated V_{CBO}, the maximum allowable base-emitter voltage designated V_{EBO}, the maximum collector current I_C, and the maximum power dissipation designated $P_{D(\text{max})}$.

For a transistor, the product of V_{CE} and I_C gives the power dissipation. Expressed as an equation:

Equation 24-8

$$P_D = V_{\text{CE}} \times I_C$$

The power dissipation P_D must never exceed the maximum power dissipation rating $P_{D(\text{max})}$. If it does, the transistor will most likely burn up.

It should be noted that $P_{D(\text{max})}$ is usually specified at a temperature of 25°C. For higher temperatures, $P_{D(\text{max})}$ decreases. Therefore, the manufacturer will usually supply a derating factor so that $P_{D(\text{max})}$ can be calculated at any temperature. The derating factor is usually specified in watts per degree Celsius (W/°C.)

Also, when $P_{D(\text{max})}$ is known, Eq. 24-8 can be arranged to solve for the maximum allowable collector current, I_C, for a specified value of collector-emitter voltage V_{CE}.

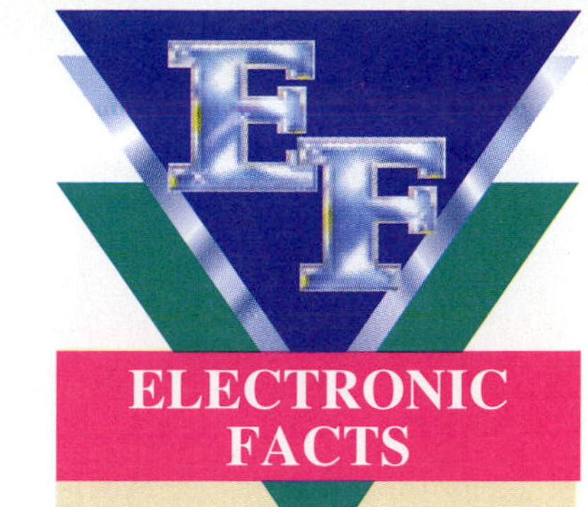

ELECTRONIC FACTS

The power dissipation rating $P_{D(\text{max})}$ decreases for higher temperatures.

REVIEW QUIZ 24-4

1. The power dissipation rating, $P_{D(\text{max})}$ ______________ for higher temperatures.
2. To calculate the power dissipation in a transistor, use the formula ______________.
3. The manufacturer usually specifies the power dissipation rating $P_{D(\text{max})}$ at ______________.

24-5 Introduction to Transistor Biasing

In order for a transistor to function properly as an amplifier, external DC voltages must be connected to provide the desired collector current I_C. The term *bias* refers to a control voltage or current. This section introduces you to two types of bias used with BJTs called base bias and voltage-divider bias.

Base Bias

Figure 24-11(*a*) on the next page shows an example of **base bias.** Together V_{BB} and R_B determine the base current I_B. The collector current I_C equals $I_B \times \beta_{\text{DC}}$. The collector resistor R_C provides the desired voltage in the collector circuit. Figure 24-11(*b*) shows the DC equivalent circuit. Because the transistor is silicon, $V_{\text{BE}} = 0.7$ volt. Notice also that the collector current source has a value dependent on both β_{DC} and I_B. In most cases, β_{DC} is assumed to have a constant value. Therefore I_C can be controlled by varying the base current I_B.

In Fig. 24-11, the base current I_B can be calculated as:

Equation 24-9

$$I_B = \frac{V_{BB} - V_{\text{BE}}}{R_B}$$

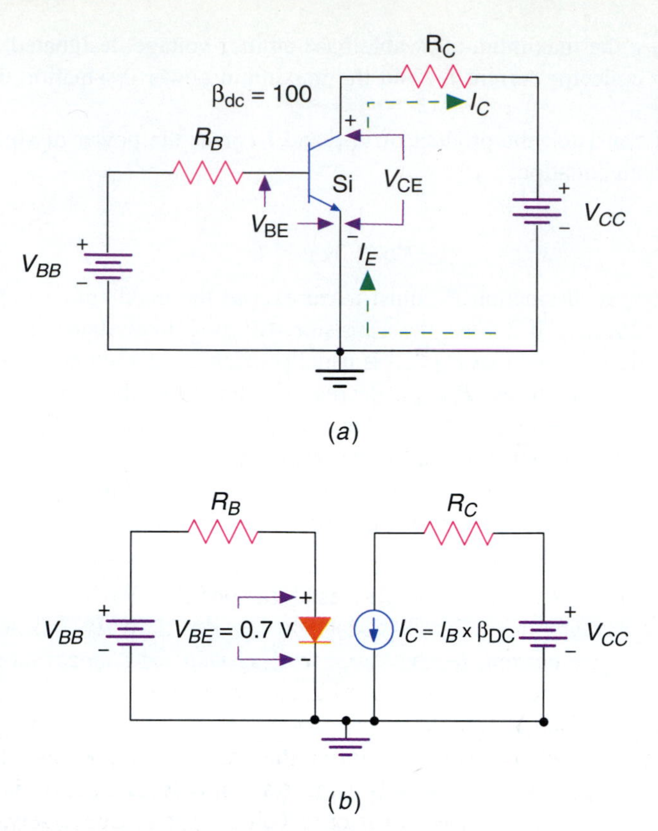

Figure 24-11 Base bias. (*a*) Circuit. (*b*) DC equivalent.

The collector current is calculated as:

Equation 24-10

$$I_C = I_B \times \beta_{\mathrm{DC}}$$

With the collector current known, the collector-emitter voltage V_{CE} can be calculated as:

Equation 24-11

$$V_{\mathrm{CE}} = V_{CC} - I_C R_C$$

Notice that the calculation for V_{CE} involves subtracting the $I_C R_C$ voltage drop from the collector supply voltage V_{CC}. If I_C increases, V_{CE} decreases owing to the increased voltage drop across R_C. On the other hand, if I_C decreases, then V_{CE} increases owing to the reduction in the voltage drop across R_C. As long as the transistor is operating in the active region, the values of R_C and V_{CC} have no effect on the value of the collector current *IC*.

Figure 24-12 shows an example of base bias that uses a single power supply V_{CC}. Notice that the power supply V_{BB} has been omitted. In this case, R_B connects directly to V_{CC}.

For the circuit in Fig. 24-12, the base current is calculated as:

Equation 24-12

$$I_B = \frac{V_{CC} - V_{\mathrm{BE}}}{R_B}$$

Notice that this is the same as Eq. 24-9 except that instead of subtracting V_{BE} from V_{BB}, we subtract it from V_{CC}.

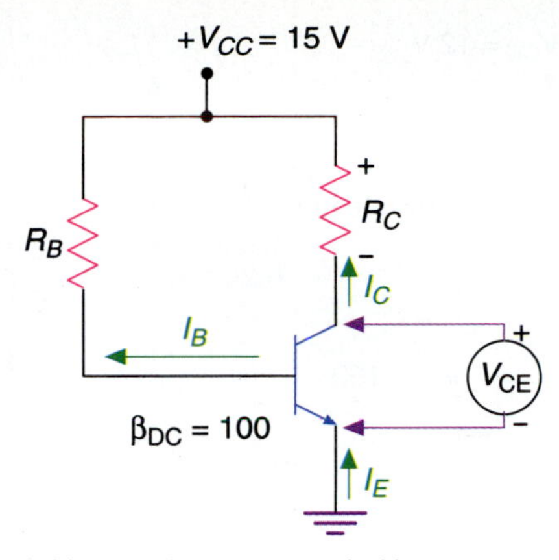

Figure 24-12 Base bias with a single power supply V_{CC}.

DC Load Line

Figure 24-13 shows a graph that includes all the possible combinations of I_C and V_{CE} for a given amplifier. The diagonal line is called a **DC load line.** For each value of collector current, I_C, the corresponding value of collector-emitter voltage V_{CE} can be found by using the load line. The end points of the DC load line are labeled $I_{C(sat)}$ and $V_{CE(off)}$. The collector current is represented by $I_{C(sat)}$ when the transistor is saturated. The collector-emitter voltage is represented by $V_{CE(off)}$ when the transistor is cut off. When the transistor is saturated, visualize the collector-emitter region of the transistor as being shorted. Conversely, when the transistor is cut off, visualize the collector-emitter region as being open. The end points $I_{C(sat)}$ and $V_{CE(off)}$ are calculated as follows:

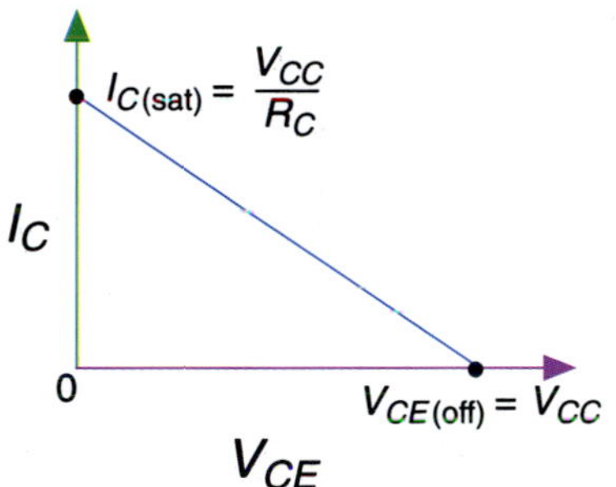

Figure 24-13 DC load line.

Equation 24-13

$$I_{C(sat)} = \frac{V_{CC}}{R_C}$$

Equation 24-14

$$V_{CE(off)} = V_{CC}$$

In most cases, the DC load line not only shows the values of $I_{C(sat)}$ and $V_{CE(off)}$, but it also shows the static values for I_C and V_{CE}. These static values of voltage and current indicate the operating point for the amplifier. The operating point is most commonly referred to as the **quiescent point,** or ***Q*** point. The term *quiescent point* refers to the DC values that exist without an AC input signal applied to the amplifier.

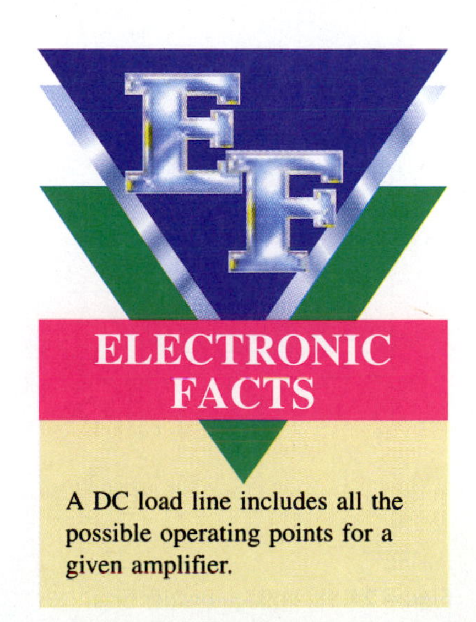

Example 24-1: For Fig. 24-14(a) on the next page, calculate the following DC quantities I_B, I_C, V_{CE}, $V_{CE(off)}$, and $I_{C(sat)}$. Also draw the DC load line showing the values for $I_{C(sat)}$ and $V_{CE(off)}$, as well as the Q point values of I_C and V_{CE}. Notice that $\beta_{DC} = 150$.

Solution: Begin by calculating I_B.

$$I_B = \frac{V_{CC} - V_{BE}}{R_B}$$

$$= \frac{12\text{ V} - 0.7\text{ V}}{270\text{ k}\Omega}$$

$$= 41.85\ \mu\text{A}$$

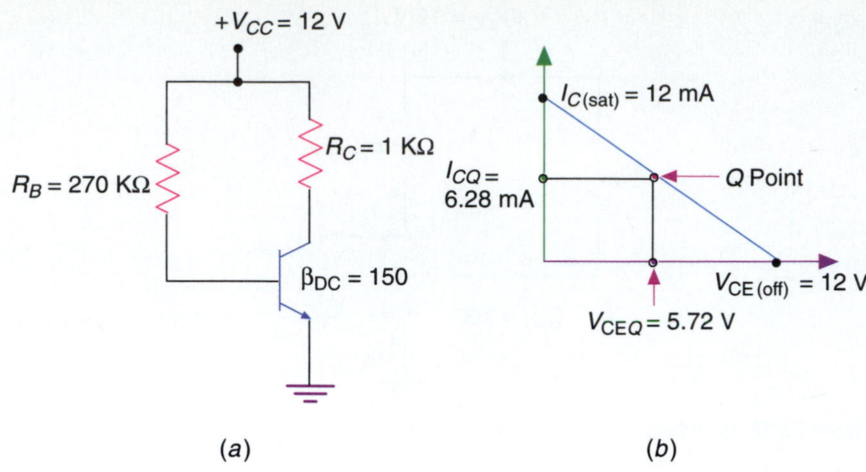

Figure 24-14 Circuit used for Example 24-1. (*a*) Circuit. (*b*) DC load line.

Next, calculate I_C. Since $\beta_{DC} = 150$, the calculation is:

$$\begin{aligned} I_C &= I_B \times \beta_{DC} \\ &= 41.85\ \mu A \times 150 \\ &= 6.28\ mA \end{aligned}$$

Finally, calculate V_{CE}.

$$\begin{aligned} V_{CE} &= V_{CC} - I_C R_C \\ &= 12\ V - (6.28\ mA \times 1\ k\Omega) \\ &= 5.72\ V \end{aligned}$$

Finally, use Eq. 24-13 and 24-14 to calculate $I_{C(sat)}$ and $V_{CE(off)}$.

$$\begin{aligned} I_{C(sat)} &= \frac{V_{CC}}{R_C} \\ &= \frac{12\ V}{1\ k\Omega} \\ &= 12\ mA \\ V_{CE(off)} &= V_{CC} \\ &= 12\ V \end{aligned}$$

The DC load line is shown in Fig. 24-14(*b*). The values of I_{CQ} and V_{CEQ} represent the operating point, or Q point, for the circuit. The letter Q in I_{CQ} and V_{CEQ} stands for quiescent, meaning static or at rest.

Base bias does have its drawbacks. The main problem is that this form of bias is especially sensitive to changes in the transistor's beta. Unfortunately, the beta of a transistor changes with its operating temperature. Therefore, if the base current I_B remains fixed, the collector current I_C can change within the transistor's operating temperature. This will cause a shift in the Q point on the DC load line. In severe cases, the Q point may shift to either saturation or cut off.

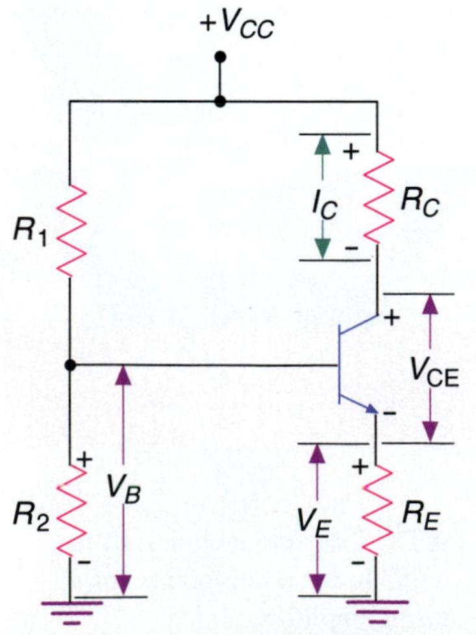

Figure 24-15 Voltage divider bias.

Voltage-Divider Bias

Figure 24-15 shows the most popular way to bias transistors, which is **voltage-divider bias.** It is given this name because R_1 and R_2 form a voltage divider to provide the desired base voltage V_B. Unlike base bias, voltage-divider bias is very stable. In fact, voltage-divider bias is almost immune to variations in the transistor's beta.

For Fig. 24-15, the following formulas are used to solve for the circuit voltages and currents.

Equation 24-15

$$V_B = \frac{R_2}{R_1 + R_2} \times V_{CC}$$

Equation 24-16

$$V_E = V_B - V_{\mathrm{BE}}$$

Equation 24-17

$$I_E = \frac{V_E}{R_E}$$

Note: Because I_B is very small, $I_E \approx I_C$, or

$$I_C = \frac{V_E}{R_E}$$

Equation 24-18

$$V_C = V_{CC} - I_C R_C$$

Equation 24-19

$$V_{\mathrm{CE}} = V_{CC} - I_C(R_C + R_E)$$

Equation 24-20

$$P_D = V_{\mathrm{CE}} \times I_C$$

The end points for the DC load line are calculated as:

Equation 24-21

$$I_{C(\mathrm{sat})} = \frac{V_{CC}}{R_C + R_E}$$

Equation 24-22

$$V_{\mathrm{CE(off)}} = V_{CC}$$

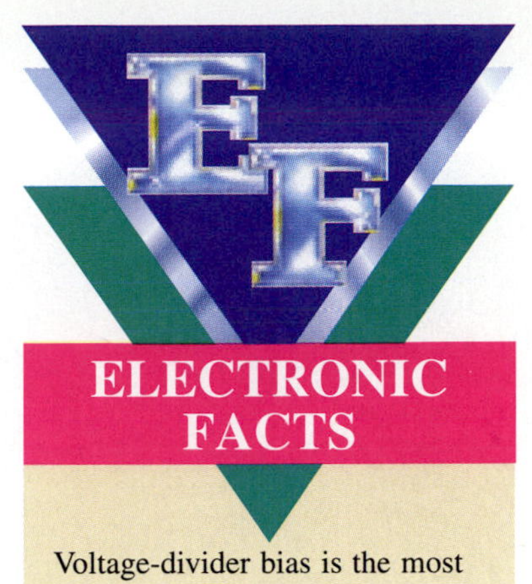

Voltage-divider bias is the most popular way to bias a transistor.

Example 24-2: In Fig. 24-16 on the next page, calculate the following DC quantities: V_B, V_E, I_{CQ}, V_C, $V_{\mathrm{CE}Q}$, $I_{C(\mathrm{sat})}$, and $V_{\mathrm{CE(off)}}$. Draw the DC load line showing the values of $I_{C(\mathrm{sat})}$, $V_{\mathrm{CE(off)}}$, I_{CQ}, and $V_{\mathrm{CE}Q}$.

Solution: Begin by calculating the base voltage V_B.

$$V_B = \frac{R_2}{R_1 + R_2} \times V_{CC}$$
$$= \frac{5.1\ \mathrm{k\Omega}}{27\ \mathrm{k\Omega} + 5.1\ \mathrm{k\Omega}} \times 15\ \mathrm{V}$$
$$= 2.38\ \mathrm{V}$$

Then, solve for V_E.

$$V_E = V_B - V_{\mathrm{BE}}$$
$$= 2.38\ \mathrm{V} - 0.7\ \mathrm{V}$$
$$= 1.68\ \mathrm{V}$$

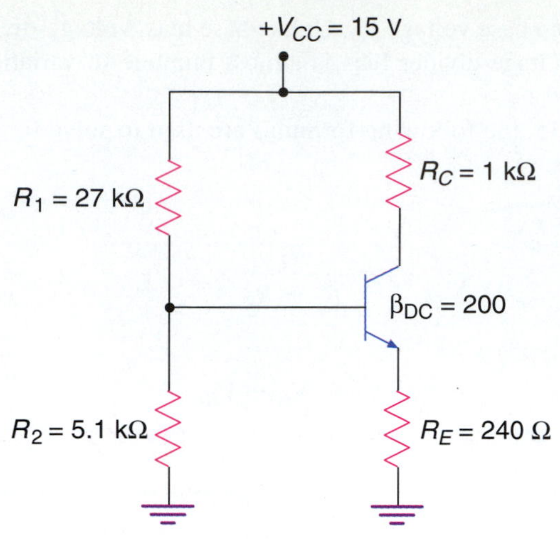

Figure 24-16 Circuit used for Example 24-2.

Next, calculate I_{CQ}.

$$I_{CQ} = \frac{V_E}{R_E} = \frac{1.68 \text{ V}}{240 \ \Omega} = 7 \text{ mA}$$

Then, solve for V_C.

$$\begin{aligned} V_C &= V_{CC} - I_{CQ}R_C \\ &= 15 \text{ V} - (7 \text{ mA} \times 1 \text{ k}\Omega) \\ &= 15 \text{ V} - 7 \text{ V} \\ &= 8 \text{ V} \end{aligned}$$

Next, calculate $V_{\text{CE}Q}$.

$$\begin{aligned} V_{\text{CE}Q} &= V_{CC} - I_{CQ}(R_C + R_E) \\ &= 15 \text{ V} - 7 \text{ mA}(1 \text{ k}\Omega + 240 \ \Omega) \\ &= 15 \text{ V} - 8.68 \text{ V} \\ &= 6.32 \text{ V} \end{aligned}$$

Finally, calculate $I_{C(\text{sat})}$ and $V_{\text{CE(off)}}$.

$$\begin{aligned} I_{C(\text{sat})} &= \frac{V_{CC}}{R_C + R_E} \\ &= \frac{15 \text{ V}}{1 \text{ k}\Omega + 240 \ \Omega} \\ &= 12.1 \text{ mA} \\ V_{\text{CE(off)}} &= V_{CC} \\ &= 15 \text{ V} \end{aligned}$$

The end points for the DC load line are shown in Fig. 24-17. Notice that the Q point values $I_{CQ} = 7$ milliamps and $V_{\text{CE}Q} = 6.32$ volts are also included.

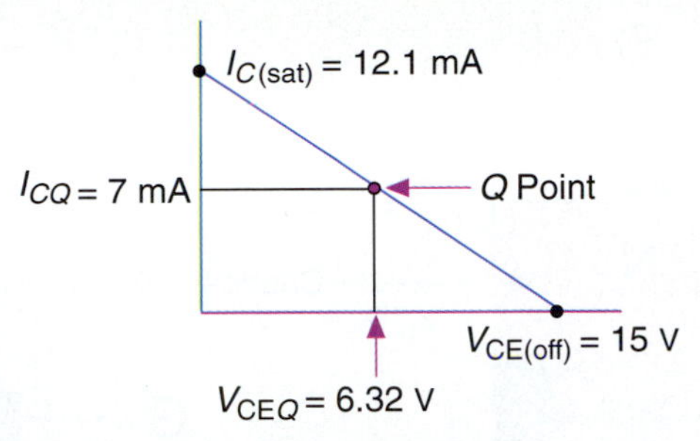

Figure 24-17 DC load line for the circuit in Example 24-2.

REVIEW QUIZ 24-5

1. Refer to Fig. 24-11(*a*). If *RB* increases, then *IC* ______________.
2. Refer to Fig. 24-11(*a*). If *RB* decreases, then V_{CE} ______________.
3. Which is a more stable form of bias: voltage-divider bias or base bias?
4. Will the collector current *IC* in Fig. 24-11(*a*) be affected if *RC* is decreased in value?

24-6 Field-Effect Transistors (FETs)

Unlike the bipolar junction transistor, the field-effect transistor (FET) is a unipolar device, meaning the FET uses only one type of charge carrier: electrons or holes. There are basically two types of FETs: the **junction field-effect transistor,** abbreviated JFET, and the metal-oxide semiconductor field-effect transistor, abbreviated MOSFET.

Unlike bipolar transistors, FETs are voltage-controlled devices. That is, an input voltage controls an output current. FETs have a very high input impedance, usually on the order of several hundred megohms. Therefore, they require very little power from the driving source.

Unlike BJTs, FETs are voltage-controlled devices, meaning that an input voltage controls an output current.

JFET

Figure 24-18(*a*) on the next page shows the basic construction of an N-channel JFET. Notice that there are three leads: the gate, (*G*), drain (*D*) and source (*S*). Notice also that the two P-type regions are internally connected to form a single gate lead. The N material between the drain and source terminals is called the channel, whereas each P-type region is called a gate.

Figure 24-18(*b*) shows the construction of a P-channel JFET. Notice that the channel between the drain and source terminals is made of P-type material, and the gate regions are made of N-type material.

For JFET, current flows in the channel between the drain and source terminals. The source current I_S and the drain current I_D are the same. The current in the channel of a JFET, however, is usually referred to as the drain current. In an N-channel JFET free electrons are the majority current carriers, whereas in the P-channel JFET holes are the majority current carriers.

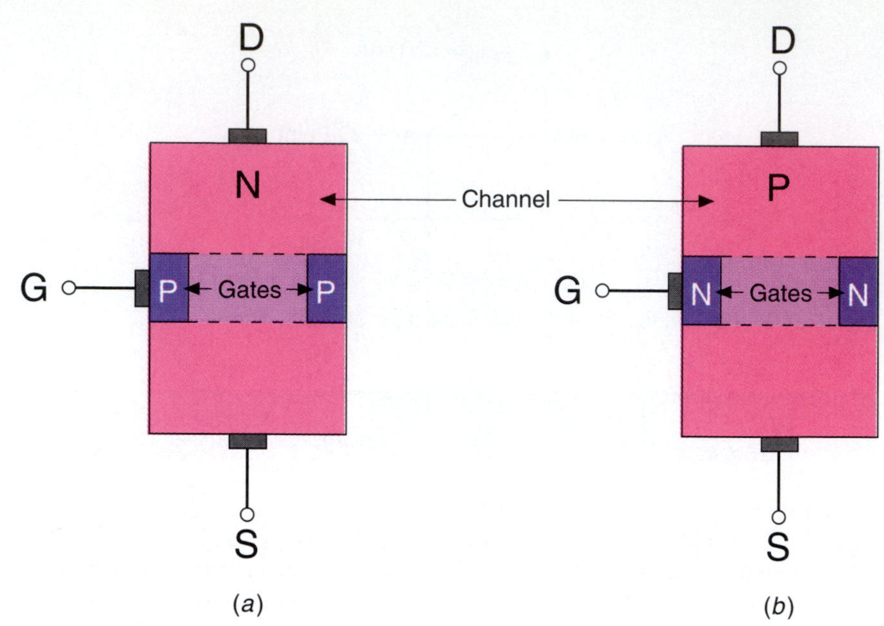

Figure 24-18 Construction of a JFET. (*a*) N-Channel JFET. (*b*) P-channel JFET.

Figure 24-19(*a*) shows the schematic symbol used for the N-channel JFET, and Fig. 24-19(*b*) shows the schematic symbol for the P-channel JFET. The thin vertical line connecting the drain and source reminds us that these terminals are connected to each end of the channel. Notice the arrow on each gate lead. For the N-channel JFET, the arrow points inward; for the P-channel JFET, the arrow points outward.

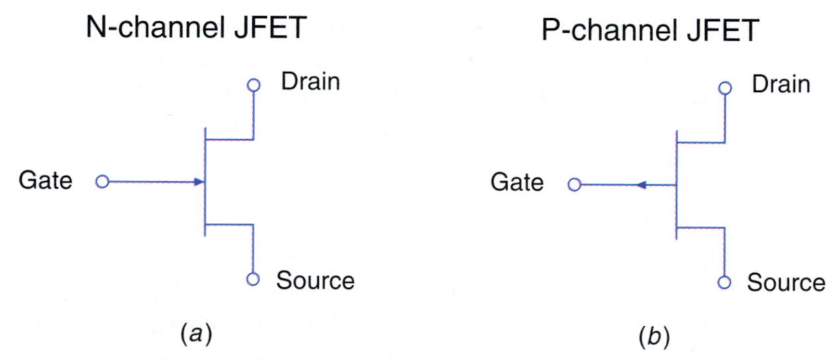

Figure 24-19 JFET schematic symbols. (*a*) N-Channel JFET. (*b*) P-channel JFET.

JFET Operation For a JFET, the gate-source junction is always reverse-biased. As a result, the gate current I_G is always zero (approximately). The reverse-biased gate-source junction has a depletion region that expands well into the channel. The depletion region narrows the channel, thereby increasing the channel's resistance. Varying the reverse-bias voltage across the gate-source junction varies the width of the depletion region, which in turn controls the drain current I_D.

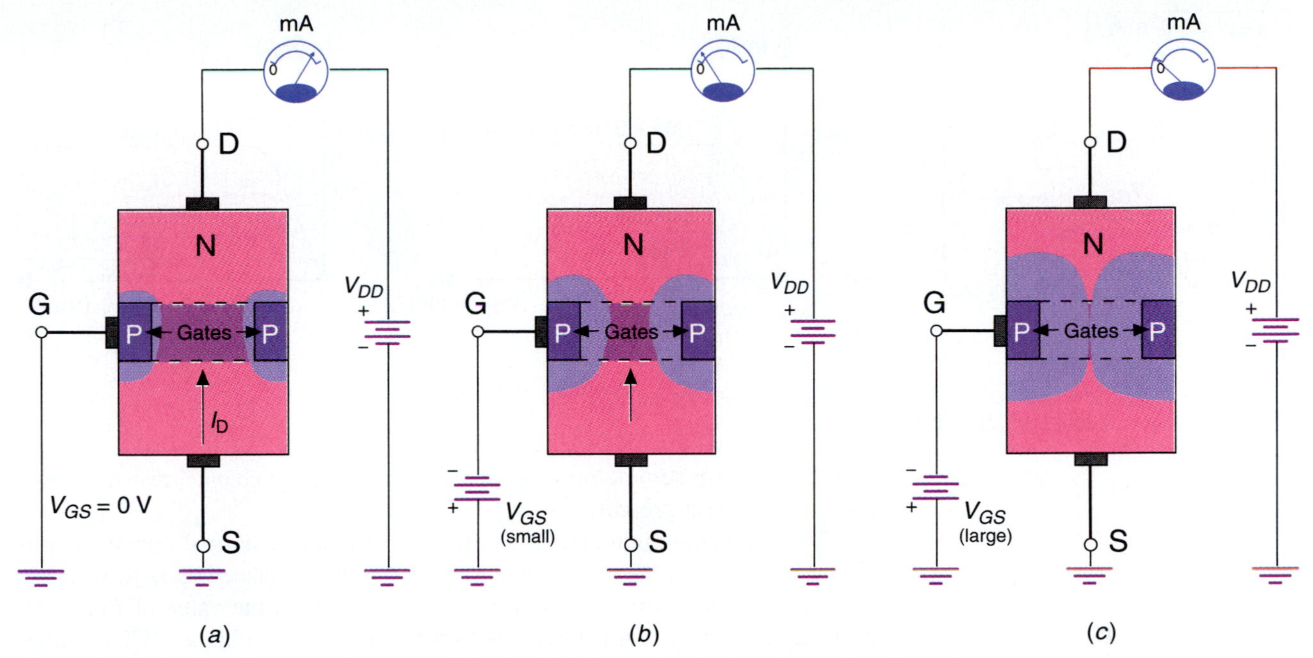

Figure 24-20 Current flow in an N-channel JFET. (*a*) With $V_{GS} = 0$ volts, the channel is wide and drain current is high. (*b*) With V_{GS} slightly negative, the channel narrows and the drain current decreases. (*c*) V_{GS} is a large negative value. The channel is pinched off, resulting in a drain current I_D of zero.

Figure 24-20 illustrates how the drain current can be controlled in an N-channel JFET by varying the negative gate-source voltage V_{GS}. The gate-source voltage in Fig. 24-20(*a*) is zero because the gate is shorted to the source. Because the drain is connected to the positive terminal of the drain supply voltage V_{DD}, the gate-source junction is always reverse-biased. Notice the wide channel between the depletion layers. This results in a significant amount of drain current I_D as indicated by the current meter. Figure 24-20(*b*) shows the channel when the gate is made slightly negative relative to the source. Because of the increase in the gate-source reverse-bias, the channel is narrower and the drain current is less when compared to Fig. 24-20(*a*). Finally, in Fig. 24-20(*c*) a very large negative voltage is applied at the gate. In this case, the depletion regions expand to a point where they touch, thereby pinching off the channel. This results in a drain current of zero. The gate source voltage required to pinch off the channel and reduce I_D to zero is called the *gate-source cutoff voltage,* designated $V_{GS(\text{off})}$.

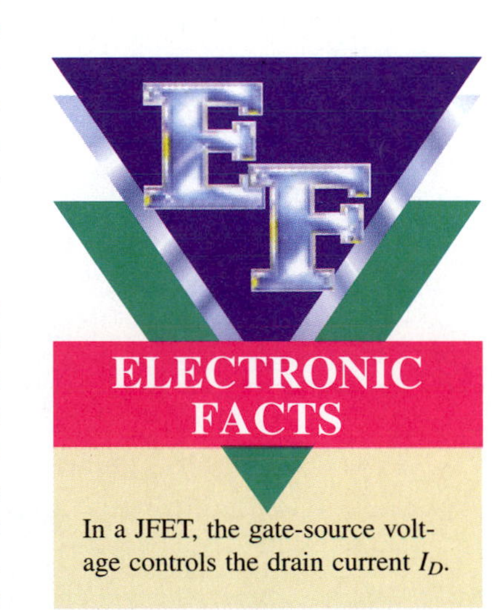

In a JFET, the gate-source voltage controls the drain current I_D.

JFET Characteristics Figure 24-21(a) on the next page shows an N-channel JFET with the proper biasing voltages. Note that the drain is positive and the gate is negative. Figure 24-21(*b*) shows the drain curve when the gate source voltage V_{GS} is zero. This graph depicts how the drain current I_D varies with the drain-source voltage V_{DS}. When V_{DS} increases from zero, I_D increases proportionally until the pinch-off voltage V_P is reached. The **pinchoff voltage** is the drain-source voltage at which the drain current levels off. The region to the left of V_P is called the *ohmic region* because I_D varies in direct proportion to V_{DS}. Because I_D levels off to the right of V_P, this region is called the *current-source*

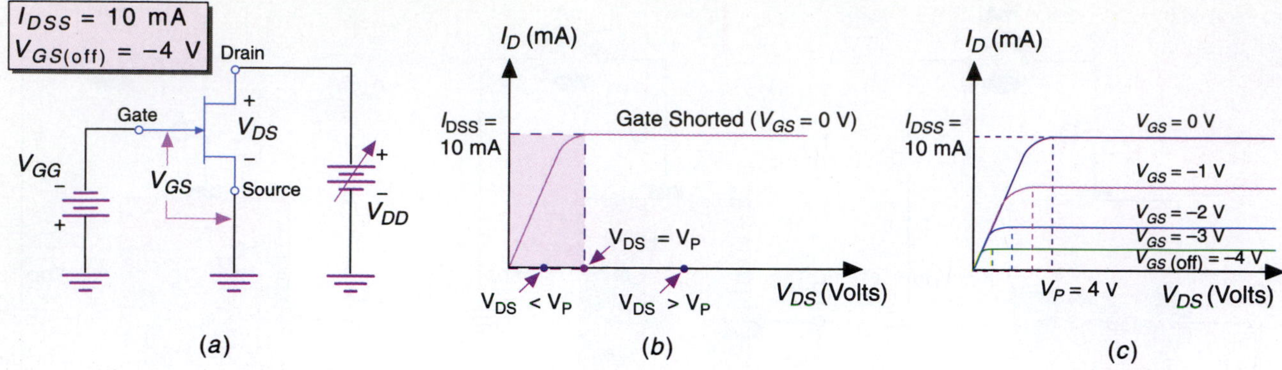

Figure 24-21 JFET drain curves. (*a*) Normal biasing voltages for an N-channel JFET. (*b*) Drain curve with V_{GS} = 0 volts. (*c*) Multiple drain curves.

region. The drain current levels off above V_P because the channel resistance R_{DS} increases in direct proportion to V_{DS}.

The maximum drain current a JFET can have under normal operating conditions occurs when $V_{GS} = 0$ volts. This current is designated as I_{DSS}, which is the drain current when $V_{GS} = 0$ volts. In Fig. 24-21(*b*), the value of I_{DSS} is 10 milliamps. Because drain current flows when $VGS = 0$ volts, a JFET is sometimes referred to as a **"normally ON"** device.

Figure 24-21(*c*) shows a complete set of drain curves. Notice that as V_{GS} is made increasingly more negative, the drain current decreases. Notice also that as V_{GS} is made more negative, I_D levels off at a lower value of drain-source voltage V_{DS}. For the JFET in Fig. 24-21, I_D equals zero when $V_{GS} = V_{GS(\text{off})}$, which is −4 volts in this case. It is interesting to note that for any JFET, the magnitudes of V_P and $V_{GS(\text{off})}$ are the same. Therefore, the following is true for any JFET: $|V_P| = |V_{GS(\text{off})}|$.

ELECTRONIC FACTS

For any JFET, the maximum drain current flows when $V_{GS} = 0$ V.

Transconductance Curve Figure 24-22 shows the **transconductance curve** of a JFET, which is a graph of I_D versus V_{GS}. This graph is nonlinear because equal changes in V_{GS} produce unequal changes in I_D.

To calculate the drain current I_D in a JFET, use Eq. 24-23.

Equation 24-23

$$I_D = I_{DSS}\left(1 - \frac{V_{GS}}{V_{GS(\text{off})}}\right)^2$$

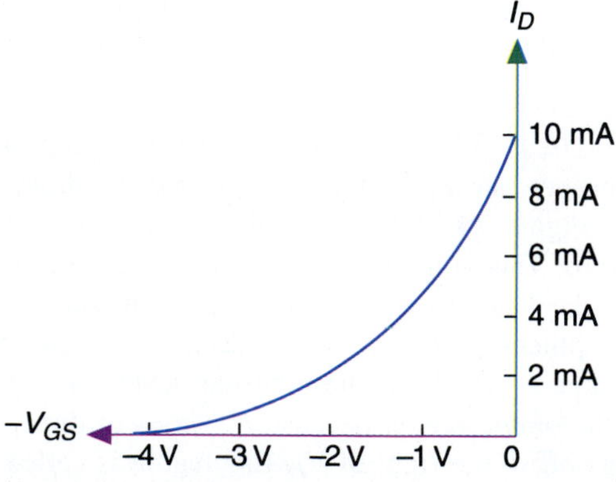

Figure 24-22 Transconductance curve.

It should be noted that the values of I_{DSS} and $V_{GS(off)}$ are usually specified on the manufacturer's data sheet.

Example 24-3: For Fig. 24-21(*a*), calculate the drain current I_D for $V_{GS} = -1.5$ volts.

Solution:
The values of I_{DSS} and $VGS_{(off)}$ in Fig. 24-21(*a*) equal 10 milliamperes and −4 volts respectively. Using Eq. 24-23, we can calculate I_D as follows:

$$\begin{aligned} I_D &= I_{DSS}\left(1 - \frac{V_{GS}}{V_{GS(\text{off})}}\right)^2 \\ &= 10 \text{ mA}\left(1 - \frac{-1.5 \text{ V}}{-4 \text{ V}}\right)^2 \\ &= 10 \text{ mA} \times 0.39 \\ &= 3.9 \text{ mA} \end{aligned}$$

MOSFET

The **metal-oxide semiconductor field-effect transistor** (MOSFET) has a gate, source, and drain just like the JFET. As in the JFET, the drain current I_D in a MOSFET is controlled by varying the gate-source voltage V_{GS}. There are two basic types of MOSFETs: the enhancement type and the depletion type. The enhancement-type MOSFET is usually referred to as an E-MOSFET and, the depletion-type MOSFET is referred to as a D-MOSFET.

The key difference between JFETs and MOSFETs is that the gate terminal in a MOSFET is insulated from the channel. Thus, MOSFETs are sometimes referred to as insulated-gate field-effect transistors, or IGFETs. Because of the insulated gate, the input impedance of a MOSFET is many times higher than that of a JFET.

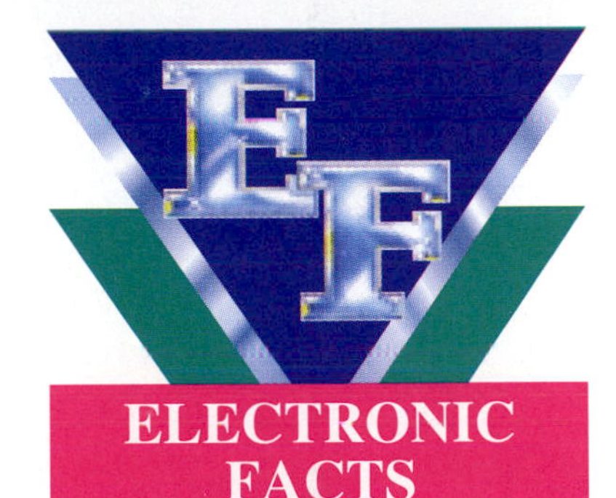

Because of the insulated gate, the input impedance of a MOSFET is many times higher than that of a JFET.

Depletion-Type MOSFET Figure 24-23(*a*) shows the basic construction of an N-channel **depletion-type MOSFET,** and Fig. 24-23(*b*) shows the schematic symbol. The drain terminal is at the top of the N material, whereas the source

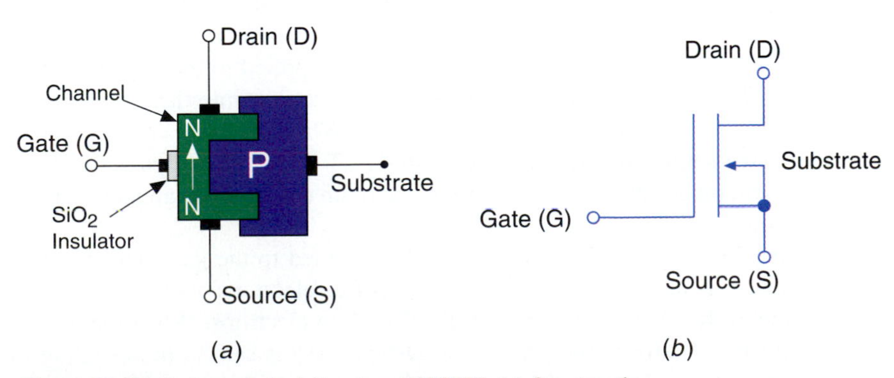

Figure 24-23 N-channel depletion-type MOSFET. (*a*) Construction. (*b*) Schematic symbol.

terminal is at the bottom. The block of P-type material forms the substrate into which the N-type material is embedded. The N-type material forms the channel. Along the left side of the N channel in Fig. 24-23(*a*) is a thin layer of silicon dioxide (SiO_2), which has been deposited to isolate the gate from the channel. Going from gate to channel we encounter metal, SiO_2, and N-type semiconductor material, in that sequence, which is how the MOSFET got its name.

Notice in Fig. 24-23(*b*) that the substrate is connected to the source. This results in a three-terminal device. That gate appears as one plate of a capacitor, and the channel acts like the other plate. The solid line connecting the source and drain terminals indicates that depletion-type MOSFETs are "normally ON" devices. This means that when V_{GS} = 0 volts, drain current I_D still flows.

Zero Gate Voltage A D-MOSFET can operate with either a positive or negative gate voltage. As shown in Fig. 24-24(*a*), the depletion-type MOSFET also conducts with the gate shorted to the source for V_{GS} = 0 volts.

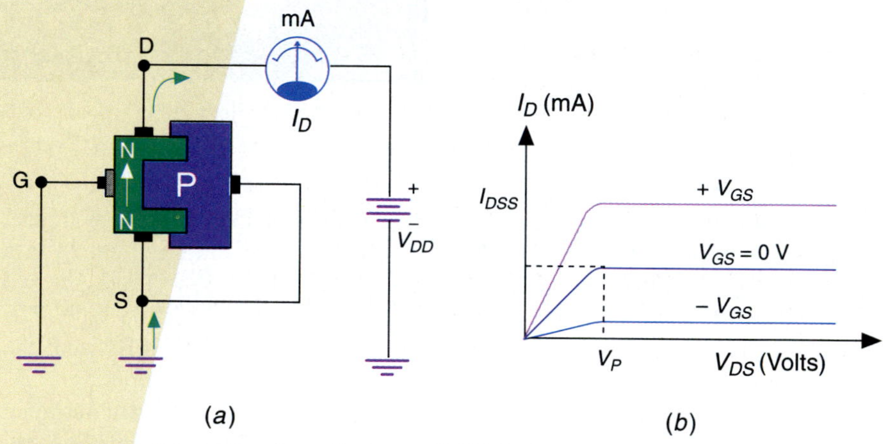

Figure 24-24 Current flow in an N-channel D-MOSFET. (*a*) Circuit with V_{GS} = 0 volts. (*b*) Drain curves.

A D-MOSFET is similar to a JFET in terms of its operating characteristics. This is shown in Fig. 24-24(*b*). Notice that for each drain curve, I_D increases linearly until the pinch-off voltage V_P is reached. When *VGS* is negative, pinch-off occurs sooner (lower values of V_{DS}), and when V_{GS} is made positive, pinch-off occurs later (higher values of V_{DS}). Notice also in Fig. 24-24(*b*) that the drain current I_D with zero gate voltage is not its maximum value. Yet I_{DSS} is still defined the same way: the drain current I_D with V_{GS} = 0 volts.

Figure 24-25(*a*) shows a positive gate voltage applied to the depletion-type MOSFET. The positive gate voltage attracts free electrons into the channel from the substrate, thereby enhancing its conductivity. When the gate is made positive relative to the source, the depletion-type MOSFET is said to be operating in the enhancement mode. When V_{GS} is positive, the drain current I_D is greater than the value of I_{DSS}.

Figure 24-25(*b*) shows a negative voltage applied to the gate. The negative gate voltage sets up an electric field that repels free electrons out of the channel. This depletes the channel of some or all of its free electrons. When the gate is made negative relative to the source, the D-MOSFET is said to be operating in the depletion mode. Making the gate negative enough will reduce the drain current *ID* to zero.

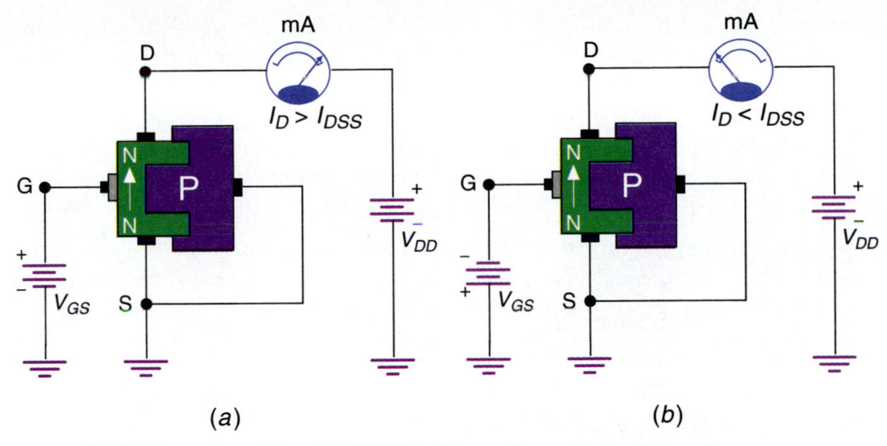

Figure 24-25 Operation of the D-MOSFET with positive and negative gate voltages. (*a*) Drain current increases with positive gate voltage. (*b*) Drain current decreases with negative gate voltage.

Figure 24-26 shows the transconductance curve for the N-channel D-MOSFET, where I_{DSS} is the drain current that flows with the gate shorted to the source ($V_{GS} = 0$ volts). It is important to note that I_{DSS} is not the maximum possible drain current. When V_{GS} is positive, the D-MOSFET operates in the enhancement mode, and the drain current I_D increases beyond the value of I_{DSS}. When V_{GS} is negative, the D-MOSFET operates in the depletion mode, and I_D is less than I_{DSS}. If V_{GS} is made negative enough, the drain current will be reduced to zero. As with JFETs, the value of gate-source voltage that reduces the drain current I_D to zero is called the *gate-source cutoff voltage,* designated $V_{GS(\text{off})}$.

Because D-MOSFETs are "normally ON" devices, the drain current I_D can be calculated using Eq. 24-23.

$$I_D = I_{DSS}\left(1 - \frac{V_{GS}}{V_{GS(\text{off})}}\right)^2$$

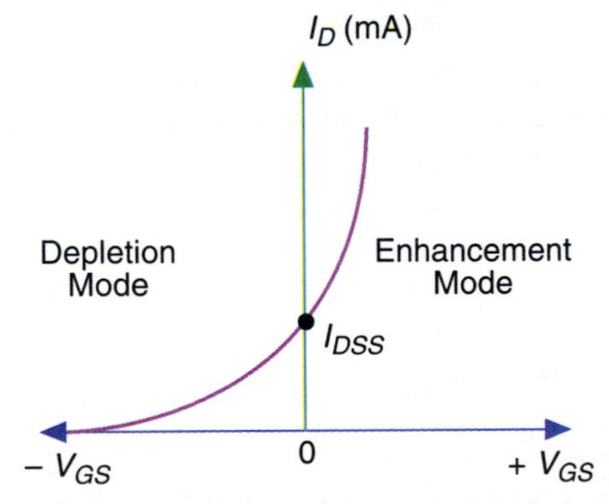

Figure 24-26 Transconductance curve for the N-channel D-MOSFET.

P-Channel Depletion-Type MOSFET Figure 24-27 on the next page shows the construction, schematic symbol, and transconductance curve for a P-channel D-MOSFET.

Figure 24-27(*a*) shows that the channel is made of P-type material, whereas the substrate is made of N-type material. Thus, P-channel D-MOSFETs require a negative drain voltage.

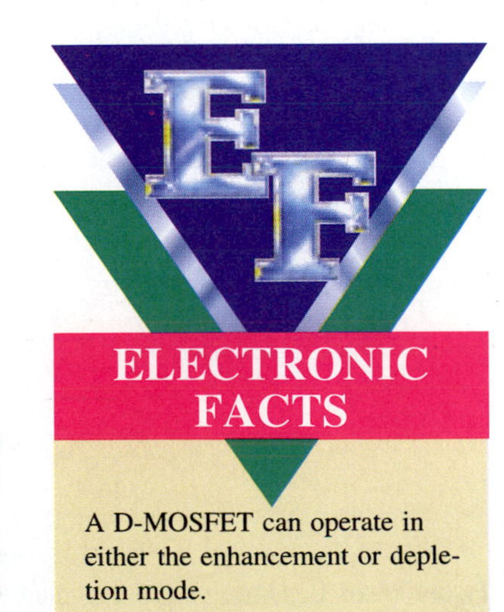

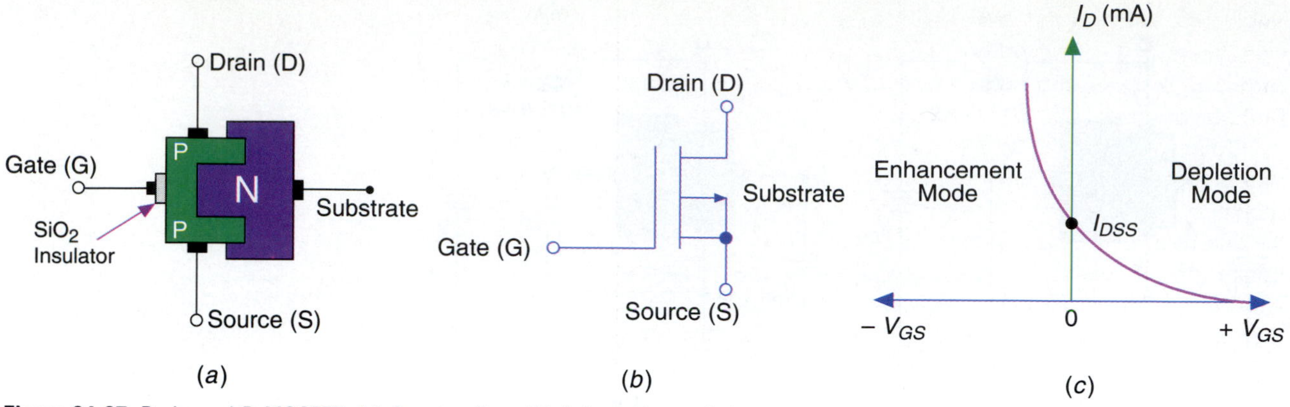

Figure 24-27 P-channel D-MOSFET. (*a*) Construction. (*b*) Schematic symbol. (*c*) Transconductance curve.

Compare the transconductance curve in Fig. 24-27(*c*) with the one in Fig. 24-26. Notice that they are opposite. The P-channel D-MOSFET operates in the enhancement mode when *VGS* is negative and in the depletion mode when *VGS* is positive. It also should be noted that holes are the majority current carriers in the P-channel.

Enhancement-Type MOSFETs Figure 24-28(*a*) shows the basic construction of an N-channel **enhancement-type MOSFET.** Notice that the P-type substrate makes contact with the SiO_2 insulator. Because of this, there is no channel for

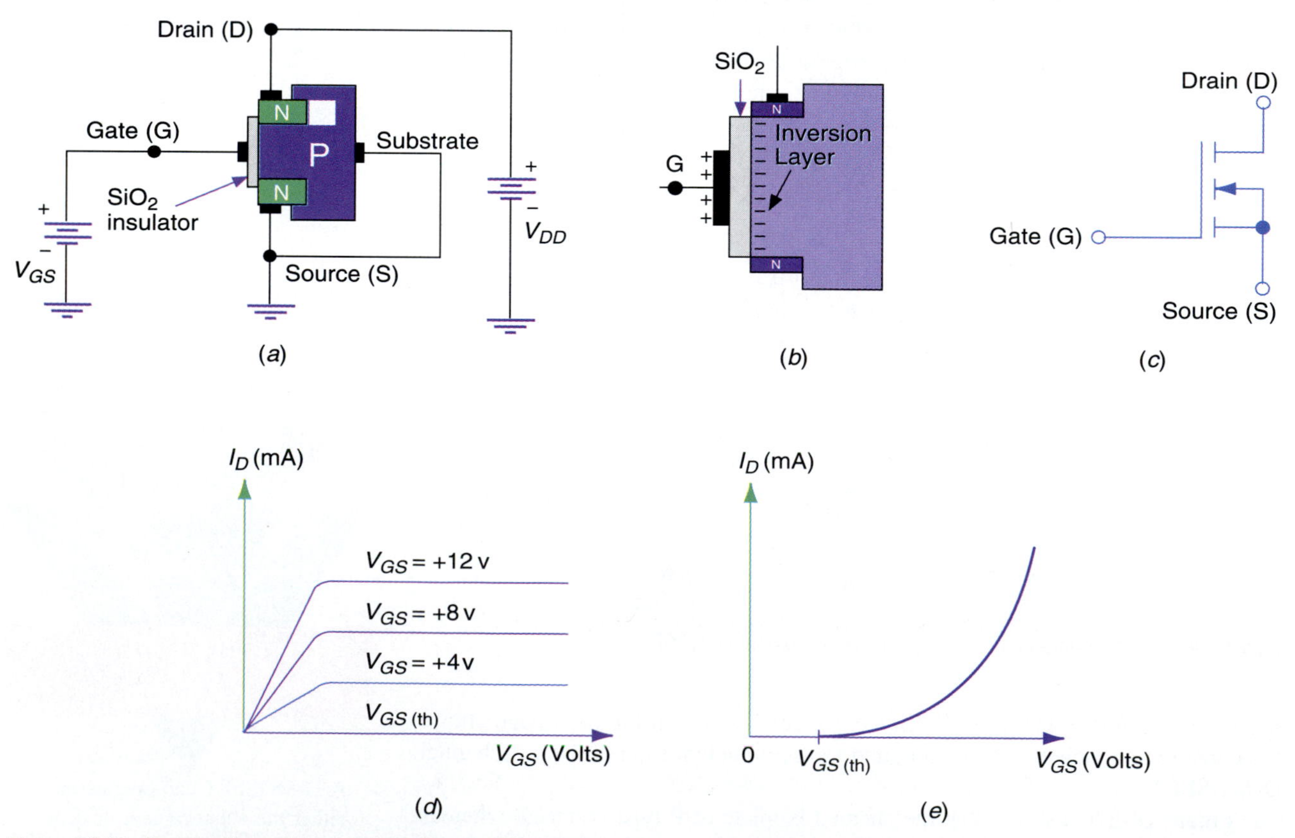

Figure 24-28 Construction and operation of an E-MOSFET. (*a*) Circuit. (*b*) Electric charges in SiO_2 insulating layer. (*c*) Schematic symbol. (*d*) Drain curves. (*e*) Transconductance curve.

conduction between the drain and source terminals. Therefore, when $V_{GS} = 0$ volts, there is no channel between the source and drain. This results in a drain current I_D of zero. Notice the polarities for the supply voltages in Fig. 24-28(*a*). Both the drain and gate are made positive with respect to the source.

To get drain current to flow, the positive gate voltage must be increased. This causes electrons to be attracted along the right edge of the SiO_2 insulator, as shown in Fig. 24-28(*b*). The P-material is said to be inverted (from P to N) at the interface between the P-doped semiconductor and the SiO_2 layer. These electrons form the channel between the source and drain. The minimum gate-source voltage V_{GS} that causes complete inversion and drain current to flow is called the **threshold voltage** and is designated as $V_{GS(\mathrm{th})}$. When the gate voltage is less than $V_{GS(\mathrm{th})}$, the drain current I_D is zero. The value of $V_{GS(\mathrm{th})}$ varies from one E-MOSFET to the next.

Figure 24-28(*c*) shows the schematic symbol for the N-channel enhancement-type MOSFET. Notice the broken channel line. The broken line represents the OFF condition that exists with zero gate voltage. Because of this characteristic, enhancement-type MOSFETS are called **"normally OFF"** devices.

Figure 24-28(*d*) shows a typical set of drain curves for the N-channel enhancement-type MOSFET. The lowest curve is the $V_{GS(\mathrm{th})}$ curve. For more positive gate voltages the drain current I_D increases. The transconductance curve is shown in Fig. 24-28(*e*). Notice that I_D is zero when the gate-source voltage is less than the threshold voltage $V_{GS(\mathrm{th})}$.

Figure 24-29(*a*) shows a P-channel E-MOSFET. Notice the arrow pointing outward, away from the P-type channel. Also notice the negative gate and drain voltages. These are the required polarities for biasing the P-channel enhancement-type MOSFET.

Figure 24-29(*b*) shows the transconductance curve for the P-channel enhancement-type MOSFET. Notice that I_D is zero until the gate-source voltage is more negative than $-V_{GS(\mathrm{th})}$.

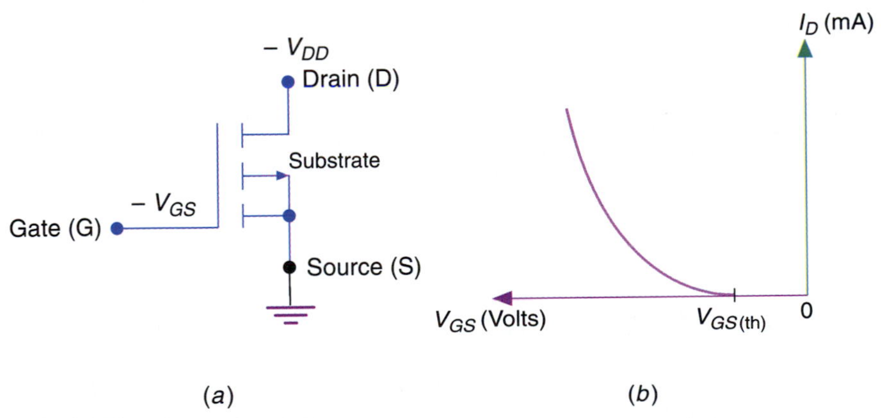

Figure 24-29 Biasing the P-channel E-MOSFET. (*a*) Proper biasing voltages. (*b*) Transconductance curve.

Handling Precautions

One disadvantage of MOSFET devices is their extreme sensitivity to **electrostatic discharge** (ESD). This sensitivity to static electricity is owing to the insulated gate-source region. The SiO_2 insulating layer is extremely thin and can be easily punctured with an electrostatic discharge.

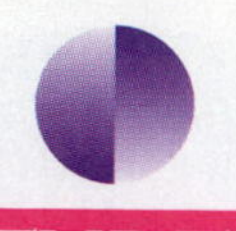

STUDENT *to* **STUDENT**

When handling MOSFET devices, always wear a grounding strap. This will prevent ESD from damaging the MOSFET that is being handled.

Typical breakdown voltages for the insulating layer can be as low as ±30 Vdc. Because MOSFETs can be easily damaged from ESD, extreme caution must be observed when handling them. The following is a list of precautions that must be observed in order to avoid damaging MOSFET devices.

1. Never insert or remove MOSFETs from a circuit with the power ON.
2. Never apply input signals when the DC power supply is OFF.
3. Wear a grounding strap on your wrist when handling MOSFET devices. The grounding strap connects to ground through a high value series resistance, such as 2.2 megohms. The grounding strap continually bleeds off static electricity (charge) from your body.
4. When storing MOSFETs, keep the device leads in contact with conductive foam, or connect a shorting ring around the leads.

Many manufacturers use protective diodes connected across the gate-source region to protect against ESD. This is shown in Fig. 24-30. The diodes are arranged so that they will conduct for either polarity of gate-source voltage V_{GS}. The breakdown voltage of the diodes is much higher than any voltage normally applied between the gate-source region, but less than the breakdown voltage of the insulating material. One drawback of using the protective diodes is that the input impedance of the device is lowered considerably.

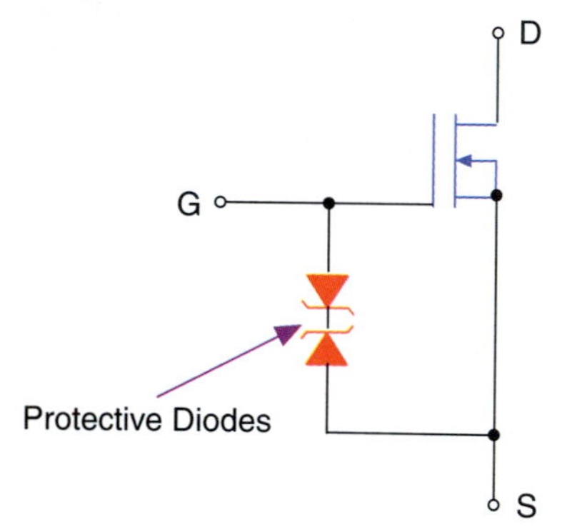

Figure 24-30 Protective diodes connected across the gate-source region protect the device against ESD.

REVIEW QUIZ 24-6

1. A JFET is a "normally ______________" device.
2. In a JFET, the gate-source junction is always ______________-biased.
3. In a JFET, the gate current I_G is ______________.
4. A JFET has the following values: I_{DSS} = 12 milliamps and $V_{GS(\text{off})}$ = −4 volts. If $V_{GS} = -2$ volts, then I_D = ______________.
5. MOSFETs are extremely sensitive to ______________.
6. For an N-channel E-MOSFET, the drain current I_D is ______________ when $V_{GS} = 0$ volts.
7. An E-MOSFET is a "normally ______________" device.
8. A D-MOSFET is a "normally ______________" device.
9. True or false? For a D-MOSFET, I_{DSS} is the maximum possible drain current.

24-7 How to Bias FETs

Many different techniques are used to bias JFET. In all cases, however, the gate-source junction is reverse-biased. Some common biasing techniques are covered in this section. They include gate bias, self-bias, and voltage-divider bias.

Gate Bias

Figure 24-31(*a*) shows an example of a JFET using **gate bias.** The negative gate voltage is applied through a gate resistor R_G, which can be any value, but it is usually quite large, such as 100 kilohms or more. Because the gate current I_G is zero in a JFET, the voltage drop across R_G is zero. The main purpose of R_G is

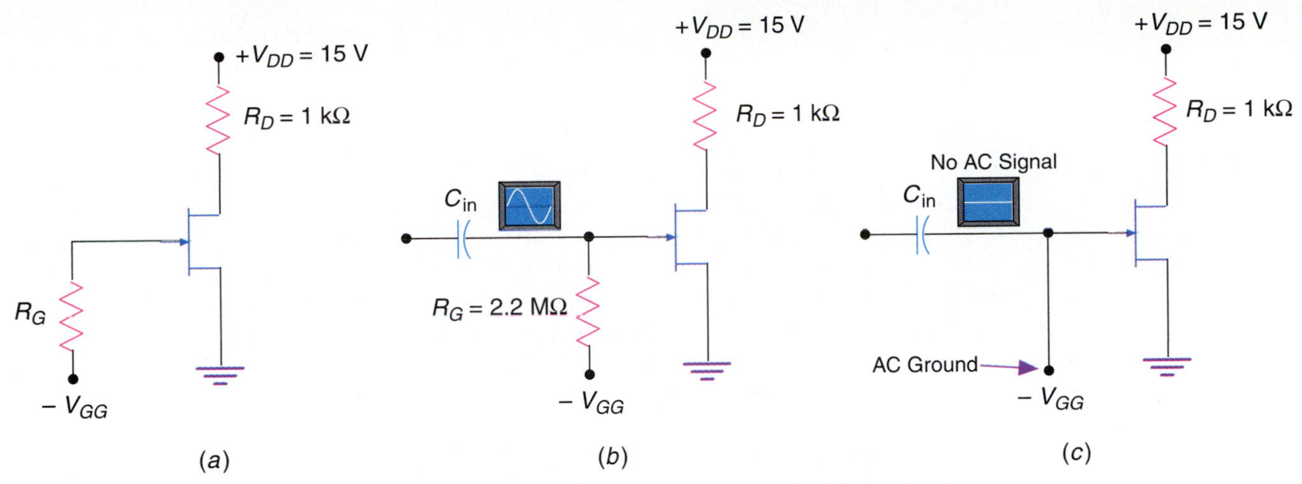

Figure 24-31 Gate bias. (*a*) Negative gate voltage applied through gate resistor R_G. (*b*) AC signal coupled to gate. (*c*) No AC signal at gate, with R_G omitted because the gate is at AC ground.

to isolate the gate from ground for AC signals. Figure 24-31(*b*) shows how an AC signal is coupled to the gate of a JFET. If R_G were omitted, as shown in Fig. 24-31(*c*), no AC signal would appear at the gate because *VGG* is at ground for AC signals. Thus, R_G is usually made equal to the value desired for the input impedance Z_{in} of the amplifier. Because the gate-source junction of a JFET is reverse-biased, its impedance is at least several hundred megohms; therefore, $Z_{in} = R_G$. In Fig. 24-31(*b*), R_G = 2.2 megohms, which means Z_{in} = 2.2 megohms.

When $V_{GS(off)}$ and I_{DSS} are known, the drain current I_D can be found for any value of V_{GS}. When the drain current is known, use Eq. 24-23 to calculate the drain-source voltage V_{DS}.

Equation 24-24

$$V_{DS} = V_{DD} - I_D R_D$$

Gate bias is seldom used with JFETs. The reason is that in mass production, the characteristics of the individual JFET, such as I_{DSS} and $V_{GS(off)}$ may vary drastically from one circuit to the next. This means that even though the gate voltage is the same from one JFET circuit to the next, the drain current I_D could be drastically different. As a result, we have an unpredictable value of drain current I_D and drain-source voltage V_{DS} with gate bias.

Self-Bias

One of the most common ways to bias JFETs is with **self-bias.** An example of self bias is shown in Fig. 24-32 on the next page. Notice that only a single power supply is used, that being the drain supply voltage V_{DD}. In this case, the voltage drop across the source resistor R_S provides the gate-source bias voltage. But how is this possible? Here's how.

When power is first applied, drain current flows, which produces a voltage drop across the source resistor R_S. For the direction of drain current shown, the source is positive with respect to ground. Also, because there is no gate current, the gate voltage V_G equals 0 volts. Therefore, V_{GS} is calculated as follows:

$$\begin{aligned} V_{GS} &= V_G - V_S \\ &= 0\text{ V} - V_S \end{aligned}$$

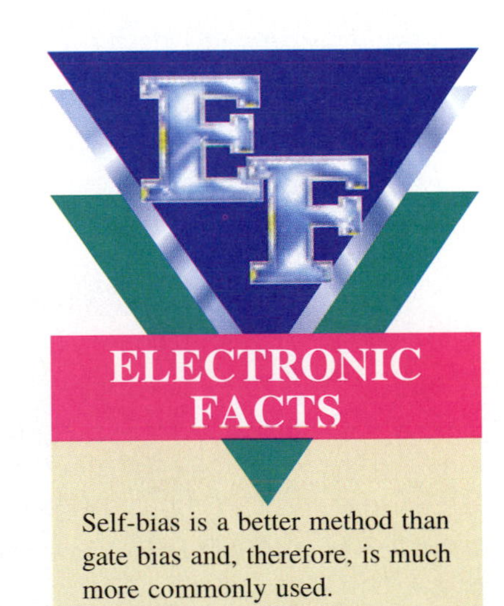

Self-bias is a better method than gate bias and, therefore, is much more commonly used.

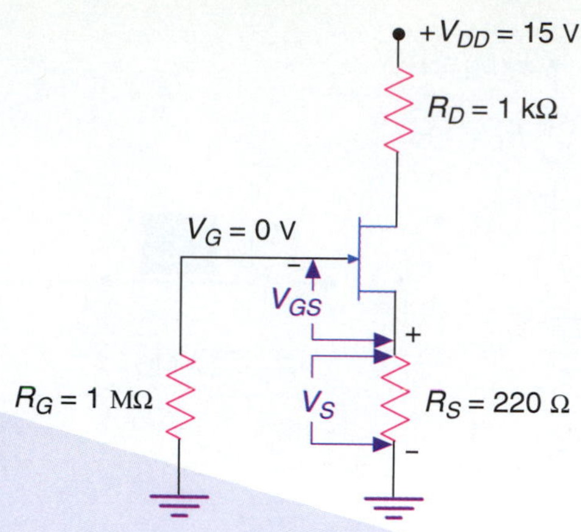

Figure 24-32 Self-biased JFET. $V_{GS} = -I_D R_S$.

Also, since $V_S = I_D R_S$, we have

Equation 24-25

$$V_{GS} = 0\text{ V} - I_D R_S \qquad \text{or}$$
$$V_{GS} = -I_D R_S$$

Self-bias is relatively stable because any increase in the drain current I_D causes V_{GS} to increase. The increase in V_{GS} causes the drain current to decrease, which partially offsets the original increase in drain current.

Likewise, a decrease in the drain current causes V_{GS} to decrease. The decrease in V_{GS} causes I_D to increase, which partially offsets the original decrease in drain current.

For the self-bias circuit in Fig. 24-32, the drain voltage V_D is calculated as follows:

Equation 24-26

$$V_D = V_{DD} - I_D R_D$$

To calculate the drain-source voltage V_{DS}, use Eq. 24-27.

Equation 24-27

$$V_{DS} = V_{DD} - I_D(R_D + R_S)$$

Example 24-4: For Fig. 24-32, assume that $I_D = 5$ milliamps. Calculate V_{GS}, V_D, and V_{DS}.

Solution:
Begin by calculating V_{GS} using Eq. 24-25.

$$\begin{aligned} V_{GS} &= -I_D R_S \\ &= -(5\text{ mA} \times 220\ \Omega) \\ &= -1.1\text{ V} \end{aligned}$$

Next, calculate V_D.

$$\begin{aligned} V_D &= V_{DD} - I_D R_D \\ &= 15\text{ V} - (5\text{ mA} \times 1\text{ k}\Omega) \\ &= 15\text{ V} - 5\text{ V} \\ &= 10\text{ V} \end{aligned}$$

Finally, solve for V_{DS}.

$$\begin{aligned} V_{DS} &= V_{DD} - I_D(R_S + R_D) \\ &= 15\ \text{V} - (5\ \text{mA} \times 1.22\ \text{k}\Omega) \\ &= 15\ \text{V} - 6.1\ \text{V} \\ &= 8.9\ \text{V} \end{aligned}$$

Voltage-Divider Bias

Figure 24-33 shows a JFET with voltage-divider bias. Because the gate-source junction has an extremely high resistance (several hundred megohms), the $R_1 - R_2$ voltage divider is practically unloaded. Therefore, the gate voltage V_G is calculated as follows:

Equation 24-28

$$V_G = \frac{R_2}{R_1 + R_2} \times V_{DD}$$

The source voltage *VS* is calculated as:

Equation 24-29

$$V_S = V_G - V_{GS}$$

Because $I_D = I_S$, the drain current is calculated as:

Equation 24-30

$$I_D = \frac{V_S}{R_S}$$

Also, the drain voltage is calculated using Eq. 24-26:

$$V_D = V_{DD} - I_D R_D$$

Voltage-divider bias does have its drawbacks. The values of V_{GS} vary from one JFET to the next, thereby making it difficult to predict the exact values of I_D and V_D for a given circuit.

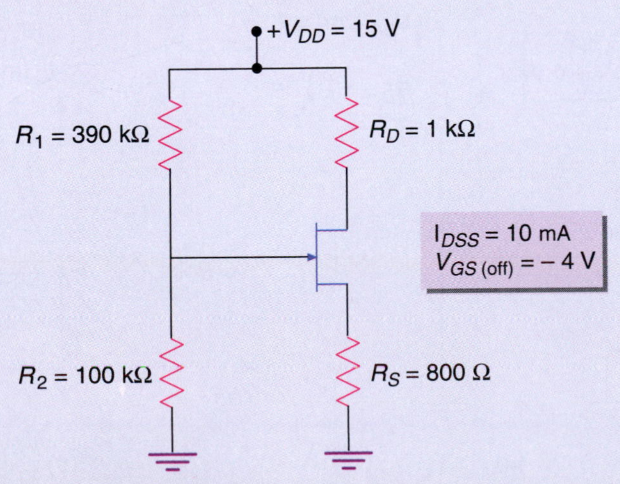

Figure 24-33 Voltage-divider bias.

Example 24-5: In Fig. 24-33, $V_{GS} = -1$ volt. Calculate V_G, V_S, I_D, and V_D.

Solution:

Begin by calculating V_G.

$$V_G = \frac{R_2}{R_1 + R_2} \times V_{DD}$$
$$= \frac{100\ \text{k}\Omega}{390\ \text{k}\Omega + 100\ \text{k}\Omega} \times 15\ \text{V}$$
$$= 3\ \text{V}$$

Next, calculate V_S.

$$V_S = V_G - V_{GS}$$
$$= 3\ \text{V} - (-1\ \text{V})$$
$$= 4\ \text{V}$$

Then calculate I_D.

$$I_D = \frac{V_S}{R_S}$$
$$= \frac{4\ \text{V}}{800\ \Omega}$$
$$= 5\ \text{mA}$$

Last, calculate V_D.

$$V_D = V_{DD} - I_D R_D$$
$$= 15\ \text{V} - (5\ \text{mA} \times 1\ \text{k}\Omega)$$
$$= 10\ \text{V}$$

MOSFET Biasing Techniques

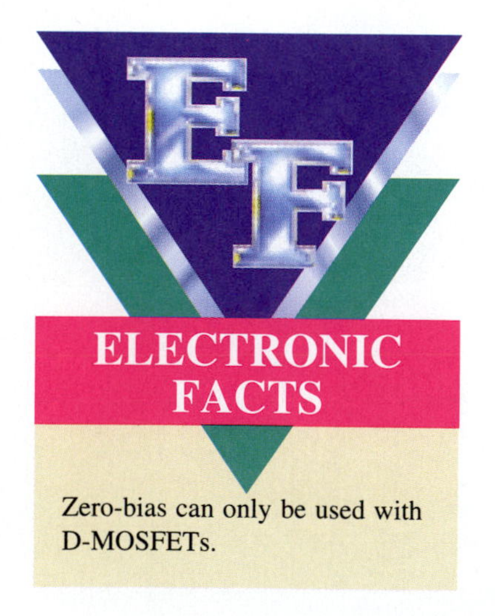

Zero-bias can only be used with D-MOSFETs.

Figure 24-34(*a*) shows a popular biasing technique that can be used only with D-MOSFETs. This form of bias is called **zero bias** because the voltage across the gate-source region is zero. With *VGS* at zero, the quiescent drain current I_D equals I_{DSS}. This is shown in Fig. 24-34(*b*). If a signal at the gate goes positive,

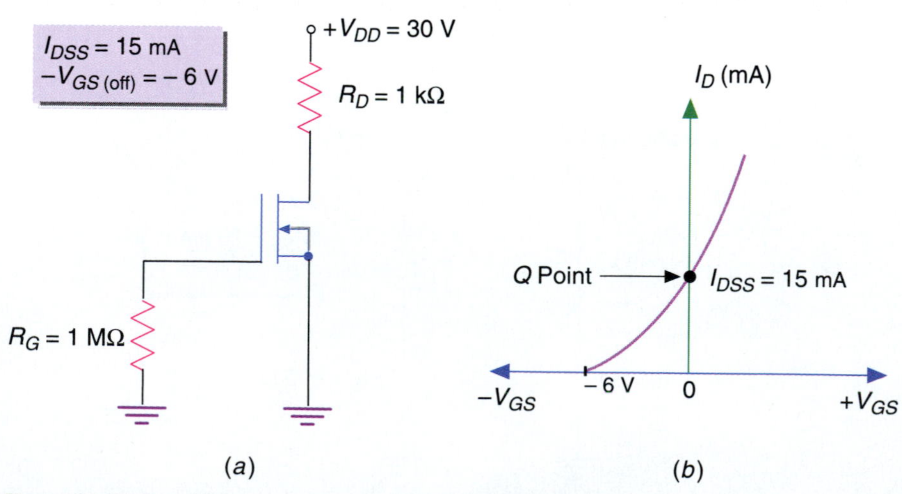

Figure 24-34 Zero bias technique used with D-MOSFET. (*a*) Circuit. (*b*) Transconductance curve.

the D-MOSFET operates in the enhancement mode, and the drain current increases. If a signal at the gate goes negative, the D-MOSFET operates in the depletion mode, and I_D decreases.

To calculate the DC voltage at the drain, simply use Eq. 24-31.

Equation 24-31

$$V_{DS} = V_{DD} - I_{DSS}R_D$$

In most cases, R_D is chosen so that $V_{DS} = V_{DD}/2$. To do this, R_D is chosen using Eq. 24-32.

Equation 24-32

$$R_D = \frac{V_{DD}}{2\,I_{DSS}}$$

It should be pointed out that even though zero bias is the most commonly used biasing technique for D-MOSFETs, other techniques can also be used. These include gate-bias, self-bias, and voltage-divider bias.

Example 24-6: In Fig. 24-34, calculate V_{DS}.

Solution:
Note that I_{DSS} = 15 milliamps. Using Eq. 24-31, the calculations are:

$$\begin{aligned} V_{DS} &= V_{DD} - I_{DSS}R_D \\ &= 30\text{ V} - (15\text{ mA} \times 1\text{ k}\Omega) \\ &= 30\text{ V} - 15\text{ V} \\ &= 15\text{ V} \end{aligned}$$

E-MOSFETs cannot be biased using either zero bias or self-bias. The reason is that V_{GS} must exceed $V_{GS(\text{th})}$ to get any drain current at all.

Figure 24-35(*a*) shows one way to bias E-MOSFETs. This form of bias is called *drain-feedback bias*. With E-MOSFETs, the manufacturer's data sheet usually specifies the value of $V_{GS(\text{th})}$ and the coordinates for one point on the transconductance curve. The quantities $I_{D(\text{on})}$, $V_{GS(\text{on})}$, and $V_{GS(\text{th})}$ are the parameters that are important when biasing E-MOSFETs. See Fig. 24-35(*b*). The transconductance curve shown in Fig. 24-35(*b*) is for a Motorola 3N169 enhancement-type MOSFET. The values shown for $I_{D(\text{on})}$, $V_{GS(\text{on})}$, and $V_{GS(\text{th})}$ are "typical" values.

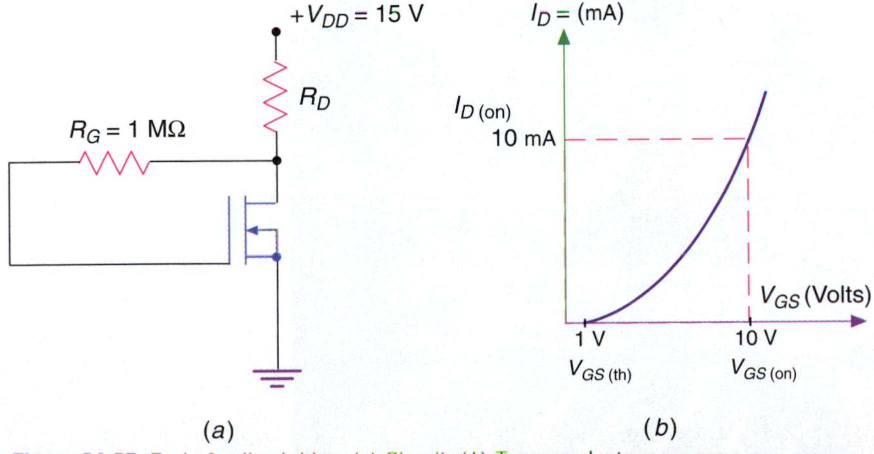

Figure 24-35 Drain feedback bias. (*a*) Circuit. (*b*) Transconductance curve.

It is somewhat strange, but the drain resistor R_D must be properly selected to provide the required bias in Fig. 24-35(*b*). The drain resistor can be calculated using Eq. 24-33.

Equation 24-33

$$R_D = V_{DD} - \frac{V_{GS(\text{on})}}{I_{D(\text{on})}}$$

Because the gate current I_G is zero, there is no voltage drop across the gate resistor R_G. Therefore, $V_{GS} = V_{DS}$.

Example 24-7: For Fig. 24-35, calculate the value of R_D required to provide an $I_{D(\text{on})}$ of 10 milliamps.

Solution:
Use Eq. 24-33 and the values from the transconductance curve in Fig. 24-35(*b*) to calculate R_D.

$$R_D = \frac{V_{DD} - V_{GS(\text{on})}}{I_{D(\text{on})}}$$
$$= \frac{15\ \text{V} - 10\ \text{V}}{10\ \text{mA}}$$
$$= 500\ \Omega$$

A 470-ohm resistor would provide the proper biasing voltage at the gate.

REVIEW QUIZ 24-7

1. True or false? Gate bias is the best way to bias a JFET.
2. True or false? With self-bias, the DC voltage measured at the gate is zero volts.
3. True or false? Zero bias is used with D-MOSFETs.
4. True or false? When a JFET uses voltage-divider bias, the source voltage V_S is 0.7 volt less than the gate voltage V_G.

SUMMARY

A bipolar junction transistor (BJT) has three leads, identified as the base (B), emitter (E), and collector (C). An NPN transistor is constructed with a thin and lightly doped P-type material sandwiched between two larger N-type regions. Conversely, a PNP transistor is constructed with a thin and lightly doped N-type material sandwiched between two larger P-type regions. The emitter region of a transistor is heavily doped. Its job is to emit or inject current carriers into the base. Most of the current carriers injected into the thin and lightly doped base diffuse across the CB junction and into the collector. The collector is the largest region within the transistor because most of the heat is dissipated in this region. In an NPN transistor, free electrons are the majority current carriers in both the emitter and collector, whereas holes are the majority current carriers in the base.

In a BJT, the emitter current $I_E = I_B + I_C$. DC alpha (α_{DC}) $= I_C/I_E$ and DC beta (β_{DC}) $= I_C/I_B$. The DC current gain of a transistor in the common-base configuration is α_{DC}, and the DC current gain of a transistor in the common emitter configuration is β_{DC}. For proper operation of a transistor, the EB junction is forward-biased and the CB junction is reverse-biased. When a transistor operates in the active region, the collector acts like a current source whose output is $I_C = I_B \times \beta_{\mathrm{DC}}$. The collector current source has a very high internal impedance. BJTs are current-controlled devices, which means that an input current controls the output current.

The power dissipation P_D of a transistor must never exceed the power dissipation rating $P_{D(\mathrm{max})}$, otherwise the transistor will burn up. $P_{D(\mathrm{max})}$ decreases for higher temperatures.

The most popular way to bias BJTs is with voltage divider bias. With this type of bias, the transistor currents and voltages remain the same even though the β_{DC} of the transistor may vary with temperature. The DC load line is a graph showing all the possible values of I_C and V_{CE} for a given transistor circuit. The end points on the DC load line are identified as $I_{C(\mathrm{sat})}$ and $V_{\mathrm{CE(off)}}$.

Field-effect transistors (FETs) have three leads identified as the gate (G), source (S), and drain (D). There are two types of FETs: the junction field-effect transistor, abbreviated JFET, and the metal-oxide semiconductor field-effect transistor, abbreviated MOSFET. All FET are voltage-controlled devices meaning that the input voltage controls the output current. For any JFET, the drain current I_D is maximum when $V_{GS} =$ 0 volts. This current is designated as I_{DSS}. The amount of gate-source voltage required to reduce the drain current I_D to zero is called the gate-source cutoff voltage. The gate-source cutoff voltage is designated as $V_{GS(\mathrm{off})}$.

The transconductance curve of a JFET is a graph of I_D versus V_{GS}. The transconductance curve is nonlinear.

There are two basic types of MOSFETs: E-MOSFETs and D-MOSFETs. In both the E-MOSFET and D-MOSFET, the gate is insulated from the channel by a thin layer of silicon dioxide (SiO_2). The input impedance of a MOSFET is many times higher than that of a JFET.

D-MOSFETs are "normally ON" electronic devices, this means that drain current flows when $V_{GS} = 0$ volts. A D-MOSFET can operate in either the depletion or enhancement mode. For D-MOSFETs, I_{DSS} is not the maximum possible drain current.

E-MOSFETs are "normally OFF" devices because there is no drain current when V_{GS} $= 0$ volts. To get drain current, the gate-source voltage must exceed the threshold voltage $V_{GS(\mathrm{th})}$.

One disadvantage of MOSFETs is their extreme sensitivity to electrostatic discharge (ESD). When handling MOSFETs, extreme care must be taken to ensure that static electricity does not puncture the thin layer of silicon dioxide separating the gate and channel.

Self-bias and voltage-divider bias are popular ways to bias a JFET. Gate bias is seldom used because it is unstable. D-MOSFETs often use a technique called zero bias. In this case, the DC drain current $I_D = I_{DSS}$. Zero bias, however, cannot be used with E-MOSFETs because the gate-source voltage must exceed $V_{GS(\text{th})}$ in order to get any drain current at all. E-MOSFETs normally use drain feedback bias.

NEW VOCABULARY

- active region
- alpha, α_{DC}
- base bias
- beta, β_{DC}
- bipolar junction transistor (BJT)
- breakdown region
- cut-off region
- depletion-type MOSFET (D-MOSFET)
- DC equivalent
- DC load line
- electrostatic discharge (ESD)
- enhancement-type MOSFET (E-MOSFET)
- gate-bias
- junction field-effect transistor (JFET)
- majority current carriers
- metal oxide semiconductor field-effect transistor (MOSFET)
- normally OFF
- normally ON
- pinch-off voltage
- quiescent point (Q point)
- saturation region
- self-bias
- threshold voltage
- transconductance curve
- voltage-divider bias
- zero bias

Questions

1. What are the three doped regions within a bipolar junction transistor?
2. How are α_{DC} and β_{DC} expressed mathematically?
3. What are the four operating regions for a transistor?
4. Why does base bias produce an unstable Q point?
5. Why is a JFET referred to as a "normally ON" device?
6. Why is the gate current IG equal to zero in a JFET?
7. Why does V_{GS} in an E-MOSFET have to exceed $V_{GS(\text{th})}$ to produce drain current?
8. Why are MOSFETs so sensitive to ESD?
9. Which is a more stable form of bias: gate bias or self-bias?
10. Can zero bias be used with E-MOSFETs? Why?
11. In a BJT, which region is: a. The most heavily doped?, b. The largest?, c. The most lightly doped?
12. Why should electronic technicians wear a grounding strap when handling MOSFETs?

Problems

1. In a transistor, I_E = 5 milliamps and IC = 4.95 milliamps. Calculate α_{DC}.
2. In a transistor, I_C = 12 milliamps and β_{DC} = 200. Calculate I_B.
3. Refer to Fig. 24-11(*a*). Assume that V_{BB} = 5 volts, V_{CC} = 18 volts, R_B = 56 kilohms, R_C = 1 kilohm, and β_{DC} = 150. Calculate I_B, I_C, V_{CE}, and P_D.
4. Refer to Fig. 24-14(*a*). Assume that V_{CC} = 24 volts, R_B = 470 kilohms, R_C = 1.5 kilohms, and β_{DC} = 200. Calculate I_B, I_C, V_{CE}, and P_D.
5. Refer to Fig. 24-15. Assume that V_{CC} = 12 volts, R_1 = 33 kilohms, R_2 = 6.2 kilohms, R_E = 500 ohms, and R_C = 2 kilohms. Calculate V_B, V_E, I_C, V_C, V_{CE}, and P_D.
6. Assume that a JFET has the following characteristics: I_{DSS} = 20 milliamps and $VGS_{(off)}$ = −5 volts. Calculate the drain current I_D for (*a*) V_{GS} = 0 volts, (*b*) V_{GS} = −1 volt, (*c*) V_{GS} = −2 volts, (*d*) V_{GS} = −3 volts, and (*e*) V_{GS} = −4 volts.
7. A D-MOSFET has the following characteristics: I_{DSS} = 15 milliamps and $V_{GS(off)}$ = −5 volts. Calculate the drain current I_D for (*a*) V_{GS} = −2 volts, (*b*) V_{GS} = 0 volts, (*c*) V_{GS} = +2 volts.
8. Refer to Fig. 24-32. Assume that V_{DD} = 12 volts, R_S = 1.5 kilohms, and R_D = 2.7 kilohms. Calculate V_{GS} and V_D if I_D = 1.6 milliamps.
9. In Fig. 24-33, V_{DD} = 18 volts. Assume that V_{GS} = −1.5 volts. Calculate V_G, V_S, V_D, and V_{DS}.
10. Refer to Fig. 24-34. If I_{DSS} = 8 milliamps and V_{DD} = 24 volts, what is the drain voltage V_D?
11. Refer to Fig. 24-35. If V_{DD} = 24 volts, calculate the value of R_D required to produce the same drain current, $I_{D(on)}$ of 10 milliamps.

Critical Thinking

1. Refer to Fig. 24-16. Assume that the BE junction of the transistor opens. Calculate the following quantities:
   V_B, V_E, I_C, V_C, and V_{CE}.
2. Refer to Fig. 24-32. If R_S opens, calculate the following quantities: I_D and V_D.
3. Refer to Fig. 24-35. If R_G is removed, calculate the drain current I_D.
4. Refer to Fig. 24-14(*a*). If R_B increases, does the collector-emitter voltage V_{CE} increase, decrease, or stay the same?

Answers to Review Quizzes

24-1
1. thin; lightly
2. electrons
3. the CB depletion layer
4. emitter

24-2
1. forward; reverse
2. current
3. small
4. 0.99
5. 250
6. 0.992

24-3
1. current source
2. 3.75 milliamps
3. breakdown
4. cut off
5. 0.7 volt

24-4
1. decreases
2. $P_D = V_{CE} \times I_C$
3. 25°C

24-5
1. decreases
2. decreases
3. voltage-divider bias
4. no

24-6
1. ON
2. reverse
3. zero
4. 3 milliamps
5. ESD
6. zero
7. OFF
8. ON
9. false

24-7
1. false
2. true
3. true
4. false

Chapter 25

Transistor Amplifiers

ABOUT THIS CHAPTER

In this chapter you will learn about the different types of BJT and FET amplifiers. For bipolar junction transistors, we will study the common-base, common-emitter, and common-collector amplifiers. For field-effect transistors, we will study the common-gate, common-source, and

After completing this chapter you will be able to:

- Calculate the voltage gain A_V of a common-base, common-emitter, and common-collector amplifier.
- Calculate the Z_{in} of a common-base, common-emitter, and common-collector amplifier.
- Determine the phase relationship between v_{in} and v_{out} in an amplifier.

common-drain amplifiers. For each of the different amplifiers we will examine the following circuit characteristics: voltage gain (A_V), input impedance (Z_{in}), and the phase relationship between v_{in} and v_{out}.

- ▽ Calculate the voltage gain A_V of a common-gate, common-source, and common-drain amplifier.
- ▽ Calculate the input impedance Z_{in} in a common-gate, common-source, and common-drain amplifier.

25-1 The Common-Emitter (CE) Amplifier

A typical **common-emitter amplifier** is shown in Fig. 25-1(*a*). Notice that voltage-divider bias is used to provide the desired circuit voltages and currents. The input signal is applied to the base, and the output signal is taken from the collector. The input coupling capacitor C_{in} has a low reactance X_C for AC signals but appears as an open to direct current. C_E is an emitter bypass capacitor. Its purpose is to bypass AC signals around the emitter resistor R_E.

The DC load line is shown in Fig. 25-1(*b*). The AC signal driving the base produces sinusoidal variations in both the base current I_B and the collector current I_C as shown. This in turn produces sinusoidal variations in the collector-emitter voltage V_{CE}.

In Fig. 25-1(*a*) the DC base voltage equals 2.5 volts. This is calculated as:

$$V_B = \frac{R_2}{R_1 + R_2} \times V_{CC}$$
$$= \frac{3.6 \text{ k}\Omega}{18 \text{ k}\Omega + 3.6 \text{ k}\Omega} \times 15 \text{ V}$$
$$= 2.5 \text{ V}$$

Furthermore, the DC emitter voltage equals 1.8 volts, calculated as:

$$V_E = V_B - V_{\text{BE}}$$
$$= 2.5 \text{ V} - 0.7 \text{ V}$$
$$= 1.8 \text{ V}$$

With V_{in} equal to 10 millivolts peak-to-peak, the base voltage varies from 2.495 to 2.505 volts as shown. The emitter bypass capacitor C_E, however, holds the emitter voltage constant at 1.8 volts. Therefore the 10 mVp–p input voltage actually appears directly across the BE junction of the transistor. When the AC base voltage is positive, the forward bias across the BE junction increases. This causes both the base current I_B and the collector current I_C to increase. Conversely, when the AC base voltage goes negative, the forward bias across the BE junction decreases. This means that the base current I_B and collector current I_C both decrease.

Before proceeding further, let's calculate the remaining unknown DC quantities in Fig. 25-1(*a*). Begin by calculating the collector current.

$$I_C = \frac{V_E}{R_E}$$
$$= \frac{1.8 \text{ V}}{240 \ \Omega}$$
$$= 7.5 \text{ mA}$$

Next, calculate the DC collector voltage.

$$V_C = V_{CC} - I_C R_C$$
$$= 15 \text{ V} - (7.5 \text{ mA} \times 1 \text{ k}\Omega)$$
$$= 15 \text{ V} - 7.5 \text{ V}$$
$$= 7.5 \text{ V}$$

Finally, calculate the DC collector-emitter voltage.

$$V_{CE} = V_{CC} - I_C(R_C + R_E)$$
$$= 15\text{ V} - 7.5\text{ mA }(1\text{ k}\Omega + 240\ \Omega)$$
$$= 15\text{ V} - 9.3\text{ V}$$
$$= 5.7\text{ V}$$

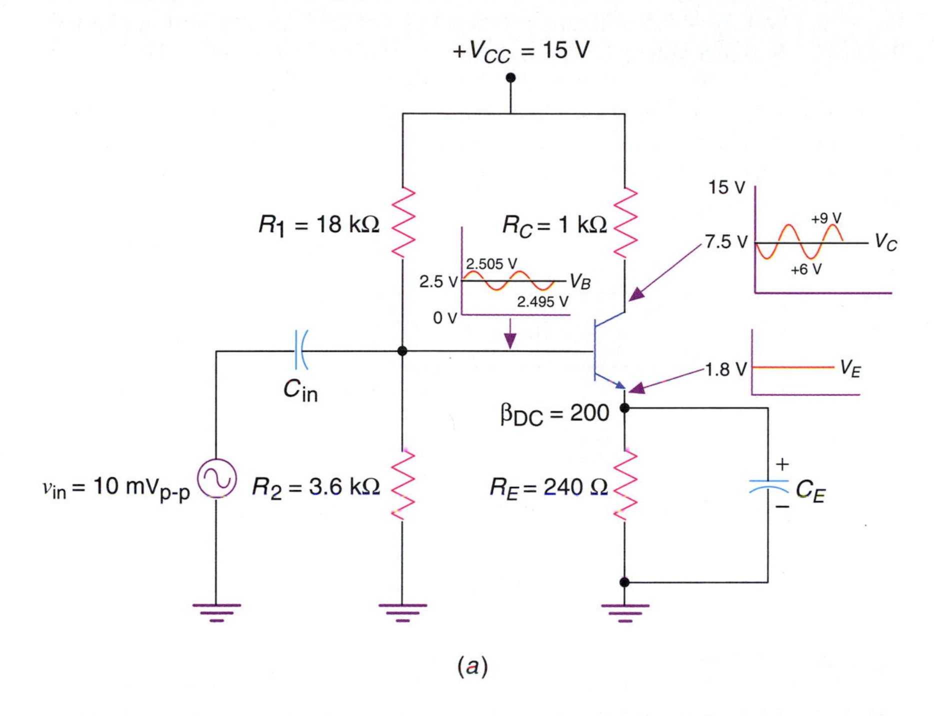

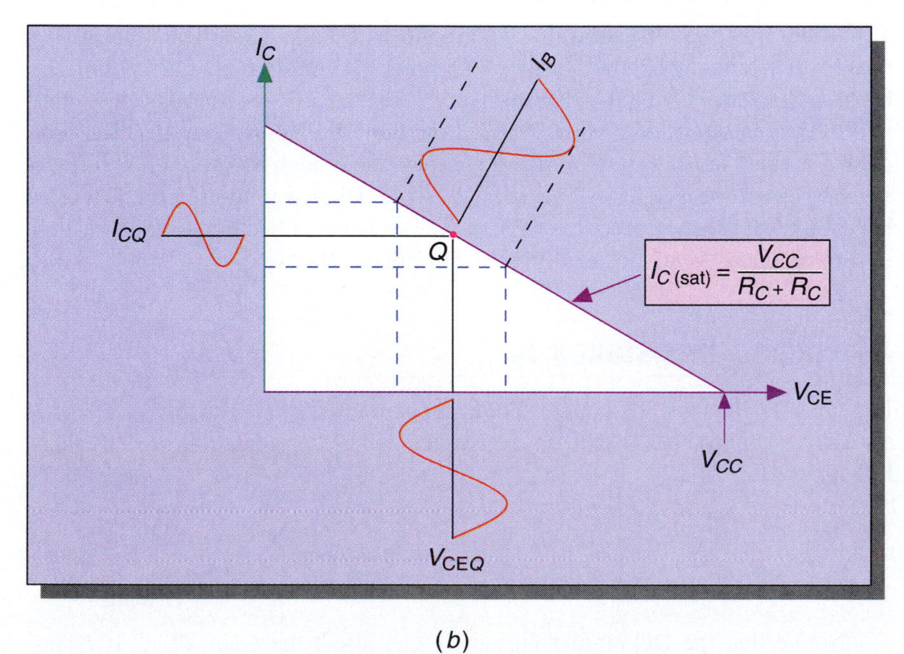

Figure 25-1 Common-emitter amplifier, where v_{in} is applied to the base and v_{out} is taken from collector. (*a*) Circuit. (*b*) DC load line.

Knowing all of the DC quantities allows us to analyze the AC operation of the common-emitter amplifier.

Amplifying the AC Signal

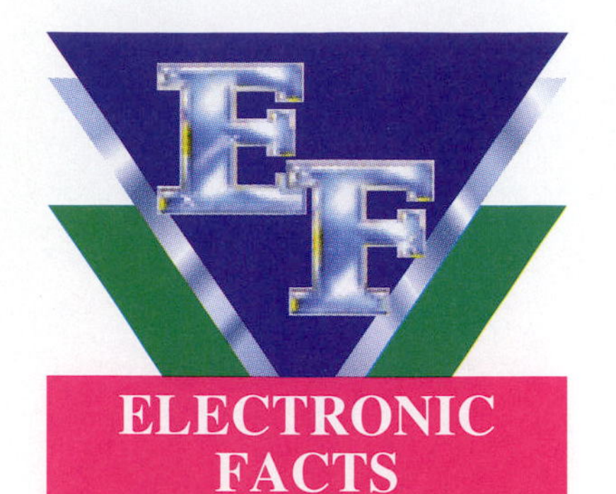

Figure 25-2 shows a graph of I_E versus V_{BE} for the transistor used in the common-emitter amplifier of Fig. 25-1. Note the following points on the graph. When $V_{BE} = 0.7$ volt, $I_E = 7.5$ milliamps. When $V_{BE} = 0.705$ volt, $I_E = 9$ milliamps. When $V_{BE} = 0.695$ volt, $I_E = 6$ milliamps.

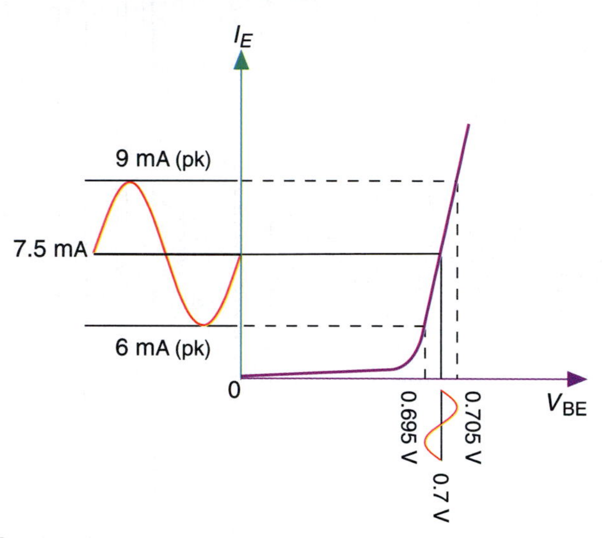

Figure 25-2 Graph of V_{BE} versus I_E for the transistor in Fig. 25-1(*a*).

When $I_E = 7.5$ milliamps, $V_C = 7.5$ volts as calculated earlier. When I_E increases to 9 milliamps, however, $V_C = 15$ volts $-$ (9 milliamps $\times$ 1 kilohm), or 6 volts. Conversely, when I_E decreases to 6 milliamps, $V_C = 15$ volts $-$ (6 milliamps $\times$ 1 kilohm), or 9 volts. Notice how the 10 mVp–p input signal is producing a much larger variation in collector voltage, which is 9 volts $-$ 6 volts $=$ 3 volts peak-to-peak. This is how simple it is. A slight variation in the forward-bias of the BE junction produces large variations in the collector current I_C, which in turn produces a large change in the collector voltage V_C.

AC Emitter Resistance r_e'

For small AC signals, the emitter diode can be treated like a resistance. This resistance is called the **AC emitter resistance** and is designated as r_e'. To calculate r_e', use Eq. 25-1.

Equation 25-1

$$r_e' = \frac{25 \text{ mV}}{I_E}$$

Notice that the DC emitter current I_E can affect the value of r_e'. If I_E increases, r_e' decreases. Conversely, if I_E decreases, r_e' increases.

The AC resistance r'_e of the emitter diode is important because it helps us analyze the AC operation of BJTs. Specifically, it is used to determine the voltage gain A_V or the input impedance Z_{in} of an amplifier.

Predicting the Voltage Gain A_V

To predict the voltage gain of a common-emitter amplifier, we must be able to write an equation for both the input and output voltages. Figure 25-3 shows the AC equivalent circuit for the common-emitter amplifier in Fig. 25-1.

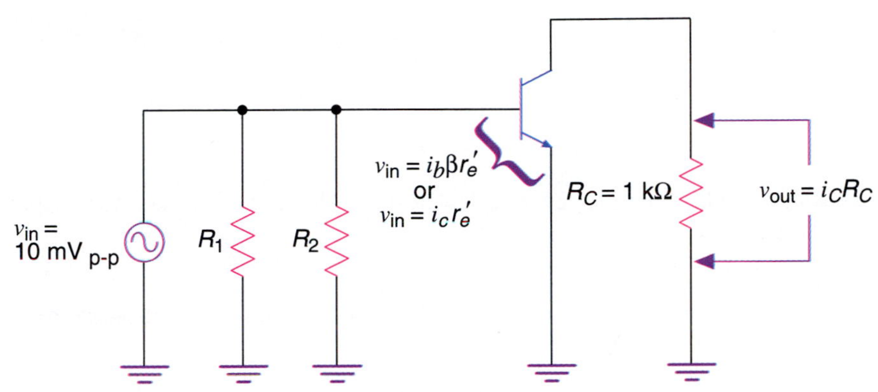

Figure 25-3 Equivalent circuit of a common-emitter amplifier as it appears to AC signals.

Notice that this circuit *does not* include C_{in}, C_E, or V_{CC}. Also, the emitter resistor R_E has been omitted. What you are looking at is the circuit as it appears to AC signals. C_{in} and C_E appear as AC short circuits and, therefore, can be eliminated. Because C_E is in parallel with R_E, R_E can be omitted as well. The voltage source V_{CC} also appears as an AC short circuit, so the top of resistors R_1 and R_C appear at ground for AC signals. The output signal is across the collector resistor R_C. An expression for the output voltage is:

Equation 25-2

$$v_{out} = i_c R_C$$

The input voltage is across the BE junction, as shown in Fig. 25-3. The input voltage v_{in} is expressed as:

Equation 25-3

$$v_{in} = i_b \beta r'_e$$

The input voltage v_{in} is directly across the BE junction of the transistor. Because base current i_b flows out the base lead, the AC emitter resistance appears larger by the factor of beta when viewed from the base. The collector current $i_c = i_b\beta$, so Eq. 25-3 can be modified as shown.

Equation 25-4

$$v_{in} = i_c r'_e$$

To calculate the **voltage gain** A_V of an amplifier, divide the output voltage by the input voltage as shown in Eq. 25-5.

Equation 25-5

$$A_V = \frac{v_{out}}{v_{in}}$$

Expanding Eq. 25-5 by using the Eqs. 25-2 and 25-4 for v_{out} and v_{in}, respectively, gives us the following:

Equation 25-6

$$A_V = \frac{i_c R_C}{i_c r'_e}$$

Which simplifies to:

Equation 25-7

$$A_V = \frac{R_C}{r'_e}$$

Example 25-1: For Fig. 25-1, calculate the voltage gain A_V and the output voltage v_{out}.

Solution:
Begin by calculating the AC emitter resistance r'_e. Because $I_E = 7.5$ milliamps, as calculated earlier, then:

$$r'_e = \frac{25 \text{ mV}}{7.5 \text{ mA}}$$

$$= 3.33\ \Omega$$

Next, use Eq. 25-7 to calculate A_V.

$$A_V = \frac{R_C}{r'_e}$$

$$= \frac{1 \text{ k}\Omega}{3.33}$$

$$= 300$$

To calculate v_{out}, multiply v_{in} by A_V.

$$v_{out} = A_V \times v_{in}$$

$$= 300 \times 10 \text{ mVp–p}$$

$$= 3 \text{ Vp–p}$$

This coincides with what was calculated earlier for Fig. 25-1.

The Effects of Connecting a Load Resistor R_L

Figure 25-4 shows the common-emitter amplifier with a load resistor R_L connected in the collector circuit. For this condition, the AC collector resistance equals the parallel combination of R_C and R_L. Therefore, the equation for the voltage gain A_V becomes:

Equation 25-8

$$A_V = \frac{r_L}{r'_e}$$

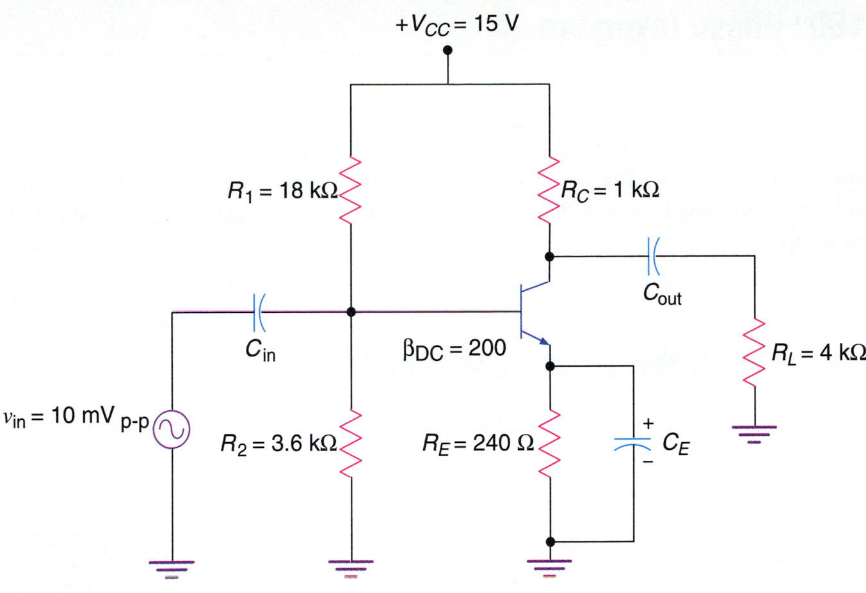

Figure 25-4 Common-emitter amplifier with load resistor R_L connected in the collector circuit.

where $r_L = \dfrac{R_C \times R_L}{R_C + R_L}$

It should be emphasized that connecting R_L does not affect the DC voltages and currents because the output coupling capacitor C_{out} appears as an open to direct current.

Example 25-2: Calculate the voltage gain and output voltage in Fig. 25-4.

Solution:
We already know that the DC emitter current I_E equals 7.5 milliamps, so r_e' still equals 3.33 ohms. The **AC collector resistance** r_L is calculated as follows:

$$\begin{aligned} r_L &= \frac{R_C \times R_L}{R_C + R_L} \\ &= \frac{1\ \text{k}\Omega \times 4\ \text{k}\Omega}{1\ \text{k}\Omega + 4\ \text{k}\Omega} \\ &= 800\ \Omega \end{aligned}$$

Now, use Eq. 25-8 to calculate A_V.

$$\begin{aligned} A_V &= \frac{r_L}{r_e'} \\ &= \frac{800}{3.33\ \Omega} \\ &= 240 \end{aligned}$$

Finally, calculate v_{out}.

$$\begin{aligned} v_{out} &= A_V \times v_{in} \\ &= 240 \times 10\ \text{mVp–p} \\ &= 2.4\ \text{Vp–p} \end{aligned}$$

Notice that a lower collector resistance means less voltage gain.

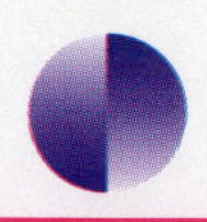

STUDENT *to* STUDENT

The CE amplifier is the only amplifier configuration that has 180° phase shift between v_{in} and v_{out}.

180° Phase Inversion

In any common-emitter amplifier, v_{in} and v_{out} are 180° out of phase. To understand why, refer back to Fig. 25-1. When v_{in} increases at the base, I_B and I_C increase. This increases the $I_C R_C$ voltage drop, which in turn reduces the collector voltage V_C. When v_{in} goes negative, I_B and I_C decrease. This decreases the $I_C R_C$ voltage drop, which means that V_C increases. Therefore, v_{in} and v_{out} are 180° out of phase.

Calculating the Input Impedance

The **input impedance** Z_{in} of a common-emitter amplifier is the impedance seen by the AC source driving the amplifier. Refer to Fig. 25-5. Notice that the input impedance $Z_{\text{in(base)}}$ is:

Equation 25-9

$$Z_{\text{in(base)}} = \beta r'_e$$

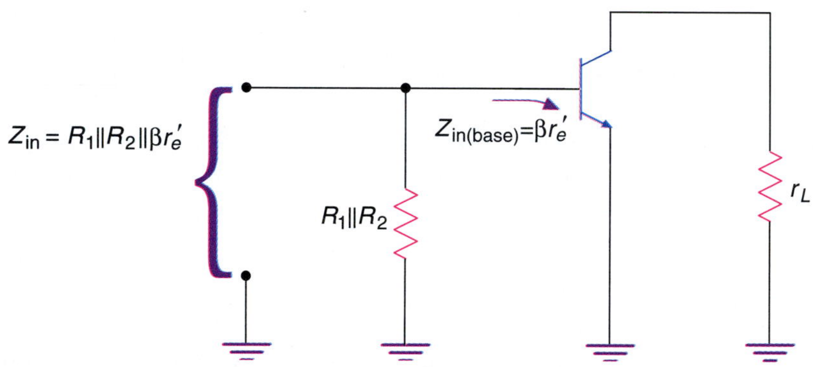

Figure 25-5 Equivalent circuit of common-emitter amplifier as seen by the AC source. The AC emitter resistance r'_e appears greater by a factor of β when viewed from the base.

This says that the AC emitter resistance r'_e appears greater by the factor of beta when reviewed from the base. The derivation of Eq. 25-9 is simple.

$$Z_{\text{in(base)}} = \frac{v_{\text{in}}}{i_{\text{in}}}$$

Because $v_{\text{in}}=i_c r'_e$ and i_{in} must equal i_b, we have:

$$Z_{\text{in(base)}} = \frac{i_c r'_e}{i_b}$$

Moreover, with $i_c/i_b=\beta$, we have

$$Z_{\text{in(base)}}=\beta r'_e$$

The input impedance Z_{in} of the amplifier includes the effects of the biasing resistors R_1 and R_2. Therefore:

Equation 25-10

$$Z_{\text{in}} = R_1||R_2||\beta r'_e$$

Example 25-3: For Fig. 25-4, calculate Z_{in}. Assume that $\beta = 200$.

Solution:
Begin by remembering that $r'_e = 3.33$ ohms. Then use Eq. 25-9 to calculate $Z_{in(base)}$.

$$Z_{in(base)} = \beta r'_e$$
$$= 200 \times 3.33\ \Omega$$
$$= 667\ \Omega$$

To calculate the input impedance Z_{in} of the amplifier, use Eq. 25-10.

$$Z_{in} = R_1 || R_2 || \beta r'_e$$
$$= 18\ \text{k}\Omega || 3.6\ \text{k}\Omega || 667\ \Omega$$
$$= 546\ \Omega$$

If the source voltage v_{in} contains any internal resistance r_i, then v_{in} will be divided between r_i and Z_{in}. Usually the higher the Z_{in} of an amplifier, the better.

Removing the Emitter-Bypass Capacitor C_E

When the **emitter-bypass capacitor** C_E is removed, both the voltage gain and input impedance are affected. The equivalent circuit is shown in Fig. 25-6. Notice that the output voltage v_{out} still equals $i_c r_L$. Now, however:

Equation 25-11

$$v_{in} \approx i_c R_E$$

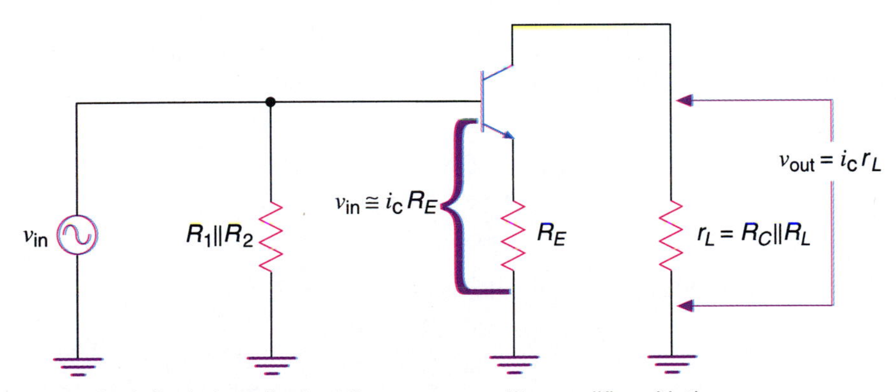

Figure 25-6 Equivalent circuit of the common-emitter amplifier with the emitter bypass capacitor C_E removed.

ELECTRONIC FACTS

When the emitter-bypass capacitor is removed, the voltage gain A_V decreases and the input impedance Z_{in} increases.

This means that most of the input voltage is now dropped across the emitter resistor R_E instead of the AC emitter resistance r'_e. With an expression for v_{in}, we can derive an equation for the voltage gain A_V.

$$A_V = \frac{i_C r_L}{i_C R_E}$$

which simplifies to:

Equation 25-12

$$A_V = \frac{r_L}{R_E}$$

Example 25-4: If $R_C = 1$ kilohm, $R_L = 4$ kilohms, and $R_E = 240$ ohms in Fig. 25-6, calculate A_V.

Solution:
Using Eq. 25-12, the calculation is:

$$\begin{aligned} A_V &= \frac{r_L}{R_E} \\ &= \frac{800\ \Omega}{240\ \Omega} \\ &= 3.33 \end{aligned}$$

where $r_L = 800\ \Omega$ as calculated earlier.

Notice the severe reduction in A_V.

To calculate $Z_{in(base)}$ without the emitter bypass capacitor, C_E, use Eq. 25-13.

Equation 25-13

$$Z_{in(base)} = \beta(r'_e + R_E)$$

In most cases, $R_E >> r'_e$, and Eq. 25-13 simplifies to:

Equation 25-14

$$Z_{in(base)} = \beta R_E$$

To calculate the input impedance Z_{in}, include the effects of the biasing resistors R_1 and R_2. Therefore:

Equation 25-15

$$Z_{in} = R_1 || R_2 || Z_{in(base)}$$

Example 25-5: For Fig. 25-6, calculate Z_{in} if $R_1 = 18$ kilohms, $R_2 = 3.6$ kilohms, $R_E = 240$ ohms, $r'_e = 3.33$ ohms, and $\beta = 200$.

Solution:

$$\begin{aligned} Z_{in(base)} &= \beta(r'_e + R_E) \\ &= 200(3.33\ \Omega + 240\ \Omega) \\ &= 48.67\ \text{k}\Omega \end{aligned}$$

Omitting the effects of r'_e gives us:

$$\begin{aligned} Z_{in(base)} &= 200\ \Omega \times 240\ \Omega \\ &= 48\ \text{k}\Omega \end{aligned}$$

To calculate Z_{in}, use Eq. 25-15. Assume that $Z_{in(base)} = 48$ kilohms.

$$\begin{aligned} Z_{in} &= R_1 || R_2 || Z_{in(base)} \\ &= 18\ \text{k}\Omega || 3.6\ \text{k}\Omega || 48\ \text{k}\Omega \\ &= 2.82\ \text{k}\Omega \end{aligned}$$

Compare this to the Z_{in} with C_E present in Example 25-3. Without C_E, $Z_{in} = 2.82$ kilohms and with C_E present in Example. 25-3, $Z_{in} = 546$ ohms.

Power Gain A_P

The **power gain** A_P of a common-emitter amplifier equals the products of its voltage gain A_V and its current gain A_i. Because the current gain A_i equals β, A_P is calculated as:

$$A_P = A_V \times A_i$$

and since $A_i = \beta$, then

Equation 25-16

$$A_P = A_V \times \beta$$

The common-emitter amplifier is the only BJT amplifier configuration that provides both a high voltage and current gain. As a result, the common-emitter amplifier has the highest power gain of all the BJT amplifier configurations.

ELECTRONIC FACTS

The CE amplifier is the only amplifier that has both voltage and current gain.

REVIEW QUIZ 25-1

1. In a common-emitter amplifier, the input signal is applied to the ______________, and the output signal is taken from the ______________.
2. In a common-emitter amplifier, v_{in} and v_{out} are ______________ phase.
3. Removing the emitter bypass capacitor C_E in a common-emitter amplifier ______________ the voltage gain.
4. The common-emitter amplifier has a(n) ______________ power gain.

25-2 The Common-Collector (CC) Amplifier

ELECTRONIC FACTS

The CC amplifier is more commonly referred to as an emitter-follower.

A typical common-collector amplifier is shown in Fig. 25-7(*a*) on the next page. Notice that the input signal v_{in} is applied to the base while the output is taken at the emitter. For the common-collector amplifier, the input and output signals are in phase. This means that when v_{in} increases, v_{out} increases. Likewise, when v_{in} decreases, v_{out} decreases. Because the output signal follows the input, the circuit is more commonly referred to as an emitter-follower.

Before analyzing the AC operation of the emitter-follower, let's calculate the DC currents and voltages in Fig. 25-7. Begin by calculating V_B.

$$V_B = \frac{R_2}{R_1 + R_2} \times V_{CC} = \frac{5.6\ \text{k}\Omega}{4.7\ \text{k}\Omega + 5.6\ \text{k}\Omega} \times 15\ \text{V} = 8.15\ \text{V}$$

Next, calculate V_E.

$$V_E = V_B - V_{BE} = 8.15\ \text{V} - 0.7\ \text{V} = 7.45\ \text{V}$$

Next, calculate I_C.

$$I_C = \frac{V_E}{R_E} = \frac{7.45\ \text{V}}{1\ \text{k}\Omega} = 7.45\ \text{mA}$$

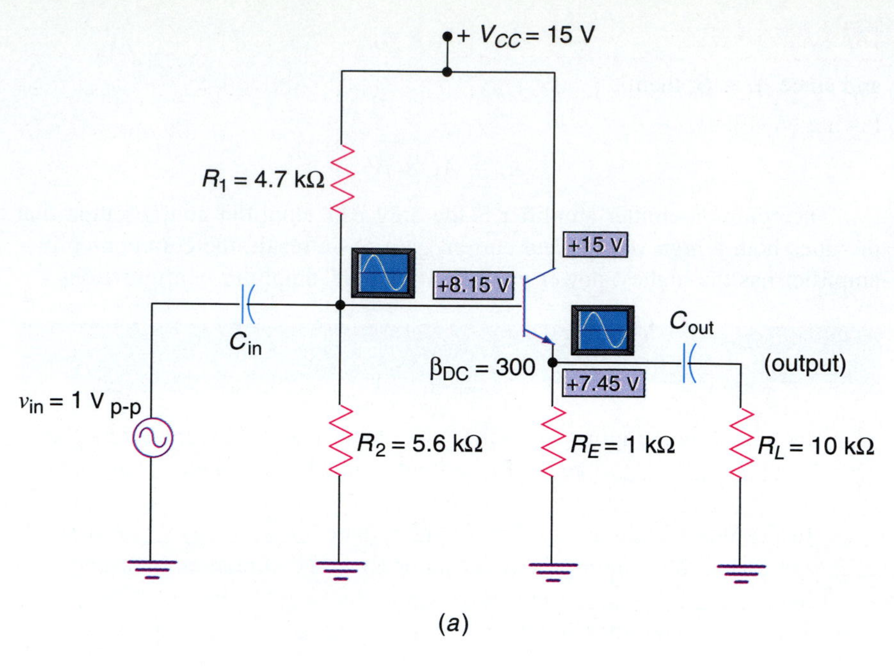

(*a*)

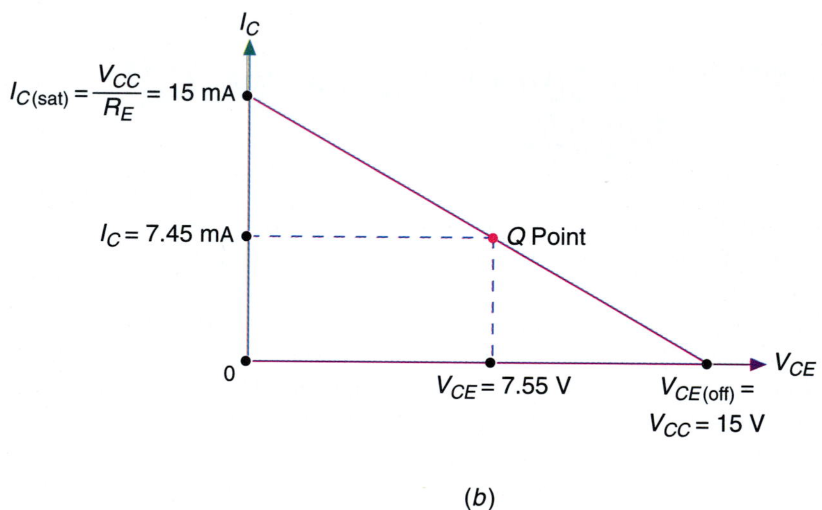

(*b*)

Figure 25-7 Common-collector amplifier. This circuit is more commonly referred to as an emitter-follower. (*a*) Circuit. (*b*) DC load line.

With the collector tied directly to V_{CC}, the collector voltage V_C equals V_{CC}.

$$\begin{aligned} V_C &= V_{CC} \\ &= 15 \text{ V} \end{aligned}$$

Finally, calculate V_{CE}.

Equation 25-17

$$\begin{aligned} V_{CE} &= V_{CC} - I_C R_E \\ &= 15 \text{ V} - (7.45 \text{ mA} \times 1 \text{ k}\Omega) \\ &= 15 \text{ V} - 7.45 \text{ V} \\ &= 7.55 \text{ V} \end{aligned}$$

DC Load Line

Figure 25-7(*b*) shows the DC load line. To calculate the end points, use the following equations.

Equation 25-18

$$I_{C(\text{sat})} = \frac{V_{CC}}{R_E}$$

Equation 25-19

$$V_{CE(\text{off})} = V_{CC}$$

In Fig. 25-7(*a*), the end points for the DC load line are calculated as follows:

$$I_{C(\text{sat})} = \frac{V_{CC}}{R_E} = \frac{15\text{ V}}{1\text{ k}\Omega} = 15\text{ mA}$$

$$V_{CE(\text{off})} = V_{CC} = 15\text{ V}$$

In Fig. 25-7(*b*), notice the Q point values for I_C and V_{CE}.

Calculating the Voltage Gain A_V

The method used to determine the voltage gain A_V of a common-emitter amplifier can also be applied to the emitter follower circuit shown in Fig. 25-7. Refer to the AC equivalent circuit shown in Fig. 25-8. Notice that the collector appears as an AC short. Also R_E and R_L are in parallel. Therefore, the output voltage v_{out} can be calculated as follows:

Equation 25-20

$$v_{\text{out}} = i_c r_L$$

where $r_L = \dfrac{R_E \times R_L}{R_E + R_L}$

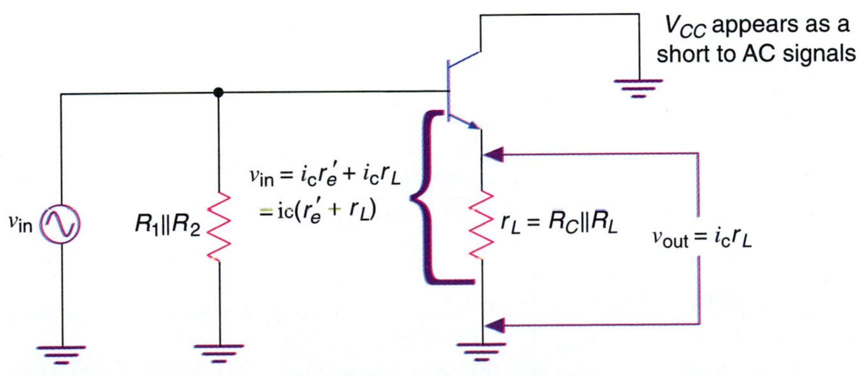

Figure 25-8 Equivalent circuit of the emitter follower as seen by AC signals. Notice the collector is at AC ground.

The AC input equals the voltage across both r'_e and r_L. This can be expressed as:

Equation 25-21

$$v_{in} = i_c(r'_e + r_L)$$

To determine A_V, we divide v_{out} by v_{in}, as shown in Eq. 25-22.

Equation 25-22

$$A_V = \frac{i_c r_L}{i_c(r'_e + r_L)}$$

which simplifies to:

Equation 25-23

$$A_V = \frac{r_L}{r'_e + r_L}$$

In most cases $r_L >> r'_e$ and Eq. 25-23 simplifies further to:

$$A_V = \frac{r_L}{r_L}$$

$$= 1 \text{ or unity}$$

ELECTRONIC FACTS

In an emitter follower, A_V is usually close to 1 or unity.

Example 25-6: Calculate A_V and v_{out} for Fig. 25-7.

Solution:
Begin by calculating r'_e. The calculation is:

$$r'_e = \frac{25 \text{ mV}}{I_E}$$

$$= \frac{25 \text{ mV}}{7.45 \text{ mA}}$$

$$= 3.35\ \Omega$$

Next, calculate the AC load resistance r_L in the emitter circuit.

$$r_L = \frac{R_E \times R_L}{R_E + R_L}$$

$$= \frac{1 \text{ k}\Omega \times 10 \text{ k}\Omega}{1 \text{ k}\Omega + 10 \text{ k}\Omega}$$

$$= 909\ \Omega$$

To calculate the exact value of A_V, use Eq. 25-23.

$$A_V = \frac{r_L}{r'_e + r_L}$$

$$= \frac{909\ \Omega}{3.35\ \Omega + 909\ \Omega}$$

$$= 0.996$$

To calculate v_{out}, proceed as shown:

$$v_{out} = A_V \times v_{in}$$

$$= 0.996 \times 1 \text{ Vp–p}$$

$$= 0.996 \text{ Vp–p}$$

Notice how close the values are for V_{in} and V_{out}. For all practical purposes, the voltage gain A_V equals 1 or unity.

Calculating the Input Impedance

An emitter-follower usually has a very high input impedance. To calculate the input impedance Z_{in} in Fig. 25-7, use Eq. 25-24.

Equation 25-24

$$Z_{in} = R_1 || R_2 || Z_{in(base)}$$

Notice that this is the same formula used for the common-emitter amplifier shown earlier.

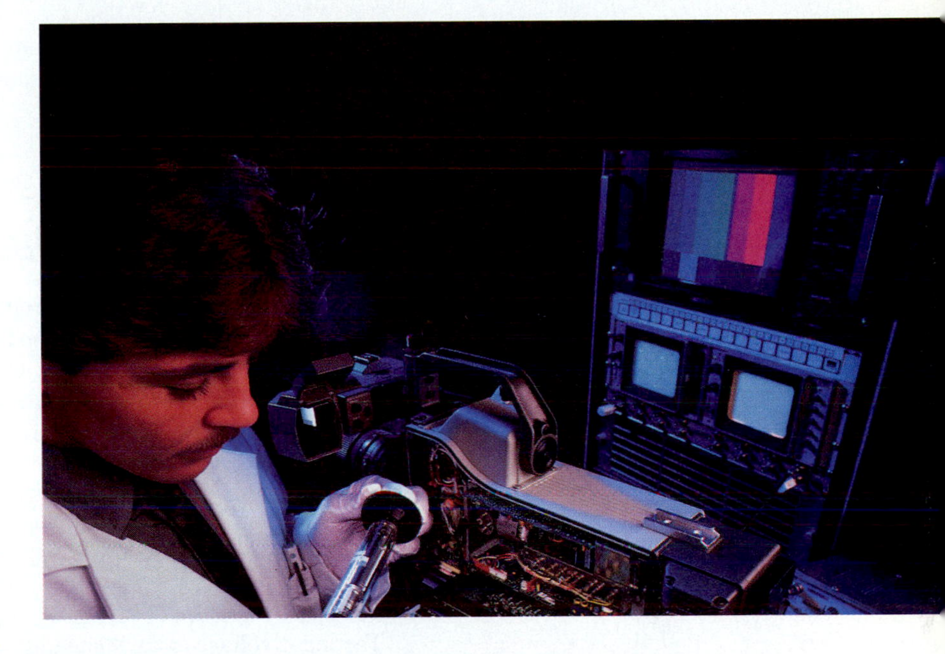

Example 25-7: Assuming that β = 300 in Fig. 25-7, calculate Z_{in}.

Solution:
Begin by calculating $Z_{in(base)}$.

$$\begin{aligned} Z_{in(base)} &= \beta(r'_e + r_L) \\ &= 300(3.35\ \Omega + 909\ \Omega) \\ &= 300 \times 912.35\ \Omega \\ &= 273.7\ \text{k}\Omega \end{aligned}$$

Next, use Eq. 25-24 to calculate Z_{in}.

$$\begin{aligned} Z_{in} &= R_1 || R_2 || Z_{in(base)} \\ &= 4.7\ \text{k}\Omega || 5.6\ \text{k}\Omega || 273.7\ \text{k}\Omega \\ &= 2.53\ \text{k}\Omega \end{aligned}$$

In many cases an emitter follower does not use the biasing resistors R_1 and R_2. Instead, the base is connected directly to the driving source, which is usually the collector of another transistor or perhaps the output of an operational amplifier (op amp). With the biasing resistors R_1 and R_2 omitted, the input impedance of the emitter-follower is stepped up considerably.

Output Impedance

The **output impedance** Z_{out} of an emitter-follower is usually quite low. In fact, 10 to 50 ohms for Z_{out} is not uncommon. Because an emitter-follower has a high Z_{in} and a low Z_{out}, it is often used in impedance matching applications. The calculation for Z_{out} are not shown here because of the complexity involved in its derivation.

Power Gain

The power gain A_P of an emitter-follower is calculated as follows:

Equation 25-25

$$A_P = A_V \times A_i$$

Because $A_V \approx 1$ and $A_i = \beta$,

Equation 25-26

$$A_P = 1 \times \beta$$
$$\approx \beta$$

REVIEW QUIZ 25-2

1. The common-collector amplifier is usually called a(n) ______________.
2. The emitter-follower has a voltage gain A_V of approximately ______________.
3. The input impedance of an emitter-follower is quite ______________.
4. The output impedance of an emitter-follower is quite ______________.

25-3 Common-Base (CB) Amplifiers

Figure 25-9 shows an example of a **common-base amplifier.** Notice that the input signal is applied to the emitter and that the output signal is taken from the collector. The base bypass capacitor C_B places the base at ground for AC signals. To determine the formula for the voltage gain A_V, refer to the AC equivalent circuit shown in Fig. 25-10. The output voltage is calculated as:

Equation 25-27

$$v_{out} = i_C r_L$$

where $r_L = \dfrac{R_C \times R_L}{R_C + R_L}$

The input voltage is calculated as:

Equation 25-28

$$v_{in} = i_C r'_e$$

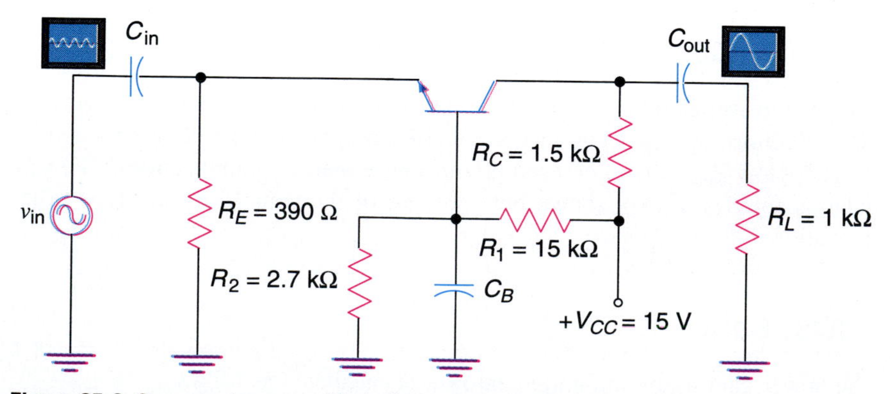

Figure 25-9 Common-base amplifier, where v_{in} is applied to the emitter while the output is taken from the collector.

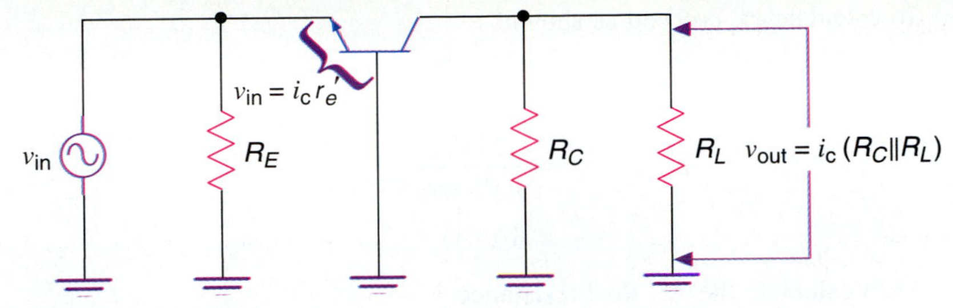

Figure 25-10 Equivalent circuit of the common-base amplifier as seen by AC signals.

Notice that v_{in} is directly across from the EB junction of the transistor.

Using the values in Eqs. 25-27 and 25-28, the voltage gain A_V can be expressed as:

Equation 25-29

$$A_V = \frac{r_L}{r_e'}$$

Example 25-8: For Fig. 25-9, calculate A_V.

Solution:
Before calculating A_V, find the AC resistance r_e' of the emitter diode. Begin by calculating all of the DC voltages and currents.

$$V_B = \frac{R_2}{R_1 + R_2} \times 15 \text{ V}$$
$$= 2.29 \text{ V}$$

Next, calculate V_E.

$$V_E = V_B - V_E$$
$$= 2.29 \text{ V} - 0.7 \text{ V}$$
$$= 1.59 \text{ V}$$

Then, solve for I_C.

$$I_C = \frac{V_E}{R_E}$$
$$= \frac{1.59 \text{ V}}{390 \ \Omega}$$
$$= 4.08 \text{ mA}$$

Then, calculate V_C.

$$V_C = V_{CC} - I_C R_C$$
$$= 15 \text{ V} - (4.08 \text{ mA} \times 1.5 \text{ k}\Omega)$$
$$= 15 \text{ V} - 6.12 \text{ V}$$
$$= 8.88 \text{ V}$$

To calculate r'_e, proceed as shown.

$$r'_e = \frac{25\text{ mV}}{I_E}$$
$$= \frac{25\text{ mV}}{4.08\text{ mA}}$$
$$= 6.13\ \Omega$$

Then calculate the AC load resistance.

$$r_L = \frac{R_C \times R_L}{R_C + R_L}$$
$$= \frac{1.5\text{ k}\Omega \times 1\text{ k}\Omega}{1.5\text{ k}\Omega + 1\text{ k}\Omega}$$
$$= 600\ \Omega$$

Knowing r_L and r'_e allows us to calculate the voltage gain.

$$A_V = \frac{r_L}{r'_e}$$
$$= \frac{600\ \Omega}{6.13\ \Omega}$$
$$= 97.9$$

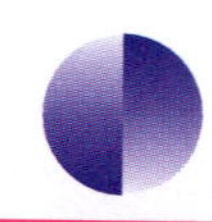

STUDENT to STUDENT

The extremely low input impedance of a CB amplifier is a major drawback, and I've been told, is the main reason it's seldom used in electronic circuitry.

Input Impedance Z_{in}

To calculate the input impedance Z_{in} of a common-base amplifier, divide the input voltage v_{in} by the input current i_{in}. Referring to Fig. 25-10, the input voltage is calculated as:

$$v_{in} = i_e r'_e$$

The input current $i_{in} = i_e = i_c$. Therefore, the input impedance Z_{in} is calculated as:

$$Z_{in} = \frac{v_{in}}{i_{in}}$$

which reduces to:

$$Z_{in} = \frac{i_C r'_e}{i_C}$$

Equation 25-30

$$Z_{in} = r'_e$$

In most cases r'_e is so much smaller than R_E that R_E can be ignored. In Fig. 25-9, $Z_{in} = r'_e$, which is 6.13 ohms in this case. The extremely low value of input impedance Z_{in} is the biggest drawback of a common-base amplifier. Because Z_{in} is so low, the common-base amplifier has only a few applications in the electronics industry.

No Phase Inversion

When v_{in} goes positive in Fig. 25-9, the emitter current I_E decreases. This in turn means that the $I_C R_C$ voltage drop decreases. As a result, the collector voltage V_C increases. Conversely, when v_{in} goes negative, the emitter current I_E increases.

This increases the $I_C R_C$ voltage drop. As a result, the collector voltage V_C decreases.

This proves that there is no phase inversion in a common-base amplifier.

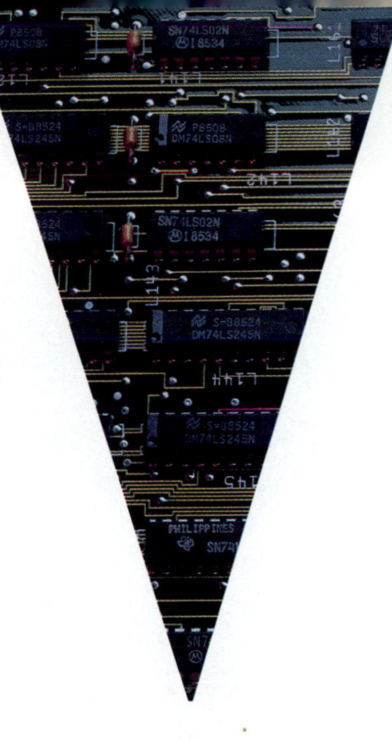

Power Gain

To calculate the power gain A_P in a common-base amplifier, use Eq. 25-31.

Equation 25-31

$$A_P = A_V \times A_i$$

$A_i \approx 1$ because the current gain of a common-base amplifier is slightly less than 1.

With $A_i \approx 1$, Eq. 25-31 can be simplified to:

Equation 25-32

$$A_P = A_V = \frac{r_L}{r_e'}$$

REVIEW QUIZ 25-3

1. A common-base amplifier has a very ______________ input impedance.
2. For a common-base amplifier, v_{in} and v_{out} are ______________ phase.
3. True or false? The common-base amplifier is used more often than the emitter-follower.

25-4 The Common-Source (CS) Amplifier

Field-effect transistors, like BJTs, can be used to amplify AC signals. FETs, however, do not provide as much voltage gain as BJTs.

Figure 25-11 on the next page shows a **common-source amplifier.** The input signal is applied to the gate, and the output is taken at the drain. The 470-ohm source resistor provides self-bias for the JFET amplifier.

Before analyzing the AC operation of the common-source amplifier, let's calculate the following DC quantities: V_G, V_{GS}, I_D, V_D, and V_{DS}.

First, $V_G = 0$ volts because the gate current I_G is zero in a JFET. Next, calculate V_{GS}. Assume that $V_S = 1.62$ volts.

$$\begin{aligned} V_{GS} &= V_G - V_S \\ &= 0\text{ V} - 1.62\text{ V} \\ &= -1.62\text{ V} \end{aligned}$$

Then calculate I_D.

$$\begin{aligned} I_D &= \frac{V_S}{R_S} \\ &= \frac{1.62\text{ V}}{470\ \Omega} \\ &= 3.45\text{ mA} \end{aligned}$$

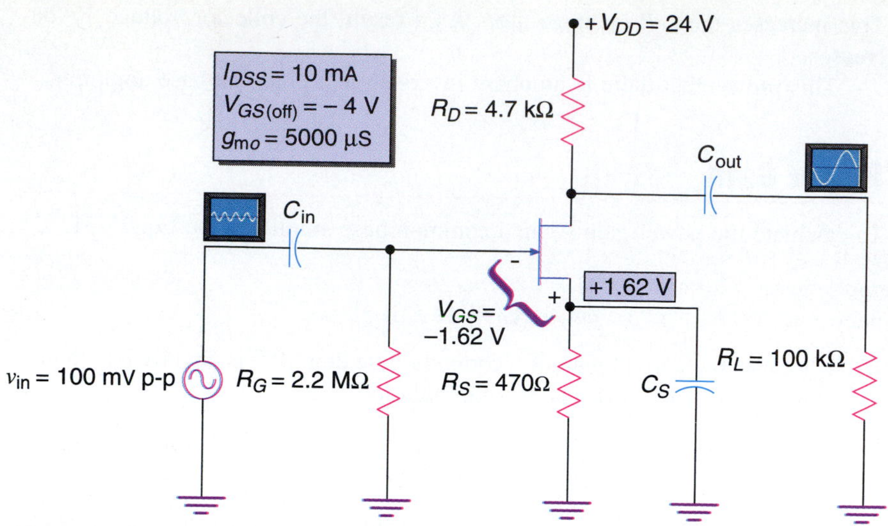

Figure 25-11 Common-source amplifier, where v_{in} is applied to the gate and v_{out} is taken from the drain.

Then solve for V_D.

$$\begin{aligned} V_D &= V_{DD} - I_D R_D \\ &= 24 \text{ V} - (3.45 \text{ mA} \times 4.7 \text{ k}\Omega) \\ &= 24 \text{ V} - 16.2 \text{ V} \\ &= 7.8 \text{ V} \end{aligned}$$

Finally, calculate V_{DS}.

$$\begin{aligned} V_{DS} &= V_{DD} - I_D (R_D + R_S) \\ &= 24 \text{ V} - (3.45 \text{ mA} \times 5.17 \text{ k}\Omega) \\ &= 24 \text{ V} - 17.84 \text{ V} \\ &= 6.16 \text{ V} \end{aligned}$$

Now, let's analyze how the common-source amplifier amplifies an AC signal.

Transconductance g_m

Remember that in a JFET, the gate-source voltage V_{GS} controls the drain current I_D. A very important JFET parameter called the **transconductance,** designated g_m, tells us how effectively V_{GS} controls I_D. Mathematically g_m is defined as:

Equation 25-33

$$g_m = \frac{I_D}{V_{GS}} \; (V_{DS} \text{ constant})$$

Because g_m is a ratio of current to voltage (which is conductance), g_m is measured in siemens (S). The transconductance g_m plays a major role in determining the voltage gain of a common-source amplifier.

When $V_{GS} = 0$ volts, the transconductance has its maximum value. At this point, the transconductance is designated as g_{mo}. The transconductance, however, is not the same for all values of V_{GS}. When g_{mo}, V_{GS}, and $V_{GS(off)}$ are known for

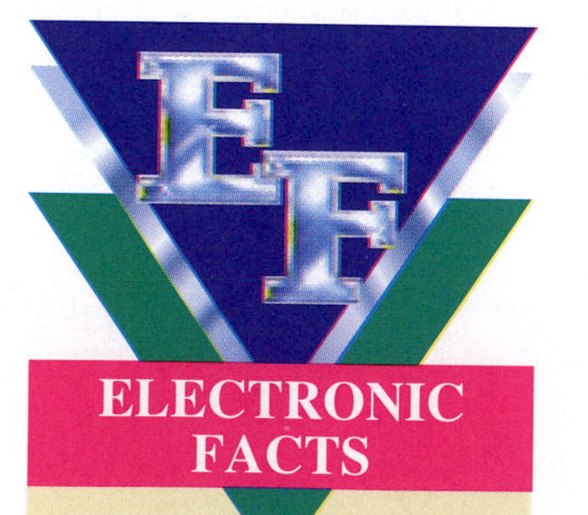

The transconductance g_m of an FET decreases as V_{GS} approaches the value of $V_{GS(off)}$.

a given JFET, the transconductance g_m can be found for any value of V_{GS} by using Eq. 25-34:

Equation 25-34

$$g_m = g_{mo}\left(1 - \frac{V_{GS}}{V_{GS(\text{off})}}\right)$$

Equation 25-34 indicates that the transconductance g_m is dependent on the value of V_{GS}. As an example, in an N-channel JFET, g_m decreases as V_{GS} is made increasingly more negative. When V_{GS} approaches $-V_{GS(\text{off})}$, the transconductance g_m approaches zero.

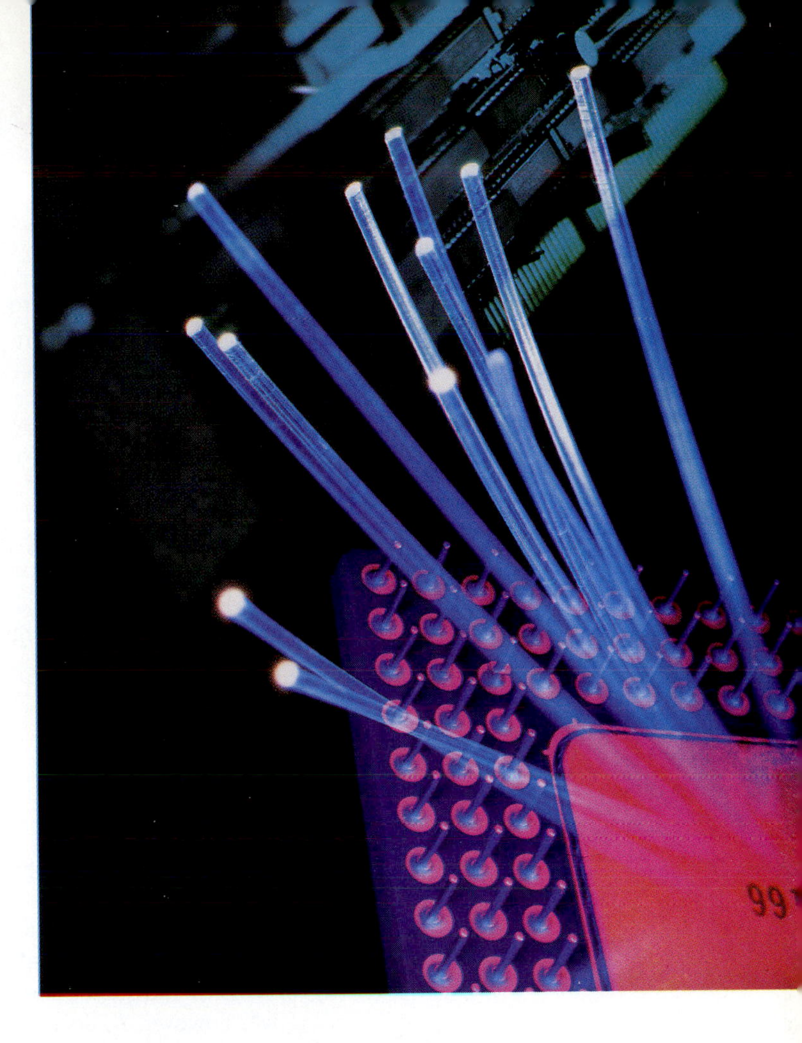

Example 25-9: For Fig. 25-11, calculate the transconductance g_m.

Solution:
Notice that $V_{GS} = -1.62$ volts, $V_{GS(\text{off})} = -4$ volts and $g_{mo} = 5000$ microsiemens. Using Eq. 25-34, the calculation for g_m is as follows:

$$\begin{aligned} g_m &= g_{mo}\left(1 - \frac{V_{GS}}{V_{GS(\text{off})}}\right) \\ &= g_{mo}\left(1 - \frac{-1.62\ \text{V}}{-4\ \text{V}}\right) \\ &= 5000\ \mu\text{S}\ (1 - 0.405) \\ &= 5000\ \mu\text{S} \times 0.595 \\ &= 2975\ \mu\text{S} \quad \text{or} \quad 2.975\ \text{mS} \end{aligned}$$

When g_m is known, A_V is calculated as follows:

Equation 25-35

$$A_V = g_m r_L$$

where $r_L = \dfrac{R_D \times R_L}{R_D + R_L}$

It should be noted that Eq. 25-35 can be used only when the source bypass capacitor C_S is present. If C_S is removed, then the voltage gain A_V can be calculated as follows:

Equation 25-36

$$A_V = \frac{g_m r_L}{1 + g_m R_S}$$

Example 25-10: For Fig. 25-11, calculate the voltage gain A_V and v_{out}.

Solution:
Begin by calculating the AC load resistance in the drain circuit.

$$\begin{aligned} r_L &= \frac{R_D \times R_L}{R_D + R_L} \\ &= \frac{4.7\ \text{k}\Omega \times 100\ \text{k}\Omega}{4.7\ \text{k}\Omega + 100\ \text{k}\Omega} \\ &= 4.49\ \text{k}\Omega \end{aligned}$$

Next, use Eq. 25-35 to calculate A_V. Because g_m was calculated earlier as 2.975 mS:

$$\begin{aligned} A_V &= g_m r_L \\ &= 2.975 \text{ mS} \times 4.49 \text{ k}\Omega \\ &= 13.36 \end{aligned}$$

To calculate v_{out}, proceed as shown.

$$\begin{aligned} v_{\text{out}} &= A_V \times v_{\text{in}} \\ &= 13.36 \times 100 \text{ mVp–p} \\ &= 1.33 \text{ Vp–p} \end{aligned}$$

Input Impedance Z_{in}

The input impedance Z_{in} of a common-source amplifier equals the value of the gate resistance R_G. Therefore:

$$Z_{\text{in}} = R_G$$

In Fig. 25-11,

$$\begin{aligned} Z_{\text{in}} &= R_G \\ &= 2.2 \text{ M}\Omega \end{aligned}$$

180° Phase Inversion

In the common-source amplifier, v_{in} and v_{out} are 180° out of phase. This is because when v_{in} goes positive, V_{GS} is less negative and I_D increases. This increases the $I_D R_D$ voltage drop and decreases the drain voltage V_D. When v_{in} goes negative, however, I_D decreases. This causes the $I_D R_D$ voltage drop to decrease and V_D increases. Therefore, v_{in} and v_{out} are 180° out of phase.

CS Amplifier Using D-MOSFETs

Figure 25-12 shows a D-MOSFET connected in the common-source configuration. Notice that the D-MOSFET uses zero bias. In this case, the DC drain-source voltage V_{DS} can be calculated as:

$$V_{DS} = V_{DD} - I_{DSS} R_D$$

The voltage gain A_V is calculated using Eq. 25-35.

Example 25-11: For Fig. 25-12, calculate the voltage gain A_V.

Solution:
Begin by calculating the AC load resistance r_L in the drain circuit.

$$\begin{aligned} r_L &= \frac{R_D \times R_L}{R_D + R_L} \\ &= \frac{1.5 \text{ k}\Omega \times 15 \text{ k}\Omega}{1.5 \text{ k}\Omega + 15 \text{ k}\Omega} \\ &= 1.36 \text{ k}\Omega \end{aligned}$$

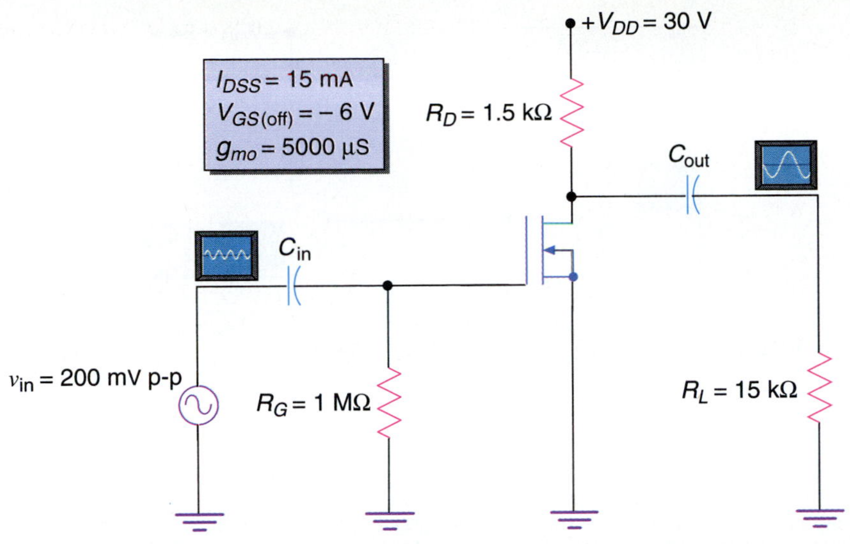

Figure 25-12 Common-source amplifier with $V_{GS} = 0$ volts. $I_D = I_{DSS} = 15$ milliamps.

Because $V_{GS} = 0$ volts, $g_m = g_{mo} = 5000$ microsiemens; therefore,

$$\begin{aligned} A_V &= g_m r_L \\ &= 5000\ \mu\text{S} \times 1.36\ \text{k}\Omega \\ &= 6.8 \end{aligned}$$

REVIEW QUIZ 25-4

1. In a common-source amplifier, v_{in} and v_{out} are ______________ phase.
2. In an N-channel JFET, the transconductance g_m ______________ when V_{GS} is made more negative.
3. In Fig. 25-12, the DC gate voltage $V_G =$ ______________ volts.
4. Removing the source bypass capacitor C_S ______________ the voltage gain.

25-5 Common-Drain (CD) and Common-Gate (CG) Amplifiers

Common-Drain Amplifier

Figure 25-13 on the next page shows a **common-drain amplifier.** Notice that v_{in} is applied to the gate and v_{out} is taken from the source. This circuit is more commonly called a source-follower because the AC signal at the source follows the AC signal at the gate.

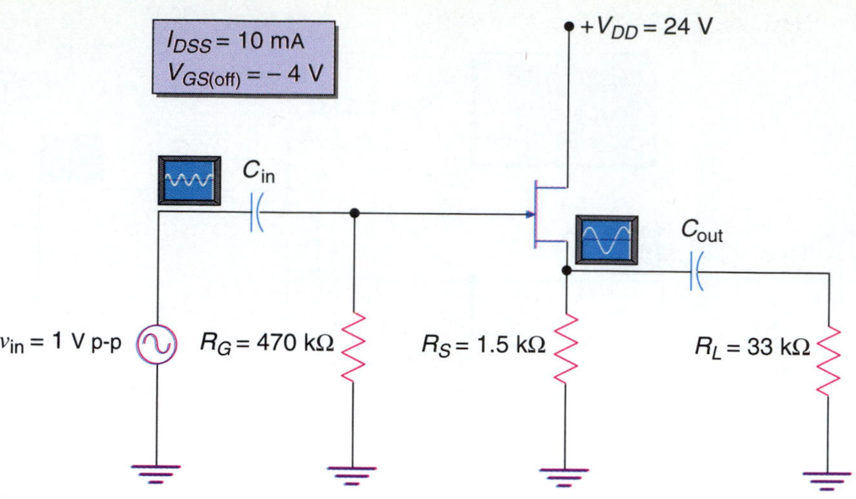

Figure 25-13 Common-drain amplifier. This circuit is more commonly called the source follower. Here, v_{in} is applied to the gate, and v_{out} is taken from the source.

Voltage Gain The voltage gain A_V of a source-follower is always less than 1. The exact calculation for A_V is:

Equation 25-37

$$A_V = \frac{g_m r_L}{1 + g_m r_L}$$

where r_L is the AC source resistance calculated as:

$$r_L = \frac{R_S \times R_L}{R_S + R_L}$$

Input Impedance The input impedance of the source-follower equals the value of the gate resistor R_G. This is expressed as:

$$Z_{in} = R_G$$

A source-follower has a low output impedance Z_{out}. To calculate Z_{out}:

Equation 25-38

$$Z_{out} = R_S || \frac{1}{g_m}$$

Example 25-12: For Fig. 25-13, calculate A_V, v_{out}, and Z_{out}. Note: $g_m = 2000$ microsiemens.

Solution:
Begin by calculating r_L.

$$r_L = \frac{R_S \times R_L}{R_S + R_L}$$

$$= \frac{1.5\text{ k}\Omega \times 33\text{ k}\Omega}{1.5\text{ k}\Omega + 33\text{ k}\Omega}$$

$$= 1.43\text{ k}\Omega$$

Next, use Eq. 25-37 to calculate A_V.

$$A_V = \frac{g_m r_L}{1 + g_m r_L}$$
$$= \frac{2000\ \mu S \times 1.43\ k\Omega}{1 + (2000\ \mu S \times 1.43\ k\Omega)}$$
$$= 0.74$$

Then solve for v_{out}.

$$v_{out} = A_V \times v_{in}$$
$$= 0.74 \times 1\ Vp\text{–}p$$
$$= 0.74\ Vp\text{–}p$$

To calculate Z_{out}, use Eq. 25-38.

$$Z_{out} = R_S || \frac{1}{g_m}$$
$$= 1.5\ k\Omega || \frac{1}{2000\ \mu S}$$
$$= 375\ \Omega$$

Common-Gate Amplifier

Figure 25-14 shows a typical **common-gate amplifier;** v_{in} is applied to the source while v_{out} is taken from the drain.

The voltage gate of a common-gate amplifier is calculated as follows:

Equation 25-39

$$A_V = g_m r_L$$

where $r_L = \dfrac{R_D \times R_L}{R_D + R_L}$

Notice that the formula for A_V is the same as that in a common-source amplifier.

The main drawback of a common-gate amplifier is its low input impedance Z_{in}. For a common gate amplifier,

$$Z_{in} = R_S \, || \frac{1}{g_m}$$

This is usually only a few hundred ohms or so.

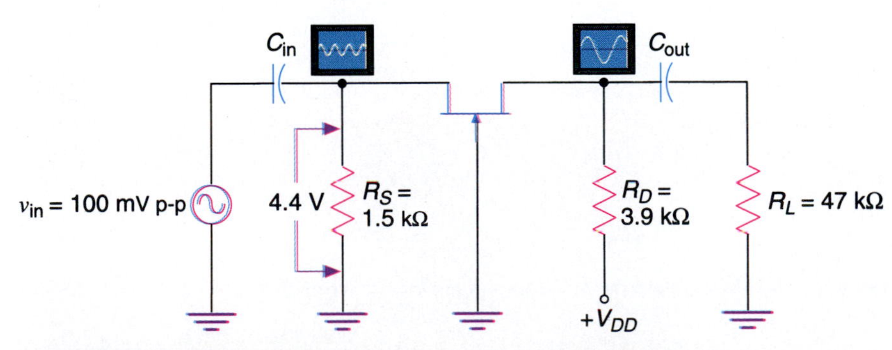

Figure 25-14 Common-gate amplifier, where v_{in} is applied to the source, and v_{out} is taken from the drain.

Example 25-13: For Fig. 25-14, calculate A_V, v_{out}, and Z_{in}. Note: $g_m = 1690$ microsiemens.

Solution:
Begin by calculating r_L.

$$r_L = \frac{R_D \times R_L}{R_D + R_L}$$

$$= \frac{3.9\ \text{k}\Omega \times 47\ \text{k}\Omega}{3.9\ \text{k}\Omega + 47\ \text{k}\Omega}$$

$$= 3.6\ \text{k}\Omega$$

Next, calculate A_V and v_{out}.

$$A_V = g_m r_L$$

$$= 1690\ \mu\text{s} \times 3.6\ \text{k}\Omega$$

$$= 6.08$$

$$v_{out} = A_V \times v_{in}$$

$$= 6.08 \times 100\ \text{mVp–p}$$

$$= 0.6\ \text{Vp–p}$$

Last, calculate Z_{in}.

$$Z_{in} = R_S || \frac{1}{g_m}$$

$$= 1.5\ \text{k}\Omega || \frac{1}{1690\ \mu\text{s}}$$

$$= 424\ \Omega$$

REVIEW QUIZ 25-5

1. True or false? In a source-follower, the voltage gain A_V is always greater than 1.
2. In the source-follower Z_{out} is quite ______________.
3. The input impedance of a common-gate amplifier is quite ______________.
4. In a common-drain amplifier, the output is taken at the ______________.

SUMMARY

The three BJT amplifier configurations are common-base, common-emitter, and common-collector. Of the three configurations, the common-emitter is the one generally used for amplifiers. With the common-emitter amplifier, v_{in} is applied to the base and v_{out} is taken from the collector. The common-emitter amplifier has both a high voltage and current gain. It is the only BJT amplifier configuration with both of these features. As a result, the common-emitter amplifier has a very high power gain. For a common-emitter amplifier, v_{in} and v_{out} are 180° out of phase. The voltage gain A_V of a common-emitter amplifier equals r_L/r'_e when the emitter bypass capacitor C_E is present. Without C_E present, the voltage gain A_V of a common-emitter amplifier equals approximately r_L/R_E. When the emitter bypass capacitor C_E is present, the input impedance seen looking into the base $Z_{in(base)}$ equals $\beta r'_e$. The input impedance of the amplifier includes the biasing resistors R_1 and R_2. In this case, $Z_{in} = R_1 \| R_2 \| \beta r'_e$. When C_E is removed, $Z_{in(base)}$ and Z_{in} increase.

A more popular name for the common-collector amplifier is emitter-follower. For an emitter-follower, v_{in} and v_{out} are in phase. The voltage gain A_V of an emitter-follower is slightly less than 1 or unity. The input impedance of an emitter-follower is usually quite high. In special cases, the biasing resistors R_1 and R_2 are omitted so that the input impedance Z_{in} is stepped up to a very high value. The output impedance Z_{out} of an emitter-follower is quite low.

The common-base amplifier has a high voltage gain. Its major drawback, however, is that its input impedance Z_{in} is quite low. For a common-base amplifier, v_{in} is applied to the emitter, and the output is taken from the collector. Because Z_{in} is so low, the common-base amplifier is seldom used.

The three FET amplifier configurations are common-source, common-drain, and common-gate. With the common-source amplifier, v_{in} is applied to the gate and v_{out} is taken from the drain. Both v_{in} and v_{out} are 180° out of phase in a common-source amplifier. The voltage gain A_V of a common-source amplifier equals $g_m r_L$ when the source bypass capacitor C_S is present. The voltage gain of a common-source amplifier is quite low compared to the voltage gain of a common-emitter amplifier. For a common-source amplifier, the input impedance Z_{in} equals the gate resistance R_G.

A more popular name for the common-drain amplifier is source-follower. A source-follower always has a voltage gain A_V that is less than 1. For a source-follower, v_{in} and v_{out} are in phase. The input impedance, Z_{in} for a source-follower equals the gate resistance R_G which is usually quite high. The output impedance Z_{out} of a source-follower is quite low. In general, the source-follower has a high input impedance and low output impedance.

For the common-gate amplifier, v_{in} is applied to the source and v_{out} is taken from the drain. Both v_{in} and v_{out} are in phase.

The common-gate amplifier has a very low input impedance, which is a big disadvantage. This is the reason why common-gate amplifiers are seldom used.

So in this chapter, we have discussed the different types of bipolar junction transistors (BJTs) and field-effect transistors (FETs). For bipolar transistors, we examined the characteristics of the common-base, common-emitter, and common-collector amplifier. For field-effect transistors, we studied the characteristics of the common-gate, common-source, and common-drain amplifier. For each we examined voltage gain, input impedance, and the phase relationship between v_{in} and v_{out}.

NEW VOCABULARY

AC collector resistance r_L
AC emitter resistance r'_e
common-base amplifier
common-collector amplifier
common-drain amplifier
common-emitter amplifier
common-gate amplifier
common-source amplifier
emitter-bypass capacitor C_E
emitter-follower
input impedance $Z_{in(base)}$ and Z_{in}
output impedance Z_{out}
phase inversion
power gain A_P
source-follower
transconductance g_m and g_{mo}
voltage gain A_V

Questions

1. What is the unit of transconductance g_m?
2. Which BJT and FET amplifier configurations have a 180° phase shift between V_{in} and V_{out}?
3. What happens to the voltage gain A_V of a common-emitter amplifier, when the emitter-bypass capacitor is removed?
4. What happens to the input impedance Z_{in} when the emitter-bypass capacitor is removed from a common-emitter amplifier?
5. Which BJT amplifier configuration has the:
   a. Lowest voltage gain A_V?
   b. Highest power gain A_P?
   c. Lowest input impedance Z_{in}?
   d. Lowest output impedance Z_{out}?
   e. Lowest current gain?
6. Does the AC resistance r'_e of the emitter diode increase or decrease when the DC emitter current I_E is increased?
7. What happens to the voltage gain A_V in a common-emitter amplifier when the AC collector resistance decreases?
8. Which will usually have a higher voltage gain: a common-emitter or common-source amplifier?
9. What is the major drawback of a common-base amplifier?
10. For FETs, is g_m always a constant value?
11. For FET amplifiers, which amplifier configuration has the:
    a. Lowest voltage gain A_V?
    b. Lowest input impedance Z_{in}?
    c. Lowest output impedance Z_{out}?

Problems

1. The DC emitter current in a transistor equals 7.5 milliamps. Calculate the AC emitter resistance r'_e.
2. A common-emitter amplifier has an input voltage v_{in} of 3 mVp–p and an output voltage v_{out} of 240 mVp–p. Calculate A_V.
3. A common-emitter amplifier has a voltage gain A_V of 100 and a current gain A_i of 150. Calculate the power gain A_P.
4. In Fig. 25-15, calculate the following DC and AC quantities: V_B, V_E, I_C, V_C, V_{CE}, r'_e, r_L, A_V, v_{out}, $Z_{in(base)}$, and Z_{in}.
5. Mentally disconnect C_E in Fig. 25-15. Calculate A_V, v_{out}, $Z_{in(base)}$, and Z_{in}.
6. In Fig. 25-16, calculate the following quantities: V_B, V_E, I_C, V_C, V_{CE}, r'_e, r_L, A_V, v_{out}, $Z_{in(base)}$, and Z_{in}.

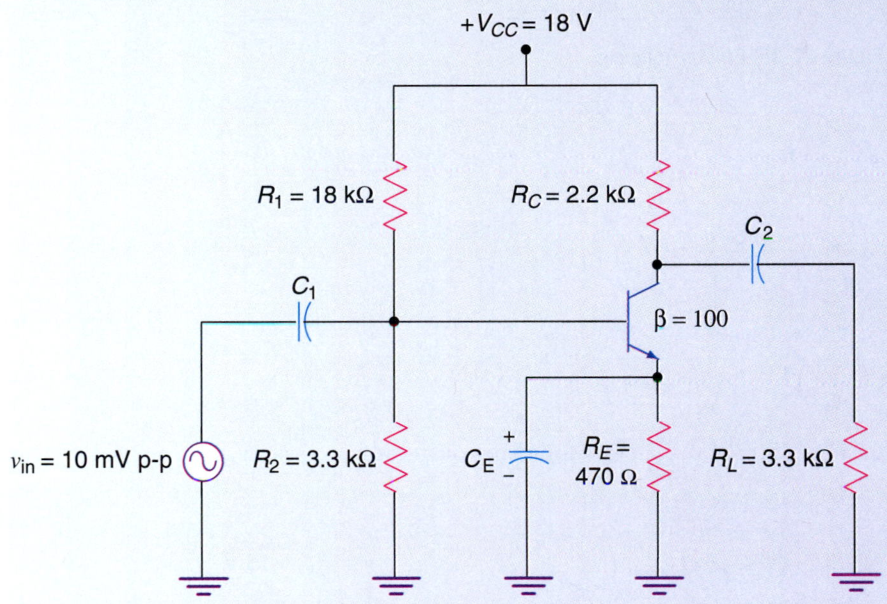

Figure 25-15 Common-emitter amplifier.

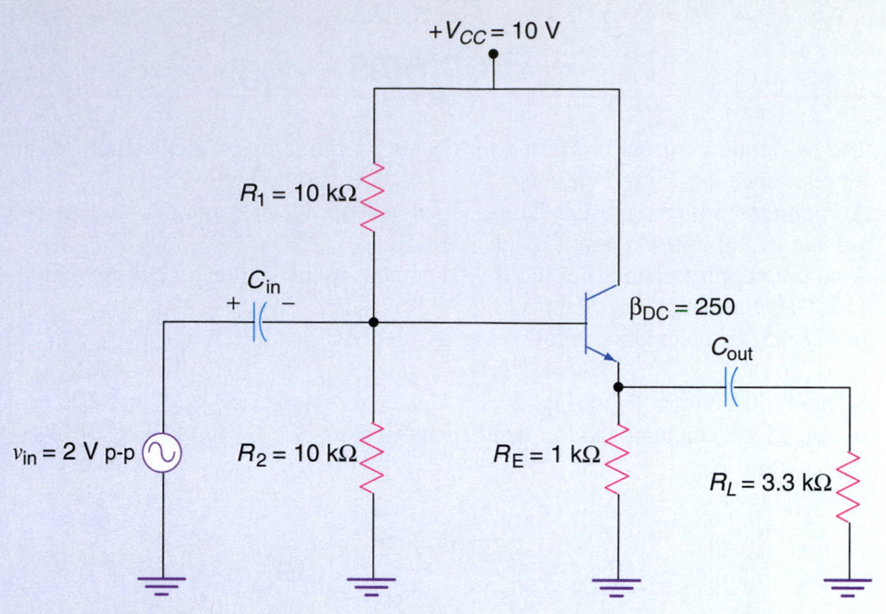

Figure 25-16 Emitter-follower.

7. In Fig. 25-17, calculate the following quantities: V_{GS}, I_D, V_D, r_L, g_m, A_V, v_{out}, and Z_{in}.

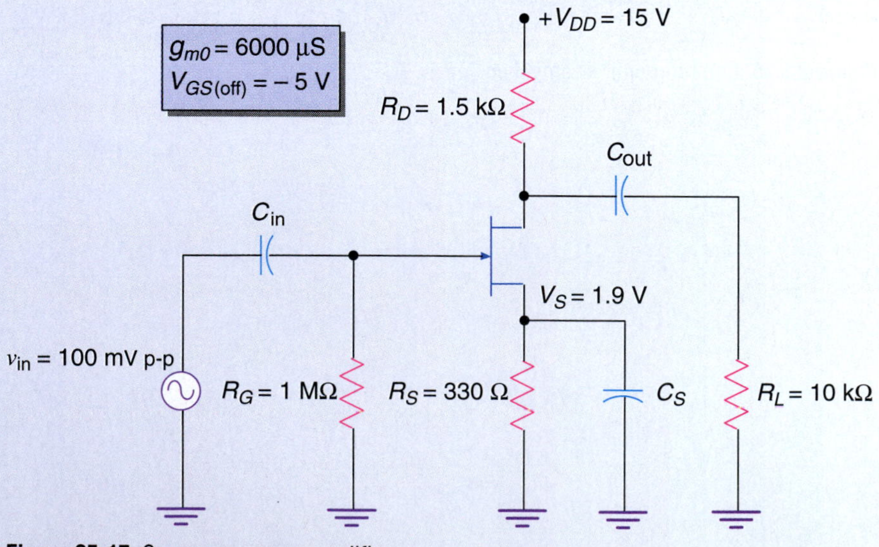

Figure 25-17 Common-source amplifier.

8. In Fig. 25-18, calculate the following quantities: V_{GS}, I_D, V_{DS}, r_L, g_m, A_V, v_{out}, Z_{in}, and Z_{out}.

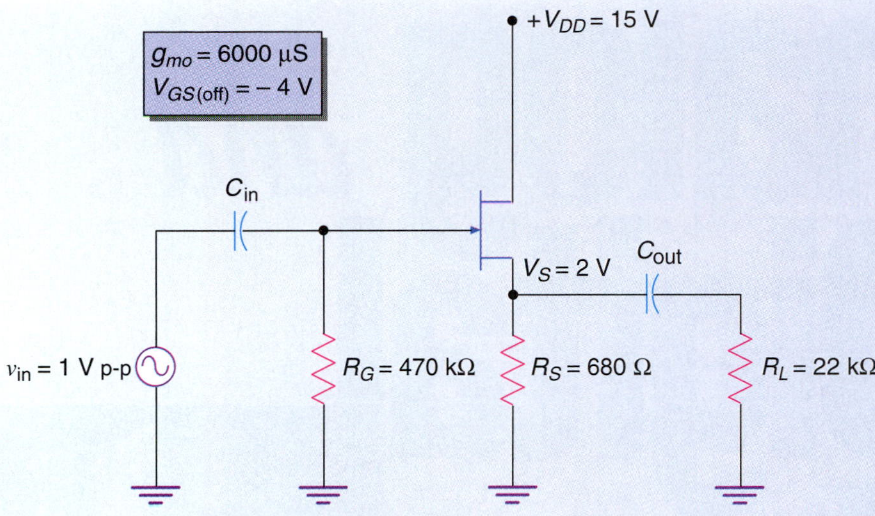

Figure 25-18 Source-follower.

Critical Thinking

1. Refer to Fig. 25-1. What is the AC signal voltage at the emitter if C_E is removed?
2. Refer to Fig. 25-1. If R_1 opens, how much AC signal voltage will appear at the collector?
3. Refer to Fig. 25-7. If R_L shorts, how much AC signal voltage is at the emitter terminal?
4. Refer to Fig. 25-9. If the base bypass capacitor C_B shorts, what effect does it have on the output voltage?
5. Refer to Fig. 25-11. What happens to the voltage gain A_V when the source bypass capacitor is removed?
6. Refer to Fig. 25-11. Is Z_{in} affected when the source bypass capacitor C_S is removed?
7. How much AC voltage would be measured at the drain terminal in Fig. 25-13?

Answers to Review Quizzes

25-1
1. base; collector
2. 180° out of
3. reduces
4. high

25-2
1. emitter follower
2. 1 or unity
3. high
4. low

25-3
1. low
2. in
3. false

25-4
1. 180° out of
2. decreases
3. 0
4. reduces

25-5
1. false
2. low
3. low
4. source

Chapter 26

Analog and Digital Integrated Circuits

ABOUT THIS CHAPTER

You have seen in earlier chapters how the operation of semiconductor diodes, BJTs, and JFETs depends on junctions that form between layers of N and P semiconductor materials. Also, you have seen how BJTs and JFETs are used in transistor amplifier circuits. But there is more

After completing this chapter you will be able to:

- Describe how to count the pins on various IC packages.
- Explain the difference between analog and digital IC circuits.
- Describe the basic function of an operational amplifier.

to a practical electronic circuit than a single transistor and its PN junctions. Practical electronic circuits require more than one transistor and many resistors. Most electronic circuits are made even more complex in order to operate at the high standards of performance required today.

This chapter describes the semiconductor technology that produces entire circuits on a single chip of silicon. Tens, hundreds, and even thousands and tens of thousands of individual PN junctions can be etched onto a single chip of silicon that is often no larger than a normal transistor. A package that includes a circuit on a single silicon chip is called an **integrated circuit,** or **IC.**

- ▽ Calculate the voltage gain of op-amp IC amplifiers.
- ▽ Recognize the symbols, truth tables, and Boolean expressions for logic functions.
- ▽ Describe the operation of a flip-flop and binary counter.

26-1 Introduction to IC Devices

Figure 26-1 provides an inside look at a typical integrated circuit (IC) device. The circuitry is microscopically photoetched in layers on the silicon base. The circuitry consists of PN junctions for the necessary diodes and transistors, lines of lightly doped P or N material that serve as resistors, and conductive tracks that interconnect the working elements and layers of etching. There is actually more space devoted to the simple electrical connections to the outside world than to the chip itself.

Electrical engineers and technicians are less concerned about the way things are assembled on the silicon chip of an IC than about the way the terminals on the IC package are interconnected with other ICs, necessary external components, and power supplies. Figure 26-2 shows the pin configurations for common types of IC package styles.

Plastic and ceramic IC packages are far more common than metal-can IC packages. And the most common types of plastic and ceramic packages are those with dual in-line pins (DIP) and surface-mount terminals. The metal IC packages with their long wire leads and the DIPs with their sturdy metal terminals are intended for mounting through holes drilled through a circuit board. The surface-mount types do not require through-holes; they are soldered directly to copper pads on the surface of a circuit board.

Figure 26-1 Internal makeup of a typical IC.

Terminals on IC devices are identified by the terminals or pins that are numbered according to a standard set of rules. IC manufacturers identify the function of each lead (amplifier input, positive voltage supply, ground connection, etc.) according to the pin numbers as shown in Fig. 26-2. The detailed view for DIPs and gull-wings shows that pin 1 is located to the left of the notch or beside a dot that is pressed into the top surface of the IC package. Pin 2 is then directly below pin 1. The counting proceeds in a counterclockwise direction through all of the pins on the IC. Most DIPs have 8, 14, 16, 18, 20, 22, 24, 28, 40, or 64 pins. Generally speaking, the greater the number of pins, the more complex the internal circuitry. A simple amplifier circuit might have only 8 pins, whereas the microprocessor in a personal computer has a minimum of 40 pins and as many as 64 when the integrated circuit is packaged in the DIP configuration.

Plastic leadless chip carrier (PLCC) packages have terminals on all four edges, so the numbering convention has to be different from that of a DIP or

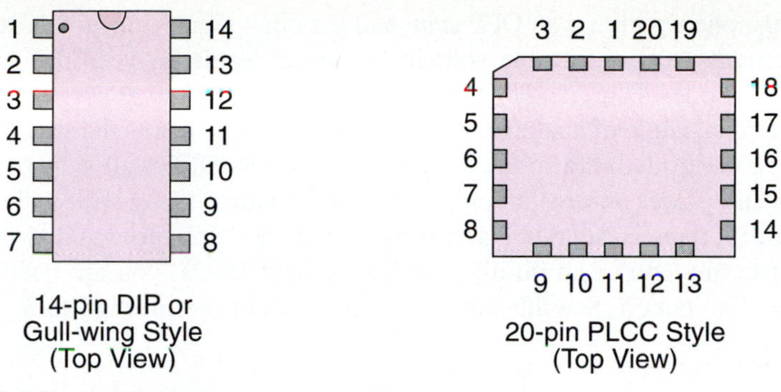

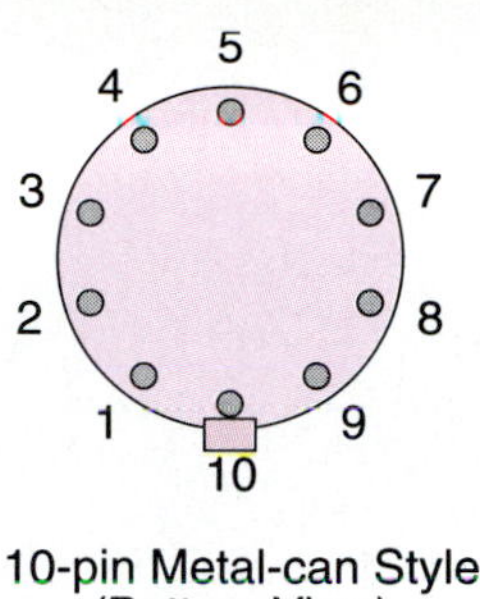

Figure 26-2 Pin configurations for popular IC package styles.

gull-wing IC package. A diagonal corner and a dot pressed into the case are used for locating pin 1 on PLCC packages. Viewing the package from the top and rotating it so that the diagonal edge is in the upper-left corner, pin 1 will be the middle pin along the top row of terminals. Pin 1 is most often identified with the dot as well. The pins are then numbered successively in the counterclockwise direction. A popular PLCC package has 20 pins, although there are examples having far more terminals than 20.

Metal-can IC packages are usually reserved for applications requiring low electrical noise levels and excellent frequency response. Unlike the plastic and ceramic IC packages, the pins are normally viewed from the bottom of the package. A tab on the case marks the location of the highest numbered pin. Pin 1 is then the first pin located clockwise from the tab.

REVIEW QUIZ 26-1

1. True or false? An IC is a set of diode and transistor junctions and resistive semiconductor materials that are integrated into a useful circuit photoetched onto a single silicon chip.
2. A plastic or ceramic IC package that has two parallel rows of pins is called a(n) ______________ package and is abbreviated as ______________.
3. PLCC is an abbreviation for ______________.
4. Viewed from the top of the package, the pins on a DIP are counted in the ______________ direction.
5. If you are viewing a 20-pin PLCC from the top, and pin 1 is located on the side directly across from you, on which side is pin 12?

26-2 Analog vs. Digital ICs

The circuits that are enclosed in IC packages can be classified as analog or digital. There are also a few IC devices that contain both classifications. Whether a particular application uses analog, digital, or a combination of analog and digital ICs mainly depends on what the circuit is expected to do.

Analog circuits are used where there are relatively smooth transitions in voltage or current levels. Digital circuits, by contrast, have only two voltage or current levels: one signifying an OFF state and the other signifying an ON state. For analog circuits, all conceivable voltage or current levels between the minimum and maximum levels are equally important.

A good example of analog and digital operations concerns the way the lighting can be controlled in a room. Where a simple ON/OFF wall switch turns on a light at full power or turns it completely off, you are dealing with a digital-like circuit. But if there is a dial in place of the switch, and you turn the dial smoothly to adjust to any one of an infinite number of light levels, you are dealing with an analog-like circuit. See the simplified circuit examples in Fig. 26-3.

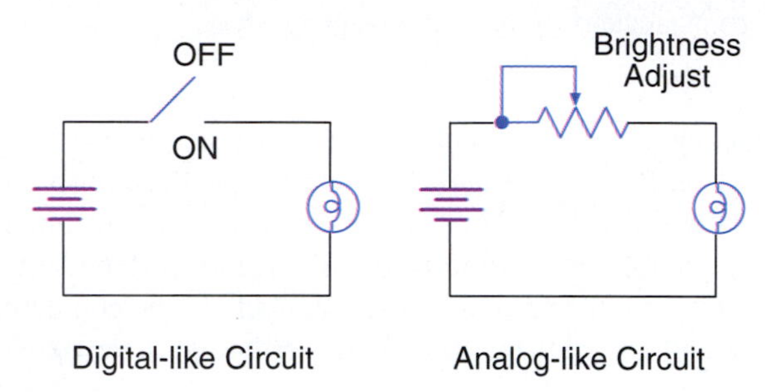

Figure 26-3 Examples of analog and digital circuits.

Linear and Nonlinear Analog Circuits

A **linear circuit** is a type of analog circuit that does not alter the basic character of the waveform it handles. A linear circuit might amplify a signal (make it larger), diminish a signal, or switch a signal from one destination to another. A linear circuit, however, does not change the essential shape and quality of the signal it processes. An audio amplifier is an example of an analog/linear circuit that must not alter the character of the original signal. Any unwanted alteration of a signal is regarded as signal distortion.

There is a class of analog circuits that is meant to transform a waveform from one shape to another. This is called a **nonlinear circuit.** An audio mixer is an example of a nonlinear analog circuit. In this instance, two or more audio signals are mixed into one signal; the resulting waveform bears little resemblance to any of the originals. Another example of a nonlinear analog circuit is one that converts a voltage level into a frequency level (a voltage-controlled oscillator, or VCO).

IC manufacturers provide a wide range of analog—linear and nonlinear—IC devices. They are mainly specified according to their input voltage sensitivity and the range of frequencies they can handle without showing significant amounts of distortion, as well as by the amount of power they can handle without overheating. The manufacturers' specifications sheets also show the types of IC packages available for each kind of circuit and the correspondence between IC pin numbers and functions within the package.

Typical power supply voltage levels for analog ICs are on the order of ±12 volts. Those that require both positive and negative supply voltages can produce meaningful output voltage levels anywhere within that power supply range. Those analog ICs that require only a positive voltage supply can produce only positive output voltages.

Combinatorial and Sequential Digital Circuits

Digital ICs are the heart of modern computers. Inputs to a computer (keyboard, disk drive, telephone modem, etc.) are all converted to sequences of ON/OFF voltage levels. Any voltage level between the full-ON and full-OFF levels is regarded as a nuisance.

Digital ICs are classified according to the types of digital functions they perform. Some perform logical and arithmetical operations, for instance. These are known as *combinatorial logic circuits.* The output of a combinatorial logic circuit depends only on the ON/OFF status of its inputs as they are at the moment. The second major classification of digital circuits is *sequential logic circuits.* The output of a sequential logic circuit may be partly determined by the input status at the moment, but mostly by the previous output status of the circuit. A digital counter is an example of a sequential logic circuit.

Figure 26-4 illustrates the difference between combinatorial and sequential logic circuits. An arithmetic adder circuit is an example of a combinatorial logic circuit. When the digital inputs represent numerals 3 and 5, the output is always a digital representation of the numeral 8. It makes no difference what was summed previously. A digital counter shows how sequential logic is different from combinatorial logic. Whenever a pulse is applied at the input of the counter, the output advances to the next higher number. If the output is a 6 when the input pulse occurs, the output changes to a 7. But the output does not change to 7 every time the input pulse occurs. The output at any given moment depends on the output prior to the occurrence of the input pulse.

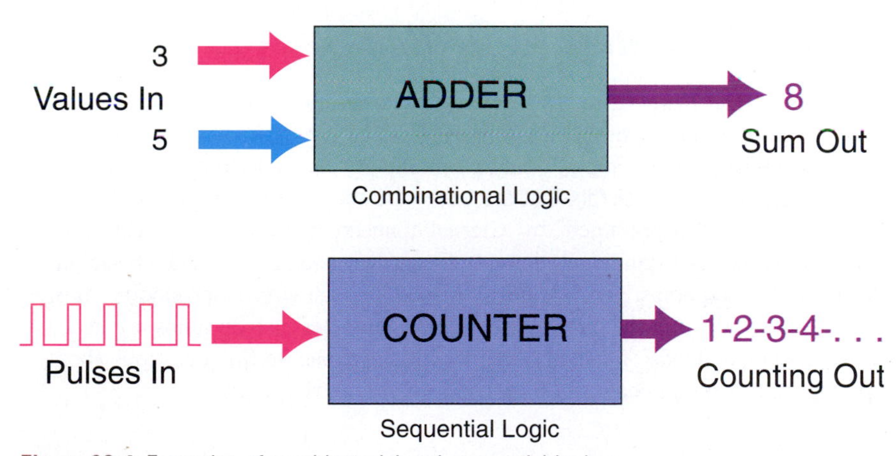

Figure 26-4 Examples of combinatorial and sequential logic.

Circuits That Combine Analog and Digital Features

Finally there are IC devices that cannot be classified as being all-analog or all-digital. In one way or another, these circuits combine the features of both. Usually this kind of circuit has an analog input and a digital output, or a digital input and an analog output. They are most often used where it is necessary to translate a signal from one of the major classifications to another.

An IC device that translates a voltage level into a set of digital pulses is called an **analog-to-digital converter,** or **A-to-D** converter. The output is a digital number that represents the amount of voltage applied at the input. The electronic digital voltmeter is a good example of how an A-to-D converter can be applied.

A **digital-to-analog converter (D-to-A converter)** accepts a digital value at its input and translates it to a voltage level that is proportional to the input. Examples of D-to-A converters can be found where a digital computer is used for controlling the amount of electric power that is applied to motors in industrial machinery and robots.

REVIEW QUIZ 26-2

1. ______________ circuits are used where every voltage level can be important.
2. ______________ circuits are used where only the ON and OFF levels are important.
3. A(n) ______________ circuit is a type of analog circuit that does not significantly alter the waveform it handles.
4. A voltage-controlled oscillator is an example of a(n) ______________ analog circuit.
5. The output of a(n) ______________ logic circuit depends on the output that existed prior to the occurrence of an input pulse.
6. An IC device that converts an analog signal to a digital signal is called a(n) ______________.

26-3 Analog IC Devices

There are a great many IC devices that are classified as analog ICs. Most, however, are built around a simple and highly versatile circuit known as an **operational amplifier** (or op amp). A single op-amp IC can replace dozens of individual transistors and resistors that are otherwise required for amplifier circuits. Op amps can be "programmed" by external connections to perform a wide variety of different amplifying operations. In fact, when you have gained a basic understanding of op amps, you will have grasped the basis for understanding most analog IC devices and the circuits built from them. Because op amps are at the heart of so many kinds of analog circuits, you can expect to find questions about op amps and their applications on industry employment exams.

NOTES

Applying an adjustable DC voltage level to an offset null terminal allows you to balance out slight internal errors in the IC. This feature is not necessary for understanding how op amps work, and many IC op amps do not have offset null terminals anyway. For this reason you will find no further discussion of them in this chapter.

Operational Amplifiers

Figure 26-5 shows the terminals that are typically available with an IC op amp. The symbol for the op amp, itself, is a triangle. Inputs are shown at the base, and the output is taken from the apex of the triangle. IC op amps have two inputs: an *inverting input* (labeled with a minus sign) and a *noninverting input* (labeled with a plus sign), and a single output. There are two power supply terminals: V_{CC} for the more positive supply voltage, and V_{EE} for the more negative supply voltage. Two other terminals, called *offset null* terminals, are used for making sure the output is zero volts when the input conditions call for zero voltage out.

Op-amp ICs are available as bipolar or MOS devices. Figure 26-6(*a*) shows the schematic and pinout diagram of the popular 741C op amp. Figures 26-6(*b*) and (*c*) show a dual (two units) op amp, and a quad (four units) op amp.

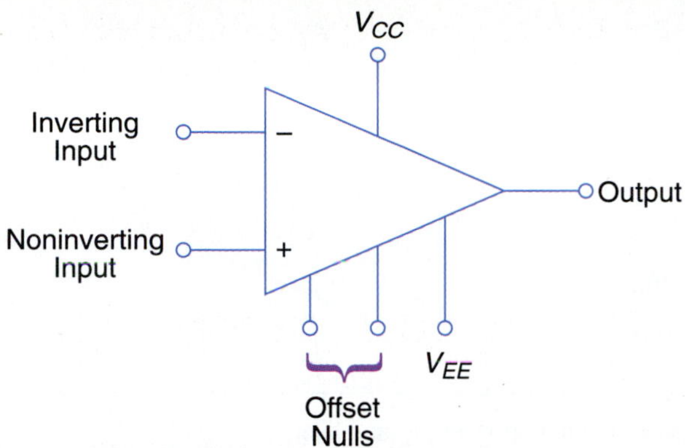

Figure 26-5 Symbol and connections for a typical IC op amp.

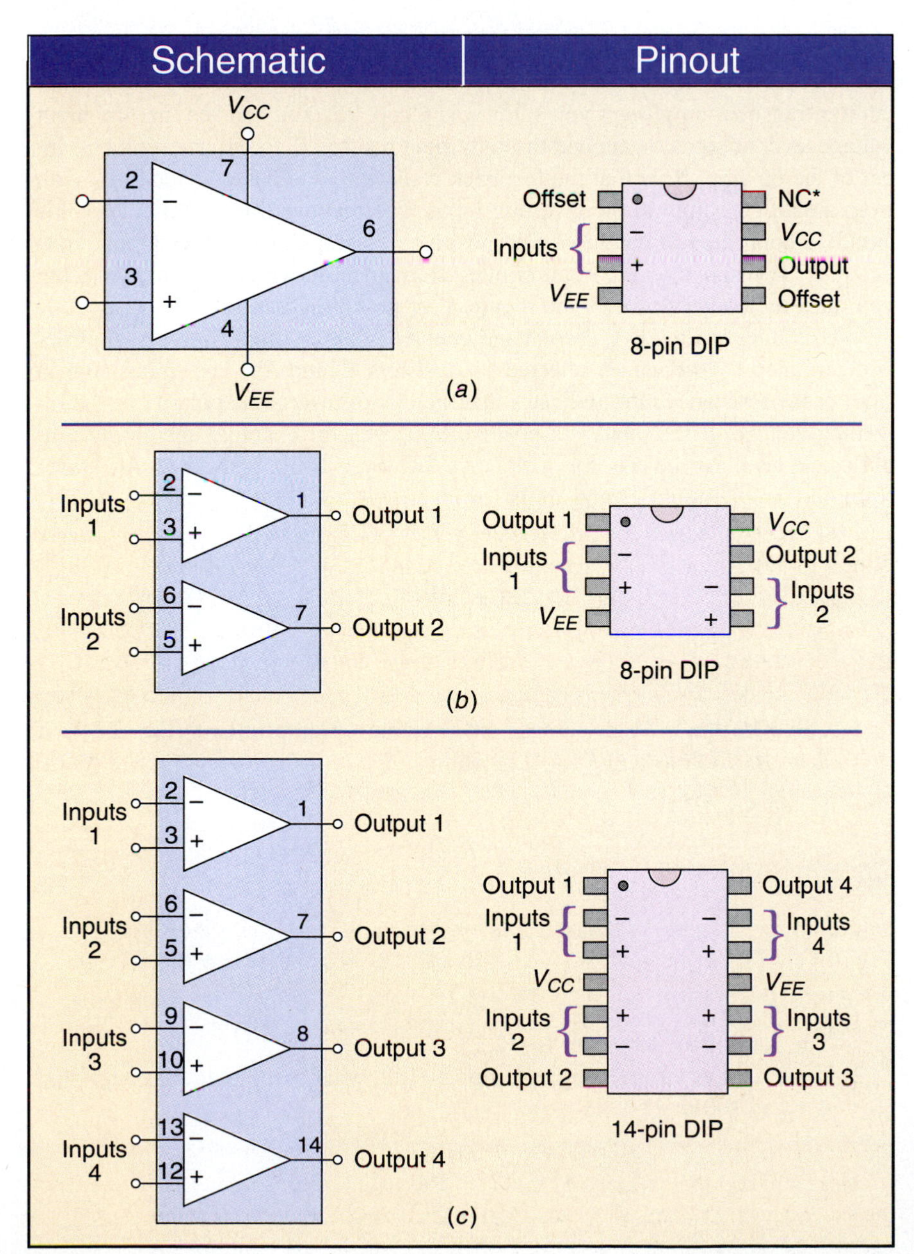

Figure 26-6 Symbols and IC pinout diagrams for op-amp ICs. (*a*) The 741C op amp. (*b*) The MC3458 dual op amp. (*c*) The LM148 quad op amp.

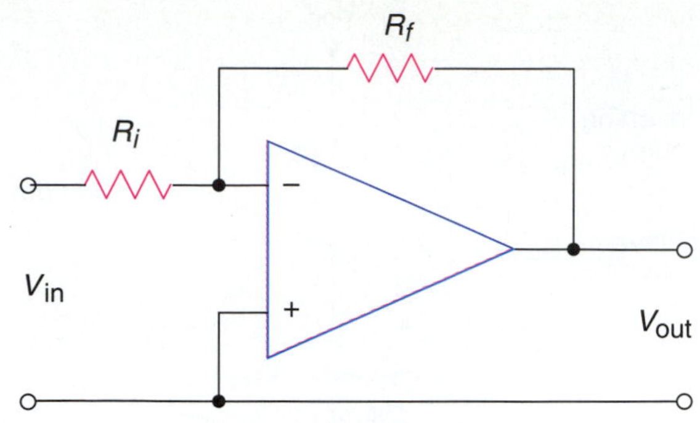

Figure 26-7 The basic op amp inverting amplifier.

Figure 26-7 shows an op amp connected as an *inverting amplifier.* It amplifies and inverts an input voltage or waveform in a manner similar to the common-emitter transistor amplifiers you studied in Chap. 25. You can see that the input voltage level or signal is applied through input resistor R_i to the inverting (−) input of the op amp. Note that the feedback resistor R_F completes a feedback path from amplified output to the inverting input. The noninverting (+) input for this circuit is connected to common. The positive (+) and negative (−) supply voltage pins (V_{CC} and V_{EE}) are often omitted from such drawings for simplicity, but you must remember that they are required for proper operation.

According to Eq. 26-1, the voltage gain A_V of an op-amp inverting amplifier is determined by the values selected for resistors R_i and R_F. The minus sign in front of the resistance ratio indicates that the circuit inverts the polarity of the incoming voltage. If the input is a positive DC level, for example, the circuit amplifies the level and inverts it to a negative DC level. If the input is an AC waveform, the circuit amplifies and shifts (or inverts) it by 180°.

Equation 26-1

$$A_V = \frac{-R_F}{R_i}$$

Example 26-1

1. Calculate the voltage gain of an inverting op-amp circuit (Fig. 27-7) if R_i = 10 kilohms and R_F = 22 kilohms.

   Using Eq. 26-1:

$$\begin{aligned} A_V &= \frac{-R_F}{R_i} \\ &= \frac{-22\text{ k}\Omega}{10\text{ k}\Omega} \\ &= -2.2 \end{aligned}$$

   The gain of the amplifier is −2.2.

2. If the gain of an inverting op amp is −2.2, what is the output voltage V_{out} if V_{in} is 100 mV?

$$\begin{aligned} V_{out} &= A_V \times V_{in} \\ &= -2.2 \times 100\text{ mV} \\ &= -220\text{ mV} \end{aligned}$$

3. Suppose you need to amplify a signal of 0.5 volt to 3 volts and you want to use an inverting op-amp circuit where R_F is fixed at 1 megohm. Calculate the necessary value for R_i.

The required voltage gain of this circuit is:

$$A_V = \frac{3\text{ V}}{0.5\text{ V}}$$
$$= 6$$

The value of R_i is given by:

$$R_i = \frac{R_f}{A_V}$$
$$= \frac{1\text{ M}\Omega}{6}$$
$$= 167\text{ k}\Omega$$

Figure 26-8 shows an op amp that is set up to operate as a *noninverting amplifier*. It amplifies, but does not invert, the input voltage or waveform. You can see that the input signal is connected directly to the noninverting input on the op amp. There is a feedback resistor R_F connected between the output and the inverting input, and there is another resistor R_1 connected between the noninverting input and ground.

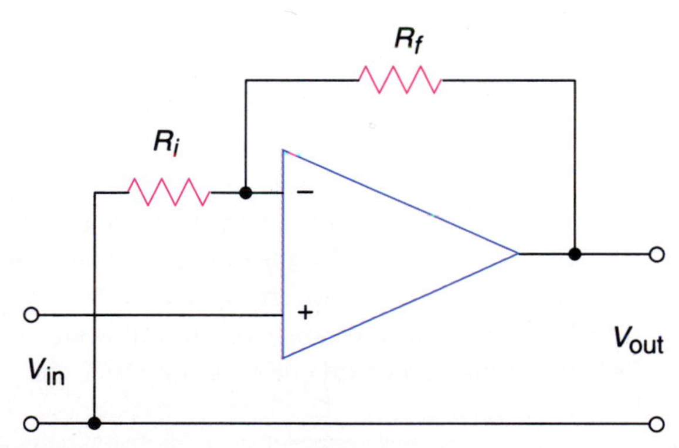

Figure 26-8 The basic op-amp noninverting amplifier.

Equation 26-2 shows that the voltage gain of this amplifier depends on the values R_F and R_1. There is no minus sign, which suggests that there is no inversion of the input signal.

Equation 26-2

$$A_V = 1 + \frac{R_F}{R_1}$$

Example 26-2

1. Calculate the voltage gain for the noninverting op-amp circuit in Fig. 26-8 if R_F = 100 kilohms and R_1 = 22 kilohms.

Using Eq. 26-2:

$$A_V = 1 + \frac{R_F}{R_1}$$
$$= 1 + \frac{100\ \text{k}\Omega}{22\ \text{k}\Omega}$$
$$= 1 + 4.5$$
$$= 5.5$$

The gain of this noninverting amplifier is 5.5.

2. Suppose you want to use a noninverting op amp to increase a voltage from 200 millivolts to 1 volt. If you have already selected a value of 1.2 megohms for R_F, what is the necessary value for R_1?

   The required voltage gain of this circuit is:

$$A_V = \frac{1\ \text{V}}{200\ \text{mV}}$$
$$= 5$$

   The value of R_1 is given by:

$$R_1 = \frac{R_F}{A_V - 1}$$
$$= \frac{1.2\ \text{M}\Omega}{5 - 1}$$
$$= \frac{1.2\ \text{M}\Omega}{4}$$
$$= 300\ \text{k}\Omega$$

A special case of a noninverting op-amp circuit is one where the value of the feedback resistor is reduced to zero ohms. (See the example in Fig. 26-9.) According to Eq. 26-2, setting R_F to zero ohms leaves A_V with a value of 1 as long as R_1 has some value greater than zero. Because the gain is one and there is no signal inversion (because this is a noninverting amplifier), the circuit is a *voltage-follower.* You learned about transistor voltage-followers, also known as emitter-followers, when studying common collector amplifier circuits.

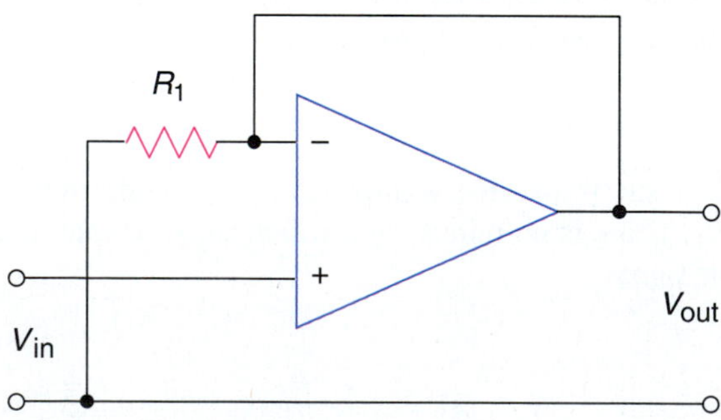

Figure 26-9 Op-amp *voltage-follower.*

Whenever a capacitor is inserted into the input or feedback paths for op amps, the circuit operates as a wave shaper and active filter. Suppose you see a series *RC* circuit in the feedback path of an inverting amplifier (Fig. 26-10). As the input frequency increases, the impedance of the feedback path decreases, thereby

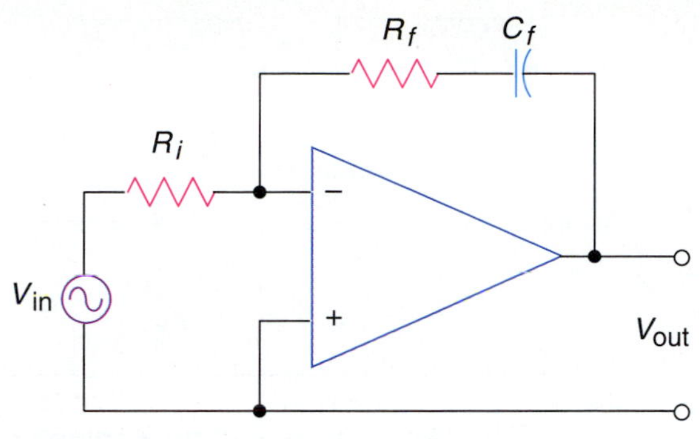

Figure 26-10 Op-amp used as an active *low-pass filter.*

decreasing the gain of the amplifier. In other words, the gain of this circuit is higher at the low-frequency end of the band. It is an active *low-pass filter* circuit.

There are literally hundreds of useful variations of op-amp circuits. For the most part, op amps have replaced discrete (stand-alone) transistor circuits. And op amps are incredibly easy to use. With just the little bit of information you have seen here, you can design a workable voltage amplifier circuit from a selection of three components.

Voltage, Current, and Power Regulators

Many of today's high-technology applications of electronic circuits require very stable and highly regulated sources of electric power. Fluctuations in the power supply voltages can seriously affect the accuracy and reliability of many different kinds of electronic equipment. Industrial electronic systems, including industrial robots, must respond accurately to signals that tell them how much power they are to apply to motors and heating elements. There are analog ICs available for regulating a wide range of voltages, currents, and power levels.

Electronic regulators work much like a thermostat system in your home. The thermostat keeps the temperature in your home as close as possible to some prescribed level, in spite of events that tend to oppose a stable temperature. When you set the thermostat to 80°F, you expect the room temperature to be established at that point, and stay there. When cooling air flows into the room from the outside, you expect the furnace to turn on, thereby opposing a drop in room temperature. Or, if it is summertime, an open door or window will tend to warm the room above the preset temperature and the air conditioner will turn on, thereby opposing an unwanted increase in room temperature.

One of the simpler and more common electronic regulator ICs is one that is used for regulating the output of a DC power supply. These fixed **voltage regulator** ICs maintain a specified output voltage level in spite of changes in the AC input voltage level and loading on the DC output. Figure 26-11 on the next page includes the block diagram for a DC power supply that includes a fixed, three-terminal voltage regulator at its output.

The figure also shows the block diagram for an *adjustable voltage regulator.* These allow the user to adjust, or trim, the output voltage level. Once it is set, the circuit maintains that voltage level in spite of fluctuations in input voltage and output loading.

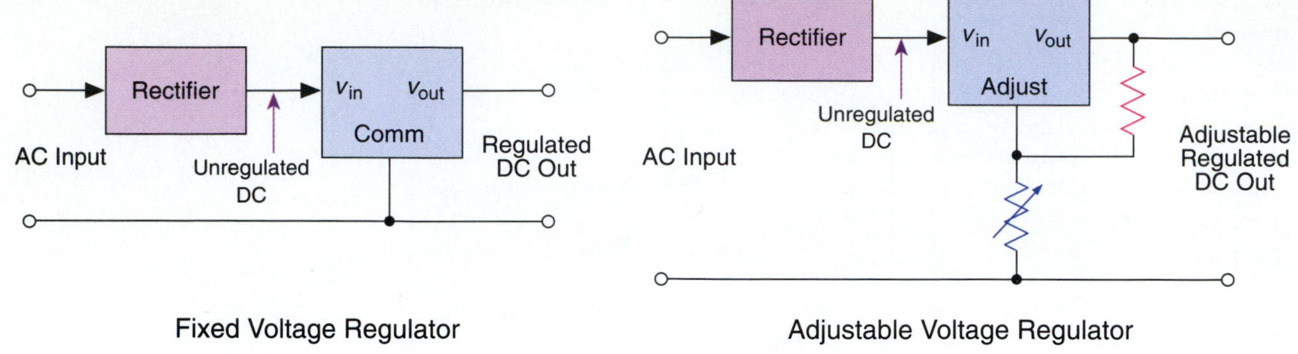

Figure 26-11 Voltage regulators used in DC power supplies.

Voltage regulators are simple, three-terminal IC devices that look like power transistors (see again Fig. 26-11). Most of them have internally integrated circuitry that uses the equivalent of two op amps and at least one BJT.

Voltage regulator ICs are specified according to their output regulating voltage level, output current-carrying capacity, and the range of input voltage levels they can handle. Table 26-1 describes a selection of fixed and adjustable voltage regulator ICs, including their part designations, output regulation voltage, input voltage range, and maximum output current.

The AC motors you can find in most home appliances run at a speed that is determined by the amount of force they have to work against. The greater the amount of load placed on the motor, the slower it runs; but if you could increase the amount of electric power applied to the motor, it could overcome the loading effect and return to a higher operating speed. So the speed of operation of such motors can be controlled through the amount of AC power applied to them. Food

Fixed Voltage Regulators

IC Designation	Output Voltage	Output Current (max)	Input Voltage Swing
LM78T05C	5.0 V	3 A	7.5 V - 35 V
MC7806	6.0 V	1.5 A	9 V - 35 V
MC78T12C	12 V	3 A	14.5 V - 35 V
LM340-15	15 V	1.5 A	17.5 V - 35 V
LM340-24	24 V	1.5 A	27.3 V - 40 V

Adjustable Voltage Regulators

IC Designation	Output Voltage	Output Current (max)	Input Voltage Swing
LM317L	1.2 V - 37 V	100 mA	5 V - 40 V
MC1723	2.0 V - 37 V	150 mA	9.5 V - 40 V
LM317M	1.2 V - 37 V	500 mA	5.0 V - 40 V
LM117	1.2 V - 37 V	1.5 A	5.0 V - 40 V
LM350	1.2 V - 33 V	3 A	5.0 V - 36 V

Table 26-1 Listing of a few fixed and adjustable DC regulator ICs.

blenders, washing machines, and certain portable electric drills have AC speed (power) controls built into them. Power control is achieved by allowing only a portion of each AC half-cycle to reach the motor windings. If the waveform is switched on 90° into each half-cycle, the motor will run at about half speed; if the waveform is switched on just 45° into each half-cycle, the motor will run at about three-fourths full speed. The inputs to a *motor speed controller* IC include a sample of the AC line's waveform, a signal that tells how fast the motor is supposed to run, and another signal that tells the actual speed of the motor. The output is a signal that specifies where on each AC half-cycle the waveform is to be switched on.

Motors that are used for setting the position of the movable parts in industrial automated machinery are classified as DC servomotors. Although these motors can be made to spin continuously like ordinary electric motors, they are rarely used in that mode. Instead, their shafts can be positioned within a fraction of a degree by applying short bursts of electric power. Signals are required for telling the control circuitry where the motor is supposed to be positioned, where the motor is actually positioned, and the number of electrical impulses that have to be applied in order to correct the difference between the desired and actual position of the shaft. All of these operations are handled today within a single IC called a *DC servomotor controller/driver.*

As long as more factories are being designed from the ground up for automation and robotic operations, the demand for more sophisticated voltage, current, and power controlling ICs will continue to grow.

Communications Circuits

Radio-frequency (RF) communications includes radio, television, radar, and all their variations and subclassifications. Radio and television, for example, are not limited to the familiar commercial versions found in our homes and automobiles. These technologies are extended to vast global and space communications links that transmit and receive computer data as well as audio and video information.

Circuits for RF communications often require the use of inductors and capacitors. It is very difficult to form capacitors of any significant value on an IC chip. At the higher frequencies of RF communications, however, the spacing between integrated components and their interconnections can serve as built-in capacitances. It is virtually impossible to work an inductance into a silicon chip. This is why you can expect to see IC devices surrounded by ordinary capacitors and inductors on the circuit boards for older and less sophisticated TVs and radios.

Rather than trying harder to integrate capacitors and inductors onto IC chips, the electronics industry decided to take an entirely different approach and develop ways to simulate the action of capacitors and inductors by means of operations that could be carried out on an IC chip. A key circuit in this regard is called a **phase-locked loop** (or PLL). PLLs, op amps, and digital control circuits are all available on IC chips and can be combined to synthesize the action of LC circuits. What is more, these vital communications circuits have been integrated on single chips to produce system-level IC devices. Here are a few examples of these RF communications ICs:

- AM stereo decoder.
- FM stereo decoder.
- Volume, balance, and bass/treble controls.

- AM broadcast receiver.
- Black-and-white TV (on a single chip).
- TV sound system.
- TV color processor.

A second major area of communications is telecommunications. This is also an information technology that deals with voice, picture, and computer data. Unlike RF communications, the information of telecommunications is confined to wires and fiber-optic strands. The ordinary telephone is the most familiar example of a piece of electronic telecommunications equipment.

The advent of the touch-tone telephone ushered in the age of telecommunications IC technology. You not only need a circuit that generates different tones when you touch the buttons (tone encoder), you also need one that translates those tones into computerized data (tone decoder) that controls IC switching units between your phone and the one you are calling. Telephone tone encoders and decoders are readily available as IC devices.

REVIEW QUIZ 26-3

1. The inputs on an IC op amp are known as ______________ and ______________.
2. The input signal for an inverting op-amp circuit is fed through the ______________ resistor to the ______________ terminal; and the feedback resistor is connected between the ______________ and ______________ terminals on the op amp.
3. One way to increase the voltage gain of an inverting op amp is to increase the value of the ______________ resistor.
4. The input signal for a noninverting op-amp circuit is fed to the ______________ terminal on the op amp; and the feedback resistor is connected between the ______________ and ______________ terminals on the op amp.
5. The output of a good DC power supply is often controlled by an IC device known as a(n) ______________.
6. The IC device that has made it possible to eliminate many LC circuits in communications equipment is called a(n) ______________.

26-4 Digital IC Devices

The origin for the term *digital* as applied to electronics technology is the same as for *digit* as applied to the fingers on your hand. When you count with your fingers, you count digitally—one, two, three, and so on. The input to a digital circuit is either ON or OFF, and nowhere in between; and so is the output of a digital circuit. Digital ICs accept digital ON-OFF inputs, perform meaningful operations on those inputs, and produce digital ON-OFF outputs.

A huge family of ICs is devoted to digital circuits and applications. Digital IC devices are constructed according to several different technologies:

- ▼ ***Transistor-transistor logic*** (TTL). A bipolar junction technology very close to that used for BJTs. It is noted for very high speed, excellent power-handling capability, reliability, ruggedness, and low cost.
- ▼ ***Complementary MOS*** (CMOS). A MOSFET technology that is noted for low power consumption, reliability, and low cost.
- ▼ ***Metal-Oxide Semiconductor*** (MOS). A MOS technology that is noted for very low power consumption and the ability to accommodate millions of elements in a single package.
- ▼ ***Emitter-Coupled Logic*** (ECL). A very high-speed, low-power bipolar technology.

The ideal digital IC is one that switches extremely rapidly from ON to OFF or vice-versa, and consumes little electric power. TTL devices are very fast, but they burn a lot of power. CMOS devices drain very little current from the power supply, but they switch slowly. Thus, there are classifications within the basic classifications of digital IC devices. Advanced low-power Schottky (ALS) ICs, for example, are basically TTL devices that are designed for very low power consumption. High-power CMOS (HC) ICs, on the other hand, are CMOS devices that are designed for higher-speed operation.

The Basic Combinatorial Logic ICs

In this section you will learn about the five most basic combinatorial logic functions and their most popular IC configurations.

According to the electrical diagram of an **AND function** shown in Fig. 26-12 on the next page, the light goes on whenever switch A *and* switch B are both closed at the same time. This is the nature of a logical AND function. All inputs must be active before the output is active. Or to put it another way, all inputs must be ON before the output is ON. It makes no difference how many switches are connected in this series arrangement—two or two hundred—the rule is the same: The output is active only when all inputs are active. If an AND circuit has two inputs (two switches in series, for instance), you can say that the output is active when switch A *and* switch B are active. If the circuit happens to have three inputs, you can say the output is active when A *and* B *and* C are active.

The circuit for the **OR function** in Fig. 26-12 shows the switches connected in parallel. The light goes on when either or both switches are closed. This defines the OR function. Again, it makes no difference how many switches are used, the rule is the same: The output is active whenever any or all inputs are active.

If an OR circuit has two inputs, you can say the output is active when A *or* B is active. It is taken for granted that the output is also active when both inputs

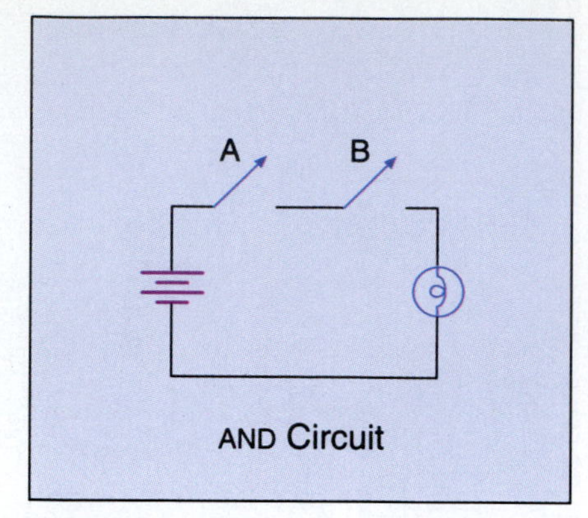

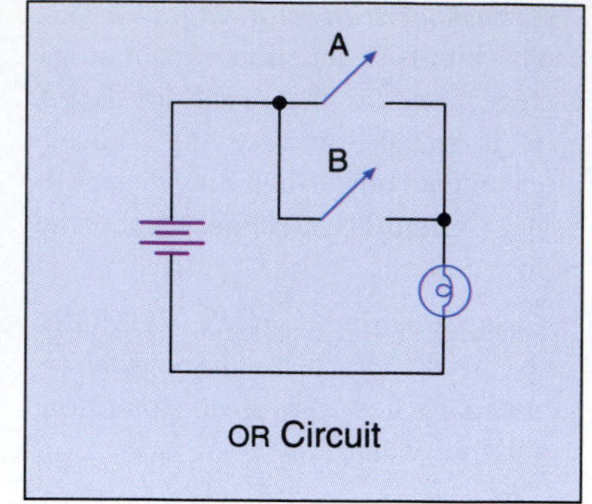

Figure 26-12 AND and OR functions represented by switches and lamps.

are active. So when you are thinking in terms of a three-input OR circuit, you can say that the output is active when A *or* B *or* C is active.

The rules for logic circuits are expressed in the forms of truth tables and logic equations. A truth table expresses logic functions in terms of the two numerals in the binary numbering system, 0 and 1. Unless specified otherwise, the 0 represents the OFF or inactive state, and the 1 represents the ON or active state. A truth table shows the output response of the logic circuit under all possible combinations of 0s and 1s at the inputs.

Figure 26-13 shows examples of truth tables for AND and OR functions. In the AND truth tables, you can see that all inputs must be at the active, logic-1 level before the output is set to logic 1. Otherwise the output is in its inactive, or logic-0, state. For the OR truth tables, however, the output is active whenever at least one of the inputs is active. A logic-1 level at any input of an OR circuit is sufficient to force its output to logic 1.

An entirely different way to portray the rules of a combinatorial logic function is by logical formulas called Boolean expressions. Named after the nineteenth-century English mathematician George Boole, these expressions are used for handling logical operations in much the same way equations are used for handling arithmetic operations.

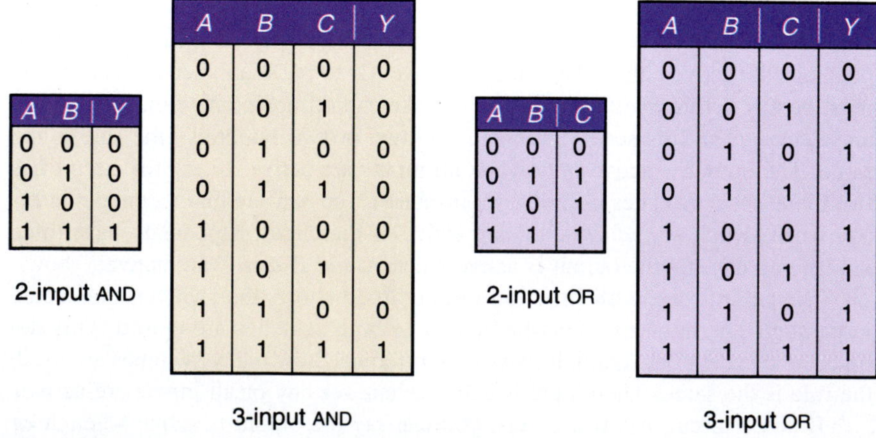

A	B	Y
0	0	0
0	1	0
1	0	0
1	1	1

2-input AND

A	B	C	Y
0	0	0	0
0	0	1	0
0	1	0	0
0	1	1	0
1	0	0	0
1	0	1	0
1	1	0	0
1	1	1	1

3-input AND

A	B	C
0	0	0
0	1	1
1	0	1
1	1	1

2-input OR

A	B	C	Y
0	0	0	0
0	0	1	1
0	1	0	1
0	1	1	1
1	0	0	1
1	0	1	1
1	1	0	1
1	1	1	1

3-input OR

Figure 26-13 A selection of AND and OR truth tables.

The Boolean sign for the AND function is a dot similar to the one sometimes used for arithmetic multiplication. Here is the Boolean expression for AND circuits having two, three, and four inputs:

$Y = A \cdot B$ Read as: "*Y* equals *A* **and** *B*"

$Y = A \cdot B \cdot C$ Read as: "*Y* equals *A* **and** *B* **and** *C*"

$Y = A \cdot B \cdot C \cdot D$ Read as: "*Y* equals *A* **and** *B* **and** *C* **and** D"

The Boolean sign for the OR function is the same plus sign that is used for arithmetic addition. The Boolean expression for OR circuits having two, three, and four inputs are as follows:

$Y = A + B$ Read as: "*Y* equals *A* **or** *B*"

$Y = A + B + C$ Read as: "*Y* equals *A* **or** *B* **or** *C*"

$Y = A + B + C + D$ Read as: "*Y* equals *A* **or** *B* **or** *C* **or** *D*"

Finally, the logic circuits themselves are represented by special symbols. It makes no difference whether the circuit is made from switches, vacuum tubes, transistors, or ICs, the logic symbol remains unchanged. Figure 26-14 shows the logic symbols for AND and OR circuits (called **logic gates** when they are drawn in this symbolic fashion).

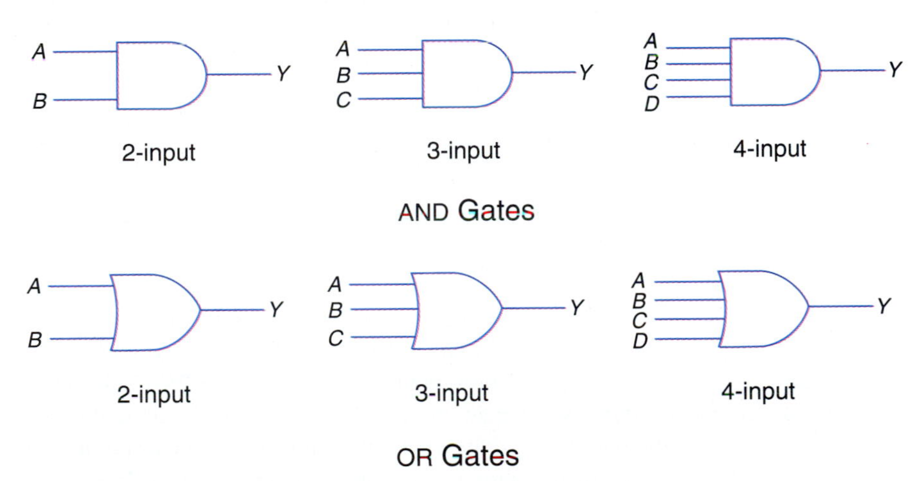

Figure 26-14 Selection of logic symbols for AND and OR logic gates.

Figure 26-15 on the next page shows some examples of how AND and OR gates are provided in 14-pin DIP ICs. The AND gate version in Fig. 26-15(*a*) is called a quad two-input AND gate. This means there are four (quad) individual two-input AND gates included in the package. The schematic diagram shows the content of the package, including the logic symbols, labeling for the inputs and outputs, and the IC pin numbers for each terminal. The IC pinout diagram shows a top view of the actual IC package and the labeling for the inputs, outputs, and power supply connections. This particular digital IC is a TTL version that is specified and catalogued as a 7408.

The OR gates device in Fig. 26-15(*b*) is a quad two-input OR gate. This is a CMOS device that is listed as a 4071. Notice there are four individual logic gates in each IC package. It is assumed that the proper supply voltage is applied to the

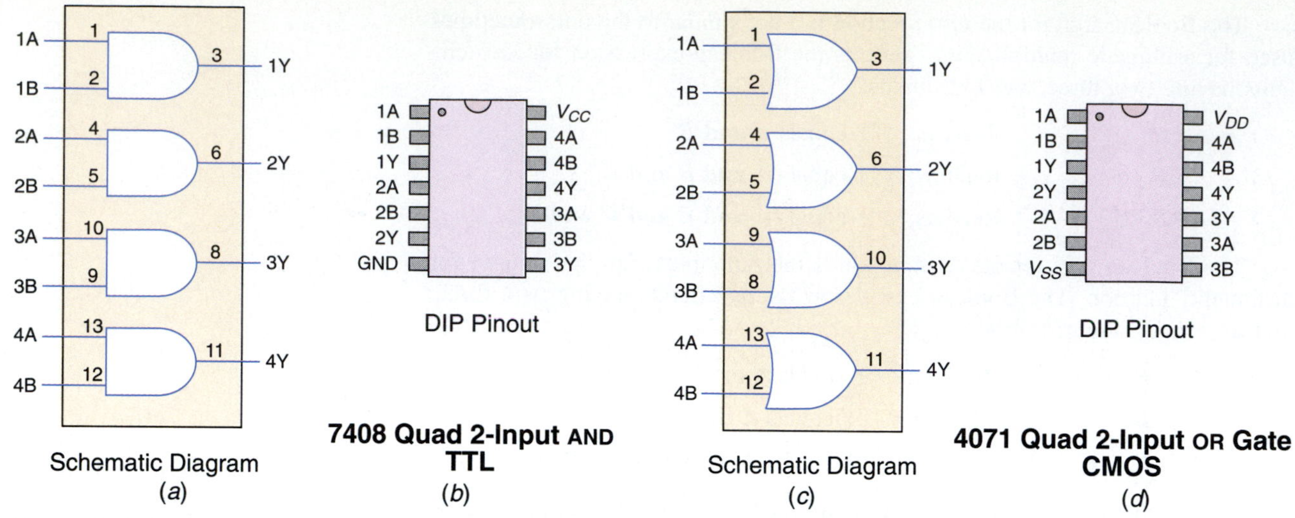

Figure 26-15 Schematics and pinout diagrams for the most popular AND-gate and OR-gate ICs.

power supply terminals: V_{CC} and GND for TTL (bipolar) digital ICs; V_{DD} and V_{SS} for MOS-type digital ICs.

As useful as they might be, AND and OR functions cannot handle all the combinatorial logic operations required by modern digital electronics. An entirely different kind of logic function, called an **INVERT or NOT function,** greatly expands the variety and usefulness of combinatorial logic circuits. As shown in Fig. 26-16, an INVERT gate simply switches (inverts) the input logic level. If the input to an INVERT gate is a logic 0, then the output is logic 1; and if the input is logic 1, the output is logic 0.

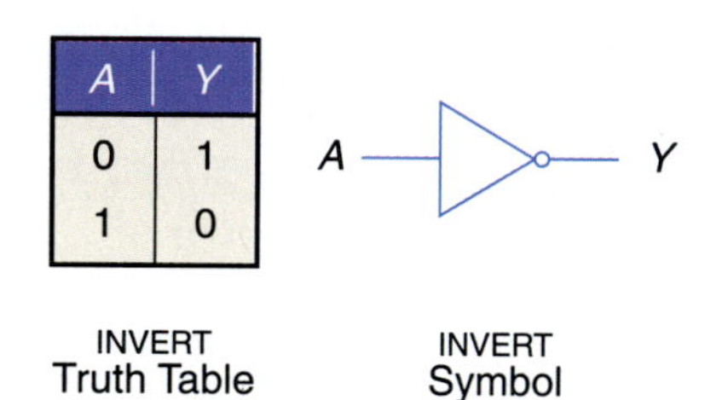

A	Y
0	1
1	0

Figure 26-16 INVERT truth table and logic symbol.

The Boolean sign for the INVERT function is an overbar, so the expression is:

$Y = \overline{A}$ Read as: "*Y* equals **not** *A*"

How can a digital function that seems as simple as an INVERT function increase the variety of digital logic circuits? Consider what happens when an INVERT gate follows the output of a two-input AND gate. Figure 26-17 illustrates this situation. For one thing, you get an entirely different kind of truth table. With an AND gate alone, the output is 1 only when all inputs are 1. But when the output is inverted, you get a logic-1 output *except* when all inputs are at logic 1. An INVERTed AND function is properly known as a **NAND function.** This name is taken from the fact that it is a NOT-AND operation.

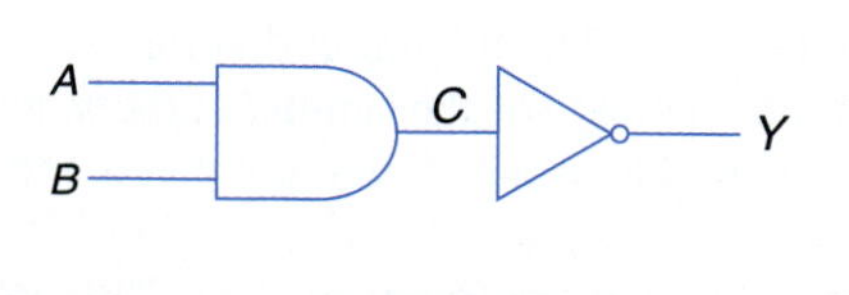

A	B	C	Y
0	0	0	1
0	1	0	1
1	0	0	1
1	1	1	0

Figure 26-17 A NAND logic diagram and truth table.

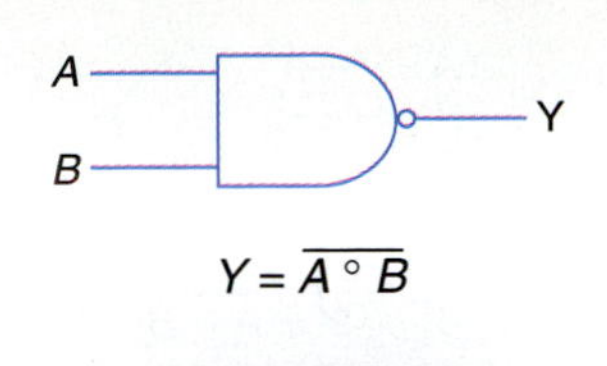

A	B	Y
0	0	1
0	1	1
1	0	1
1	1	0

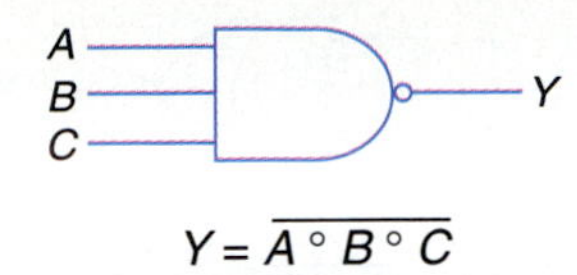

$Y = \overline{A \circ B \circ C}$

A	B	C	Y
0	0	0	1
0	0	1	1
0	1	0	1
0	1	1	1
1	0	0	1
1	0	1	1
1	1	0	1
1	1	1	0

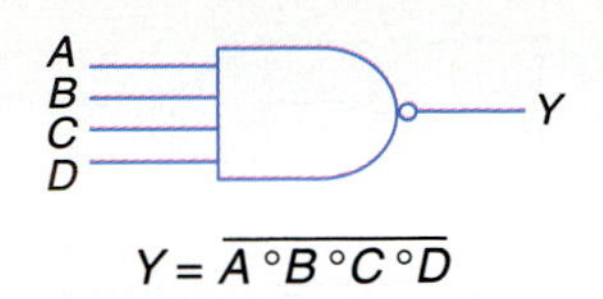

$Y = \overline{A \circ B \circ C \circ D}$

A	B	C	D	Y
0	0	0	0	1
0	0	0	1	1
0	0	1	0	1
0	0	1	1	1
0	1	0	0	1
0	1	0	1	1
0	1	1	0	1
0	1	1	1	1
1	0	0	0	1
1	0	0	1	1
1	0	1	0	1
1	0	1	1	1
1	1	0	0	1
1	1	0	1	1
1	1	1	0	1
1	1	1	1	0

Figure 26-18 Logic symbols, Boolean equations, and truth tables for two-input, three-input, and four-input NAND gates.

The NAND function is an important digital operation in its own right. Figure 26-18 shows truth tables and symbols for three common NAND-gate IC devices. Notice that the symbol for a NAND gate looks like an AND symbol with a "bubble" located at the output. This "bubble" signifies that the logic level is inverted at that point. The Boolean equations look like the AND equations, but with bars across the input terms.

Just as a basic AND function can be INVERTed to produce a NAND function, a basic OR function can be INVERTed to produce a **NOR function** (NOT-OR). This is shown in Fig. 26-19 on page 512.

Figure 26-20 on pages 512–514 shows some examples of INVERT, NAND, and NOR IC packages.

Sequential Logic ICs

The fundamental device in sequential digital electronics is called a **flip-flop.** The simplest kind of flip-flop device has one input and one output. The input is a trigger pulse. Each time the trigger pulse occurs, the output of the circuit "flips" from the present logic level to the opposite logic level. If an input trigger pulse is applied while the output is at logic 0, the output is flipped to a logic-1 level. On the other hand, if the trigger pulse occurs when the output *(cont. on page 515)*

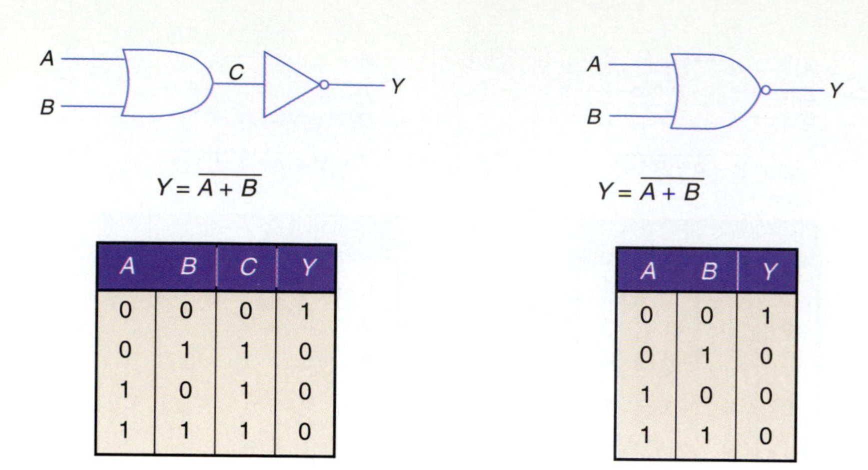

A	B	C	Y
0	0	0	1
0	1	1	0
1	0	1	0
1	1	1	0

A	B	Y
0	0	1
0	1	0
1	0	0
1	1	0

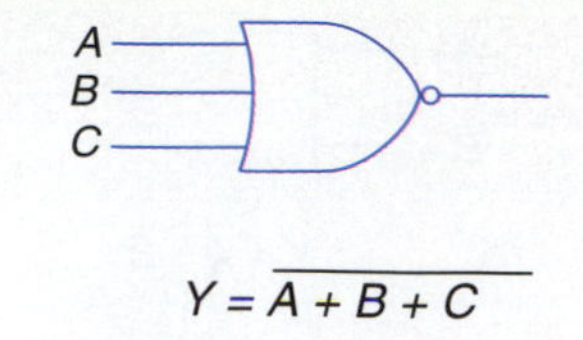

$Y = \overline{A + B + C}$

A	B	C	Y
0	0	0	1
0	0	1	0
0	1	0	0
0	1	1	0
1	0	0	0
1	0	1	0
1	1	0	0
1	1	1	0

Figure 26-19 Logic symbols, Boolean equations, and truth tables for a NOR circuit, a two-input, and a three-input NOR gate.

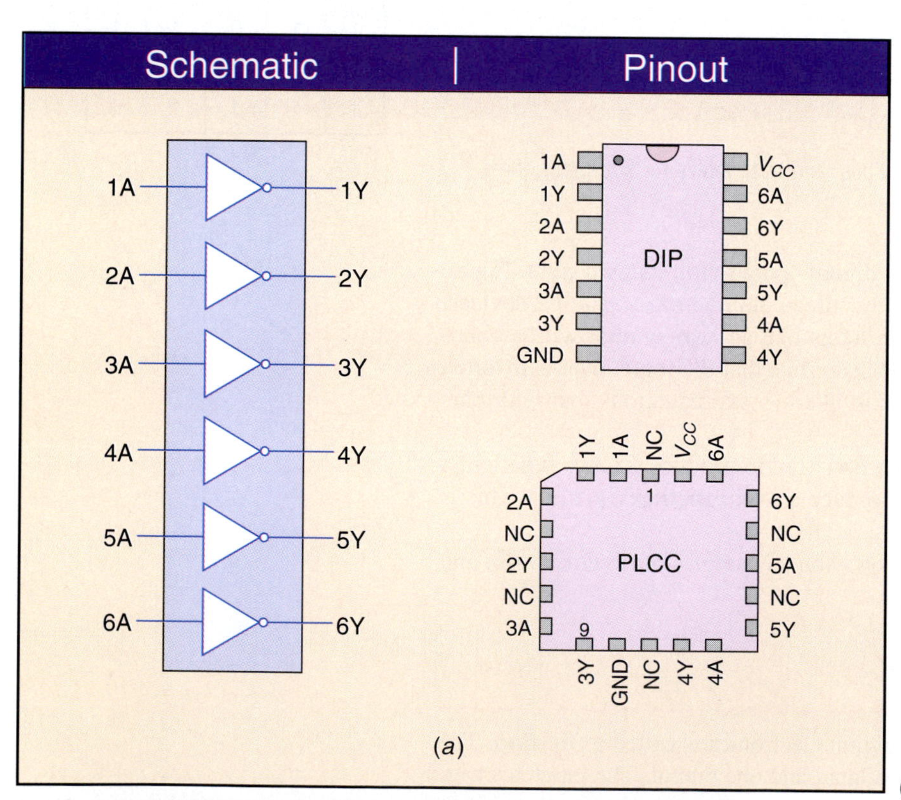

(*a*) 7404 hex inverter

Figure 26-20 Schematic diagrams and selected IC package pinout diagrams for (*a*) 7404 hex inverter, (*b*) 7400 quad two-input NAND gate, (*c*) 7410 triple three-input NAND gate, (*d*) 7420 dual four-input NAND gate, (*e*) 7402 quad two-input NOR gate. (Continues following pages)

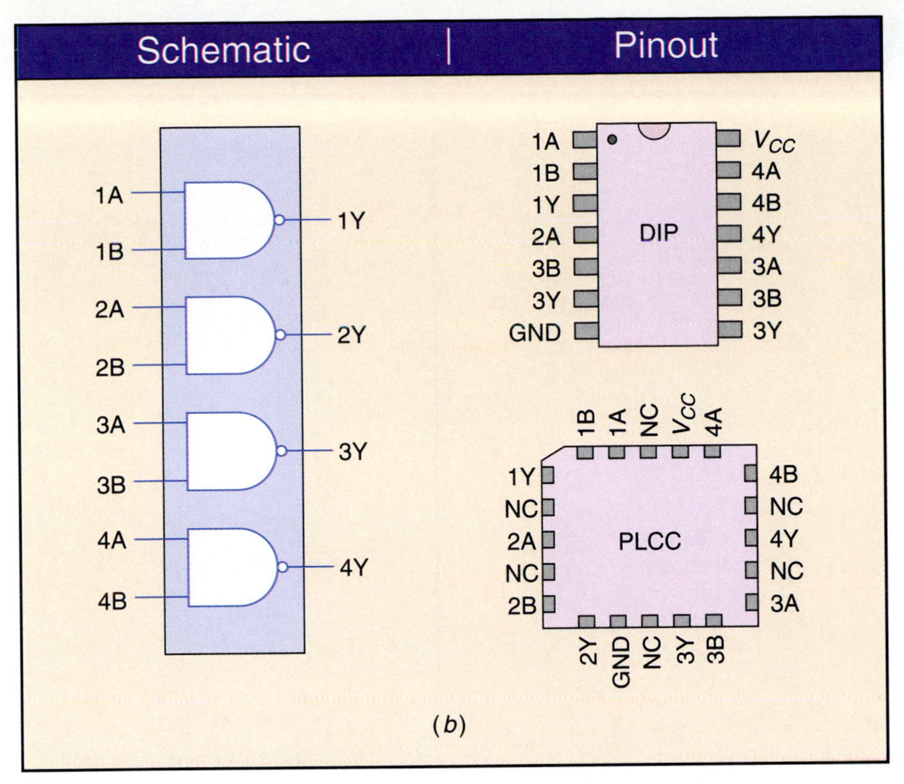

(*b*) 7400 quad two-input NAND gate.

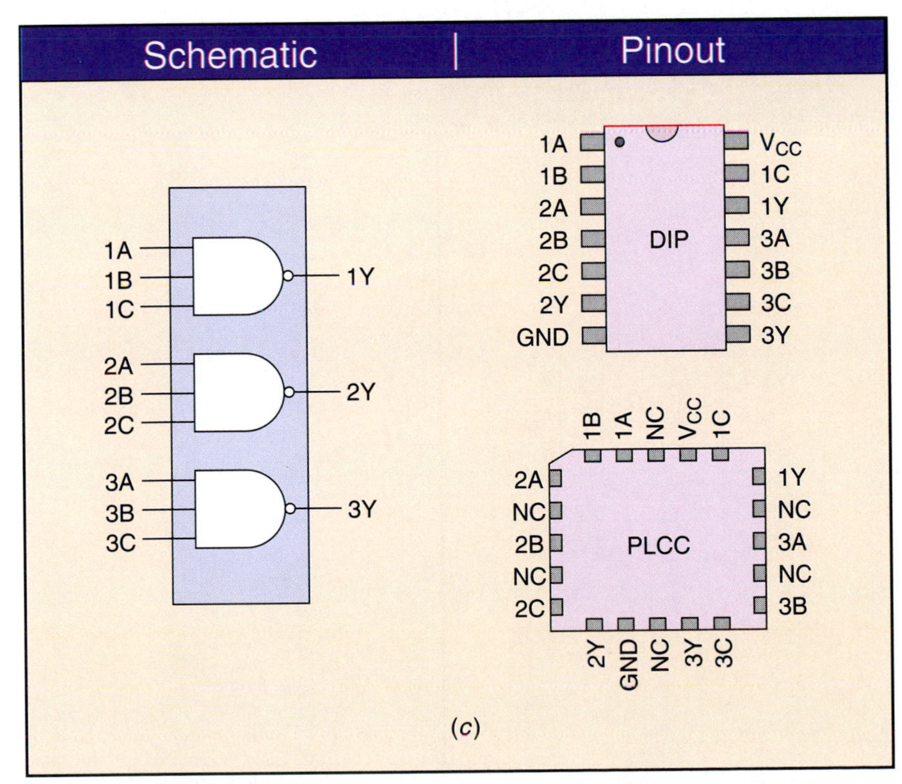

(*c*) 7410 triple three-input NAND gate.

Figure 26-20 continued.

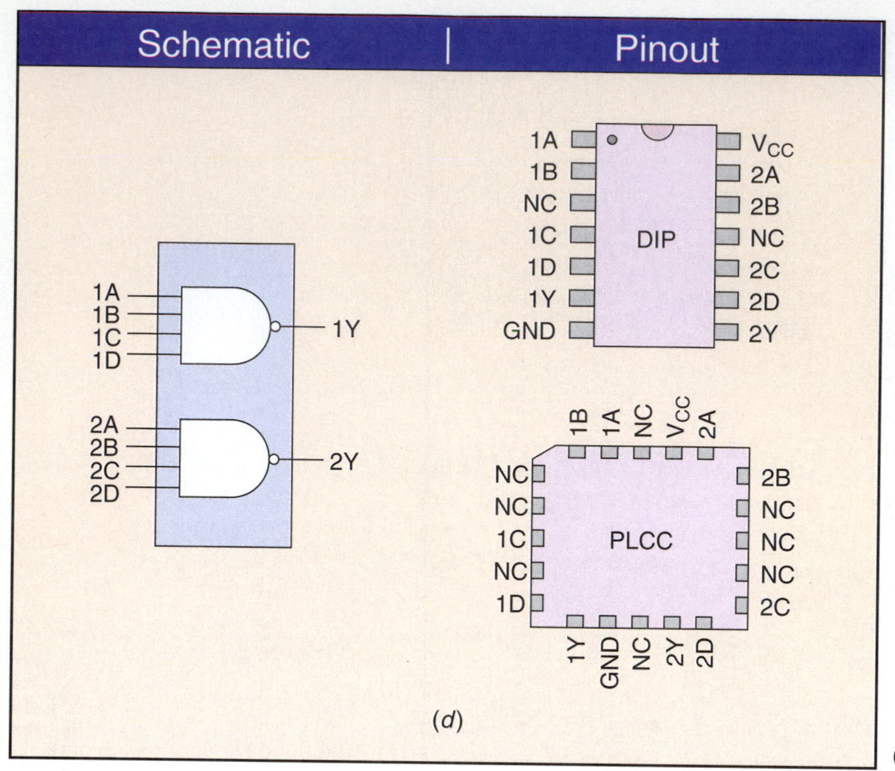

(*d*) 7420 dual four-input NAND gate.

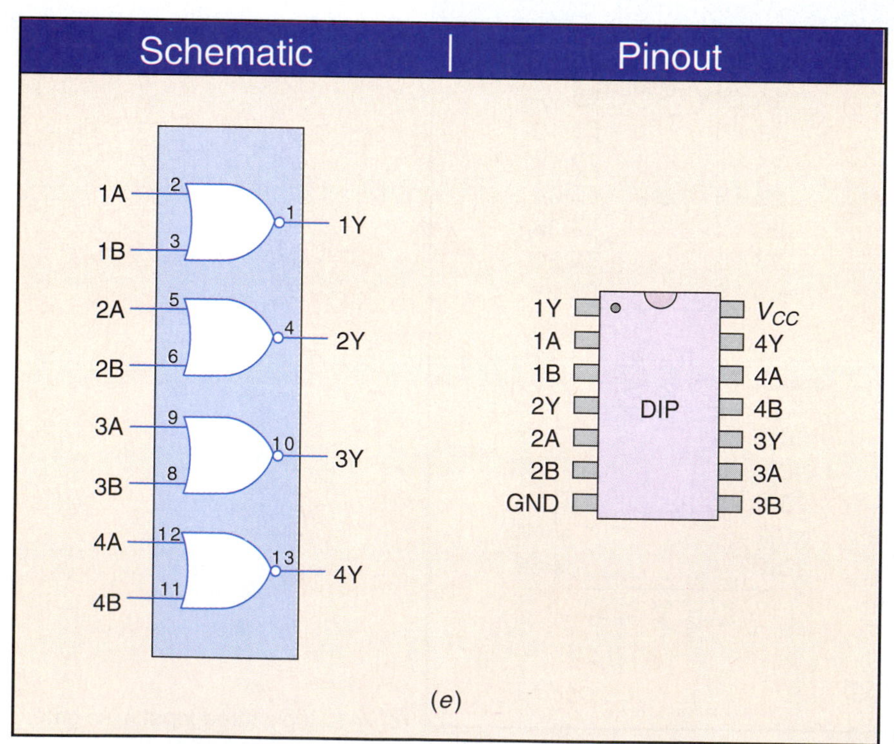

(*e*) 7402 quad two-input NOR gate.

Figure 26-20 continued.

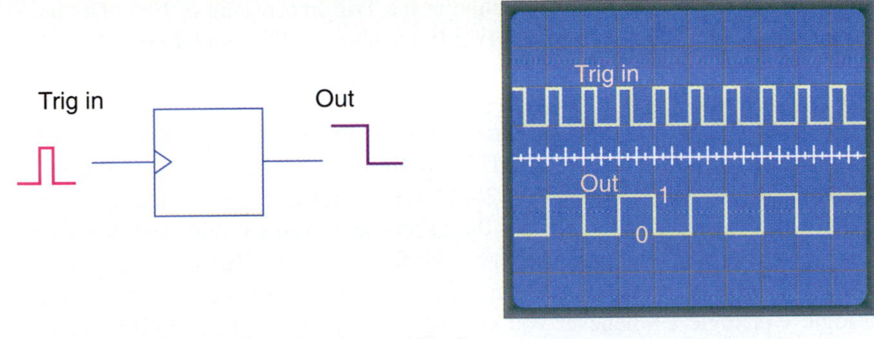

Figure 26-21 Basic action of a triggered flip-flop.

is at logic 1, the output is flopped to logic 0. The waveforms in Fig. 26-21 show how a sequence of trigger pulses causes the output of the device to flip and flop between 0 and 1.

A flip-flop is considered a sequential logic circuit because you cannot say whether the output will go to a logic 0 or logic 1 when the next trigger pulse occurs, unless you know whether the output is currently at 0 or 1. The sequence of output levels for a flip-flop is 0-1-0-1-0-1-, and so on.

The binary, or base 2, number system is one that is especially well suited for the two possible states (ON or OFF) of digital electronics. Digital flip-flops can be cascaded (connected one after the other) to produce longer binary counting sequences. Figure 26-22 shows four basic flip-flops that are cascaded to produce a 4-bit binary counting sequence.

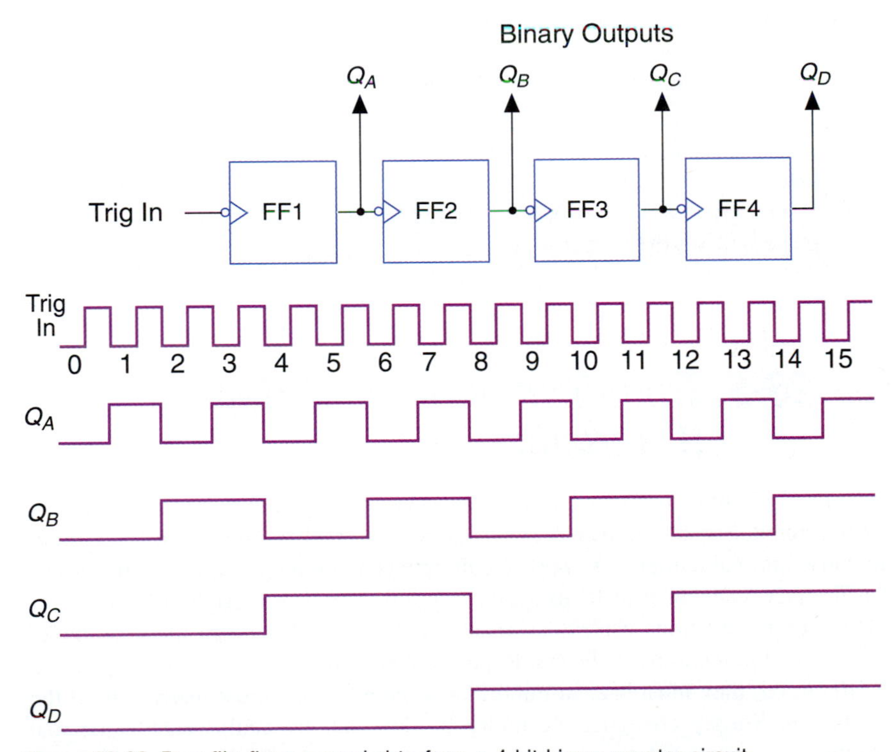

Figure 26-22 Four flip-flops cascaded to form a 4-bit binary counter circuit.

The original trigger pulse is applied at the Trig In terminal of FF1. The output of FF1 is then connected to the Trig In terminal of FF2, and so on down the line. Each time a Trig In on one of the flip-flops sees a 1-to-0 transition, that particular flip-flop switches state.

Figure 26-23 shows the IC schematic pinout drawing for a popular dual (two in a package) flip-flop called the 7476 *J-K* flip-flop. Comparing the 7476 with the basic flip-flop described in Fig. 26-23, the trigger is applied to the CLK terminal, and the output is taken from the Q terminal. You can also see, however, that the 7476 dual *J-K* flip-flop offers more features than a basic flip-flop. Generally speaking, these additional features allow you to preset or clear the outputs to logic 1 or logic 0 whenever you choose (by using the PRE and CLR inputs) or force the outputs to 0 or to 1 whenever a trigger pulse is applied (*J* and *K* inputs).

Other sequential logic devices that are based on flip-flop principles include shift registers and data latches. Certain types of computer memories are actually a form of register or data latch. Microprocessors (40-pin IC packages that handle most of the fundamental operations of personal computers) include hundreds of sequential as well as combinatorial logic functions.

REVIEW QUIZ 26-4

1. TTL is an abbreviation for ______________, and CMOS is an abbreviation for ______________.
2. A(n) ______________ logic gate produces a logic-1 output only when all inputs are at the logic-1 level.
3. A(n) ______________ logic gate produces a logic-1 output only when all inputs are at the logic-0 level.
4. A NOR function is a(n) ______________-OR function.
5. A dual four-input NAND gate IC contains ______________ individual ______________-input NAND gates. A quad two-input NAND gate IC contains ______________ individual ______________-input NAND gates.
6. If the output of a flip-flop is currently at logic 0, the next input trigger pulse will set the output to ______________.

26-5 Troubleshooting and Repairing IC Circuits

Integrated circuits are extremely reliable devices; they rarely "go bad" on their own account. Something has to go wrong in the circuitry outside an IC in order to cause internal damage. A short circuit between conductors on a circuit board, for instance, can cause an IC to overheat and destroy itself. High-voltage surges along communications cables can reach MOS ICs and damage them by puncturing the thin metal oxide layers. Repair technicians will indeed find bad ICs in a circuit, but they must bear in mind that something else most likely caused the IC to fail. Simply changing the faulty IC does not normally remedy the real trouble.

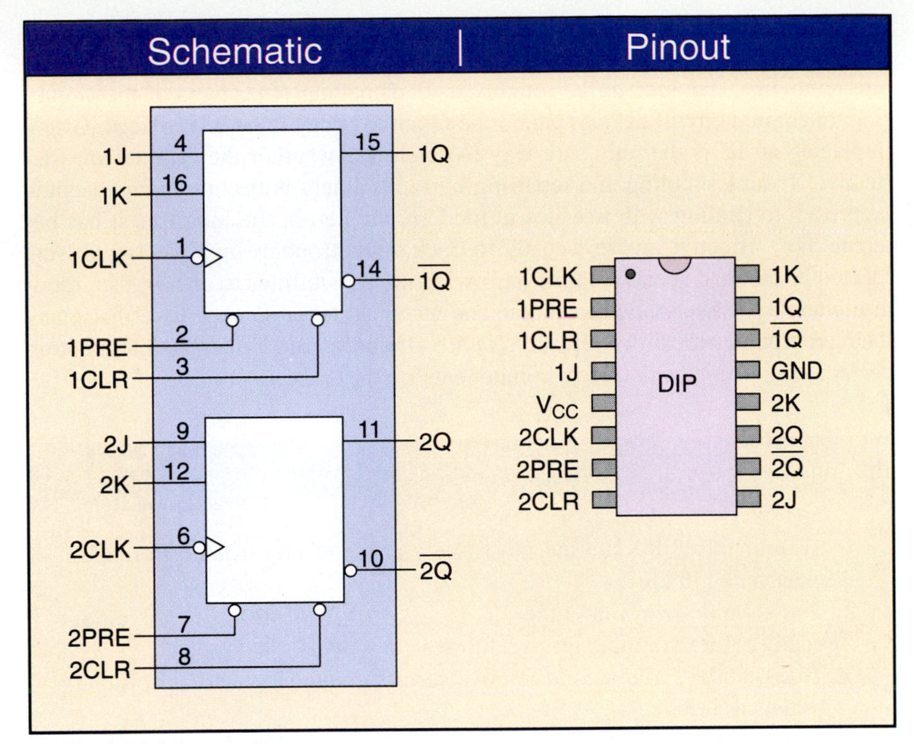

Figure 26-23 Schematic and pinout of a 7476 dual J-K flip-flop.

So there are two problems to be solved at the same time when you are troubleshooting a circuit constructed with IC devices: determine which IC is defective and determine how it got that way. Using your understanding of how the circuit is supposed to work, narrow down the number of likely candidates for bad ICs. Track down the flow of signals through the circuit to reduce the number of possibly bad ICs even further. Then consider what might have caused the suspected ICs to go bad. A close visual inspection of the circuit board sometimes turns up the main cause of the trouble—bits of metal lying across conductive tracks, cracked conductive tracks, solder bridges between the pins on an IC. Look for resistors, diodes, and transistors that show signs of burning or melting. If you cannot locate any physical evidence of circuit damage, check the power supply voltages to the suspected ICs and use an oscilloscope to take a more thorough look at the flow of signals.

When it is time to replace an IC with a good one, first take note of how the IC is oriented on the board. DIP and gull-wing packages can be fit into place in two ways, PLCCs can be installed in four different directions, and you can fit most metal-can ICs onto a circuit board in ten different positions—but only one position is correct!

Remove the suspected IC with the proper removal tools. Carefully avoid possible heat and pressure damage to surrounding conductive tracks, discrete components, and neighboring ICs. Make sure your desoldering procedure does not release tiny beads of solder that can roll along the board and become trapped between the terminals on other ICs, thereby shorting the terminals together.

Important: Do not attempt to remove surface-mount devices that are soldered in place without using the desoldering tools that are designed for that purpose.

ICs that are not soldered to the board, but pressure-fit into chip carriers, should also be removed carefully, even if you intend to throw away the IC you are removing. Handling an IC roughly risks damaging other parts of the circuit.

Integrated circuit devices cannot be repaired. They must be replaced. Often, replacing an IC is the only sure way to determine whether the original was defective. Troubleshooting and repairing by replacement is becoming an accepted approach to dealing with troubles in modern circuits. In the long run, it has become more effective and less costly to track down troubles by replacing all suspected ICs than to spend the time narrowing the possibilities to one. In fact, modern circuits are becoming so reliable and technical labor so expensive that many large companies repair by replacing entire circuit boards, rather than attempting to locate and replace defective components on the boards.

REVIEW QUIZ 26-5

1. True or false? ICs are the most likely cause of any trouble in modern electronic circuits.
2. You should always note the ______________ of an IC on the circuit board before you attempt to remove it.
3. True or false? You should remove surface-mount ICs with an ordinary soldering iron.
4. Tiny beads of solder can ______________ together the terminals on an IC.
5. It is becoming more economical to troubleshoot and repair circuits by ______________ the suspected ICs.

SUMMARY

Integrated circuits (ICs) are semiconductor devices that combine two or more devices to form a circuit that is integrated onto a single silicon chip. It is not unusual that an IC will contain the equivalent of a hundred transistors and diodes, a corresponding number of resistances, and all of the necessary interconnections. The most sophisticated ICs include in excess of a million components.

IC technology has made it possible to develop cheap and reliable versions of older electronic devices such as AM radio receivers, television sets, and stereo amplifiers. It has also made possible new consumer items such as personal computers, fax machines, cellular phones, and video games.

Integrated circuits tend to be one of two main types: analog or digital. Analog ICs are used where it is important to work with all possible voltage levels that exist between two extremes. The most significant analog IC device is the operational amplifier (op amp). An op amp can be "programmed" by means of external components and connections to perform a wide variety of different analog tasks.

Digital ICs, on the other hand, operate on an ON/OFF basis. Digital ICs are then further classified as combinatorial logic or sequential logic devices. Combinatorial logic devices produce an output that depends only upon the status of the inputs at the moment. Examples of combinatorial IC logic circuits are AND, OR, INVERT, NAND, and NOR logic gates. The output from a sequential logic circuit depends on the previous state of the output as well as the present state of the inputs. Flip-flops are the leading examples of sequential logic devices.

NEW VOCABULARY

- A-to-D converter
- AND function
- complementary MOS (CMOS)
- D-to-A converter
- emitter-coupled logic (ECL)
- flip-flop
- integrated circuit (IC)
- INVERT function
- linear circuit
- logic gates
- NAND function
- nonlinear circuit
- NOR function
- operational amplifier (op amp)
- OR function
- phase-locked loop (PLL)
- transistor-transistor logic (TTL)
- voltage regulator

Questions

1. In which direction are the terminals counted on an IC chip when viewing it from the top?
2. What are the main differences and similarities between DIP and gull-wing IC devices?
3. A typical op amp has how many input terminals? Power supply terminals?
4. How many separate op amps are included in a quad op-amp package? A dual op-amp package?
5. Is DC voltage regulation accomplished with analog or digital IC devices?
6. Is AC motor control accomplished with analog or digital IC devices?
7. What analog IC device has made it possible to eliminate many of the LC circuits normally required for communications circuits?
8. Which is a better example of a sequential logic device, a flip-flop or NOR gate?
9. What is a dual four-input NAND gate?
10. What basic digital circuit forms the heart of a binary counter IC?

Problems

1. What is the voltage gain of an amplifier when the input is 1 volt and the desired output level is 7 volts?
2. What is the voltage gain of an inverting op amp when R_F is 10 kilohms and R_i is 10 kilohms?
3. What is the voltage gain of a noninverting op amp when $R_1 = R_F$?
4. If R_F for an inverting op-amp circuit is 2.2 megohms and the voltage gain is to be -10, what is the necessary value for the input resistor?
5. If R_1 for a noninverting op-amp circuit is 1 kilohm, the input voltage level is 0.24 volt and you want to amplify the signal to 2.5 volts, what is the calculated value for R_F?
6. Write the Boolean equation for a four-input AND gate.
7. Write the Boolean equation for a four-input NOR gate.

Critical Thinking

1. Which type of IC circuit—analog or digital—would most likely feed the input to an A-to-D converter?
2. Which type of IC circuit—analog or digital—would most likely follow the output of a D-to-A converter?
3. Why do DIP and PLCC IC packages always have an even number of terminals?
4. What pin is directly across from pin 6 on a 16-pin DIP IC?
5. What pin is directly across from terminal 1 on a 20-terminal PLCC IC?
6. If you can make an active low-pass filter by connecting a series *RC* circuit in the feedback path of an inverting op amp, what kind of active filter will you have if the series *RC* circuit is connected in the input path instead?
7. Which type of basic op-amp configuration should be used for making a voltage follower circuit?

8. If you can produce a NAND function by following an AND gate with an INVERTER, what type of function is created by following a NAND gate with INVERTER?
9. Which type of digital IC—TTL or CMOS—will drain more power from the power supply?
10. How many different combinations of output logic levels can be found from a counter composed of two cascaded flip-flops?

Answers to Review Quizzes

26-1
1. true
2. dual in-line; DIP
3. plastic leadless chip carrier
4. counterclockwise
5. on the side toward you

26-2
1. analog
2. digital
3. linear
4. nonlinear
5. sequential
6. analog-to-digital converter

26-3
1. inverting; noninverting
2. input; inverting; output; inverting
3. feedback
4. noninverting; output; inverting
5. voltage regulator
6. phase-locked loop

26-4
1. transistor-transistor logic; complementary metal-oxide semiconductor
2. AND
3. NOR
4. inverted
5. two; four; four; two
6. logic 1

26-5
1. false
2. orientation
3. false
4. short
5. replacing

Chapter 27

Career Opportunities *and* Industry Requirements

ABOUT THIS CHAPTER

Electronics is a very broad field that offers exciting and diverse opportunities to those who are willing to accept its technical challenges. It offers one of the broadest ranges of physical and mental requirements to be found in any occupational category. Jobs range from sitting at

After completing this chapter you will be able to:

- Identify the physical requirements of several major occupational categories available in electronics.
- Identify the level of technical training required by each occupational category.

a comfortable workstation in an air-conditioned environment to installing antennas atop two-hundred-foot-tall towers on mountain peaks in the middle of January, which adds a whole new meaning to "air" and "conditioned." The purpose of this chapter is to provide some insight into the various occupational categories available to electronics professionals. It focuses on the physical requirements, technical training levels, relative earning potential, and the occupation's growth potential.

- ▼ Identify what supplemental training may be required for a specific occupational category.
- ▼ Make an informed decision as to the potential a specific occupational category may hold.

27-1 Occupations With Short-Term Training Requirements

A typical training program for electronics technicians is about two years in length. The topics of DC theory, AC theory, electronic devices, and digital electronics are prerequisites to the specific training necessary to become employable in most occupational categories. Not all occupations, however, require an in-depth knowledge of component and circuit operation.

Electronic Assembly

The manufacture of electronic equipment requires that components be soldered onto printed circuit boards and that the boards be installed into their final assemblies. Installing components onto a circuit board is called **stuffing**. The stuffing of circuit boards does not necessarily involve soldering, as that is often done by machine. Figure 27-1 shows a **wave-soldering machine** in action. The circuit board is prepared for this process by stuffing it with components, trimming their leads, and applying a suitable flux. The solder is then applied to the circuit board by wave action in a molten solder bath. After being soldered, the board is cleaned of any residual flux and contaminates. The process is overseen by technicians who are skilled in the operation of the machines and the processes involved. No knowledge of electronics is required.

A very high percentage of the components used in computers and other digital based devices are of the surface-mount variety and are both placed and soldered onto circuit boards by machine. Surface mount components do not have leads in the normal sense and are not mounted into holes. The components used in digital circuitry are well suited for surface-mount technology. The assembly of circuit boards using surface mount technology is mostly done by machine, but this technology also requires high volumes to justify the cost of the equipment.

Figure 27-1 Components are often soldered onto printed circuit boards by machines employing wave-soldering techniques.

Repairing circuit boards populated with surface-mount components is performed by technicians or repair persons who are highly skilled in the use of specialized soldering and desoldering equipment. Such equipment may use specialized soldering pads shaped to heat all the pins of a complex integrated circuit at once or use small jets of hot air or nitrogen to melt the solder. Parts are resoldered using special creams of **eutectic solder**. Eutectic solders, which melt at lower than normal temperatures, tend to pull in toward themselves, preventing unwanted bridging of closely spaced pins.

Stuffing circuit boards with conventional components is frequently done by professional assemblers who also solder them into place and trim their leads to the desired length. Such hand assembly is especially likely when production runs are small. Many manufacturers find it economically advantageous to contract assembly work out to companies that specialize in hand assembly. These companies hire many individuals with excellent soldering skills who are able to recognize components on sight, identify their values, and follow printed directions. The charges made by the assembly houses are usually based on a per-hole cost, so great emphasis is placed on speed while maintaining accuracy. Assembly is also conducive to piecework, which may be farmed out to individuals with the necessary skills and equipment.

In many cases, assemblers must attend special soldering classes and obtain appropriate certification before they are employable. Certification is almost always mandatory for those assembling military or aerospace electronics.

In many cases semitechnical assembly jobs include the construction of specialized cable assemblies and installing circuit boards or equipment into their final housings. These jobs often require proficiency in the use of hand tools such as screwdrivers, wire cutters, wire strippers, connectorization tools, and various wrenches. A knowledge of small power tools such as a power drill and various small saws may also be required. Subassemblies consisting of circuit boards or complete pieces of equipment can be thought of as electrical components being tied together into a system.

Electronics assembler jobs have the following pros and cons:

Pros

▲ Training period is relatively short.

▲ Employment with a desirable company may present advancement or further training opportunities.

▲ Piecework may provide desired flexibility while providing a supplemental income.

▲ Opportunity to gain experience and familiarity with a wide range of electronic equipment.

▲ Light physical requirements and a pleasant work environment.

Cons

▼ Wages are lower than in other areas of electronics.

▼ Work may not be steady.

▼ Work is very repetitive and banal.

▼ Automation, improved technology, and foreign manufacturers are eroding job opportunities.

Fiber-Optics Technician

The technology behind fiber optics is sophisticated and fascinating. Fiber highways are revolutionizing communications and expanding at an incredible rate. The performance of fiber over any form of copper cable is so great that a comparison quickly becomes an exercise in futility. But the virtues of fiber do not translate into the job market as much as might be imagined. Working with fiber optics falls into the areas of installation, connectorization, optical loss testing, system design, and system troubleshooting.

The installation of fiber-optic cable is usually done by electrical contractors, contractors who specialize in fiber-optic cable, or telephone linemen. The procedures are similar to those used with copper wire and copper cable. In the case of telecommunication companies, the fiber-optic cable is routed along existing right-of-ways on telephone poles or underground. There is a great deal of physical effort associated with the installation of any type of wire or cable, along with some knowledge of building construction, building codes, and use of appropriate power tools.

Telecommunications companies buy fiber-optic cable in five-mile lengths and use a process called **fusion splicing** to join the sections. Fusion splicing melts the glass fibers together so that optically they appear to be one very long cable. The fusion splicing machines only require a short time to master.

Once the fiber-optic cable is in place, the connectorization process begins. There are several standard connector types that require varying degrees of manual dexterity and patience. The actual fibers are usually smaller than a human hair and are very brittle. Crews specially trained in connectorization follow the installation crews and install the connectors. It is not uncommon for an installation such as a large computer network to have 25,000 to 50,000 connections. The physical requirements for those installing the connectors are less than for those installing the cable. However considerable climbing and physical mobility are required in both jobs.

The connectorization crews also perform optical loss testing. This process ensures that the light gets from one end of the fiber-optic cable to the other with minimum loss. A thorough knowledge of the test equipment is required to ensure the integrity of the system.

Fiber-optic networks can be damaged like any other form of cable. Technicians are required to locate the fault in the system and correct it. These technicians must be familiar with all aspects of fiber-optic installation, connectorization, loss testing, and system operation. This aspect of maintaining a fiber-optic system also requires good physical mobility, but as might be expected, this area pays more because of the technical knowledge required.

Fiber-optic system design can be placed in the hands of qualified technicians. This responsibility requires that the designer be a draftsperson capable of creating the blueprints the cable installers will follow. Physical requirements are modest and the pay can be excellent.

The photo in Fig. 27-2(*a*) shows an installation crew pulling fiber-optic cable. Figure 27-2(*b*) shows a fiber-optic connectorization crew performing an assembly line operation.

Fiber-optics technician jobs have the following pros and cons:

Pros

- ▲ Excellent growth potential.
- ▲ Short training period.

(*a*)

(*b*)

Figure 27-2 Fiber-optic systems are replacing wire and coaxial systems as fast as technology can move.

▲ Reasonable pay.

▲ Light to moderate physical requirements. (Only light lifting requirements, but good physical mobility needed.)

▲ Good stepping stone into more technical areas.

Cons

▼ Work is heavily dependent on new construction.

▼ Job market is somewhat limited.

Cable Television Technician

Cable television has taken America by storm and continues to grow at an amazing rate. Several cable companies have merged with telecommunications companies in preparation for the proposed "digital highway." Cable television companies hire technicians in three general categories: installer, field technician, and bench technician. The first two categories require only a limited training period.

Installation crews dig ditches to bury coaxial cable and run the cable into and throughout homes, offices, and apartments. The installation technicians use power tools and equipment to dig ditches in which to bury their cable. They must also have knowledge of general building construction and overall system operation. They run the cable, connectorize the cable, and hook up equipment. If the equipment doesn't work, they call in more highly skilled technicians who are familiar with the test equipment necessary to determine where the signal is lost, interfered with, or distorted in some way.

The pros and cons of jobs as a **cable television technician** (installation technician and field technician) are as follows:

Pros

▲ An in-depth knowledge of electronic devices and circuitry is not required.

▲ Stable industry with good growth potential.

▲ Upward mobility with outside study.

▲ Good wages.

Cons

▼ Demands physical mobility and stamina.

▼ Little opportunity for technical self-improvement.

▼ Direct contact with the public can be demanding.

Calibration Technician

Manufacturers of electronic equipment must often calibrate and align that equipment before it is ready for the marketplace. This situation is particularly true for the manufacturers of test instruments. More often than not, the first job a would-be technician will be given is in calibration. This helps familiarize the technician with the equipment and the company. As a general rule, movement into more technical areas is just a matter of time, assuming one has the technical skills to warrant such a move.

The pros and cons of being a **calibration technician** are as follows:

Pros

▲ Opportunity to become familiar with the company and its products so that an internal move is an informed one.

▲ Employer training is provided that is helpful in gaining upward mobility and higher wages.

▲ Stable and pleasant work environment.

Cons

▼ Wages are relatively low.

▼ Limited opportunity to use technical skills.

▼ There is the potential to become trapped in this position or bypassed for advancement since promotion is not always based on performance.

Line Repair or Troubleshooting Technician

Line repair technicians work for manufacturers of electronic equipment. A certain amount of newly manufactured equipment is dead or nonoperable off the assembly line. Technicians must quickly determine the cause of the problem or declare the unit unsalvageable. This is definitely an entry-level position, and quotas can seem demanding. The work is very repetitive and requires only limited technical skill because most problems are mechanical in nature. Electronics manufacturers, however, also have positions at the other end of the scale in the engineering department.

The pros and cons of being a line repair technician are as follows:

Pros

▲ Because most companies hire from within through promotions, these jobs can lead to some exciting careers.

- ▲ Light work in a pleasant work environment.
- ▲ Technical comfort level is high after only a short time on the job.
- ▲ Opportunities to learn complex test equipment and testing procedures.

Cons

- ▼ Relatively low pay.
- ▼ High production quotas can be excessively demanding.

27-2 Major Occupational Categories

Employment in the major occupational categories requires an excellent knowledge of basic electronics and formal training in the occupation. These jobs require a high degree of troubleshooting and problem-solving skills.

Consumer Electronics Technician

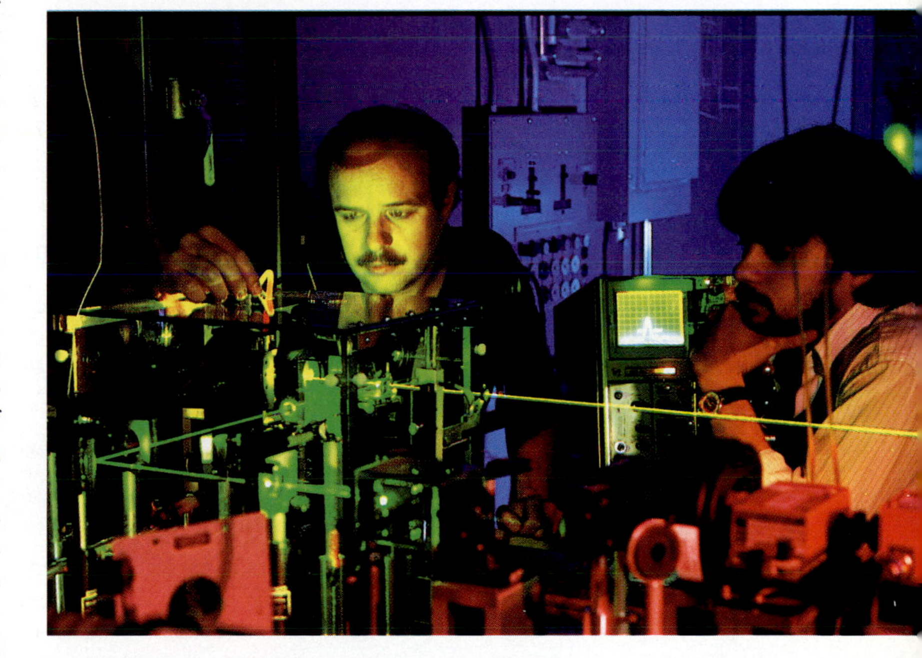

Technicians in consumer electronics repair home entertainment electronics such as televisions, stereos, VCRs, CD players, and automotive electronics. At one time this was an apprenticeable trade and may still be in some areas of the country. The technical demands are heavy because home entertainment electronics requires an extremely broad knowledge of electronic circuits and mechanical devices. Technicians in this area can find employment almost anywhere.

This field has had some down-sizing over the last few years, but it continues to remain viable in many areas. The sales of high-end home entertainment devices such as projection televisions, camcorders, expensive car electronics, and editing VCRs have stabilized the business. This is an area where component-level troubleshooting is alive and well. With HDTV and interactive television on the horizon, this occupational area could become much stronger. Most shops are small, employing from 1 to 50 technicians, which is conducive to an almost familylike relationship with the employer.

Some pros and cons of working as a **consumer electronics technician** are as follows:

Pros

- ▲ Employment is available almost anywhere in the United States.
- ▲ Chance to use a broad range of technical skills.
- ▲ A very defined occupational category with good industry support.
- ▲ Wages are moderate to high.

Cons

- ▼ Industry often has an image problem.
- ▼ Direct contact with the public, who, in many cases, are angered that they are forced to seek your services.
- ▼ Constantly increasing technology at lower prices have made many pieces of equipment impractical to repair and have increased the pressure on technicians for more production.

Industrial Electronics Technician

Manufacturers of almost everything rely on electronic equipment to keep the factories rolling. Equipment life can easily span 40 or more years, requiring industrial technicians to be very resourceful. Industrial technicians must also have skills in other trade areas. It is very helpful to have some skills as a welder, machinist, plumber, and mechanic. This job category requires a high degree of mobility and the ability to work under extreme pressure. When a factory is running well, there may be little to do, but when something breaks and production stops the pressure is on.

Here are some pros and cons of working as an **industrial electronics technician**.

Pros

- ▲ High wages.
- ▲ Job security and stability.
- ▲ An opportunity to use a wide range of skills.
- ▲ Excellent job satisfaction.

Cons

- ▼ Can be high pressure at times.
- ▼ Requires a fair level of proficiency on several other job classifications.
- ▼ Can require physical strength as well as mobility often under unpleasant working conditions.
- ▼ Industry has down-sized and become less defined over several decades in keeping with the general decline in manufacturing in the United States. This field, however, shows some signs of reversing as manufacturing seems to be on the rise.

Electronics Communications Technician

Radio-frequency communication has never been so prevalent. Every state, county, and city government is directly or indirectly a potential employer. Like consumer electronics, this is a well-defined occupation requiring a high degree of technical skill and resourcefulness. Employment opportunities are very broad and available almost anywhere in the United States. Many opportunities are also available in other countries. Communications technicians find employment anywhere from telecommunications companies to commercial radio and television broadcast stations. Digital communications and data networks are growing rapidly. Wages and

Figure 27-3 Electronics communications technicians work with very complex circuitry that is both digital and analog.

benefits in the communications industry are generally excellent. Physical mobility is important in most areas of communications. Most technicians start as installers, or they may do the physical tasks that the more senior technicians would rather not do. Figure 27-3 shows a technician working in a communications shop.

Some pros and cons of being an **electronics communications technician** are as follows:

Pros

- ▲ Has some of the highest wages in the electronics industry for nondegree technicians.
- ▲ Strong growth potential in digital communications over fiber-optic networks.
- ▲ Wide range of skill requirements and geographical locations.
- ▲ Strong industry certification programs.

Cons

- ▼ Equipment is very complex and often difficult to work on because of its compactness.
- ▼ Physical requirements can be demanding in many areas because so much equipment is remote or mobile.

Medical Electronics Technician

Medical electronics technicians are always factory trained. The liabilities are too great otherwise. A firm knowledge of basic electronics and digital circuitry is required, along with hydraulics, pneumatics, and mechanics. Once those requirements are met, an employment application can be made. Hospitals frequently employ certified technicians who work in a close relationship with the manufacturer of the equipment. Medical equipment manufacturers will often provide up to a year of training with the trainee on the payroll. These jobs require a professional appearance and demeanor because it is necessary to work closely with medical professionals. Pros and cons are as follows.

Pros

- ▲ Very prestigious job requiring a close working relationship with medical professionals.
- ▲ Good to excellent wages, benefits, and working conditions.
- ▲ Training provided by employer.

Cons

- ▼ Industry is perceived to value youth and appearance above technical ability.
- ▼ Job opportunities are limited.

Aerospace Electronics Technician

The aerospace industry employs many technicians in areas of high technical responsibility. Technicians are often responsible for the design, testing, and limited production of specialized electronic devices used to test and evaluate aircraft systems or manufacturing procedures. This is an area where technicians can use all of their skills and creativity to the upmost. Electronic technicians in the aerospace industry are generally unionized, enjoy high wages, have excellent benefits, and have higher than average job security. Figure 27-4 shows an aerospace technician at work.

Figure 27-4 Aerospace technicians work with limited production runs of very specialized equipment.

A very high level of proficiency must be obtained to hold these positions. In addition to electronic theory, a working knowledge of metal fabrication and other manufacturing processes is required.

The pros and cons of being an **aerospace electronics technician** are as follows:

Pros

- ▲ High wages and excellent benefits.
- ▲ Opportunity to be creative and industrious.
- ▲ Job presents new challenges on a regular basis.
- ▲ These are usually union protected positions.
- ▲ Opportunity to work with advanced technologies.

Cons

- ▼ Best jobs are not obtained at entry level.
- ▼ The perception that politics can be a factor in promotions and job placement.
- ▼ Impersonal employment.
- ▼ High fluctuation in employment levels puts more recently hired employees in a vulnerable position when down-sizing occurs, regardless of job performance or technical ability.

Computer Repair Technician

Computer repair has become big business over the past few years. The level of technical expertise required to repair computers is not in keeping with the technology they contain. The various circuit boards are treated as plug-in components. Some of the peripherals, however, especially monitors, lend themselves to component level repair. A large part of the personal computer repair industry is customer relations. Computer users often do not understand the machines they are working with or the proper use of the software the computers are running. A computer technician must understand both the software and hardware better than the customer.

Most wholesale and retail dealers must have a repair facility in order to sell certain brand names. The computer manufacturers require servicing dealers to send their technicians to factory schools for training. At the completion of that training, a certificate is issued. When looking for a job as a computer repair technician, the excuse for not being hired is that the applicant is not certified in the repair of the equipment. When that certification is sought, the would-be technician is told that he or she cannot be admitted to the certification school because the technician does not work for an authorized dealer. This apparently impossible situation can be remedied by dealers if they are in need of a technician. The dealer can hire you and then send you to school. If you are told that you cannot be hired because you do not have a certificate often really means that the dealer is not hiring.

Computer technicians must often go to the customer's workplace to service the computer. Field service is a big part of the business. Being able to deal with uninformed customers can be very challenging.

The pros and cons of being a **computer repair technician** are as follows:

Pros

▲ There is a lot of work.

▲ Wages are good to excellent.

▲ The work is light and usually takes place in a pleasant work environment.

Cons

▼ Computer technology changes so rapidly that it is difficult to become as familiar with the equipment as a technician should be.

▼ Service information is often nonexistent because of the numerous equipment which is manufactured in other countries.

▼ Being familiar with a wide range of operating systems and customer software can be very difficult, if not impossible.

27-3 Automotive Electronic Technician

Today's automobiles contain an extensive amount of on-board computers that control not only the engine management system but the rest of the vehicle as well. As on-board electronic control system complexity increases every new-car year, future job opportunities are endless for a certified technician with a strong background in analog and digital electronics and solid, logical troubleshooting skills. Although this field is one of the toughest industries to work in, every day contains a new challenge parked in the diagnostic/repair bay. The first few years are spent mastering the basics in mechanical and service repair, acquiring the needed personal tools that a technician must provide, and gaining experience in diagnosing a complex, expensive investment.

Technicians are certified by the Institute for **Automotive Service Excellence,** or **ASE.** The tests, categorized by systems, require technicians to understand and diagnose problems they may face on a daily basis. With the introduction of a national inspection and maintenance program sponsored by the government, the need for well-trained technicians will be the highest in the industry's history.

The pros and cons of being an **automotive electronic technician** include:

Pros

▲ Job security is excellent.

▲ Employment can be obtained anywhere there are cars.

▲ Pay is above average to excellent for a well trained and certified technician.

▲ Job satisfaction can be very high.

Cons

- ▼ The initial outlay of time required to learn the basics is equal to a two-year electronics degree program. Often much of the learning must be self-taught, because automotive electronic training is not up to date in some automotive technology programs.
- ▼ Entry-level pay is below average.
- ▼ The personal tools required to reach high-level positions average $10,000 to $20,000.
- ▼ Support from automotive manufacturers with documentation (control system operation, adequate and understandable schematics, etc.) is not the best.
- ▼ Some textbooks and study guides covering automotive electronics are not up to date.
- ▼ Community acceptance of an automotive technician is still low; many are still thought of as "mechanics" or rip-off artists. But this is changing as car owners learn more about what it takes in knowledge and specialized equipment to repair high-technology vehicles. The day of the driveway or backyard mechanic is gone forever.

27-4 Advanced Positions

One of the most sought after jobs in electronics is that of engineering technician. The engineering technician may, in part, be responsible for a large part of the design phase. Designs from the engineer must be prototyped and tested by the engineering technician. These positions are few and require about six years of experience. A two-year degree is also recommended. Engineering technicians must provide written reports on their findings and justify their suggestions. Most employers like to see some college experience to guarantee a certain level of proficiency in written and oral communication.

The pros and cons of being an **engineering technician** are as follows:

Pros

- ▲ The highest position a technician can attain.
- ▲ Pay and prestige are as high as it gets.
- ▲ Provides the opportunity to use all of your skills and develop new ones.
- ▲ Work is performed at a comfortable pace in a pleasant work environment.
- ▲ There is seldom any heavy lifting involved.

Cons

- These positions are usually not available as entry-level positions, regardless of ability, without at least a two-year degree.
- Positions in military electronics or aerospace companies are undergoing a strong down-sizing trend. The bottom has not been reached.

SUMMARY

The choice of what area of electronics to enter as a profession is very personal. Some positions require more people skills than technical skills, which may appeal to many. But others may be more at home with the creativity afforded by working closely with the design process and watching something take form. Electronics offers jobs whose physical requirements range from sitting in a chair to literally climbing mountains. Wages in the industry are generally above average. In the areas of industrial, communications, and engineering, wages for electronics technicians are on-par with the highest paying construction trades, but they come without the strenuous physical labor and feast or famine cycle so common to the construction trades. The information contained in this chapter is, of necessity, very limited, but it does provide some insight into the career opportunities offered by the electronics field.

NEW VOCABULARY

aerospace electronics technician
automotive electronic technician
cable television technician
calibration technician
computer repair technician
consumer electronics technician
electronics communications technician
electronics assembler
engineering technician
eutectic solder
fiber-optics technician
fusion splicing
industrial electronics technician
Institute for Automotive Service Excellence (ASE)
line repair technician
medical electronics technician
stuffing
wave-soldering machine

STEAM TURBINE

GLOSSARY

A-to-D converter (analog-to-digital converter) Converter that translates a voltage level into a set of digital pulses.

AC collector resistance The type of resistance in an AC circuit which is found using the formula

$$r_L = \frac{R_c \times R_L}{R_c + R_L}$$

AC emitter resistance The type of resistance in an AC circuit which is found using the formula

$$r_e' = \frac{25\text{ mV}}{I_E}$$

ACOS Alternate function for COS on a calculator.

active region Region in which the collection curves of a transistor are nearly horizontal.

aerospace electronics technician Person responsible for design, testing, and limited production of specialized electronic devices to test and evaluate aircraft systems or manufacturing.

alternating current (AC) Current whose direction reverses at regularly occurring intervals.

alternators AC generators using slip rings.

American Wire Gauge (AWG) Units of standardized wire sizes in the U.S. that range from the smallest, number 40, to the largest, 4/0, or "four aught."

ammeter Instrument to measure current.

ampere Basic unit of current. One ampere is the rate of flow of electric charge equivalent to one coulomb of charge moving past a given point in one second.

amplitude Peak values of voltage, current, or power.

analog circuits Circuits that use continuous variations in current or voltage.

AND function Basic combinatorial logic function in which all inputs must be on before the output is active.

anode Positive side of a diode.

antilog The number that corresponds to a logarithm number. For example, the number that corresponds to the logarithm number 2.3142 is 206.15791.

antiresonance Another term for *parallel resonance.*

apparent power The voltage and current, expressed in volt-amperes, drawn by a transformer.

ARC Calculator function that will ascertain angle theta if SIN, COS, or TAN is known.

armature A moveable part of a relay.

ASIN Alternate function for SIN on a calculator.

ATAN Alternate function for TAN on a calculator.

atom Fundamental unit of all matter.

attentuation Condition that exists as frequency increases and voltage decreases in a low-pass filter.

autoformer An inexpensive transformer; usually a tapped coil of wire used to step up or step down voltage.

automotive electronics technician Repair specialist who deals with on-board computers that manage today's cars.

autoranging feature Feature of digital multimeters that can automatically select the best scale for a selected function.

avalanche The free flow of current when a diode breaks down.

B-H curve A graph that shows the effects of hysteresis.

back emf Another term for *counter electromotive force.*

band-rejection Another term for *band-stop filter.*

band-stop filter Filter that prevents a band of frequencies from passing to a load.

bandwidth A range of frequencies with a resonant effect in an *L-C-R* circuit.

battery A group of electrically connected cells that change chemical energy into electrical energy.

BCS theory Superconductivity theory of American physicists John Bardeen, Leon N. Cooper, and J. Robert Schrieffer; states that mobile electrons travel in pairs.

bias Average DC level of amplifier voltage or current to set operating characteristics.

bipolar junction transistor (BJT) Transistor that has three leads identified as the base (B), emitter (E), and collector (C); can be NPN or PNP type.

bleeder Resistor that allows current to drain in order to reduce shock hazard.

breakdown region Region in which V_{ce} exceeds breakdown voltage rating of the CB junction.

bridge circuit A diamond-shaped circuit that is divided by a bridge resistor that dissects the circuit into two triangles. Also called a double delta.

cable television technician Installer, field technician, or bench technician in the cable industry.

capacitance The ability to store energy in the electrostatic field between two electrically isolated conductors with dissimilar charges.

capacitive reactance A capacitor's opposition to current.

capacitor Device used to concentrate or store electricity. Formerly called a *condenser.*

cathode Negative side of a diode.

cell A battery component.

Celsius Temperature scale, also called *centigrade,* that uses zero degrees for the freezing point of water and 100 degrees for the boiling point of water.

centigrade Temperature scale, also called *Celsius,* that uses zero degrees for the freezing point of water and 100 degrees for the boiling point of water.

cgs One of two major divisions of the metric system; uses centimeter-gram-second units of measure.

characteristic The part of a logarithm to the left of the decimal point.

charge A property of particles of matter. A charged particle can attract or repel other particles.

choke Inductor with ability to choke off high frequencies.

circuit (chassis) ground Also called *hot ground.*

clampers Device used to fix, or clamp, a signal to a particular level. Also called a *DC restorer.*

clipper Device used to control or remove part of a signal. Also called a *limiter.*

coefficient of coupling The decimal value between 0 and 1 of the degree of magnetic coupling between two coils.

coercive force Magnetomotive force created by a coil because it forces atoms to align in opposition to their normal patterns.

common-base amplifier Amplifier in which the input signal is applied to the emitter and the output signal is taken from the collector.

common-collector amplifier Amplifier in which the input signal is applied to the base while the output signal is taken at the emitter.

common-drain amplifier Amplifier in which the input signal is applied to the gate and the output signal is taken from the source.

common-gate amplifier Amplifier in which the input signal is applied to the source while the output signal is taken from the drain.

common ground A return to earth for AC power lines.

common-source amplifier Amplifier in which the input signal is applied at the gate and the output signal is taken at the drain.

commutator A device that converts alternating current to direct current.

complementary MOS (CMOS) Digital IC device noted for its reliability, low power consumption, and low cost.

complex number Mathematical expression that completely defines a phasor diagram. Also called a *phasor notation.*

computer repair technician Person specializing in the repair of computers.

condenser Device used to concentrate or store electricity. Now known as a *capacitor.*

conductance The ease with which current can flow through a material, component, circuit, or device.

conductor Current-carrying component, usually made from metal.

constant k An analysis technique used to calculate the L and C components of a passive filter.

consumer electronics technician Person specializing in the repair of home entertainment electronics and automotive electronics.

Cooper pairs Electron pairs formed when electrons interact with mechanical vibrations in atoms.

cosecant The secant of the complement of an angle or arc.

cosine (COS) In a right triangle, the function of an acute angle that is expressed as the ratio of the side adjacent to the angle to the hypotenuse.

cotangent The tangent of a complement of an angle or arc.

coulomb (C) Fundamental unit of electric charge.

Coulomb's law The electrical force of repulsion or attraction between two charged bodies is directly proportional to the strength of their charges and inversely proportional to the square of the distance between their centers.

counter emf Electromagnetic force in a coil that flows back against the source voltage, retarding an increase of current.

current divider relationship The state in which currents through two parallel resistors are inversely proportional to their values.

cutoff frequency Frequency at which higher or lower frequencies do not pass through a filter.

cutoff region Plotted graphically, the region in which a transistor has only a very small collector current.

cycle One complete set of values for a repeating waveform.

D-to-A converter Device that accepts a digital value at its input and translates the value to a voltage level that is proportional to the input.

damping A decrease in amplitude of each successive oscillation in a wave train.

d'Arsonval/Weston meter Once the most popular analog design, it is now surpassed by digital designs, although it continues to offer faster response to changes in voltage.

DC component Determined by the average amplitude value of the periodic wave.

DC load line A graph showing all the possible values of I_c and V_{ce} for a given temperature.

DC restorer A circuit that adds a fixed bias voltage to a signal.

decibel (dB) One-tenth of a bel. A logarithmic unit that represents a power or voltage ratio.

degree Unit that divides a circle into 360 equal parts.

delta Fourth letter of the Greek alphabet, triangular in shape. Lends its name to delta and double-delta circuits.

depletion region The region in a diode where some of the free electrons are captured and are no longer available to support current flow.

diamagnetics Materials that do not normally produce any magnetism but are slightly repelled by a magnet because their magnetic properties change when under the influence of an external field. They include beryllium, copper, silver, gold, germanium, and bismuth.

dielectrics Another name for *insulators.*

digital multimeter (DMM) Device that measures voltage, current, and resistance, and displays those values automatically.

direct current (DC) Current that flows in only one direction.

divider current Current diverted from the load current in a circuit.

domains Microscopically small arrangements in ferromagnetic materials that produce magnets.

doping Adding impurity elements to semiconductors to improve their electrical characteristics.

double subscript notation System for identifying voltage across two points. The first subscript is the reference, and the second is the polarity.

duty cycle On time of a waveform divided by the total time period, which is the sum of the on time and off time. The ratio is expressed as a percentage.

eddy currents Circulating currents created in the iron core of an inductor.

electric field The lines of force that allow action at one point in space to cause an action at another point, with no physical connection between the two actions.

electric polarities An electric charge that is either positive (electron deficiency) or negative (excessive electrons).

electricity The class of phenomena involving electric charges and their effects when at rest or in motion.

electrolysis The production of a chemical change by passing electricity through a liquid.

electrolyte A liquid conductor.

electrolytic capacitor A capacitor that contains a liquid conductor or electrolyte; must be operated from direct current.

electromagnetism The force exerted by magnetic fields created by electric currents flowing through coils of wire.

electromotive force (emf) Voltage that produces a current in a circuit.

electronic communications technician Person specializing in fields such as radio, television, or telecommunications.

Electronic Industries Association (EIA) The association that sets standards for the industry including establishing color code and numerical equivalents to identify resistor value.

electronics assembler Person who installs components onto printed circuit boards.

electrons Basic units of negative charge that orbit the nucleus in an atom.

electroscope An instrument that detects the presence and polarity of an electric charge.

electrostatic deflection Movement of electron beams using the interaction of electric fields.

electrostatic discharge The discharge of static electricity that can destroy sensitive electronic devices.

electrostatic induction The ability of a body to acquire the static charge of another when in proximity.

emitter bypass capacitor A capacitor that has very low reactance in a parallel path.

emitter-coupled logic Family of high-speed, low-power bipolar technology.

energy The ability to do work.

engineer technician Person involved in electronics design, prototyping, and testing.

erg Unit for energy in the cgs system.

ether Contrived substance the ancient Greeks said allowed light waves to travel through seemingly void spaces.

eutectic solder A type of solder that melts at lower than normal temperature and is used in the repair of circuit boards.

exponent A small figure, symbol, or number placed at the upper right of another figure to determine how many times the larger figure is to be multipled by itself; also called the *power* or *logarithm.*

Fahrenheit An American system of temperature measurement that uses 32 degrees as the freezing point of water and 212 degrees as the boiling point of water.

farad (F) Unit of capacitance expressed in charge/voltage ratio. One farad is the SI unit of capacitance that will store one coulomb of charge when one volt is applied.

Faraday's law The law of electromagnetic induction. It involves the generation of current in a wire and also charges in electrolytic cells.

ferrite core Also called *powdered iron core.* It is indicated by dashed lines alongside a core symbol.

ferrites Magnetic oxides.

ferromagnetic Pertaining to substances that can be strongly magnetized. Such elements include iron, nickel, cobalt, and gadolinium.

fiber-optics technician Person involved in installation, connectorization, optical loss testing, system design, and system troubleshooting in fiber optics.

field A region characterized by electric, magnetic, or gravitational lines of force. Fields interact only with other fields of their type.

field intensity Strength of magnetomotive force measured in oersted units at a coil's center.

first harmonic Another term for *fundamental frequency.*

flip-flop Sequential logic circuit where the input trigger impulse flips output from the present logic level to the opposite logic level.

flux The number of lines of force that pass through any closed surface.

flux density (B) Strength of a magnetic field gauged by the number of lines of force in a specified area.

flywheel effect The ability of an *L-C-R* to hand energy back and forth between the inductor and capacitor after the source has been removed.

free electrons Highly conductive valence electrons that move freely within metals.

frequency Number of times per second a waveform repeats.

frequency domain The way in which waveforms are displayed with a spectrum analyzer.

fundamental frequency The first harmonic in a waveform.

fuse Metal conductor that melts from excessive current to open a circuit.

fusion splicing The joining of miles-long lengths of fiber-optic cables by melting them together using a fusion master machine.

Galvanism The belief that living organisms converted chemical energy into a separate form of electricity apart from that of metallic contact.

gate bias A technique used to bias field effect transistors, in which a negative gate voltage is applied through a gate resistor.

gauss (G) Unit of measure for flux in the cgs system. One gauss is equal to one maxwell per square centimeter.

Gauss's law Law that states that flux is directly proportional to the amount of charge enclosed by a surface.

generators Devices that transform mechanical energy into electricity.

giga (G) One billion, or 10 to the 9th power.

gilbert (Gb) The cgs unit for magnetic force. One gilbert is equal to 0.796 ($10 \div 4\pi$) ampere-turns.

grad Unit of angular measurement in Europe. A grad divides a circle's circumference into 400 equal parts.

half-wave symmetry The condition in which a part of a waveform during the first half of its period mirrors the portion of a waveform during the second half.

Hall device Forerunner of direct current magnetic probes.

Hall voltage Small voltage generated when a current-carrying material is placed in a magnetic field with the current and field at right angles.

harmonics Multitudes of fundamental frequency.

henry (H) Fundamental unit of inductance. A current of one ampere per second induces one volt across an inductance of one henry.

hertz (Hz) Unit of frequency equal to one cycle per second.

hole Positive charge created when an electron is removed from a valence shell. The hole can be filled by another electron.

horsepower Unit of mechanical power. One horsepower equals 550 footpounds lifted one foot in one second; equivalent to about 746 watts.

hot ground Circuit ground.

hypotenuse The side of a right triangle that is opposite the right angle.

hysteresis The hesitation of a material's atoms or molecules to follow changes in an external magnetic field.

impedance Total opposition to current in a circuit regardless of the cause.

in phase The state in which there is no difference between the voltage across a resistor and the current through it.

incandescent Pertaining to a device that can produce light from heat.

inductance The ability of any circuit to oppose any change in current.

inductive reactance The opposition an inductor presents to an AC source. It is different from resistance in that inductive reactance does not consume power from the source.

inductors Coils of wire with inductance.

industrial electronics technician Person who repairs electronic manufacturing equipment.

input impedance Impedance seen by the AC source driving the amplifier.

Institute for Automotive Service Excellence (ASE) Organization that certifies automotive electronic technicians.

insulators Materials that provide significant opposition to electric current.

integers Whole numbers.

integrated circuit (IC) A circuit that contains transistors, diodes, resistors, and capacitors in one package.

INVERSE Calculator function that will ascertain angle theta if SIN, COS, or TAN is known.

ion An atom that has become electrically unbalanced by gaining or losing an electron.

j term With the real term, it forms the two terms in the rectangular form of a complex number.

JFET Junction field-effect transistor.

joule (J) Unit of energy in the SI and mks systems.

junction barrier A neutralized zone in a semiconductor where negative electrons have been attracted across to the P material side.

kelvin (K) SI unit for temperature. The range starts at zero kelvins, called *absolute zero,* and increases in the same size units as the Celsius, or centigrade, system.

kickback voltage Voltage generated by a conductor when its magnetic field is collapsing that exceeds the source voltage if the source is removed.

kilogram (kg) A metric unit of mass equal to 1,000 grams.

kilowatt-hour One thousand watts of power used for one hour.

kinetic energy Energy of motion.

Kirchhoff's voltage law The sum of voltages around a closed path equals zero.

L-type filter A low-pass filter whose response is not as sharp as a *T*-type filter.

ladder One of the most common types of network configurations.

law of charges Charles Du Fay's law that states that unlike kinds of electricity attract and like kinds repel.

law of inverse squares The electric attraction or repulsion between two bodies varies inversely as the square of the distance between them.

law of magnetic poles Like poles repel and opposite poles attract.

left-hand motor rule When the thumb, first finger, and second finger of the left hand are held at right angles to one another, the thumb will indicate the direction of motion, the first finger will show the current path, and the second finger will point in the north-south direction of the magnetic field. This applies when magnetic polarities and direction of electron flow are known.

light-dependent resistor (LDR) Resistive element that changes value when exposed to light. Most LDRs are made from cadmium sulfide or silicon.

limiter Another term for *clipper.*

linear circuit Analog circuit that does not alter the basic character of the waveform it handles.

load That which consumes power from a voltage source in an electric circuit.

load current Current consumed from a voltage source.

logarithm The power, or exponent, to which 10 is raised to equal a given number.

logic gates Logic circuit with two or more inputs but one high or low output.

m-derived filter A variation of the constant-k filter, named because the values of inductance and capacitance are multiplied by the common factor m.

magnetic circuit Channels for magnetic lines of force through magnetic materials.

magnetic deflection Movement of electron beams using the interaction of magnetic fields.

magnetic flux The number of lines of force passing through a magnetic field.

magnetic induction The magnetizing of ferrous materials by introducing a magnet.

magnetomotive force (mmf) Force created by an electromagnet.

magnetostriction The change in the length of a material with a change in magnetic condition.

magnet Ferromagnetic material that produces a magnetic field outside itself.

magnitude term With the angle term, it forms the two terms in the polar form of a complex number.

mantissa The part of a logarithm to the right of a decimal point.

maxwell (Mx) Unit for magnetic flux in the English system. One maxwell is one line of force.

medical electronics technician Factory-trained technician who works on medical equipment.

mega (M) One million, or 10 to the 6th power.

Meissner effect The complete expulsion of magnetic fields by superconductors when their temperature is dropped below transition temperature.

mental approximations Circuit calculations made without using a calculator.

mesh One of three major network theorems.

mesh analysis Reliable method of calculating the loop currents in a simple two-dimensional network.

meters (m) Primary unit of length used in SI. A meter is equal to 39.37 inches.

micro (μ) One millionth, or 10 to the -6 power.

micro-microfarad Capacitor value that states a millionth of a millionth of a farad.

milli (m) One-thousandth, or 10 to the -3 power.

mks One of two major divisions of the metric system; uses meter-kilogram-second units of measure.

molecule The smallest amount of a material that can exist and still maintain the properties of that material.

MOSFET Metal-oxide semiconductor field-effect transistor.

motor A device, also called a *solenoid,* that can convert electricity into mechanical energy.

multimeter Device that measures voltage, current, and resistance.

mutual inductance Ability of one coil to induce voltage in neighboring coils.

N-type semiconductor A negatively charged semiconductor containing a donor impurity.

NAND function An inverted AND function. Name is taken from the fact that it is a not-AND operation.

nano (n) One billionth, or 10 to the −9 power.

negative temperature coefficient (NTC) Characteristic of a material whose resistance decreases with an increase in temperature. Semiconductors, electrolytes, and insulators have NTC.

network A group of circuits whose parameters cannot be found using Ohm's law and basic circuit rules.

neutrons Basic units without electrical charge that orbit the nucleus in an atom.

newton (N) Unit of force. One newton can accelerate one kilogram mass one meter per second.

nichrome Metal alloy of copper, nickel, and iron that can carry large amounts of current.

nonlinear circuit A type of analog circuit meant to transform a waveform from one form to another.

NOR function Inverted OR function. Name is taken from the fact that it is a not-OR operation.

normally OFF Another name for *EMOSFET.* No drain current flows when V_{GS} equals zero volts.

normally ON Another name for *JFET.* Drain current flows when V_{GS} equals zero volts.

octave The doubling of the frequency, or an interval between two frequencies that have a 2:1 ratio.

oersted (Oe) The cgs unit of field intensity. The oersted is one gilbert per centimeter.

ohm (Ω) Unit of resistance.

ohmmeter Device that measures resistance by forcing current through a component, circuit, or device being tested.

operational amplifier (op amp) Amplifier that can be programmed to perform a variety of amplifying operations.

OR function Logic function in which output is active when any or all inputs are active.

oscilloscope Electronic instrument that graphically displays voltage over time with a small point of light moving across a screen.

output impedance The load impedance which, when connected to the output terminals, results in the optimum transfer of power.

P-N junction The intersection of P and N material in a diode.

P-type semiconductor A positively charged semiconductor containing an acceptor impurity.

parallel circuit Basic circuit type in which resistors are connected directly across one another.

parallel R-C circuit A circuit containing a resistor and a capacitor connected in parallel where voltage is the same across the resistor and capacitor.

parallel series circuit A circuit that has a series of two impedances in one of its branches.

passband Range of frequencies between the response of a filter and the desired frequency.

peak-to-peak Pertaining to the total range of voltage between the positive and negative peaks of an AC cycle.

period Time consumed by a waveform before it repeats.

permeability Extent to which magnetic forces can be concentrated.

permittivity The ease with which a dielectric permits electrostatic lines of force to exist in comparison to a vacuum.

phase The difference in time between the starting points of each sine wave forming the composite waveform; measured in degrees or radians.

phase inversion The inversion of the waveform that is done by single-ended amplifier circuits or transformers.

phase lock loop An electronic circuit for locking an oscillator in phase with an input signal.

phasor notation Another term for *complex number.*

phi (ϕ) Greek letter used in identifying the sides of a right triangle to determine the hypotenuse.

photovoltaic effect When light energy is converted into electrical energy using solar cells.

pi One of the most common types of networks; shaped like the Greek letter pi (π).

pi-type (π-type) filter A low-pass filter.

pico (p) One trillionth, or 10 to the -12 power.

pinch-off voltage The drain-source voltage at which the drain current levels off.

plates Metal surfaces of a capacitor that face each other but do not touch.

polar form A type of complex-number notation that specifies a phasor diagram in terms of the length of its hypotenuse and its phase angle.

positive temperature coefficient (PTC) Characteristic of a material if its resistance increases with an increase in temperature. Metals and most alloys have PTC.

potential energy Stored energy.

potentiometer A three-terminal resistor, also known as a *pot,* connected as a voltage divider.

power The rate of doing work, measured electrically in watts.

power gain (A_P) $A_P = A_V \times A_I$.

printed circuit board Modern-day replacement for metal chassis on which a circuit image is printed.

prodivisum Mathematical derivative of reciprocals equation used when two resistors are connected in parallel. One resistor value is multiplied by the other, and the product is divided by the sum of the two values.

protons Basic units of positive charge that orbit around the nucleus in an atom.

protractor Semicircular instrument with a graduated arc used to measure angles.

pulsating direct current Voltage that varies in amplitude and periodically drops to zero; common in digital logic circuits.

pure direct current Steady current or voltage whose amplitude does not change; usually produced by batteries and solar cells.

Pythagorean theorem The sum of the squares of two sides of a right triangle is equal to the square of the hypotenuse.

Q factor Factor of quality for a resonant circuit inversely proportional to bandwidth.

quadrant One-quarter of a 360-degree rotation of a conductor in a magnetic field. The four 90-degree quadrants are numbered 1 to 4 in counterclockwise direction.

quadrature A configuration in which two waveforms are 90 degrees out of phase.

R-over-N equation When resistors are all the same value, resistance of the circuit is inversely proportional to the number of resistors.

radian A line through the length of a circle's radius that is curved to fit its circumference.

reactance Voltage created by inductors that opposes a source.

real term With the j term, it forms the two terms in a rectangular form of a complex number.

reciprocal equation Basic equation used for finding the equivalent resistance of a parallel circuit.

rectifier Diode used to change, or rectify, alternating current into direct current, mostly in power supplies.

relay Electric switch operated by a current in a coil.

resistivity Specific resistance of a material.

resonance Condition in which the current through a series *L-C-R* circuit is maximum, being limited only by the resistance of the circuit and the applied voltage.

resonant filters Devices used to pass a specific band of frequencies to a load or to stop certain frequencies from reaching a load.

rheostat Resistor with two terminals to vary current.

rho (ρ) Greek letter that is the symbol for resistivity.

ripple The undesired modulation of a signal or power source.

saturation The condition that occurs when the maximum possible numbers of atoms or molecules are brought into magnetic alignment by an external magnetic field. Once a material is saturated, the field intensity cannot be increased.

saturation region Vertical portion of the collector characteristic curves near the origin.

scientific notation Numeric shorthand that expresses numbers as a power of 10.

secant A line from the center of a circle through the circumference to another line tangent to the circle.

Seebeck effect Two dissimilar metals in contact with each other produce an electrical potential when heated.

self-bias The means of providing effective gate bias in a field-effect transistor.

self-inductance Ability of a coil to induce a voltage into itself.

semiconductor A material whose conductivity lies between that of a conductor and that of an insulator.

series-aiding Pertaining to the situation in which the positive terminal of one source is connected to the negative terminal of another.

series circuit A circuit formed by components connected end-to-end, with the same current flowing through the components when connected across a voltage source.

series-opposing Pertaining to positive-to-positive or negative-to-negative terminal connection in a series.

series-parallel Basic circuit created when series and parallel circuits are combined.

series-parallel circuit A series circuit that has a set of parallel impedances.

series R-C circuit Circuit containing a resistor and a capacitor connected in a series across an AC source.

short circuit Situation that occurs when a small resistance approaches zero.

shunt A metal strip or low-value resistor connection.

shunt (clipper) A limiter that shunts output voltage.

shunt capacitor A short that channels undesired high frequencies away from the load.

SI Stands for "International System of Units," a system of measurements.

siemens Unit of conductance.

sine (SIN) In a right triangle, the function of an acute angle that is expressed as a ratio of the side opposite the angle to the hypotenuse.

sine wave The condition in which the amplitude of a wave at any point is determined using the trigonometric sine function.

sinusoidal Having the characteristics of a sine wave; having a curving, snakelike motion.

skin effect When difference in measure Q of a circuit is less than the calculated Q, high-frequency alterations of current through the inductor force electrons to the surface of the wire.

slope Steepness of a filter's response curve.

smoothing capacitor Rectifier that smooths pulsating direct current into a steady value of voltage.

solenoid A motor that converts electricity into mechanical energy.

source That which develops power in an electric circuit.

specific resistance The ohmic value of a material at room temperature; also called *resistivity.*

spectroscopy The study of the distribution of light through the use of an instrument that splits visible light into wavelengths by means of a split and a prism.

spectrum analyzer Device used to view frequency components of a waveform.

square wave Waveform created by pressing the momentary contact switch.

static electricity Electricity without motion that is produced whenever dissimilar nonmetallic materials are rubbed together.

stop band Those frequencies prohibited from passing to a load by a stop-band filter.

superconductors Materials that lose all electrical resistance when cooled to temperatures near absolute zero.

superposition Procedure to calculate voltage of a network with multiple sources.

switch The simplest controlling device for power in an electric circuit.

T network Network in which three components are connected with one end in a common connection and the other ends to three lines; same as *wye network.*

T-type filter A low-pass filter with a sharper response than the L-type filter. Also called a π-type filter.

tangent (TAN) In a right triangle, the function of an acute angle expressed as the ratio of the side opposite the angle to the side adjacent to the angle.

tank circuit A parallel resonant circuit with the ability to store energy.

temperature coefficient The measure of how voltage varies with a change in temperature.`

tera (T) One trillion, or 10 to the 12th power.

tesla (T) Unit of flux density in the mks system and in SI. One tesla is equal to one weber per square meter, or 10,000 gauss.

test point Terminal in a circuit that extends above the circuit board where voltage can be measured.

thermistor A resistor that changes value with changes in temperature.

thermocouple A device that converts heat energy into electrical energy.

theta (θ) Greek letter used to identify the major reference point when naming sides of a right triangle.

Thévenin's theorem Formula used to determine the current through a selected resistor somewhere in a network.

time constant After a sudden change in voltage or current, the time required to change equals 63 percent.

time domain Oscilloscope's two-dimensional plot of how voltage varies with time.

toroidal electromagnet Electromagnet with a doughnut-shaped magnetic core.

transconductance (g_m and g_{mo}) The factor in a FET which specifies how the gate voltage V_{GS} controls the drain current I_D. The formulas used are:

$$g_m = \frac{I_D}{V_{GS}} \text{ } (V_{DS} \text{ constant}) \text{ and } g_{mo} = \frac{I_D}{V_{GS}} \text{ when } V_{GS}=0$$

transconductance curve A graph of I_D versus V_{GS}. The graph is nonlinear because equal changes in V_{GS} produce unequal changes in I_D.

transformer A device with two or more coil windings used to step up or step down AC voltage.

transistor-transistor logic (TTL) An IC device with a bipolar junction technology very close to that of a BJT; noted for high speed, power-handling, reliability, and low cost.

trimmer capacitor A capacitor that can make small adjustments in stored electric energy.

true power The product of the applied voltage and the current through a resistor; also called *resistive power.*

universal charge curve Curve formed when plotting the increase in voltage across a capacitor as it charges.

universal discharge curve A graph of the output voltage plotted against the first five constants as a capacitor discharges; inverted image of the charge curve.

watt (W) Unit of electrical power.

work Force multiplied by distance.

valence Outermost shell of all atoms.

variac Transformer that can select any voltage between zero and the source.

varying direct current Voltage whose amplitude changes but never drops to zero; common in circuits of amplifying devices.

vector A line that indicates magnitude and direction.

volt-ampere (VA) Unit of apparent power equal to the product of current and voltage.

volt-ampere curve A graph of the elctrical characteristics of any conductor, component, or circuit.

volt-ampere reactive (VAR) The volt-ampere measurement of pure inductance which always has a voltage-to-current phase relationship of 90 degrees.

volt-ohmmeter (VOM) Device that measures voltage.

voltage Current times resistance.

voltage-dependent resistor (VDR) Resistive component that changes value with changers in applied voltage.

voltage-divider bias The situation in which transistor current remains the same even though the β_{DC} of the transistor may vary with temperature.

voltage dividers Series circuits made from resistors that distribute voltage across each resistor.

voltage regulator A device that maintains a specified output voltage level despite changes in input voltage levels or output load current.

voltmeter Instrument that measures in volts the potential difference between two points.

waveform A mathematical plot of how the amplitude of an electrical quantity changes with time.

weber (Wb) The mks and SI unit of magnetic flux. One weber is equal to 100 million lines of force.

Wheatstone bridge A balanced circuit used for measuring resistance or to compare one resistance to another.

windings The inductors of a transformer.

wye network A network in which three components are connected at one end while their other ends are connected to three lines; same as *T network.*

zener Semiconductor diode used to regulate voltage; used in reverse direction.

zero bias The situation in which voltage across a gate-source region is zero.

INDEX

A

B

C

D

E

F

G

I

Q

R

S

Z

PHOTO CREDITS

XVI, Greg Pease/Tony Stone Images
2–3, Michael Gilbert/Science Photo Library/Photo Researchers
3, (tl) © Science Source/Photo Researchers, (tr) Michael Gilbert/Science Photo library/Photo Researchers, (br) © 1990 David A. Wagner/The Stock Market
10, Colin Cuthbert/Nawrocki Stock Photo
17, Henry Dakin/Science Photo Library/Photo Researchers
18–19, Michael Gilbert/Science Photo Library/Photo Researchers
19, (tl) © Science Source/Photo Researchers, (tr) Michael Gilbert/Science Photo Library/Photo Researchers, (br) © 1990 David A. Wagner/The Stock Market
30, Courtesy of The Light Brigade, Kent, WA
31, (tl) Peter Scholey/Nawrocki Stock Photo, (tr) Tony Craddock/SPL/Photo Researchers, (bl) John Mead/Science Photo Library/Photo Researchers, (br) © Francois Gohier/Photo Researchers, Inc.
32, (t) © 1990 Craig Tuttle/The Stock Market, (b) Peter Lamberti/Tony Stone Images
36–37, Michael Gilbert/Science Photo Library/Photo Researchers
37, (tl) © Science Source/Photo Researchers, (tr) Michael Gilbert/Science Photo Library/Photo Researchers, (br) © 1990 David A. Wagner/The Stock Market
47, David Parker/Science Photo Library/Photo Researchers
51, Lonnie Duka/Tony Stone Images
56–57, Michael Gilbert/Science Photo Library/Photo Researchers
57, (tl) © Science Source/Photo Researchers, (tr) Michael Gilbert/Science Photo Library/Photo Researchers, (br) © 1990 David A. Wagner/The Stock Market
72–73, Michael Gilbert/Science Photo Library/Photo Researchers
73, (tl) © Science Source/Photo Researchers, (tr) Michael Gilbert/Science Photo Library/Photo Researchers, (br) © 1990 David A. Wagner/The Stock Market
79, Chemical Design Ltd./Science Photo Library/Photo Researchers
85, Ben Swedowsky/The Image Bank
89, © 1992 Robert Essel/The Stock Market
92–93, Michael Gilbert/Science Photo Library/Photo Researchers
93, (tl) © Science Source/Photo Researchers, (tr) Michael Gilbert/Science Photo Library/Photo Researchers, (br) © 1990 David A. Wagner/The Stock Market
94, Dr. Jeremy Burgess/Science Photo Library/Photo Researchers
100, Science Photo Library/Photo Researchers
105, © 1992 Larry Keenan/The Image Bank
106–107, Michael Gilbert/Science Photo Library/Photo Researchers
107, (tl) © Science Source/Photo Researchers, (tr) Michael Gilbert/Science Photo Library/Photo Researchers, (br) © 1990 David A. Wagner/The Stock Market
109, Courtesy Central Scientific Company
112, Courtesy of Simpson Electric Company
113, Courtesy of Tektronix, Inc.
117, File photo
122–123, Michael Gilbert/Science Photo Library/Photo Researchers
123, (tl) Science Source/Photo Researchers, (tr) Michael Gilbert/Science Photo Library/Photo Researchers, (br) © 1990 David A. Wagner/The Stock Market
127, File Photo
133, © 1992 Larry Keenan/The Image Bank
137, Ken Cooper/The Image Bank
140–141, Michael Gilbert/Science Photo Library/Photo Researchers
141, (tl) © Science Source/Photo Researchers, (tr) Michael Gilbert/Science Photo Library/Photo Researchers, (br) © 1990 David A. Wagner/The Stock Market
147, Andrzej Dodzinski/Science Photo Library/Photo Researchers
149, © 1992 Larry Keenan/The Image Bank
150, Simon Fraser, Welwyn Electronics/Science Photo Library/Photo Researchers
154–155, Michael Gilbert/Science Photo Library/Photo Researchers
155, (tl) © Science Source/Photo Researchers, (tr) Michael Gilbert/Science Photo Library/Photo Researchers, (br) © 1990 David A. Wagner/The Stock Market
157, Simon Fraser, Welwyn Electronics/Science Photo Library/Photo Researchers
166, Kim Steele/The Image Bank
172–173, Michael Gilbert/Science Photo Library/Photo Researchers
173, (tl) © Science Source/Photo Researchers, (tr) Michael Gilbert/Science Photo Library/Photo Researchers, (br) © 1990 David A. Wagner/The Stock Market
179, © Ken Biggs/Photo Researchers
187, Rick Altman/Nawrocki Stock Photo
190–191, Michael Gilbert/Science Photo Library/Photo Researchers
191, (tl) © Science Source/Photo Researchers, (tr) Michael Gilbert/Science Photo Library/Photo Researchers, (br) © 1990 David A. Wagner/The Stock Market
199, © Will & Deni McIntyre/Photo Researchers
203, © 1992 Larry Keenan/The Image Bank

204, Alvis Upitis/The Image Bank
208–209, Michael Gilbert/Science Photo Library/Photo Researchers
209, (tl) © Science Source/Photo Researchers, (tr) Michael Gilbert/Science Photo Library/Photo Researchers, 209, (br) © 1990 David A. Wagner/The Stock Market
217, © 1992 Larry Keenan/The Image Bank
221, © 1992 Richard Wahlstrom/The Image Bank
226–227, Brett Froomer/The Image Bank
228–229, © Science Source/Photo Researchers, Inc.
229, (tl) © 1990 David A. Wagner/The Stock Market, (tr) © Science Source/Photo Researchers, (br) Michael Gilbert/Science Photo Library/Photo Researchers
242, © 1992 Uniphoto, Inc.
247, Andy Sacks/Tony Stone Images
250–251, © Science Source/Photo Researchers, Inc.
251, (tl) © 1990 David A. Wagner/The Stock Market, (tr) © Science Source/Photo Researchers, (br) Michael Gilbert/Science Photo Library/Photo Researchers
262, Science Photo Library/Photo Researchers
266–267, © Science Source/Photo Researchers
267, (tl) © 1990 David A. Wagner/The Stock Market, (tr) © Science Source/Photo Researchers, (br) Michael Gilbert/Science Photo Library/Photo Researchers
276, Charles Thatcher/Tony Stone Images
283, © 1992 Larry Keenan/The Image Bank
289, Kim Steele/The Image Bank
290–291, © Science Source/Photo Researchers
291, (tl) © 1990 David A. Wagner/The Stock Market, (tr) © Science Source/Photo Researchers, (br) Michael Gilbert/Science Photo Library/Photo Researchers
293, © 1992 Larry Keenan/The Image Bank
296, © Telegraph Colour Library/FPG International
305, David Parker/Science Photo Library/Photo Researchers
311, (l) © 1992 Larry Keenan/The Image Bank, (r) © Telegraph Colour Library/FPG International
312–313, © Science Source/Photo Researchers
313, (tl) © 1990 David A. Wagner/The Stock Market, (tr) © Science Source/Photo Researchers, (br) Michael Gilbert/Science Photo Library/Photo Researchers
315, SuperStock
323, Tony Stone Images
326, SuperStock
330–331, © Science Source/Photo Researchers, Inc.
331, (tl) © 1990 David A. Wagner/The Stock Market, (tr) © Science Source/Photo Researchers, (br) Michael Gilbert/Science Photo Library/Photo Researchers
337, Astrid and Hanns-Frieder Michler/Science Photo Library/Photo Researchers
343, © 1990 Stephen Hunt/The Image Bank
344–345, © Science Source/Photo Researchers
345, (tl) © 1990 David A. Wagner/The Stock Market, (tr) © Science Source/Photo Researchers, (br) Michael Gilbert/Science Photo Library/Photo Researchers
349, Alfred Pasieka/Science Photo Library/Photo Researchers
353, © Alvis Upitis/The Image Bank
358–359, © Science Source/Photo Researchers
359, (tl) © 1990 David A. Wagner/The Stock Market, (tr) © Science Source/Photo Researchers, (br) Michael Gilbert/Science Photo Library/Photo Researchers
365, Ken Whitmore/Tony Stone Images
367, Alan Levenson/Tony Stone Images
380–381, © Science Source/Photo Researchers
381, (tl) © 1990 David A. Wagner/The Stock Market, (tr) © Science Source/Photo Researchers, (br) Michael Gilbert/Science Photo Library/Photo Researchers
383, © Joseph Nettis/Photo Researchers
396, © 1992 Larry Keenan/The Image Bank
403, © 1990 Joe Robbins/FPG International
404–405, © 1990 David A. Wagner/The Stock Market
405, (tl) © Science Source/Photo Researchers, (tr) © 1990 David A. Wagner/The Stock Market, (br) Michael Gilbert/Science Photo Library/Photo Researchers
412, 414, 423, SuperStock
424–425, © 1990 David A. Wagner/The Stock Market
425, (tl) © Science Source/Photo Researchers, (tr) © 1990 David A. Wagner/The Stock Market, (br) Michael Gilbert/Science Photo Library/Photo Researchers
433, Dr. Jeremy Burgess/Science Photo Library/Photo Researchers
445, Larry Keenan Associates/The Image Bank
449, © 1992 Larry Keenan/The Image Bank
456, © 1988 Alvis Upitis/The Image Bank
460–461, © 1990 David A. Wagner/The Stock Market
461, (tl) © Science Source/Photo Researchers, (tr) © 1990 David A. Wagner/The Stock Market, (br) Michael Gilbert/Science Photo Library/Photo Researchers
475, Thompson & Thompson/Tony Stone Images
479, SuperStock
481, Ralph Mercer/Tony Stone Images
491, Andy Sacks/Tony Stone Images
492–493, © 1990 David A. Wagner/The Stock Market
493, (tl) © Science Source/Photo Researchers, (tr) © 1990 David A. Wagner/The Stock Market, (br) Michael Gilbert/Science Photo Library/Photo Researchers
494, © 1986 Dick Luria/FPG International
507, SuperStock
521, © Robert A. Issacs/Photo Researchers
522–523, © 1990 David A. Wagner/The Stock Market
523, (tl) © Science Source/Photo Researchers, (tr) © 1990 David A. Wagner/The Stock Market, (br) Michael Gilbert/Science Photo Library/Photo Researchers
524, Courtesy of Meteor Communications Corp., Kent, WA
527, (l) and (r) Courtesy of the Light Brigade, Kent, WA
529, © Hank Morgan/VHSID Lab/ECE Dept., U. of MA./Science Source/Photo Researchers
531, Courtesy of DTS Electronics, Tacoma, WA
532, Courtesy of Meteor Communications Corp., Kent, WA
533, Greg Pease/Tony Stone Images
535, Mel Lindstrom/Tony Stone Images
537, Franz Edson/Tony Stone Images